Methods in Molecular Biology

Series Editor
John M. Walker
School of Life and Medical Sciences
University of Hertfordshire
Hatfield, Hertfordshire, AL10 9AB, UK

For further volumes:
http://www.springer.com/series/7651

The Mitotic Spindle

Methods and Protocols

Edited by

Paul Chang

Massachusetts Institute of Technology, Cambridge, MA, USA

Ryoma Ohi

Department of Cell & Developmental Biology, Vanderbilt University, Nashville, TN, USA

Humana Press

Editors
Paul Chang
Massachusetts Institute of Technology
Cambridge, MA, USA

Ryoma Ohi
Department of Cell & Developmental Biology
Vanderbilt University
Nashville, TN, USA

ISSN 1064-3745 ISSN 1940-6029 (electronic)
Methods in Molecular Biology
ISBN 978-1-4939-3540-6 ISBN 978-1-4939-3542-0 (eBook)
DOI 10.1007/978-1-4939-3542-0

Library of Congress Control Number: 2016938221

Printed on acid-free paper

This Humana Press imprint is published by Springer Nature
The registered company is Springer Science+Business Media LLC New York

Preface

The mitotic spindle is a large macromolecular structure that is both beautiful and functionally important. It has long been a focus of intense investigation due to its role as a guardian of ploidy, responsible for the equal segregation of genetic material into daughter cells during cell division. Not surprisingly, the spindle has also been a target of therapeutic inhibition for diseases that involve misregulation of cell division, such as cancer. Indeed, inhibitors of spindle function, such as tubulin poisons, are among the earliest-developed forms of cancer therapy and remain first-line therapies for many forms of cancer. More recently, the spindle has been shown to play equally important roles in other aspects of cell function and human disease. These advances have relied on the development of advanced technology, and as the tools used to study the spindle have evolved so has our understanding of it.

Innovations in many areas have impacted mitosis research. Modern mitosis studies began in the 1950s, alongside key developments in light microscopy and the introduction of the electron microscope. Improvements in light microscopy facilitated real-time analysis of cell division and demonstrated the dynamic nature of mitosis. Electron microscopy revealed ultrastructural details of the dividing cell that were critical to the identification of key functional components of the spindle, including the kinetochore. Biochemical discoveries resulted in the identification of tubulin as the building block of the mitotic spindle. Further studies then identified key proteins important for the regulation of microtubule function and have derived key principles of microtubule polymerization dynamics, both of which are fundamental to mitosis. "In vitro" reconstitution of spindle assembly using *Xenopus laevis* egg extracts provided a simple yet effective assay to identify key components of the spindle required for function. This led to the realization that physics, as well as biology, is central to spindle assembly and the coordination of spindle function. Lastly, knowledge of core cell cycle control mechanisms has allowed us to begin to connect mechanical aspects of mitosis with the physiological state of the cell.

Today, methods to analyze mitotic spindle assembly and function are impressively sophisticated and draw from diverse disciplines that include physics, chemistry, and computational modeling. Protein labeling techniques allow us to visualize key spindle components with high spatial and temporal resolution. Chemically induced dimerization gives us control over when and where proteins interact within the cell. Improvements in the isolation and biochemical reconstitution of complex protein assemblies have enabled a thorough documentation of their components and are permitting powerful activity-based studies.

In this volume of Methods in Molecular Biology, we have collected a series of protocols from leading investigators in the field whose expertise centers on mitotic spindle function. These methods cover a broad range of techniques, from basic microscopy to the study of spindle functions relevant to cancer. We include methods that can be applied to diverse model systems, which range from the cell-free *Xenopus* egg extract system to the moss *Physcomitrella patens*, in an effort to demonstrate the key contributions made by researchers using multiple model organisms. Finally, our chapters integrate cutting-edge technologies, which have only become available due to cross-disciplinary efforts, e.g., analog-sensitive inhibition of kinases.

It is our hope that these chapters will be informative for researchers new to mitosis, as well as for those that are already expert in the field.

Cambridge, MA, USA *Paul Chang*
Nashville, TN, USA *Ryoma Ohi*

Contents

Contributors

NAOWRAS AL-OBAIDI • *Department of Biology, Konstanz Research School Chemical Biology (KoRS-CB), University of Konstanz, Konstanz, Germany*

SAI K. BALCHAND • *Department of Biology and Program in Molecular and Cellular Biology, University of Massachusetts Amherst, Amherst, MA, USA*

EDWARD R. BALLISTER • *Department of Biology, University of Pennsylvania, Philadelphia, PA, USA*

MICHAEL D. BLOWER • *Department of Molecular Biology, Massachusetts General Hospital, Boston, MA, USA; Department of Genetics, Harvard Medical School, Boston, MA, USA*

DANIELA CIMINI • *Department of Biological Sciences, Biocomplexity Institute, Virginia Tech, Blacksburg, VA, USA*

TRISHA N. DAVIS • *Department of Biochemistry, University of Washington, Seattle, WA, USA*

JENNIFER G. DELUCA • *Department of Biochemistry and Molecular Biology, Colorado State University, Fort Collins, CO, USA*

KEITH F. DELUCA • *Department of Biochemistry and Molecular Biology, Colorado State University, Fort Collins, CO, USA*

AMY B. EMERMAN • *Department of Molecular Biology, Massachusetts General Hospital, Boston, MA, USA; Department of Genetics, Harvard Medical School, Boston, MA, USA*

STEFAN FLORIAN • *Department of Systems Biology, Harvard Medical School, Boston, MA, USA*

KIMBERLY K. FONG • *Department of Biochemistry, University of Washington, Seattle, WA, USA*

CINDY FONSECA • *Department of Molecular Physiology and Biophysics, University of Vermont College of Medicine, Burlington, VT, USA*

TATSUO FUKAGAWA • *Department of Molecular Genetics, National Institute of Genetics and Graduate University for Advanced Studies (SOKENDAI), Shizuoka, Japan; Graduate School of Frontier Biosciences, Osaka University, Suita, Japan*

NEIL J. GANEM • *Division of Hematology and Oncology, Department of Pharmacology, Boston University School of Medicine, Boston, MA, USA; Department of Experimental Therapeutics and Medicine, Boston University School of Medicine, Boston, MA, USA*

MATTHEW C. GOOD • *Department of Cell and Developmental Biology, University of Pennsylvania, Philadelphia, PA, USA; Department of Bioengineering, University of Pennsylvania, Philadelphia, PA, USA; Cell and Molecular Biology Graduate Group, University of Pennsylvania, Philadelphia, PA, USA*

GOHTA GOSHIMA • *Division of Biological Science, Graduate School of Science, Nagoya University, Nagoya, Japan*

BETH GRACZYK • *Department of Biochemistry, University of Washington, Seattle, WA, USA*

AARON C. GROEN • *Department of Systems Biology, Harvard Medical School, Boston, MA, USA*

BIN HE • *Department of Biological Sciences, Biocomplexity Institute, Virginia Tech, Blacksburg, VA, USA*

JACOB A. HERMAN • *Department of Biochemistry and Molecular Biology, Colorado State University, Fort Collins, CO, USA*

PRASAD V. JALLEPALLI • *Molecular Biology Program, Memorial Sloan-Kettering Cancer Center, New York, NY, USA*
ASHWINI JAMBHEKAR • *Department of Molecular Biology, Massachusetts General Hospital, Boston, MA, USA; Department of Genetics, Harvard Medical School, Boston, MA, USA*
MATHEW J.K. JONES • *Molecular Biology Program, Memorial Sloan-Kettering Cancer Center, New York, NY, USA*
VLADIMIR JOUKOV • *Department of Biological Chemistry and Molecular Pharmacology, Harvard Medical School, Boston, MA, USA*
JOHANNA KASTL • *Department of Biology, Konstanz Research School Chemical Biology (KoRS-CB), University of Konstanz, Konstanz, Germany*
MATTHEW KING • *Department of Molecular Biology, Princeton University, Princeton, NJ, USA*
TOMOMI KIYOMITSU • *Division of Biological Science, Graduate School of Science, Nagoya University, Nagoya, Japan; Precursory Research for Embryonic Science and Technology (PRESTO) Program, Japan Science and Technology Agency, Saitama, Japan*
GEERT J.P.L. KOPS • *Hubrecht Institute – KNAW (Royal Netherlands Academy of Arts and Sciences), Utrecht, The Netherlands; Cancer Genomics Netherlands, University Medical Center Utrecht, Utrecht, The Netherlands*
MIJUNG KWON • *Department of Pediatric Oncology, Howard Hughes Medical Institute, Dana-Farber Cancer Institute, Boston, MA, USA; Department of Cell Biology, Harvard Medical School, Boston, MA, USA*
MICHAEL A. LAMPSON • *Department of Biology, University of Pennsylvania, Philadelphia, PA, USA*
NICOLAS LECLAND • *Centre Biologie du Développement, UMR 5547 CNRS-Université Paul Sabatier, Toulouse, France*
JENS LÜDERS • *Cell and Developmental Biology Programme, Institute for Research in Biomedicine (IRB Barcelona), The Barcelona Institute of Science and Technology, Barcelona, Spain*
BARBARA J. MANN • *Department of Biology and Program in Molecular and Cellular Biology, University of Massachusetts Amherst, Amherst, MA, USA*
THOMAS U. MAYER • *Department of Biology, Konstanz Research School Chemical Biology (KoRS-CB), University of Konstanz, Konstanz, Germany*
MATTHEW D.D. MIELL • *Department of Biochemistry, Stanford University School of Medicine, Stanford, CA, USA*
TOMOHIRO MIKI • *Division of Biological Science, Graduate School of Science, Nagoya University, Nagoya, Japan*
TIMOTHY J. MITCHISON • *Department of Systems Biology, Harvard Medical School, Boston, MA, USA*
DANIEL J. NEEDLEMAN • *School of Engineering and Applied Sciences, Harvard University, Cambridge, MA, USA; Departments of Applied Physics, and Molecular and Cellular Biology, Harvard University, Cambridge, MA, USA*
ARCANGELA DE NICOLO • *Department of Medicine, Harvard Medical School, Boston, MA, USA; Department of Cancer Biology, Dana-Farber Cancer Institute, Boston, MA, USA*
TATSUYA NISHINO • *Department of Molecular Genetics, National Institute of Genetics and Graduate University for Advanced Studies (SOKENDAI), Shizuoka, Japan; Department of Biological Science and Technology, Graduate School of Industrial Science and Technology, Tokyo University of Science, Tokyo, Japan*
SABINE PETRY • *Department of Molecular Biology, Princeton University, Princeton, NJ, USA*

ADRIAN T. SAURIN • *Division of Cancer Research, Medical Research Institute, Jacqui Wood Cancer Centre, Ninewells Hospital and Medical School, University of Dundee, Dundee, UK*
ELIZABETH M. SHENK • *Division of Hematology and Oncology, Department of Pharmacology, Boston University School of Medicine, Boston, MA, USA; Department of Experimental Therapeutics and Medicine, Boston University School of Medicine, Boston, MA, USA*
RAJESH KUMAR SONI • *Cell Biology Program, Memorial Sloan-Kettering Cancer Center, New York, NY, USA*
AARON F. STRAIGHT • *Department of Biochemistry, Stanford University School of Medicine, Stanford, CA, USA*
JASON STUMPFF • *Department of Molecular Physiology and Biophysics, University of Vermont College of Medicine, Burlington, VT, USA*
MENG-FU BRYAN TSOU • *Cell Biology Program, Memorial Sloan-Kettering Cancer Center, New York, NY, USA*
PATRICIA WADSWORTH • *Department of Biology, Program in Molecular and Cellular Biology, University of Massachusetts Amherst, Amherst, MA, USA*
JOHANNES C. WALTER • *Department of Biological Chemistry and Molecular Pharmacology, Harvard Medical School, Howard Hughes Medical Institute, Boston, MA, USA*
MOÉ YAMADA • *Division of Biological Science, Graduate School of Science, Nagoya University, Nagoya, Japan*
TAE YEON YOO • *School of Engineering and Applied Sciences, Harvard University, Cambridge, MA, USA*
MARIJA ZANIC • *Department of Cell and Developmental Biology, Vanderbilt University School of Medicine, Nashville, TN, USA; Department of Chemical and Biomolecular Engineering, Vanderbilt University, Nashville, TN, USA*

Part I

Methods Focused on Cytology

Chapter 1

Using Fluorescence Microscopy to Study Mitosis

Sai K. Balchand, Barbara J. Mann, and Patricia Wadsworth

Abstract

Fluorescence microscopy is one of the most important approaches in the cell biologist's toolbox for studying the mitotic spindle. In fact, many of the key insights into our understanding of mitosis have been enabled by the visualization of mitotic processes using fluorescence microscopy. Here, we summarize some of the important considerations for imaging mitosis using fluorescence microscopy. Because light can damage live cells, we emphasize the importance of minimizing cellular damage while obtaining informative images.

Key words Spinning disc confocal microscopy, Confocal laser scanning microscopy, Mitosis

1 Introduction

The impact of fluorescence microscopy on cell biology in general, and on the field of mitosis in particular, is profound. This major impact has occurred due to the confluence of the development of fluorescent reporter molecules that can be delivered to and expressed in live cells and advances in equipment used for imaging. Over the past few decades, major improvements in detectors, light sources, filters, objective lenses, and the increased availability of turnkey systems for confocal microscopy have provided powerful tools to the researcher seeking to image the events of mitosis. Researchers have taken advantage of sophisticated new equipment to image cells that have been engineered to express mitotic proteins tagged with genetically encoded fluorescent proteins, FRET sensors and other fluorescent markers. Methods for introducing engineered proteins into cells have been greatly improved, enabling the researcher to control level of expression [1] and to perform near simultaneous imaging of multiple fluorescent proteins in the spindle. Moreover, recent introduction of gene editing protocols for mammalian cells promises to revolutionize imaging of tagged proteins by permitting modification of the endogenous genetic

Paul Chang and Ryoma Ohi (eds.), *The Mitotic Spindle: Methods and Protocols*, Methods in Molecular Biology, vol. 1413,
DOI 10.1007/978-1-4939-3542-0_1, © Springer Science+Business Media New York 2016

locus with fluorescent tags [2]. These new developments will undoubtedly provide unanticipated insights into mitosis.

Obtaining meaningful, quantitative data with the fluorescence light microscope requires a thorough understanding of the many variables that must be controlled during imaging [3]. Perhaps the most important consideration is that the cells must be kept healthy and dividing throughout the acquisition process [4]. Unfortunately, light is not inert and will cause damage to the live cells if not used judiciously. Fortunately, many key parameters of the experimental design for successful live cell imaging using fluorescence microscopy have been worked out [5]; similarly, protocols for generating cell lines expressing tagged proteins are also available [6, 7]. Here, we outline methods for fluorescence microscopy of mitosis in mammalian cultured cells, including maintaining cell viability. With minor modifications, similar methods can also be used for other cultured cells from other species (e.g., *Drosophila* S2 cells) and other model systems (e.g., *C. elegans* embryos).

2 Materials

1. Non-CO_2 medium: culture medium without phenol red indicator dye and without bicarbonate buffer and containing 20 mM HEPES, pH adjusted to 7.2 with potassium hydroxide, and supplemented with 10 % fetal bovine serum (FBS).
2. Oxygen scavenging system, e.g., Oxyrase.
3. Sterile glass coverslips, thickness #1.5.
4. Chambers for holding coverslips or glass bottom culture dishes.
5. Thermistor probe.
6. Vacuum grease or mineral oil.
7. Lens cleaner.

3 Methods

3.1 Maintenance of Mitotic Cell Health

It is of utmost importance that cell health is maintained during imaging. In the case of imaging mitotic cells, this can be easily evaluated by comparing the progress through mitosis of the imaged cells with appropriate control cells that were not exposed to fluorescent light. Successful completion of mitosis, with kinetics similar to the control cells, is a good indication that the imaging protocol has not caused cellular damage. Whenever possible, mitotic progression should be evaluated for the imaging conditions used for your experiments.

3.1.1 Protocol 1: Maintaining Mitotic Progression During Imaging

1. Plate the cells in appropriate dishes or on coverslips for use in closed chambers (*see* **Note 1**). Coverslips should be flame-sterilized or autoclaved prior to use; #1.5 glass is the proper thickness for high numerical aperture objective lenses (*see* **Note 2**). Cells are typically plated 1–2 days prior to experimentation.
2. Choose an appropriate stage/microscope heater system and pre-warm to the desired temperature (*see* **Note 3**).
3. Remove the cell culture medium and replace with pre-warmed Non-CO_2 medium that lacks bicarbonate buffer and contains 20 mM HEPES, pH 7.2 with potassium hydroxide, and supplemented with 10 % FBS. Non-CO_2 culture medium lacks phenol red, which is autofluorescent.
4. Add an oxygen scavenging system to retard photobleaching. Because Oxyrase contains bacterial membrane fragments, it is added to the medium, and then filtered, using a 0.22 μm syringe filter, prior to adding to the chamber or dish containing the cells.
5. Bring the cells to the microscope for imaging.

3.2 Fluorescence Microscopy of Mitotic Cells

3.2.1 Choosing an Imaging System

For a number of years, imaging mitotic cells by spinning disc confocal microscopy (SDCM) has been the gold standard in the field, primarily because the rapid acquisition possible with this type of confocal system enables the dynamics of the mitotic spindle to be captured [8]. However, laser scanning confocal microscopes (LSCM) can also capture dynamic events, especially instruments equipped with resonant scanning mirrors. LSCMs have additional advantages, including the ability to change pinhole size, to increase magnification using electronic zoom, spectral detectors to enable imaging with a wide range of fluorophores and built-in capacity for photobleaching and photoactivation experiments [3]. For a discussion of some of the major differences between these types of confocal systems, and the option of imaging using wide-field fluorescence followed by deconvolution, *see* **Note 4**.

3.2.2 Protocol 2: Imaging Mitosis Using Spinning Disc Confocal Microscopy

1. Turn on microscope system and follow the manufacturer's recommendation for warming up the lasers. This protocol assumes familiarity with the basic software commands for the system you are using.
2. Bring a dish of cells to the microscope in pre-warmed non-CO_2 medium and secure the dish to the microscope stage. Because cells are not under ideal conditions, i.e., they are in a non-CO_2 buffered environment, switch to a new dish of cells after about 1 h of imaging. Long-term imaging is best accomplished with an environmental chamber that controls humidity, gas, and temperature.

3. Select an appropriate objective lens. Using an objective lens with high numerical aperture and corrected for chromatic aberration gives the best possible results (*see* **Note 2**). Most SDCM systems have pinholes that are optimized for higher magnification lenses; newer systems have a choice of pinholes, enabling imaging at a range of magnifications.
4. Select a suitable cell for imaging. Use phase-contrast, DIC, or bright-field optics to find a cell at the appropriate stage of mitosis. Avoid using fluorescence light, unless absolutely necessary, so as to reduce photobleaching prior to image acquisition (*see* **Note 5**).
5. Set the laser power for each wavelength you will use, and select an exposure time, using the imaging software. Go to live imaging mode, and collect a single image of one fluorophore; adjust focus if needed. Adjust exposure time and/or laser power as if needed so that you can just detect the fluorescent signal(s) of interest. Observe the histogram of pixel intensity distribution and avoid saturating the pixels in the imaging field. Repeat for each fluorophore as needed.
6. If you are imaging a single *Z*-plane in the cell, engage the focus control system to automatically maintain focus (*see* **Note 6**).
7. For time-lapse acquisition of a single focal plane, set the interval between images and the duration of the time-lapse sequence. If you want to record the progression through mitosis, use an interval between images that is relatively long to avoid damage to the cell (e.g., 1 image per ~30–60 s) (*see* Fig. 1b). Using these settings, confirm that the cell progresses through mitosis, and that the fluorescent signal is not photobleached. If better time resolution is needed, try taking images more frequently, again confirming that the cell is not damaged by exposure to the fluorescent light (Fig. 2a).
8. To acquire a *Z*-series at a single time point, use the microscope and software controls to set the top and bottom limits for the *Z*-stack. This is typically performed in live imaging mode. Use a lower exposure time to reduce bleaching as you set up the *Z*-stack parameters, and reset the exposure to the desired time for the acquisition. Set the number of *Z*-planes you wish to collect. Acquire the stack.
9. To acquire a *Z*-series at multiple time points, set the parameters for *Z*-image acquisition and the parameters for a time-lapse sequence (*see* **Note 7**).
10. For multicolor imaging, use the software to collect images of multiple fluorophores during a time-lapse sequence or *Z*-series (*see* **Note 8**).

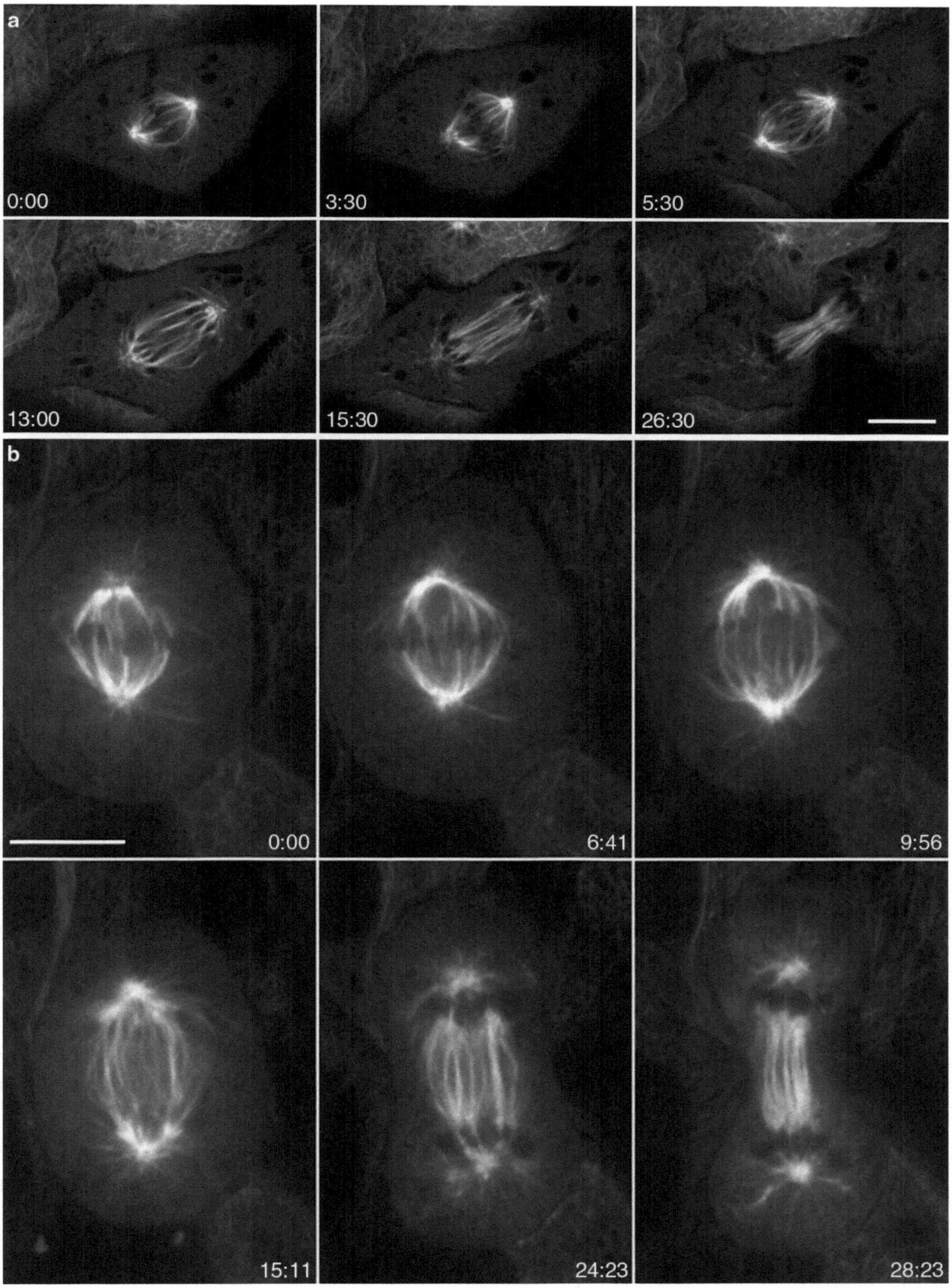

Fig. 1 Live mitotic LLC-Pk1 cells expressing GFP-tubulin. (**a**) Images acquired using a laser scanning confocal microscope, 60× objective lens. (**b**) Images acquired using a spinning disc confocal microscope, 100× objective lens. (**a**, **b**) are selected images of time series of images acquired at 30 s intervals. Bars = 10 μm. Time in min:s

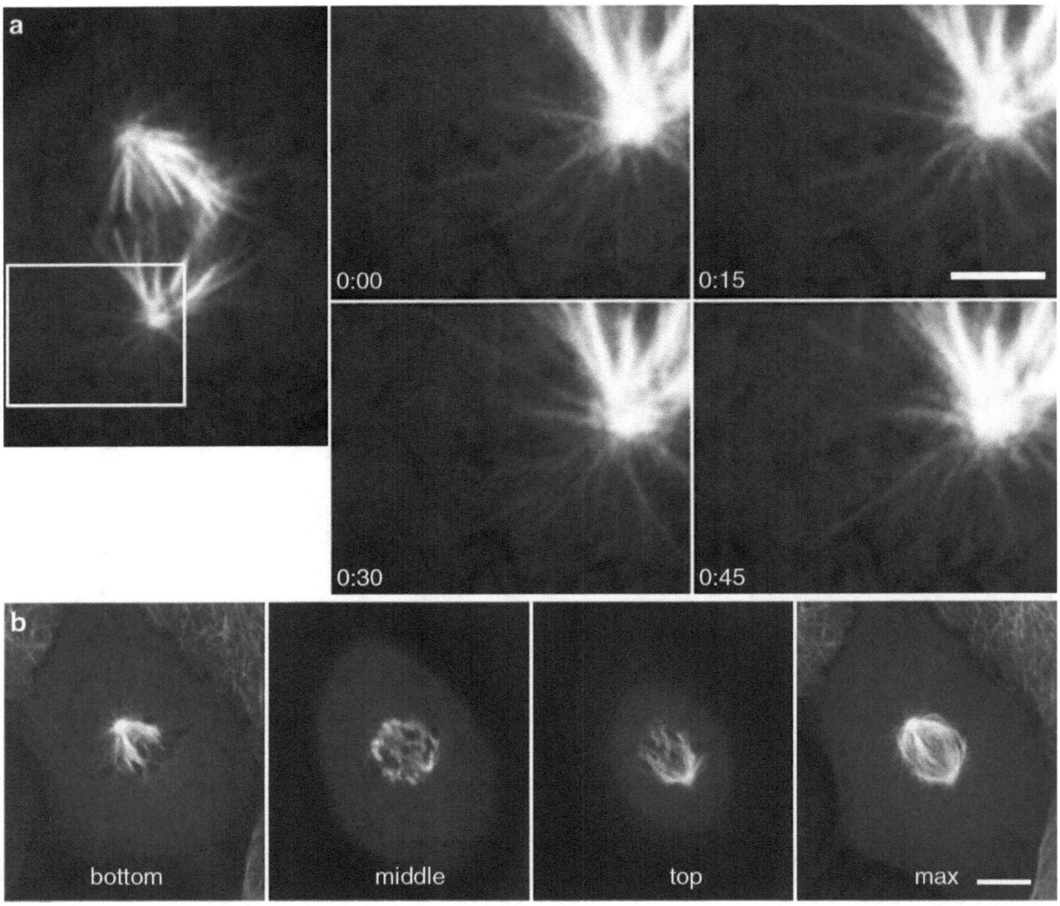

Fig. 2 Live LLC-Pk1 cells expressing GFP-tubulin, images acquired using spinning disc confocal microscopy, 100× objective lens. (**a**) Astral microtubule dynamics. *Left* shows single plane of a prometaphase cell; *right panels* show selected images at 15 s intervals. (**b**) *Z*-stack of images of a cell with a tilted mitotic spindle. *Bottom*, *middle*, and *top planes* are shown; *right panel* shows maximum intensity projection of complete *Z*-series. Bar = 5 μm in (**a**) and 10 μm in (**b**)

11. Shut down and clean up. Turn off all parts of the system in the order recommended by the manufacturer. Archive data from imaging computer.
12. Clean the objective lens; oil can damage lens elements and must be removed thoroughly at the end of the imaging session (*see* **Note 9**).

3.2.3 Protocol 3: Imaging Live Mitotic Cells with Laser Scanning Confocal Microscopy (LSCM)

Follow **steps 1–4** from protocol 2.

1. Using software controls, set the scan speed and laser power. Acquire a single scan. Adjust parameters so that the fluorescent structure of interest is visible. Observe the histogram of pixel intensity distribution to check for saturated pixels; adjust imaging parameters using the live scan mode so that no pixels are

saturated. If necessary, use averaging to decrease noise in the image (*see* **Note 10**).

2. Use microscope software to increase the magnification using the zoom feature. Use Nyquist sampling to obtain the theoretical resolution limit for the objective.
3. Collect a time-lapse sequence for the duration of mitosis (*see* Fig. 1a). See Protocol 2, **step 7**.
4. To collect a *Z*-stack of images, use the microscope software to mark the top and bottom of the region you wish to image. See Protocol 2, **step 8**. Select the desired number of *Z*-planes to collect. Acquire the *Z*-stack (Fig. 2b).
5. Most laser scanning microscopes allow simultaneous acquisition of multiple wavelengths. However, it must be noted that during simultaneous acquisitions, it is possible for the emission signal to bleed over into the red shifted wavelengths and create artifacts. To avoid this, it is best to acquire images sequentially if the fluorophores used overlap in their excitation/emission profiles.
6. Shut down and clean up. See Protocol 2, **steps 11, 12**. Turn off all parts of the system in the order recommended by the manufacturer. Archive data from imaging computer. *See* **Note 9**.

4 Notes

1. Two basic chamber designs are typically used for imaging cells by fluorescence microscopy: glass bottom cell culture dishes and closed chambers that hold a coverslip [4]. Of the two, the dish with a coverslip bottom (e.g., MatTek dishes) is the simplest. Cells are plated (at the desired density) in the dishes, which come in various sizes, 1–2 days prior to experimentation. The main drawback of the dishes is evaporation of the medium and the resulting condensation on the lid. Evaporation can be reduced by placing a layer of mineral oil on the top of medium in the dish or by using vacuum grease to seal the lid to the dish bottom. Another drawback is that the dishes are not always perfectly flat or stable depending on the stage holder, and the lack of stability raises the possibility of spills from the dish onto the microscope stage or objective lens. The advantage of the dishes is primarily the ease of use and the fact that inhibitors, or other reagents, can be added by simply lifting the top of the dish and exchanging the medium.

   As an alternative to an open dish, we and others use closed chambers. One example is the Rose Chamber [4, 9], which consists of two metal plates and a middle spacer that have a central hole. The coverslip is placed on the bottom plate, covered with the spacer, the top plate is added and the entire assembly is screwed together. Medium is added and a top coverslip (circular, 25 mm diameter) which acts at the lid. (Alternatively, the top coverslip can be placed beneath the top plate and medium can be added and removed using two syringes and needles inserted into the spacer). All parts of the chamber can be autoclaved if it is important to maintain sterility. We find that cell health is improved in these closed chambers and that the metal, once warmed to 37 °C, retains heat better than a plastic dish, resulting in reduced temperature fluctuations. Additionally, the chambers rarely leak. The chambers can be placed in an incubator, after the experiment, for subsequent observation, and the top lid can be removed for adding inhibitors or fixative. The chambers can be home fabricated in a machine shop; similar chambers are commercially available. Although there is up-front expense, the chambers can be reused for years, eliminating the need for disposable plastic dishes.

2. Selection of objective lens. In fluorescence microscopy, image brightness varies as the fourth power of the objective numerical aperture and is inversely proportional to the square of objective magnification https://www.microscopyu.com/articles/formulas/formulasimagebrightness.html.

   For imaging mitosis, a 60× or 100× oil immersion objective lenses of NA 1.4 is ideal. Using an objective lens that is corrected for both spherical and chromatic aberrations (plan apochromatic lenses have both corrections) is also important, especially if co-localization of multiple fluorophores will be performed. Microscope objectives are designed for use with coverslips with a thickness of 0.17 mm, corresponding to a number 1.5 coverglass. Refractive index differences between the lens immersion medium and the sample contribute to spherical aberration; if live cells will be imaged, a lens designed to be used with different refractive index media (i.e., water) can be used [10].

3. Mammalian cells are typically maintained at 37 °C during imaging, although cell division will continue, more slowly, at somewhat lower temperatures (30–36 °C). Microscope manufacturers provide a number of options for regulating temperature, ranging from environmental chambers that enclose all or a large portion of the microscope to smaller stage heaters, or air heating devices [4]. Environmental chambers that also regulate humidity and gas level, which need to be controlled when imaging for long periods, can also be used. If such an environmental chamber is not available or desired (some users find that the large enclosure impedes access to the microscope), we find that a stage heater is an excellent choice, especially when

used along with an objective lens collar that warms the lens so that it does not function as a heat sink. To measure the temperature during imaging, use a Thermistor probe with a microsensor that can be placed in the culture dish. This will provide a much more accurate measure of the temperature in the dish than the read-out of the stage heater.

4. Selecting a Microscope system. Because several different types of fluorescence microscope systems can be used for imaging mitotic cells, understanding their principle features is important to make an informed choice. The first issue is to determine if a conventional wide field or a confocal microscope is required. The principle advantage of confocal microscopy is the capacity to remove fluorescence light coming from focal planes above and below the plane of focus. By removing this light, the ability to detect fine structure is greatly improved. However, it is important to recognize that confocal imaging per se does not significantly change the resolution limit of the light microscope, assuming all other factors are the same, e.g., objective lens NA. Because the mitotic spindle is a three-dimensional structure, the advantage of using a confocal microscope to remove out of focus fluorescence is significant [3]. That being said, a similar improvement can be accomplished by collecting a *Z*-series of images with a conventional wide field microscope, followed by deconvolution to remove out of focus fluorescence [11]. In fact some would argue that wide field microscopy and deconvolution offer important advantages over confocal, including speed of acquisition and the fact that essentially all the light is collected, whereas in confocal, some light is rejected at the pinhole. For additional information the reader is directed to http://micro.magnet.fsu.edu/primer/.

   A laser scanning confocal microscope (LSCM) is built on a conventional optical microscope stand. To generate an image, laser light, focused by the objective lens, is raster scanned across the sample. Fluorescence light emitted from the sample is collected, also by the objective lens, and passes through a pinhole, which eliminates fluorescence coming from above and below the plane of focus. The detector is a sensitive photomultiplier tube (PMT). The scanning, collection and display of the resulting image is under computer control. To obtain information from different *Z* depths of the sample, the microscope has a motorized *Z* control (accomplished by moving the objective lens or the stage). Thus, the user can generate a *Z*-stack of images; the thickness of each plane in *Z* is controlled by the diameter of the pinhole and the number of discrete images taken.

   Modern, turnkey LSCM systems are equipped with multiple solid-state or gas lasers, each generating a specific wavelength (or wavelengths) of light. In practice this means that multiple fluorophores can be excited and imaged, either sequentially or in some cases simultaneously. Additionally, many LSCM systems are equipped with spectral detectors that can distinguish fluorophores by their emission spectra. Thus an important advantage of such systems is their use in experiments that require use of several probes or for experiments such as FRET where there is some spectral overlap.

Because the laser light scans the sample, LSCM systems are well suited for both photoactivation and photobleaching experiments. Using software controls, local activation or bleaching can be easily performed by selective scanning of a predetermined region at the appropriate laser wavelength and power.

Obtaining the best images with a LSCM system requires understanding of the various parameters that can be controlled during imaging. For example, by controlling the laser scan speed the amount of excitation light illuminating (and collected from) the sample can be varied; slower scanning results in more light collected, but more bleaching. Rapid scan rates typically generate noiser images; to reduce noise, multiple scans can be averaged. The laser power can be regulated independently of scan speed and averaging. Finally, LSCM systems have a zoom feature that allows the user to decrease the area that is scanned, essentially allowing for a range of magnifications from a single objective lens. Generally, for live cell imaging, less light (lower laser power) and faster acquisition (rapid scan speed) is preferred to keep the cells dividing, and typically this must be determined empirically.

The major alternative to a LSCM is a Spinning Disc confocal microscope (SDCM). The major difference between the two types of confocals is that in a SDCM multiple points of laser light scan the sample simultaneously, thus typically reducing the time required to collect a single image [8]. For example, using SDCM, images of microtubules in mitotic cells can be acquired at approximately one second intervals for many minutes with minimal photobleaching [8, 12]. Another difference in the two types of system is that a SDCM uses a camera as the detector, not a PMT. EMCCD cameras are frequently used for SDCM systems because of their high sensitivity, although the newer CMOS technology allows faster readout and smaller pixels than most EMCCD cameras, thus potentially increasing resolution and speed [13]. SDCM systems with ports for more than one camera are also available so that the user can select the camera best suited for the particular experiment. The microscope has optics that can magnify the image from the microscope onto the camera chip, to achieve Nyquist sampling. Although speed is the major advantage of the SDCM, the pinhole size is fixed in the microscope design, so the user cannot vary this parameter. Both LSCM and SDCM systems are available with multiple lasers for imaging a range of fluorescent probes.

A third alternative for imaging mitosis is to use a wide-field fluorescence microscope and perform deconvolution to remove out-of-focus fluorescence [11]. Turnkey systems are available for this mode of microscopy (e.g., DeltaVision). The major advantages include the speed of acquisition (no scanning is used) and the fact that a greater percentage of the emitted light is collected, because there is no pinhole. However, images must be processed post-acquisition and depending on the system, this can be slow.

5. Photobleaching of the fluorescent molecule of interest is perhaps one of the most frustrating issues in fluorescence microscopy. Possible remedies include using multiple fluorophores (e.g., triple GFP tag, [14]), switching to a fluorophore that is brighter and/or bleaches more slowly [15]; using genetically encoded tags that can be modified with a bright organic dye (e.g., Halo tag and Halo ligand). Adjusting the imaging parameters (i.e., exposure time, light intensity) can also help reduce photobleaching; the optimum settings need to be determined empirically for your sample.
6. During a time-lapse acquisition, the focal plane can change. Typically this results from drift of microscope components due to thermal motion—which is particularly notable when the stage is being heated for live cell imaging. Newer equipment overcomes this issue by holding a set focus using a far red laser beam reflected off the lower surface of the coverslip. In addition to drift resulting from thermal motion, mitotic cells round up to some degree during mitosis. In this case, the structure of interest may go out of focus significantly as the cell changes shape and necessitates changing the focus to keep the structure of interest in focus. Although this can be frustrating, especially during long acquisitions, with practice the user can learn to refocus the cell during the imaging sequence.
7. Typically, progression through mitosis is adversely impacted if a complete *Z*-stack is acquired at multiple, consecutive time points. One solution is to collect the fewest number of *Z*-planes that are required to obtain the necessary information [16], and to increase the time interval between acquisitions. Alternatively, if the information in *Z* is essential, a complete *Z*-stack can be taken at different time points, using different cells if necessary.
8. Collecting multicolor images will expose the mitotic cell to additional light. Using the software, you can acquire images of

each fluorophore at each time point, or take fewer images of one (or more) of the fluorophores, depending on the time resolution that is needed for each.

9. Cleaning oil off of the objective lens is particularly important for inverted microscopes and when the microscope is heated. Under these conditions the warm oil will flow down the sides of the objective lens. One solution is to use a fabric-coated elastic band (e.g., ponytail holder) placed around the lens; the fabric absorbs the oil and the elastic secures it to the lens.
10. We find that to preserve cell health, faster scan time, lower laser power, and 2× averaging work well.

References

1. Poser I, Sarov M, Hutchins JRA, Heriche J-K, Toyoda Y, Pozniakovsky A, Weigl D, Nitsche A, Hegemann B, Bird AW, Pelletier L, Kittler R, Hua S, Naumann R, Augsburg M, Sykora MM, Hofemeister H, Zhang Y, Nasmyth K, White KP, Dietzel S, Mechtler K, Durbin R, Stewart AF, Peters J-M, Buchholz F, Hyman AA (2008) BAC transgeneOmics: a high-throughput method for exploration of protein function in mammals. Nat Methods 5:409–415
2. Munoz IM, Szyniarowski P, Toth R, Rouse J, Lachaud C (2014) Improved genome editing in human cell lines using CRISPR method. Plos One 9:e109752
3. Pawley JB (2006) Handbook of biological confocal microscopy, 3rd edn. Springer, New York, NY
4. Wadsworth P (2010) Studying mitosis in cultured mammalian cells. In: Goldman RD, Swedlow JR, Spector DL (eds) Live cell imaging a laboratory manual. Cold Spring Harbor Laboratory Press, Cold Spring Harbor, NY, pp 571–582
5. Swedlow JR (2010) In vivo imaging of mammalian cells. In: Goldman RD, Swedlow JR, Spector DL (eds) Live cell imaging a laboratory manual. Cold Spring Harbor Laboratory Press, Cold Spring Harbor, NY, pp 317–331
6. Goodson HL, Dzurisn JS, Wadsworth P (2010) Methods for expressing and analyzing GFP-tubulin and GFP-microtubule-associated proteins. In: Goldman RD, Swedlow JR, Spector DL (eds) Live cell imaging: a laboratory manual. Cold Spring Harbor Laboratory Press, Cold Spring Harbor, NY, pp 605–622
7. Wadsworth P, Rusan NM, Tulu US, Fagerstrom C (2005) Stable expression of fluorescently tagged proteins for studies of mitosis in mammalian cells. Nat Methods 2:981–987
8. Maddox P, Moree B, Canman JC, Salmon ED (2003) Spinning disk confocal microscope system for rapid high-resolution, multimode, fluorescence speckle microscopy and green fluorescent protein imaging in living cells. Methods Enzymol 360:597–617
9. Rose GG, Pomerat CM, Shindler TO, Trunnel JB (1958) A cellophane strip technique for culturing tissue in multipurpose culture chambers. J Biophys Biochem Cytol 4:761–764
10. Waters JC (2009) Accuracy and precision in quantitative fluorescence microscopy. J Cell Biol 185:1135–1148
11. Swedlow JR, Sedat JW, Agard DA (1997) Deconvolution in optical microscopy. In: Jansson PA (ed) Deconvolution of images and spectra. Academic, New York, pp 284–309
12. Rusan NM, Wadsworth P (2005) Centrosome fragments and microtubules are released and transported asymmetrically away from division plane in anaphase. J Cell Biol 168:21–28
13. Winter PW, Shroff H (2014) Faster fluorescence microscopy: advances in high speed biological imaging. Curr Opin Chem Biol 20:46–53
14. Lee W-L, Oberle JR, Cooper JA (2003) The role of the lissencephaly protein Pac1 during nuclear migration in budding yeast. J Cell Biol 160:355–364
15. Rizzo MA, Davidson MW, Piston DW (2010) Fluorescence protein tracking and detection. In: Goldman RD, Swedlow JR, Spector DL (eds) Live cell imaging a laboratory manual. Cold Spring Harbor Laboratory Press, Cold Spring Harbor, New York, pp 3–34
16. Magidson V, O'Connell CB, Loncarek J, Paul R, Mogilner A, Khodjakov A (2011) The spatial arrangement of chromosomes during prometaphase facilitates spindle assembly. Cell 146:555–567

Chapter 2

Using Photoactivatable GFP to Study Microtubule Dynamics and Chromosome Segregation

Bin He and Daniela Cimini

Abstract

Mitosis is a highly dynamic process during which the genetic material is equally distributed between two daughter cells. During mitosis, the sister chromatids of replicated chromosomes interact with dynamic microtubules and such interactions lead to stereotypical chromosome movements that eventually result in chromosome segregation and successful cell division. Approaches that allow quantification of microtubule dynamics and chromosome movements are of utmost importance for a mechanistic understanding of mitosis. In this chapter, we describe methods based on activation of photoactivatable green fluorescent protein (PA-GFP) that can be used for quantitative studies of microtubule dynamics and chromosome segregation.

Key words Photoactivation, Mitosis, Microtubules, Chromosomes, Segregation, Poleward flux

1 Introduction

Accurate segregation of chromosomes during mitosis is critical for ensuring equal partitioning of the genetic material between the two daughter cells. Errors in mitotic chromosome segregation can lead to the formation of daughter cells with abnormal chromosome numbers, a condition known as aneuploidy, which is a leading cause of miscarriage and genetic disorders [1] and is ubiquitously found in cancer [2–4]. Mitotic cell division is a highly dynamic process resulting from the concerted dynamic behavior of different mitotic apparatus components. The microtubules of the mitotic spindle and the chromosomes, aided by numerous microtubule- and chromosome-associated molecular motors and non-motor proteins, are the major players in this process. A key event in chromosome segregation is the establishment of microtubule–chromosome interactions, with the major point of interaction between microtubules and chromosome being the kinetochore. During prometaphase, establishment of kinetochore–microtubule attachment leads to chromosome congression to the spindle equator. In a mature,

Paul Chang and Ryoma Ohi (eds.), *The Mitotic Spindle: Methods and Protocols*, Methods in Molecular Biology, vol. 1413,
DOI 10.1007/978-1-4939-3542-0_2, © Springer Science+Business Media New York 2016

metaphase mitotic spindle, the chromosomes are lined up at the spindle equator, forming the so-called metaphase plate. The two sister chromatids are bound to microtubules, with each sister kinetochore bound to microtubules from one pole (i.e., sister kinetochores bound to opposite poles). The bound microtubules remain dynamic, undergoing poleward flux and dynamic instability at the plus end (facing the kinetochore). These microtubule dynamics produce chromosome movement, so that in many cell types the aligned chromosomes oscillate back and forth about the metaphase plate. Moreover, kinetochore-bound microtubules turn over while maintaining stable kinetochore attachment. In other words, metaphase kinetochores are stably bound to bundles of microtubules (k-fibers), but individual microtubules within such k-fibers can detach from kinetochores (and disassemble) with certain probabilities, and this results in measurable turnover rates.

A number of perturbations can cause changes in microtubule dynamics, and when such changes are specific to kinetochore microtubules, they can result in kinetochore–microtubule attachment errors and chromosome mis-segregation. Thus, methods to quantify chromosome and microtubule dynamics in mitosis represent powerful tools for the identification of specific chromosome segregation errors and the identification of the responsible mechanisms.

Here, we specifically focus on the use of photoactivatable GFP to study microtubule dynamics, including poleward flux and turnover, and chromosome segregation. We describe experimental methods as well as data acquisition and analysis, and briefly discuss some of the caveats and precautions related to these methods.

2 Materials

2.1 Cells

1. PA-GFP-tub PtK1; PtK1 cells stably expressing photoactivatable-GFP-tubulin (*see* **Note 1**).
2. H2B-PA-GFP PtK1; PtK1 cells stably expressing photoactivatable-GFP-tagged H2B (*see* **Note 2**).

2.2 Media, Coverslips, Dishes, Chambers, Taxol

1. Ham's F-12 medium complemented with 10 % fetal bovine serum, antibiotics, and antimycotic. Occasionally, cells (particularly the PA-GFP-tub PtK1 cells) may be cultured for a few days in the presence of antibiotics (1 mg/ml G418 for PA-GFP-tub PtK1 cells and 0.1 μg/ml puromycin for H2B-PA-GFP PtK1 cells).
2. Acid-washed 22 mm × 22 mm, # 1.5 glass coverslips (*see* **Note 3**).
3. Sterile 35 mm polystyrene petri dishes or 6-well plate.

Fig. 1 Modified Rose chamber. (**a**) Image of disassembled Rose chamber with bottom plate (*left*), silicone gasket and screws (*center*), and top plate (*right*). (**b**) Assembled Rose chamber. The top plate shown in this picture is cut off to allow for easier access of microinjection needles or for viewing/imaging with short working distance condenser. For most applications, a cutoff top plate is not required

4. Modified Rose chambers, including top and bottom plates, silicone gaskets, and screws (Fig. 1; *see* **Note 4**).
5. 21 Gauge (21G) needles and 5 ml syringes.
6. Glass-bottom 35 mm petri dishes.
7. Imaging medium: Leibovitz's L-15 with 1.822 g/l HEPES, and 4.5 g/l glucose. Adjust pH to 7.2 with NaOH, and complement with 10 % fetal bovine serum and antibiotics/antimycotic solution.
8. Taxol, 5 mM stock solution in DMSO.

3 Methods

3.1 Microtubule Photoactivation

3.1.1 Background

Different populations of microtubules coexist within the bipolar mitotic spindle. The two major classes of microtubules relevant to this procedure are the kinetochore microtubules and the non-kinetochore microtubules present in the region between the two spindle poles. These two classes of microtubules display different dynamic behavior [5]. Indeed, attachment to the kinetochore results in microtubule stabilization, which causes kinetochore microtubules to be less dynamic than non-kinetochore microtubules. Importantly, a number of perturbations associated with mitotic defects cause changes in kinetochore microtubule dynamics, and abnormal kinetochore microtubule dynamics are associated with chromosomal instability in cancer cells [6–8]. Early studies measured microtubule poleward flux and/or microtubule turnover through a variety of methods, including fluorescence activation of injected caged-fluorescein tubulin [5, 9], a combination of microtubule depolymerizing drugs and electron microscopy [10], and fluorescent speckle microscopy [11]. In recent years, the emergence of photoactivatable and photoswitchable fluorescent proteins [12, 13] has allowed the generation of cell lines that express tubulin fused to a fluorescent protein that can be "switched on" at any given time and at any given location within the cell. Here, we describe the method used to measure rates of kinetochore microtubule poleward flux and spindle microtubule turnover in PtKl cells expressing tubulin fused to photactivatable green fluorescent protein (PA-GFP-tub PtKl).

3.1.2 Sample Preparation

1. If using glass-bottom dishes, plate PA-GFP-tub-PtKl cells in the dishes and replace culture medium with pre-warmed imaging medium prior to the experiment.
2. If planning to use modified Rose chambers, place sterile coverslips in 35 mm petri dishes or 6-well plate. To do so, use clean forceps to remove one acid-washed coverslip (*see* **Note 3**) from the ethanol-containing jar, drain the excess ethanol on the edge of the jar, and flame the coverslip prior to depositing it in a petri dish/well. As a cautionary measure, keep the flame source and the flaming coverslip away from the ethanol-containing jar.
3. Repeat the procedure to prepare as many coverslip-containing petri dishes/wells as needed for the experiment. Then, plate the cells on the sterile coverslips inside the petri dishes/wells. Perform the whole procedure under sterile conditions (i.e., biosafety cabinet).
4. Prior to the experiment, remove the coverslip and assemble a modified Rose chamber (described in [14]) with a top coverslip. Briefly, deposit the coverslip on the Rose chamber bottom plate, cell-side up, making sure to center it on the round hole; place the silicon gasket on top of the coverslip; place a top coverslip (again making sure to center it on the round hole); place the top plate; secure the chamber with screws (do not tighten too much at this stage).
5. To fill the chamber with imaging media, prepare a syringe with pre-warmed imaging media and a 21G needle. Then, insert a 21G needle (exhaust) into the silicone gasket on one side of the chamber and insert the needle of the syringe on the opposite side. Hold the chamber vertically, with the syringe at the bottom and the exhaust needle at the top and fill the chamber with imaging media making sure that as the media goes in, all the air bubbles are expelled out of the chamber through the exhaust needle. When the chamber is filled, further tighten the screws and clean the bottom coverslip with a Kimwipes moist with deionized water and then with a Kimwipes moist with 70 % ethanol.
6. Once ready, place the chamber or the glass-bottom dish on the stage of an inverted microscope equipped with a system for temperature control (*see* **Note 5**) and appropriate photoactivation (*see* **Note 6**) and imaging light sources.

3.1.3 Data Acquisition

1. After placing your sample on the microscope stage, use transmitted light and a DIC or phase-contrast 60× objective to locate a mitotic cell (typically in prometaphase or metaphase) and acquire pre-activation GFP fluorescence (480/40 nm illumination in our setup) and transmitted light images. Using the phase-contrast image on the screen, draw a ~0.5 μm-wide line perpendicular to the spindle long axis and located on one side of the metaphase plate. Using the appropriate software command, activate the fluorescence (1–2 s with our setup) within the selected region and immediately acquire post-activation fluorescence and phase-contrast images.
2. Acquire images (*see* **Note 7**) again after 10–15 s and then every 15–20 s thereafter for 5–10 min (or until the fluorescence signal disappears at the spindle pole due to microtubule poleward flux; Fig. 2, top, last frame).
3. Acquire data sets under each experimental condition as well as in samples pretreated for ~1 h with 10 μM of the microtubule-stabilizing drug [15] taxol (in growth medium) and maintained in 10 μM taxol (in imaging medium) during imaging. The taxol data will be used to calibrate for photobleaching (*see* **Note 8**).
4. Given that each cell is typically imaged for no longer than 10 min, multiple cells can be activated and imaged on the same coverslip one after the other both in taxol-treated and experimental samples.

3.1.4 Quantification of Microtubule Turnover

1. For each photoactivated cell (both experimental samples and taxol-treated samples), fluorescence dissipation must be quantified after background subtraction. To do this, draw a rectangular region around the area of activation and an equivalent rectangular region on the opposite, non-activated, side of the spindle. Measure the total integrated fluorescence within those two regions at each time point using any image analysis software (we use either NIS elements image analysis package or the open source program Fiji/ImageJ). For cells displaying normal microtubule dynamics, the fluorescent mark will move poleward over time. Consequently, the rectangular regions (for fluorescence and background measurements) must be moved at each time frame to follow the poleward movement of the activated mark. For taxol-treated cells, the mark is expected to be static, and therefore, quantification of background fluorescence and fluorescence intensity of the activated region can be performed without moving the rectangular quantification regions.
2. Fluorescence intensity at each time point will be obtained by subtracting the background value from the value of the fluo-

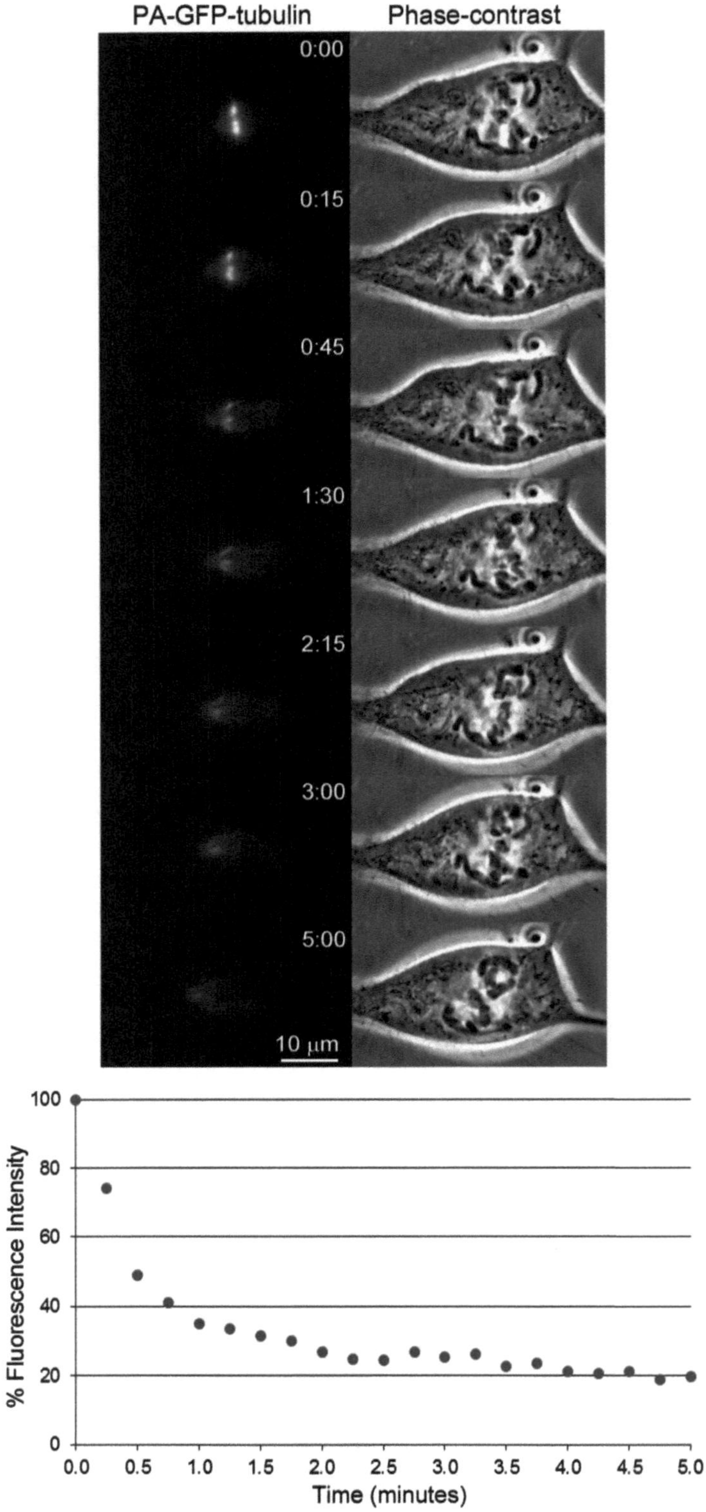

Fig. 2 Microtubule photoactivation. The images (stills from a time-lapse movie) show photoactivation of PA-GFP-tubulin (*left column*) in a prometaphase PtK1 cell (chromosomes are visible in the phase-contrast images on the *right column*). A narrow rectangular region across the left half spindle was activated and fluorescence and phase-contrast images were acquired nearly simultaneously immediately after activation and every 15 s thereafter. Elapsed time is shown in min:sec. The graph at the bottom reports the change in fluorescence intensity over time within the activated region (see text for details on how to quantify the fluorescence intensity)

rescent mark. Quantification of fluorescence dissipation over time in taxol-treated cells will represent the loss of fluorescence due to photobleaching.

3. To calculate the average photobleaching-dependent fluorescence loss, collect data from 10 or more taxol-treated cells and average the fraction of fluorescence loss over time. The fluorescence intensity will be normalized to 1 at t_0 (time point immediately after photoactivation) and the fluorescence intensity at subsequent time points will be a fraction of the fluorescence intensity at t_0, and we will refer to them as photobleaching values. The data from taxol-treated cells are expected to display a linear, but very slow (how slow will depend on the imaging system, exposure times, etc., which will affect how much photobleaching occurs at each time point), decay.
4. After quantifying the fluorescence intensity for the experimental data (fluorescence intensity of activated mark minus background), divide the values obtained by the "photobleaching values." These final values will represent the fluorescence intensity values after background subtraction and photobleaching correction and should be represented as percent fluorescence intensities, where the t_0 value is set to 100 % (see graph in Fig. 2).
5. Data from multiple cells are typically averaged [16, 17] for representation and for quantitative analysis (see details below).
6. To obtain quantitative information about microtubule dynamics, the data should be subject to nonlinear regression analysis. The best fit for this type of data is a double exponential curve [5], with equation $y = A_1 \times \exp(-k_1 \times t) + A_2 \times \exp(-k_2 \times t)$, where A_1 and A_2 are the percentages of the total fluorescence contributed by, and hence the percentages of, non-kinetochore and kinetochore microtubules, respectively; k_1 and k_2 are the rate constants of turnover/fluorescence dissipation for non-kinetochore and kinetochore microtubules, respectively; and t is time after photoactivation. Microtubule half-lives ($t_{1/2}$) can be calculated as follows: $t_{1/2} = \ln 2/k_1$ for non-kinetochore microtubule; $t_{1/2} = \ln 2/k_2$ for kinetochore microtubules. Thus, this method allows quantification of both the percentage of microtubules incorporated into k-fibers vs. unbound microtubules and turnover rates of these two populations of microtubules.

3.1.5 Quantification of Microtubule Poleward Flux

Due to the bundling of microtubules within k-fibers, the fluorescence marks after activation look particularly well-defined along k-fibers. Quantifying microtubule poleward flux can easily be performed by tracking the poleward movement of such marks on the k-fibers. In PtK1 cells, a phase-contrast image of good quality allows visualization of the spindle pole. Alternatively, at the end of the experiment, the whole spindle can be activated and imaged, and this image can be used as a reference for distances from the spindle poles. Poleward flux can be quantified by simply measuring the mark-to-pole distance at subsequent time points (either manually or with simple MatLab scripts) and then calculating the rate of flux as d/t, where d is the distance (in μm) traveled by the mark over a given period of time, t (in minutes), which will give a rate in μm/min.

3.2 Chromosome Photoactivation

3.2.1 Background

Accurate chromosome segregation is an absolute requirement for maintenance of cell and organismal function. Indeed, chromosome segregation errors result in aneuploidy, which is a leading cause of miscarriage and genetic disorders in humans [1] and has been long recognized as a hallmark of cancer [2–4]. In addition to being aneuploid, most cancer cells also display high rates of chromosome mis-segregation [7, 18–21], a phenotype typically referred to as chromosomal instability, or CIN [21]. Given the dire consequences of chromosome segregation errors, many methods have been developed over the years to examine chromosome segregation during mitosis. These methods, including both fixed and live-cell approaches, allow labeling of one or a few specific chromosomes or labeling of all chromosomes. Fixed-cell analysis of anaphase cells, possibly combined with immunostaining of kinetochore proteins [22, 23], or live imaging of cells expressing histone-GFP [24, 25] are common methods for visualization of all chromosomes. These methods, however, preclude observation of certain chromosome segregation errors, such as that occurring when two sister chromatids segregate to the same spindle pole. To obviate this limitation, fluorescence in situ hybridization with chromosome specific probes has been performed on ana-telophase cells [24, 26, 27], but this approach can be very laborious and is unlikely to provide insight into the mechanism responsible for the mis-segregation events. Alternatively, individual chromosomes have been visualized in live cells using the LacO-LacI system, which relies on the integration of a Lac operator array at a specific locus and expression of a fluorescent protein-tagged Lac repressor [28]. This approach has been extensively used in yeast and occasionally to study chromosome segregation in vertebrate cells [29, 30]. However, with this method the site of integration along vertebrate chromosomes cannot be controlled. Recent efforts have attempted to overcome this limitation through the development of CRISPR-based methods [31, 32]. Photoactivation of H2B-PA-GFP is a relatively simple method that offers many advantages over other currently used approaches. Indeed, any chromosome can be marked and different chromosomes can be marked in different mitotic cells. Moreover, several chromosomes within the same cell can be activated if necessary and the mark can be created in any position along the chromosome.

Most importantly, the decision of which chromosome or which region of a chromosome to mark and follow can be made at the time of imaging rather than at the time of experiment set up or establishment of the cell line.

3.2.2 Sample Preparation

For sample preparation, plate H2B-PA-GFP PtK1 cells following the instructions provided in Subheading 3.1.2.

3.2.3 Data Acquisition

1. After placing your sample on the microscope stage, use transmitted light and a DIC or phase-contrast 60× objective to locate a mitotic cell (typically in early prometaphase) and acquire a transmitted light image.
2. Using the phase-contrast image on the screen, select a region (*see* **Note 9**) on one or two chromosomes of choice for photoactivation (*see* **Note 6**). Using the appropriate software command, activate the fluorescence within the selected region and acquire post-activation fluorescence and phase-contrast images immediately after activation and thereafter at regular time intervals.
3. The activation time varies depending on the goal of the experiment (as well as expression level of PA-GFP in the cell), and it will be determined accordingly (*see* **Note 10**).
4. The time intervals for imaging of photoactivated chromosomes should be chosen in such a way that will allow imaging through mitosis without causing excessive bleaching of the activated signal (*see* **Note 11**). One way to keep the cell healthy and the fluorescence visible long enough is to monitor or image the cell by phase-contrast and only periodically acquire a fluorescence image, until the cell reaches the phase of interest (e.g., metaphase-anaphase transition if the goal is to examine chromosome segregation). At that point, GFP and phase images can be acquired simultaneously and at shorter time intervals. With our setup we can follow cells through anaphase (~20 min) imaging at 30 s intervals. However, the total imaging time and time intervals can vary widely depending on the purpose of the experiment and the type of microscope setup.

3.2.4 Data Analysis

The type of analysis will depend on the goal of the experiment, and we will describe here three different types of experiments and their respective data analysis. A small region on one or two chromosomes can be marked in early prometaphase cells. Alternatively, we have marked a single region on the last unaligned chromosome in prometaphase cells. In both cases, analysis of the time-lapse video will provide information on the segregation of the marked chromosome(s). In addition to this qualitative information, the mark on individual chromosomes could also be used to determine the dynamics of the chromosome. Particularly, if the mark is located

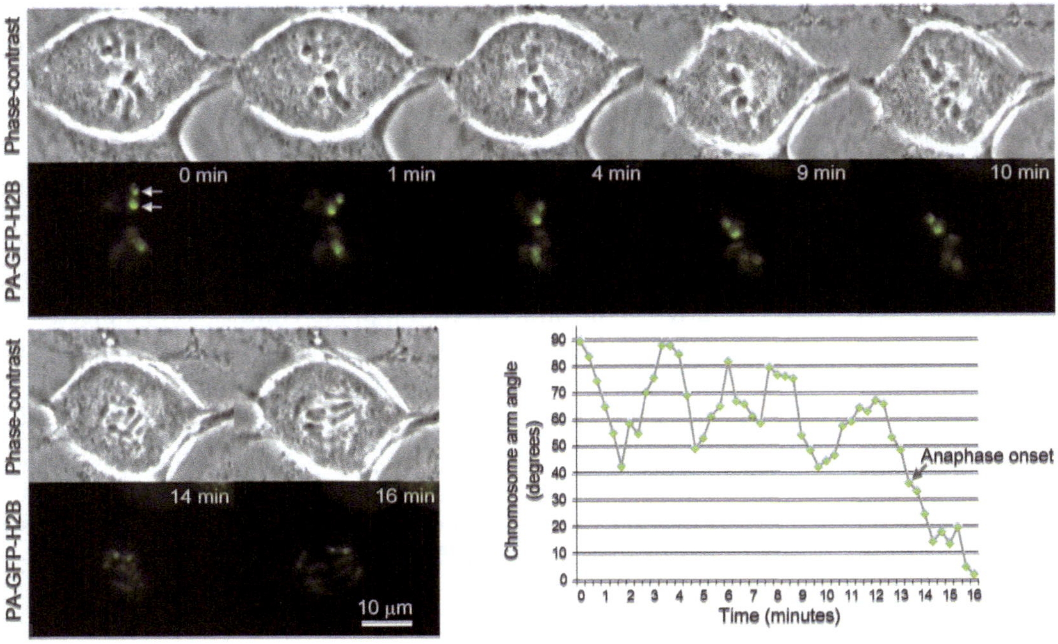

Fig. 3 Chromosome photoactivation. The images (stills from a time-lapse movie) show photoactivation of two points (*arrows* in 0 min image) on the same chromosome arm of a PtK1 cell. The chromosomes can be viewed in the top row of phase-contrast images, whereas the photoactivated GFP signals are shown in the bottom row of images. The cell was imaged from metaphase into anaphase, with anaphase onset occurring at ~13 min, as indicated in the graph. During image analysis, the movement of the two activated points was tracked (using a MatLab script) in order to describe the behavior of the chromosome arm. Using these data, we calculated the angle between the chromosome arm and the spindle long axis. The change of this angle over time is shown in the graph at the bottom right. From the data reported in the graph it is evident that the orientation of the chromosome arm displays an oscillatory pattern resulting from spindle-dependent forces acting on the chromosome. After anaphase onset, the marks are only visible on one sister chromatid (due to activation at a single focal plane), and as this chromatid starts its poleward movement, the angle approaches 0°, as expected as a result of the typical V-shaped orientation of segregating chromosomes, in which individual chromosome arms become nearly parallel to the spindle axis

in proximity of the centromere, tracking its movement in metaphase will allow quantification of directional instability [33]. A MatLab script [34], commercial image analysis software, or open source programs, such as Fiji/ImageJ, can be used for tracking. If the mark is located away from the centromere on a chromosome arm (Fig. 3), the arm dynamics can also be quantified. By marking two spots on a single chromosome arm (Fig. 3), we were able to quantify the changes in orientation of the chromosome arm with respect to the spindle axis. Interestingly, this type of experiment revealed that in PtK1 cells the arms of peripheral metaphase chromosomes undergo extensive oscillations (Fig. 3), even though their centromeres/kinetochores were previously shown to not oscillate [34–37]. This method could be used to investigate arm dynamics for chromosomes at various spindle locations, which would, in turn, reveal possible variations in the forces exerted on chromosomes in different regions of the mitotic spindle. Finally, activation of a whole chromosome-containing micronucleus in late prophase makes it possible to distinguish that given chromosome from all the others upon nuclear envelope breakdown [38]. Analysis of the time-lapse movie enables studies of the detailed behavior, through mitosis, of the micronucleus chromosome in terms of segregation and position relative to other chromosomes or the spindle [38].

4 Notes

1. In principle, the methods described here could be applied to any adherent cell line. However, we described methods that were specifically optimized for variants of the PtK1 cell line, a stable epithelial cell line derived from the kidney of a normal adult female rat kangaroo (*Potorous tridactylus*). PtK1 cells are easily maintained in culture and have been used as a model of choice for many studies of mitosis, particularly those based on imaging approaches. Two key features, remaining flat during mitosis and possessing a small number of large chromosomes, make these cells particularly suited for microscopy-based studies and allow for easy identification and/or tracking of individual chromosomes/kinetochores and individual k-fibers. Two minor drawbacks with this cell line are that the *Potorous tridactylus* genome has not been sequenced and that PtK1 cells do not transfect well with standard lipid-based methods or electroporation. Nevertheless, several groups have been able to sequence a number of PtK1 genes or to infer sufficient sequence information to design efficient siRNAs ([39, 40]; Shah and Berns, personal communication). Moreover, both lipid-based transfection methods and retroviral gene transduction have been successfully employed to generate PtK1 or PtK2 (equivalent cell line derived from male *Potorous tridactylus*) cell lines stably expressing a variety of gene products [24, 36, 41, 42]. The large size of these cells and the fact that they remain flat during mitosis has also enabled the use of microinjection in mitotic cells as a method to inhibit specific target proteins, introduce dominant-negative mutant proteins, or introduce fluorescently labeled proteins [17, 37, 43–47]. The PA-GFP-tub PtK1 cell line was initially generated by Alexey Khodjakov [36] using lipid-based transfection and has been previously used in several studies to quantify microtubule poleward flux and turnover rates [16, 17, 36].
2. The H2B-PA-GFP PtK1 cell line was generated by Daniela Cimini (while still at UNC, Chapel Hill) using retrovirus-mediated gene transduction. These cells are currently being used

to uncover novel mechanisms of chromosome mis-segregation and to investigate the mitotic behavior of micronucleus-derived chromosomes (He and Cimini, unpublished).

3. Acid washed coverslips are prepared as follows. First, separate coverslips (from 2 to 4 packages) from one another and place them into a glass beaker containing the amount of water necessary to make a 1 M HCl solution that will cover the coverslips. Add the required amount of 37 % HCl to make a 1 M solution. Loosely cover the glass beaker with a watch glass and heat the 1 M HCl solution at 50–60 °C for 4–16 h. Cool to room temperature, pour the HCl solution out, and rinse out the 1 M HCl with deionized water. Fill container with deionized water and sonicate in ultrasonic cleaner for 30 min. Pour the water out, fill the beaker with fresh deionized water, and sonicate in ultrasonic cleaner for 30 min. Repeat this last step one more time. Pour the water out, fill the beaker with 50 % ethanol, and sonicate in ultrasonic cleaner for 30 min. Pour the 50 % ethanol out, fill the beaker with 70 % ethanol, and sonicate in ultrasonic cleaner for 30 min. Pour the 70 % ethanol out, fill the beaker with 95 % ethanol, and sonicate in ultrasonic cleaner for 30 min. Pour the 95 % ethanol out, transfer the coverslips to a glass jar with screw cap, fill with 95 % ethanol, and store for later use.
4. Rose chambers (Fig. 1) are typically custom-made by in-house machine shops. Note that the plates should not be painted/varnished (the paint/varnish will peel off), but anodized. The modified Rose chambers should be disassembled immediately after use and silicone gaskets, top and bottom plates should be immersed in deionized water to prevent the media from drying out on the plates and causing corrosion. Top and bottom plates and silicone gaskets can be washed in bulk and this should be done manually using an appropriate detergent, such as Sparkleen 1 (Fisher Scientific). To wash, immerse the plates and gaskets in a large container (e.g., a dish pan) with a solution of deionized water and Sparkleen 1, scrub each piece with a glassware brush and transfer it into a secondary large container filled with deionized water. Rinse one more time and conclude by rinsing under running deionized water. Let the parts dry before use.
5. We typically perform our experiments at ~35–36 °C either on a microscope stage heated with an air stream incubator (Nevtek) or in a Tokai Hit stage top incubator, and we make sure that the system has reached the desired temperature prior to starting the experiment. We also wait 5–10 min after placing the dish/Rose chamber on the stage, to make sure the sample has reached the correct temperature prior to the beginning of the experiment. Different groups use slightly different temperatures [5, 9, 16, 36], but temperatures slightly below 37 °C

appear to protect cells from photodamage [48]. Importantly, because temperature is known to affect the rates of tubulin polymerization/depolymerization, it is important to maintain the temperature constant for all experimental conditions.

6. We perform our photoactivation experiments on an epifluorescence microscope (Nikon Ti-E) equipped with a Mosaic Photoactivation System (Photonic Instruments/Andor) consisting of digital diaphragm optical head with micromirror array and optics with transmission spectra <370–700 nm. For activation, we use a 100 W Olympus mercury lamp and a dichroic mirror that transmits light at 365–435 nm and reflects above 435 nm (a 405 nm laser can also be used for activation of PA-GFP). The Mosaic system allows for activation of user-defined regions of interest of any shape, size, and complexity.

   Other methods, besides the Mosaic digital diaphragm, can be used for fluorescence photoactivation. In earlier studies, microtubule fluorescence activation was achieved by focusing the activation light through a narrow slit introduced into the light path of a standard epifluorescence microscope [9, 49, 50]. In other more recent studies, photoactivation of a pseudo-Gaussian line region was achieved with a 408 nm laser focused by a cylindrical lens [16, 17, 36]. The problem with these methods is that the activation line/bar will have a fixed orientation, which requires either exclusive selection of cells whose spindle has the right orientation (i.e., spindle long axis perpendicular to the activation line/bar) or the use of a rotating stage [9]. Finally, another recent commercial system that has been used by some groups to study kinetochore microtubule dynamics [6, 51] is the Micropoint System (Photonic Instruments). Whereas this system shares some of the advantages offered by the Mosaic System, the area of activation is limited to a spot instead of a user-selected region. Thus, this system could be easily employed to activate small regions along chromosome arms, but would limit microtubule activation to a small region of the spindle rather than a region spanning the spindle width (i.e., all the k-fibers on a given focal plane). In fact, a bar could be created by activating multiple aligned spots, but this may compromise the accuracy of the quantification, given that different points along the line would be activated at slightly different time points.

7. When choosing image acquisition settings for photoactivated tubulin, the first point that should be considered is the level of PA-GFP-tubulin expression in the cell of interest that should be high enough to allow generation of a mark that fluoresces significantly above the background. To generate a mark with good signal-to-noise ratio, however, the exposure time for

post-activation image acquisition must also be carefully selected. Therefore, we suggest performing pilot experiments in which the variability in levels of PA-GFP-tubulin expression, as well as the optimal exposure times are evaluated and the experimental settings are optimized.

8. Due to imaging of the fluorescently labeled tubulin, some of the sample fluorescence will photobleach, and this photobleaching should be accounted for. To this end, it is important to collect a set of data under experimental conditions in which fluorescence dissipation is only due to photobleaching and not to microtubule turnover, which is achieved by performing tubulin photoactivation in taxol-treated cells, as described in Subheading 3.1.3.
9. For chromosome photoactivation, one important point to consider is the size and position of the activation region, which will vary depending on the goal of the experiment. We have successfully used photoactivation to mark chromosomes in several different ways. For instance, we labeled whole chromosome-containing micronuclei to study the segregation of such chromosomes at the mitosis following micronucleus formation [38]. In other experiments, we activated a small chromosome region close to the centromere in early prometaphase chromosomes (He and Cimini, unpublished). This results in labeling of both sister chromatids and allows for analysis of segregation of the two sisters, including segregation of two sisters to the same pole. We also performed a number of control experiments in which we labeled either entire nuclei or a portion of a nucleus and found that photoactivation *per se* does not affect chromosome segregation [38]. Finally, in some trial experiments, we generated two small marks on the same chromosome arm, which allowed us to measure chromosome arm movement (Fig. 3).
10. For chromosome photoactivation, the activation time will vary depending on the goal of the experiment (as well as expression level of PA-GFP in the cell). For example, when the goal is to mark a small region on a prometaphase chromosome, one second illumination results in good signal, although in some cases one of the two sisters may not be marked as well as the other. Depending on the mitotic stage and how condensed the chromosomes are, activation may be performed on more than one focal plane to make sure the whole chromosome is activated. However, this should be avoided when possible, because it may result in unwanted activation of other chromosomes. On the other hand, for activation of the bulk of chromatin within an entire micronucleus, we routinely use multiple 100 μs illumination pulses over five focal planes (spaced 0.6 μm of each other), including two focal planes above and two below the center focal plane.

11. One problem we encounter with activated H2B-PA-GFP is a general low-level fluorescence activation on all the chromosomes over time. This appears to be caused by slight photoactivation resulting from repetitive exposure to the GFP excitation wavelength (480/40 nm for our filter set) during imaging. This, combined with bleaching of the activated mark, leads to a reduction of the signal-to-noise ratio over time, thus limiting the time and frequency of imaging, despite PA-GFP initially being shown to remain stable for days [52]. Therefore, although we believe that labeling of chromosomes/chromosome regions by photoactivation has great potential in the study of chromosome segregation, the imaging set up must be carefully evaluated. Pilot experiments should be performed to determine exposure and imaging times compatible with collection of data that will address the desired experimental question(s). Moreover, as described in the methods Subheading 3.2.2, the fluorescent mark can be preserved by restricting GFP imaging in time. This can be achieved if the cell of interest is imaged by phase-contrast microscopy until it approaches the time window of interest, at which point imaging of the fluorescent signal is added.

Acknowledgements

We would like to thank Lisa Cameron (Dana Farber Cancer Institute) for critical reading of the manuscript. Funding in the Cimini lab provided by NSF grants MCB-0842551 and MCB-1517506 and HFSP grant RGY0069/2010.

References

1. Hassold T, Hunt P (2001) To err (meiotically) is human: the genesis of human aneuploidy. Nat Rev Genet 2(4):280–291
2. Cimini D (2008) Merotelic kinetochore orientation, aneuploidy, and cancer. Biochim Biophys Acta 1786(1):32–40
3. Mitelman F, Johansson B, Mertens F (2014) Mitelman database of chromosome aberrations and gene fusions in cancer. http://cgap.nci.nih.gov/Chromosomes/Mitelman
4. Weaver BA, Cleveland DW (2006) Does aneuploidy cause cancer? Curr Opin Cell Biol 18(6):658–667
5. Zhai Y, Kronebusch PJ, Borisy GG (1995) Kinetochore microtubule dynamics and the metaphase-anaphase transition. J Cell Biol 131(3):721–734
6. Bakhoum SF, Genovese G, Compton DA (2009) Deviant kinetochore microtubule dynamics underlie chromosomal instability. Curr Biol 19(22):1937–1942
7. Bakhoum SF et al (2014) The mitotic origin of chromosomal instability. Curr Biol 24(4): R148–R149
8. Bakhoum SF et al (2009) Genome stability is ensured by temporal control of kinetochore-microtubule dynamics. Nat Cell Biol 11(1):27–35
9. Mitchison TJ (1989) Polewards microtubule flux in the mitotic spindle: evidence from photoactivation of fluorescence. J Cell Biol 109(2):637–652
10. Cassimeris L et al (1990) Stability of microtubule attachment to metaphase kinetochores in PtK1 cells. J Cell Sci 96(Pt 1):9–15

11. Waterman-Storer CM et al (1998) Fluorescent speckle microscopy, a method to visualize the dynamics of protein assemblies in living cells. Curr Biol 8(22):1227–1230
12. Lippincott-Schwartz J, Patterson GH (2008) Fluorescent proteins for photoactivation experiments. Methods Cell Biol 85:45–61
13. Zhou XX, Lin MZ (2013) Photoswitchable fluorescent proteins: ten years of colorful chemistry and exciting applications. Curr Opin Chem Biol 17(4):682–690
14. Rieder CL, Hard R (1990) Newt lung epithelial cells: cultivation, use, and advantages for biomedical research. Int Rev Cytol 122:153–220
15. Schiff PB, Horwitz SB (1980) Taxol stabilizes microtubules in mouse fibroblast cells. Proc Natl Acad Sci U S A 77(3):1561–1565
16. Cimini D et al (2006) Aurora kinase promotes turnover of kinetochore microtubules to reduce chromosome segregation errors. Curr Biol 16(17):1711–1718
17. DeLuca JG et al (2006) Kinetochore microtubule dynamics and attachment stability are regulated by Hec1. Cell 127(5):969–982
18. Nicholson JM, Cimini D (2011) How mitotic errors contribute to karyotypic diversity in cancer. Adv Cancer Res 112:43–75
19. Nicholson JM, Cimini D (2013) Cancer karyotypes: survival of the fittest. Front Oncol 3:148
20. Thompson SL, Compton DA (2008) Examining the link between chromosomal instability and aneuploidy in human cells. J Cell Biol 180(4):665–672
21. Lengauer C, Kinzler KW, Vogelstein B (1997) Genetic instability in colorectal cancers. Nature 386(6625):623–627
22. Cimini D et al (2003) Histone hyperacetylation in mitosis prevents sister chromatid separation and produces chromosome segregation defects. Mol Biol Cell 14(9):3821–3833
23. Cimini D et al (2001) Merotelic kinetochore orientation is a major mechanism of aneuploidy in mitotic mammalian tissue cells. J Cell Biol 153(3):517–527
24. Cimini D et al (2002) Merotelic kinetochore orientation versus chromosome mono-orientation in the origin of lagging chromosomes in human primary cells. J Cell Sci 115(Pt 3):507–515
25. Kanda T, Sullivan KF, Wahl GM (1998) Histone-GFP fusion protein enables sensitive analysis of chromosome dynamics in living mammalian cells. Curr Biol 8(7):377–385
26. Cimini D, Tanzarella C, Degrassi F (1999) Differences in malsegregation rates obtained by scoring ana-telophases or binucleate cells. Mutagenesis 14(6):563–568
27. Torosantucci L et al (2009) Aneuploidy in mitosis of PtK1 cells is generated by random loss and nondisjunction of individual chromosomes. J Cell Sci 122(Pt 19):3455–3461
28. Robinett CC et al (1996) In vivo localization of DNA sequences and visualization of large-scale chromatin organization using lac operator/repressor recognition. J Cell Biol 135(6 Pt 2):1685–1700
29. Thompson SL, Compton DA (2010) Proliferation of aneuploid human cells is limited by a p53-dependent mechanism. J Cell Biol 188(3):369–381
30. Thompson SL, Compton DA (2011) Chromosome missegregation in human cells arises through specific types of kinetochore-microtubule attachment errors. Proc Natl Acad Sci U S A 108(44):17974–17978
31. Chen B et al (2013) Dynamic imaging of genomic loci in living human cells by an optimized CRISPR/Cas system. Cell 155(7): 1479–1491
32. Ma H et al (2015) Multicolor CRISPR labeling of chromosomal loci in human cells. Proc Natl Acad Sci U S A 112(10):3002–3007
33. Skibbens RV, Skeen VP, Salmon ED (1993) Directional instability of kinetochore motility during chromosome congression and segregation in mitotic newt lung cells: a push-pull mechanism. J Cell Biol 122(4):859–875
34. Wan X et al (2012) The coupling between sister kinetochore directional instability and oscillations in centromere stretch in metaphase PtK1 cells. Mol Biol Cell 23(6):1035–1046
35. Civelekoglu-Scholey G et al (2013) Dynamic bonds and polar ejection force distribution explain kinetochore oscillations in PtK1 cells. J Cell Biol 201(4):577–593
36. Cameron LA et al (2006) Kinesin 5-independent poleward flux of kinetochore microtubules in PtK1 cells. J Cell Biol 173(2):173–179
37. Cimini D, Cameron LA, Salmon ED (2004) Anaphase spindle mechanics prevent mis-segregation of merotelically oriented chromosomes. Curr Biol 14(23):2149–2155
38. He B et al (2014) Chromosomes mis-segregated into micronuclei cause chromosomal instability by further mis-segregating at subsequent mitoses. Mol Biol Cell 25:P1839
39. Guimaraes GJ et al (2008) Kinetochore-microtubule attachment relies on the disordered N-terminal tail domain of Hec1. Curr Biol 18(22):1778–1784

40. Stout JR et al (2006) Deciphering protein function during mitosis in PtK cells using RNAi. BMC Cell Biol 7:26
41. Dumont S, Salmon ED, Mitchison TJ (2012) Deformations within moving kinetochores reveal different sites of active and passive force generation. Science 337(6092):355–358
42. Shah JV et al (2004) Dynamics of centromere and kinetochore proteins; implications for checkpoint signaling and silencing. Curr Biol 14(11):942–952
43. Canman J, Salmon E, Fang G (2002) Inducing precocious anaphase in cultured mammalian cells. Cell Motil Cyto 52:61–65
44. Canman JC, Hoffman DB, Salmon ED (2000) The role of pre- and post-anaphase microtubules in the cytokinesis phase of the cell cycle. Curr Biol 10(10):611–614
45. Howell BJ et al (2000) Visualization of Mad2 dynamics at kinetochores, along spindle fibers, and at spindle poles in living cells. J Cell Biol 150(6):1233–1250
46. Shannon KB, Canman JC, Salmon ED (2002) Mad2 and BubR1 function in a single checkpoint pathway that responds to a loss of tension. Mol Biol Cell 13(10):3706–3719
47. De Antoni A et al (2005) The Mad1/Mad2 complex as a template for Mad2 activation in the spindle assembly checkpoint. Curr Biol 15(3):214–225
48. Schulze E, Kirschner M (1988) New features of microtubule behaviour observed in vivo. Nature 334(6180):356–359
49. Mitchison TJ, Salmon ED (1992) Poleward kinetochore fiber movement occurs during both metaphase and anaphase-A in newt lung cell mitosis. J Cell Biol 119(3):569–582
50. Salmon ED et al (2007) A high-resolution multimode digital microscope system. Methods Cell Biol 81:187–218
51. Maffini S et al (2009) Motor-independent targeting of CLASPs to kinetochores by CENP-E promotes microtubule turnover and poleward flux. Curr Biol 19(18):1566–1572
52. Patterson GH, Lippincott-Schwartz J (2002) A photoactivatable GFP for selective photolabeling of proteins and cells. Science 297(5588):1873–1877

Part II

Methods Focused on Microtubules and the Mitotic Spindle

Chapter 3

Purification and Fluorescent Labeling of Tubulin from *Xenopus laevis* Egg Extracts

Aaron C. Groen and Timothy J. Mitchison

Abstract

For many years, microtubule research has depended on tubulin purified from cow and pig brains, which may not be ideal for experiments using proteins or extracts from non-brain tissues and cold-blooded organisms. Here, we describe a method to purify functional tubulin from the eggs of the frog, *Xenopus laevis*. This tubulin has many benefits for the study of microtubules and microtubule based structures assembled in vitro at room temperature. Frog tubulin lacks many of the highly stabilizing posttranslational modifications present in pig brain-derived tubulin, and polymerizes efficiently at room temperature. In addition, fluorescently labeled frog egg tubulin incorporates into meiotic spindles assembled in egg extract more efficiently than brain tubulin, and is thus superior as a probe for Xenopus egg extract experiments. Frog egg tubulin will provide excellent opportunities to identify active nucleation complexes and revisit microtubule polymerization dynamics in vitro.

Key words Microtubules, Tubulin, Cell-free cytoplasm, *Xenopus laevis*

1 Introduction

Undiluted, cell-free cytoplasm prepared from meiotic *Xenopus laevis* eggs, commonly called "CSF extract" or "crude egg extract," reconstitutes many biological phenomena, including cell cycle progression, meiotic spindle assembly, and cytokinesis signaling [1–3]. This cytoplasm contains over 100 mg/ml of protein, including abundant ribosomes and organelles, and it is in a normal physiological state, making it a good source for biochemical purifications [4]. Here we describe a protocol for purifying frog egg tubulin from the cytoplasm isolated from the eggs of *Xenopus laevis*. The resulting "frog tubulin" has several advantages over bovine or porcine brain tubulin; it polymerizes at room temperature and does not contain posttranslational modifications that could interfere with in vitro polymerization dynamics [5, 6]. Labeled frog egg tubulin incorporates more efficiently than labeled bovine brain tubulin into meiotic spindles assembled in frog egg extract (Fig. 5).

Paul Chang and Ryoma Ohi (eds.), *The Mitotic Spindle: Methods and Protocols*, Methods in Molecular Biology, vol. 1413,
DOI 10.1007/978-1-4939-3542-0_3,

We suspect this is because the frog tubulin probe polymerizes more efficiently at room temperature, but posttranslational modification might also play a role. With such a relatively simple method for purification of native, unmodified tubulin, we anticipate future advances in understanding the biochemistry of microtubules.

Our purification method is based on conventional polymerization-depolymerization cycles using high salt (PIPES) to elute the contaminating microtubule associated proteins [7]. An alternative purification approach, based on affinity to a CH-TOG domain, was recently described [8]. The affinity approach is ideal for purification of tubulin from lysates where it is too dilute to polymerize efficiently. However, tubulin polymerizes efficiently in frog egg extract under the conditions we describe, which makes purification by polymerization easy and efficient.

In the protocol below, we provide a detailed description of how to prepare a conventional "CSF extract," using methods similar to those described previously [2]. We then describe our tubulin purification technique using the concentrated extract, and a method to label frog tubulin on lysine residues with fluorescent probes using NHS-ester chemistry. Depending on need, we typically label a portion of each frog egg tubulin preparation with several different dyes.

2 Materials

2.1 General Equipment

18G 1½ gauge needles.

Syringes (Model 1001 LT-Syringe 1 ml Luer-tip).

50 ml Falcon tube.

Amicon Ultra-concentrator 50 kDa cutoff (Amicon Ultra-15 Centrifugal Filter Unit Ultracel-50 membrane).

SP-Sepharose beads.

2.2 Reagents

PMSG (gonadotropin from pregnant mare serum).

HCG (chorionic gonadotropin human).

Cysteine (l-cysteine).

Leupeptin.

Pepstatin.

Chymostatin.

Cytochalasin D.

DMSO.

2.3 Extract Buffers

MMR (Marc's Modified Ringer's; 1×: 0.1 M NaCl, 2 mM KCl, 1 mM $MgSO_4$, 2 mM $CaCl_2$, 5 mM HEPES, pH 7.8, 0.1 mM EDTA).

Dejelly (1× XB salts; 2 % l-cysteine) *400 ml.*

XB Salts (20×: 2 M KCl, 20 mM $MgCl_2$, 1 mM $CaCl_2$).

CSF-XB (1× XB salts, 10 mM K-HEPES, pH 7.7, 5 mM EGTA, pH 8, 50 mM sucrose) *2 l.*

CSF-XB plus protease inhibitors (1 μg/ml leupeptin, 1 μg/ml pepstatin, 1 μg/ml chymostatin) *200 ml.*

2.4 Quick Buffer Recipe for Extract Buffers

2 l MMR: (80 ml 25× MMR).

800 ml Dejelly: (40 ml 20× XB salts, 8 g L cysteine, 3.6 ml 10 N KOH).

2 l CSF-XB: (100 ml 20× XB salts, 20 ml 0.5 M EGTA, pH 8.0, 20 ml 1 M K-HEPES, pH 7.7, 2 ml 1 M $MgCl_2$, 20 ml 2 M sucrose, 220 μl 10 N KOH).

2.5 Tubulin Prep Buffers

Dilution buffer (30 % glycerol in 1× BRB-80) *300 ml.*

Cushion buffer (40 % glycerol in 1× BRB-80) *100 ml.*

Depolymerization buffer (50 mM K-MES, 1 mM $CaCl_2$, pH 6.6) *50 ml.*

PIPES buffer (1 M K-PIPES, 10 mM $MgCl_2$, 20 mM EGTS, pH 6.9) *50 ml.*

BRB-80 (80 mM K-PIPES, 1 mM $MgCl_2$, 1 mM EGTA, pH 6.8) *50 ml.*

High pH cushion (0.1 K-HEPES, pH 8.6, 1 mM $MgCl_2$, 1 mM EGTA, 60 % glycerol).

Low pH cushion (1× BRB-80 in 60 % glycerol).

Labeling buffer (0.1 M K-HEPES, 1 mM $MgCl_2$, 1 mM EGTA, 40 % glycerol).

Quench (2× BRB-80, 100 mM K-glutamate, 40 % glycerol).

3 Methods

3.1 Preparation of Crude Xenopus laevis Egg Extracts Without Intact Actin

Maintain frogs and buffers and perform the entire preparation at 16 °C. Ten frogs should be sufficient to yield ~5 mg of frog egg tubulin from ~20 ml of crude extract generated (*see* Fig. 1). Prep time: ~1 h. (*See* Fig. 2 for flowchart of protocol.)

1. Inject each *Xenopus laevis* frog with 100 international units (IU) of pregnant mare gonadotropin (PMSG) at least 2 days before extract prep. (PMSG is diluted in sterilized ddH_2O).
2. One day prior to the preparation inject frogs with 500 IU of human chorionic gonadotropin (HCG) and placed in a single container containing 2 l MMR. (HCG is diluted in sterilized ddH_2O.)
3. Collect and pool eggs 24 h after HCG injection.
4. Remove poor quality eggs (activated or unequal animal and vegetal poles) throughout collection and washing procedure.
5. Eggs are washed 3× in MMR (~300 ml per wash) then 3× in dejelly solution (~200 ml per wash). For the last wash, incu-

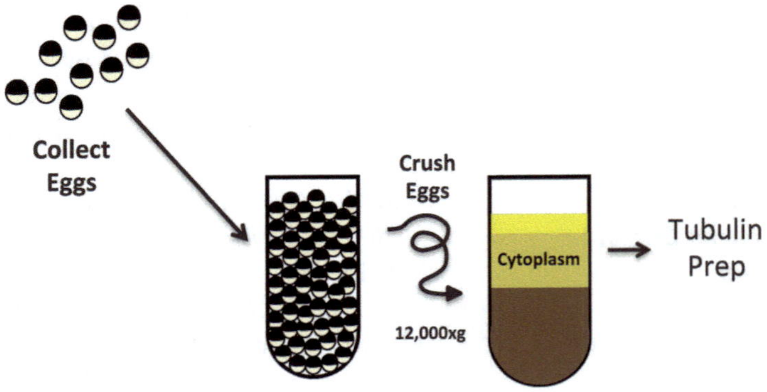

Fig. 1 *Xenopus laevis* egg extract preparation

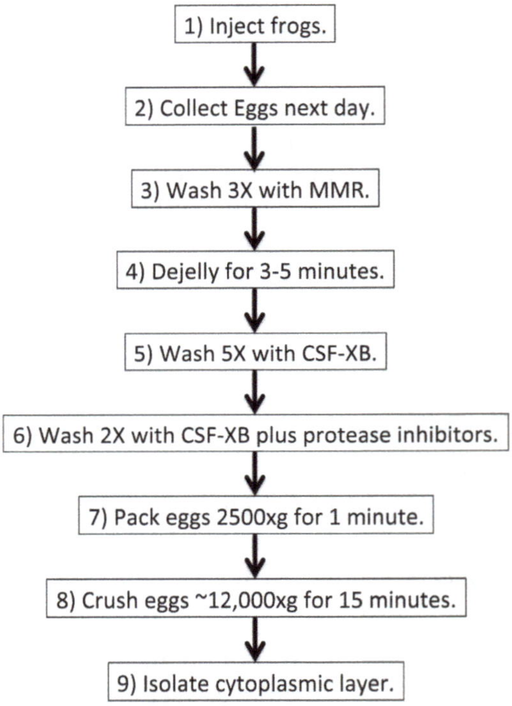

Fig. 2 Flow diagram of extract protocol

bate eggs in dejelly solution for 3–5 min or until eggs are tightly packed at the bottom of the dish. Gently swirling eggs helps to remove jelly and decreases incubation time.

6. Wash dejellied eggs 5× in CSF-XB (~400 ml per wash). Gently swirl eggs during each wash to prevent lysis.
7. Wash eggs 2× in CSF-XB plus protease inhibitors (~100 ml per wash).

8. Using a transfer pipet with the tip cut off, transfer eggs to Beckman Ultra-Clear 13×51 mm tubes containing 1 ml of CSF-XB plus protease inhibitors and cytochalasin D (10 μg/ml).
9. Remove extra buffer by packing eggs via centrifugation (2500×*g* for 1 min). Remove buffer using vacuum aspirator.
10. Eggs are crushed at ~12,000×*g* in a swinging bucket rotor for 15 min (SW50.1 at 10,000 rpm or JS13.1 at 9700 rpm). The rotor is kept at room temperature, while the centrifuge is maintained at 16 °C.
11. Extract layer (straw colored layer) is isolated with a 18G 1½ gauge needle attached to a 1 ml syringe (*see* **Note 1**; Fig. 2). Collect the dense straw colored layer above the black egg shell layer by rotating the bevel of needle in this area as you remove the extract with the syringe.)
12. Pool extract from each tube and add 1 μg/ml LPC, 1 μg/ml cytochalasin D, and sucrose to 50 mM.
13. Extract is now ready for tubulin purification (*see* **Note 2**).

3.2 Purification of Xenopus laevis Tubulin

Tubulin polymerization steps are performed at room temperature (18–24 °C), whereas depolymerization steps occur at 0–4 °C. Prep time: ~5–6 h (Modified from Refs. [2, 7, 9]; *See* Fig. 3 for flow-chart of protocol.)

1. Place crude extract (this prep is for 20 ml) in a 50 ml Falcon tube and add 10 % DMSO (v/v). Cap Falcon tube and slowly invert three to four times. Incubate for 30 min at room temperature. This induces efficient but reversible polymerization of the egg tubulin.
2. Aliquot 5 ml of polymerization reaction to new Falcon tubes and dilute tenfold in dilution buffer resulting in 50 ml extract solution. Glycerol in the dilution buffer prevents microtubule depolymerization.
3. Layer each 50 ml reaction over 2 ml of cushion buffer and spin for 30 min at 27,000×*g* at 23 °C. (We use Beckman JA-17 for this spin.). The cushion prevents pelleting of the abundant organelles in the extract.
4. Keep tubes at room temperature. Remove 50 ml above the cushion layer and wash cushion interface 3× with 1× BRB-80.
5. Remove cushion carefully without disturbing the pellet.
6. Resuspend each pellet in 0.5–1 ml cold depolymerization buffer (Pellet 1; Fig. 4a). The calcium in the buffer greatly enhances depolymerization compared to cold temperature alone.
7. Place on ice for 20 min and triturate every 2 min.

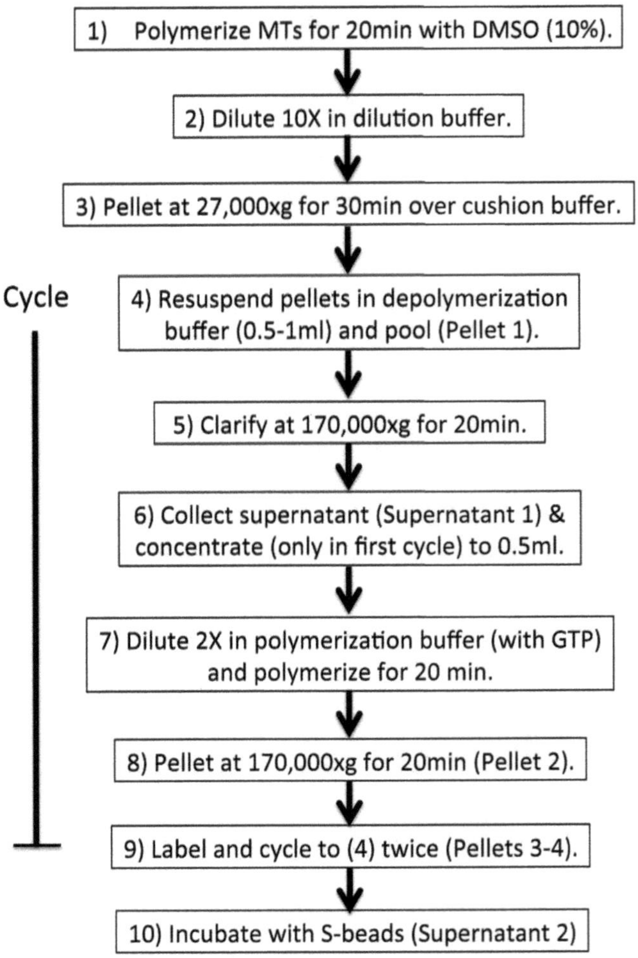

Fig. 3 Flow diagram of the purification of tubulin from *Xenopus laevis* egg extracts

8. Pool all pellets. Spin at 170,000 × *g* for 20 min at 4 °C. (We use an MLA-80.). This spin pellets ribosomes, organelles, keratin, and actin and intermediate filaments that co-pelleted with the microtubules.
9. Collect supernatant (Supernatant 1; Fig. 4a) and concentrate to 0.5 ml in an Amicon Ultra-concentrator 50 kDa cutoff. The final protein concentration should be in the range of 30–50 mg/ml. Concentrating at this step makes the subsequent polymerization more efficient.
10. Dilute 2× in polymerization buffer and add ATP (1 mM), GTP (5 mM), DMSO (10 %), and glycerol (10 %). Incubate for 20 min at room temperature to polymerize.

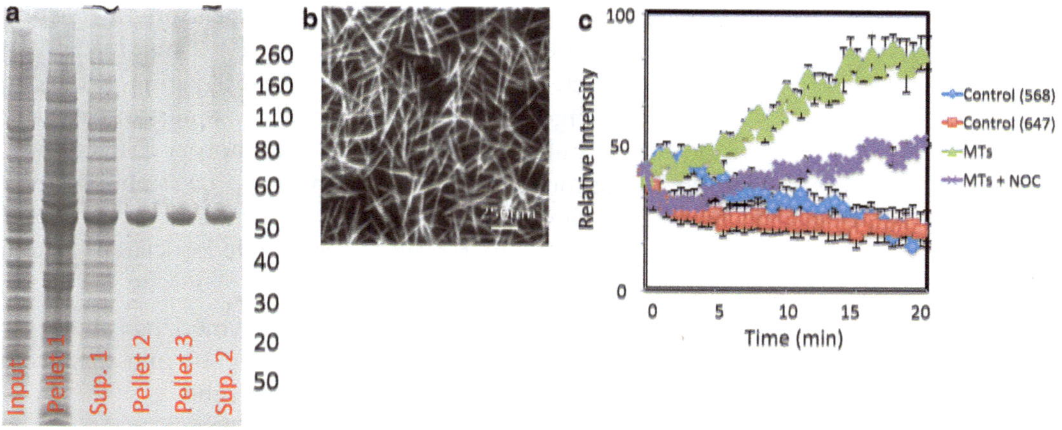

Fig. 4 Purification of tubulin from cytoplasm. (**a**) Coomassie stained SDS-PAGE showing isolated fractions during tubulin prep. (**b**) 60× TIRF image of *Xenopus laevis* tubulin polymerized in vitro on plain glass. Scale Bar = 250 mm. (**c**) Analysis of microtubule polymerization using FRET between A568/A647 on a fluorescence spectrophotometer. Control (568) and Control (647) are in vitro microtubule polymerization reactions containing frog tubulin labeled with either Alexa 568 or Alexa 647. MTs and MTs + NOC (+10 μm nocodazole) are microtubule polymerization reactions containing frog tubulin with tubulin labeled with both A568 and A647. Relative intensity of fluorescence observed with 578Ex/665Em. $N = 3$. Error bars = Standard error

11. Spin at 170,000 × *g* for 20 min at 23 °C over a 3 ml high-pH cushion to make fluorochrome-labeled tubulin, or use a pH 6.8 cushion buffer to make unlabeled tubulin; Pellet 2. Elevated pH is required for efficient microtubule labeling using NHS-ester chemistry.
12. Keep at room temperature. Remove 1 ml above cushion layer and wash interface 3× with labeling buffer. For unlabeled tubulin wash interface with 1× BRB-80; for labeled tubulin skip to Subheading 3.3).
13. Resuspend the pellet in 200 μl of cold depolymerization buffer. Incubate on ice for 20 min and triturate every 2 min.
14. Spin at 170,000 × *g* for 20 min at 4 °C. (We use TLA-100.1.)
15. Collect supernatant and dilute 2× in polymerization buffer then add ATP (1 mM), GTP (5 mM), DMSO (10 %), and glycerol (10 %). Incubate for 20 min at room temperature to polymerize.
16. Place polymerized microtubules over 1 ml of cushion buffer and spin at 170,000 × *g* for 20 min at 23 °C. (Pellet 3; We use TLA-100.3.)
17. Resuspend in 200–300 μl cold 1× BRB-80 in a 1.5 ml eppendorf tube. Incubate on ice for 20 min.
18. Wash 100 μl of 50 % slurry of SP-Sepharose Fast Flow beads with 1× BRB-80 three times. Completely dry beads by picking

a small hole near the bottom of the eppendorf tube with a 27-gauge needle and spinning 2000 × *g* for 2 min into a 2 ml eppendorf (to collect the excess BRB-80) using a tabletop microcentrifuge. SP-Sepharose efficiently binds and removes the microtubule binding proteins and motors that co-purify with microtubules. It serves the same purpose as phosophocellulose in a conventional brain tubulin purification protocol. It removes some of the residual microtubule associated proteins present after the high salt wash.

19. Add dry beads with a spatula to the depolymerized tubulin and incubate for 1 h on ice. Flick tube every 10 min to prevent beads from settling to the bottom of the tube.
20. Spin mixture at 4 °C for 1 min at 10,000 × *g* and collect as much of the supernatant as possible without collecting beads.
21. To collect remaining tubulin from the beads, pick a small hole near the bottom of the eppendorf tube with a 27-gauge needle and spin into a 2 ml eppendorf using a tabletop microcentrifuge. Pool both samples.
22. Spin tubulin at 170,000 × *g* for 20 min at 4 °C for a final clarifying spin (Supernatant 2).
23. Determine tubulin concentration and stoichiometry of labeling (*see* **Note 3**).
24. Flash-freeze in liquid nitrogen in 1–10 μl aliquots and store at −80 °C. In our experience tubulin is stable for at least 1 year at −80 °C although warming to higher temperatures during storage can inactivate it.

3.3 Fluorescent Labeling of Frog Egg Tubulin

1. Resuspend pellet in 0.4 ml of labeling buffer and add the fluorescent probe NHS ester to 1 mM final concentration, in two steps separated by 15 min. We use the Alexa conjugated NHS esters which are stored at −20 °C as 50 mM DMSO stock solutions. Other NHS esters, including long-chain biotin derivatives, can also be used. Incubate for 30 min at room temperature and triturate every 2 min for the lysine labeling reaction to proceed.
2. Add 100 μl of quench buffer. This inactivates remaining NHS esters, which protects the tubulin from further lysine modification after depolymerization.
3. Spin at 170,000 × *g* for 20 min at 23 °C over a 3 ml low-pH cushion. There should be a small, intensely colored pellet, and a much large volume of dye solution over the cushion.
4. Keep at room temperature. Remove 0.5 ml above cushion and wash interface 3× with BRB-80 to remove as much unbound dye as possible. Proceed to **step 13** in Subheading 3.2.

3.4 Confirmation of Polymerization Activity of Purified Frog Tubulin

3.4.1 Microscopy: Confirming Microtubule Polymerization Activity

1. Make tubulin polymerization reaction (1× BRB80, 1 mM GTP, 1 mM DTT, tubulin) above the critical concentration for polymerization (~1 mg/ml) and incubate on ice for 5 min (inspired by Refs. [10, 11].
2. Incubate at 23 °C for 20 min. Do not dilute microtubules below the critical concentration unless Taxol is added in a equimolar concentration to tubulin [12].
3. Directly image microtubules using light microscopy—wide-field, TIRF, DIC microscopy (Fig. 4b). Microtubules should appear as single long filaments.

3.4.2 Fluorescence Spectrophotometry: Measuring Polymerization Kinetics

1. Make two different tubulin polymerization reactions (1× BRB80, 1 mM GTP, 1 mM DTT, tubulin) labeled with different fluorophores capable of FRET (i.e., A647 and A568). The stoichiometry of label must be close to 1 to measure FRET and the polymerization reaction must contain an equal concentration of each labeled tubulin.
2. Combine both tubulin polymerization reactions and measure on a fluorescence spectrophotometer.
3. Polymerization kinetics should include a brief lag phase followed by an increase in FRET signal, reaching a maximum in approximately 20 min. Any deviations suggest tubulin purification was not successful.

3.4.3 Physiological Confirmation of Microtubule Polymerization Activity

1. Assemble meiotic Xenopus egg extract spindles as previously described [2].
2. Add directly labeled frog egg tubulin to meiotic spindles (~50 μg/ml for general viewing or 10 ng/μl for single molecule speckle microscopy).
3. Image using light microcopy (wide-field or spinning disc confocal microscopy; Fig. 5). Note that frog egg tubulin incorporates (relative to background) to a greater extent than bovine tubulin (Fig. 5).

4 Notes

1. The straw colored layer is fragmented into many parts. Collect everything, including the denser yellow layer right above the black layer.
2. For larger tubulin preps, extract can be frozen for storage. Freeze in liquid nitrogen in 1 ml aliquots and store in −80 °C.

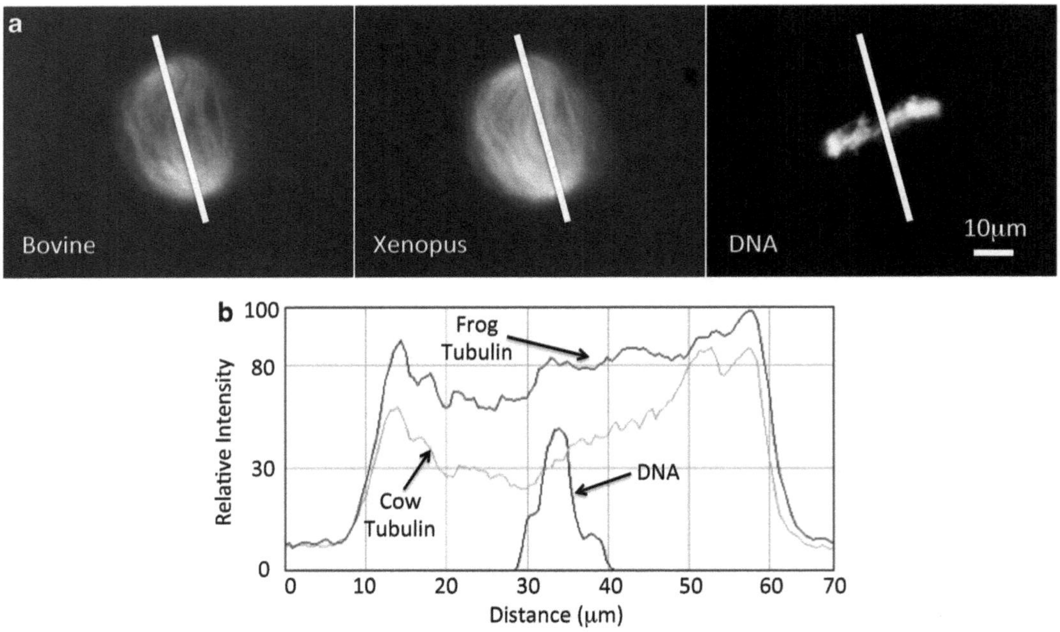

Fig. 5 Frog tubulin incorporation in meiotic extract spindles is greater than bovine tubulin. (**a**) Meiotic extract spindles incorporating A568 bovine tubulin (*left*), A647 Xenopus tubulin (*middle*), and DNA (DAPI). (**b**) Intensity profile of line shown in (**a**). Note: more Xenopus tubulin (relative to the background) incorporates into the microtubules within the spindle than bovine tubulin

3. To determine the concentration of tubulin and stoichiometry of label: (a) *Stoichiometry of dye* = (*Concentration of dye*)/(*Concentration of tubulin*). (b) *Concentration of Tubulin* = [Absorbance (at 280 nm) − (Absorbance at excitation wavelength of dye) × (Absorbance of dye at 280 as fraction of peak absorbance or CF)] × Dilution Factor/Extinction Coefficient of tubulin (115,000). (c) *Concentration of fluorophore* = Absorbance (at excitation wavelength of dye) × dilution factor/extinction coefficient of dye.

Acknowledgements

Our work was funded primarily by NIH grant GM23928. I would like to thank Dr. Mary Badon, MD/MBA, for her guidance and contributions to the figures.

References

1. Lustig KD, Stukenberg PT, McGarry TJ et al (1997) Small pool expression screening: identification of genes involved in cell cycle control, apoptosis, and early development. Methods Enzymol 283:83–99
2. Desai A, Murray A, Mitchison TJ, Walczak CE (1999) The use of Xenopus egg extracts to study mitotic spindle assembly and function in vitro. Methods Cell Biol 61:385–412
3. Nguyen PA, Groen AC, Loose M et al (2014) Spatial organization of cytokinesis signaling reconstituted in a cell-free system. Science 346:244–247. doi:10.1126/science.1256773
4. Groen AC, Ngyuen PA, Field CM et al (2014) Glycogen-supplemented mitotic cytosol for analyzing Xenopus egg microtubule organization. Methods Enzymol 540:417–433. doi:10.1016/B978-0-12-397924-7.00023-6
5. Song Y, Brady ST (2014) Post-translational modifications of tubulin: pathways to functional diversity of microtubules. Trends Cell Biol. doi:10.1016/j.tcb.2014.10.004
6. Podolski M, Mahamdeh M, Howard J (2014) Stu2, the budding yeast XMAP215/Dis1 homolog, promotes assembly of yeast microtubules by increasing growth rate and decreasing catastrophe frequency. J Biol Chem 289:28087–28093. doi:10.1074/jbc.M114.584300
7. Castoldi M, Popov AV (2003) Purification of brain tubulin through two cycles of polymerization-depolymerization in a high-molarity buffer. Protein Expr Purif 32:83–88. doi:10.1016/S1046-5928(03)00218-3
8. Widlund PO, Podolski M, Reber S et al (2012) One-step purification of assembly-competent tubulin from diverse eukaryotic sources. Mol Biol Cell 23:4393–4401. doi:10.1091/mbc.E12-06-0444
9. Hyman A, Drechsel D, Kellogg D et al (1991) Preparation of modified tubulins. Methods Enzymol 196:478–485
10. Correia JJ, Wilson L (2013) Microtubules, in vitro. Academic Press, Amsterdam
11. Cammack R, Atwood T, Campbell P et al (2006) Oxford dictionary of biochemistry and molecular biology, 2nd edn. Oxford University Press, Oxford
12. Parness J, Horwitz SB (1981) Taxol binds to polymerized tubulin in vitro. J Cell Biol 91:479–487

Chapter 4

Measuring the Effects of Microtubule-Associated Proteins on Microtubule Dynamics In Vitro

Marija Zanic

Abstract

Microtubule dynamic instability, the process by which individual microtubules switch between phases of growth and shrinkage, is essential for establishing the architecture of cellular microtubule structures, such as the mitotic spindle. This switching process is regulated by a complex network of microtubule-associated proteins (MAPs), which modulate different aspects of microtubule dynamic behavior. To elucidate the effects of MAPs and their molecular mechanisms of action, in vitro reconstitution approaches with purified components are used. Here, I present methods for measuring individual and combined effects of MAPs on microtubule dynamics, using purified protein components and total-internal-reflection fluorescence (TIRF) microscopy. Particular focus is given to the experimental design, proper parameterization, and data analysis.

Key words Cytoskeleton, Reconstitution, Microtubule, Tubulin, Microtubule dynamics, Microtubule-associated proteins

1 Introduction

Microtubules are cytoskeletal polymers essential for cell division, motility, and intracellular transport. To accomplish many cellular functions, such as building the mitotic spindle, the microtubule network has to rapidly remodel in response to cellular cues. This remodeling is achieved through tight regulation of microtubule dynamic instability, the process of switching between growth and shrinkage of individual microtubule polymers [1]. Although microtubules reconstituted in vitro from purified tubulin alone exhibit this fascinating switching behavior, microtubule dynamics in vivo are far more complex due to the action of many microtubule-associated-proteins (MAPs) that regulate microtubule behavior [2]. The network of MAPs is large and heavily interconnected [3], and thus the effects of any individual MAP are hard to discern by the standard depletion and overexpression approaches in cells.

Paul Chang and Ryoma Ohi (eds.), *The Mitotic Spindle: Methods and Protocols*, Methods in Molecular Biology, vol. 1413, DOI 10.1007/978-1-4939-3542-0_4, © Springer Science+Business Media New York 2016

In recent years, due to advances in recombinant protein expression and purification, in vitro reconstitution approaches have become essential in elucidating the mechanisms and characterizing the effects of MAPs on dynamic microtubule behavior [4]. Not only individual MAPs, but also combinations of MAPs can now be studied, increasing the level of complexity in reconstitution systems. With the growing interest in the use of reconstitution approaches in microtubule studies, a set of experimental methods that include guidelines on measuring and parameterizing microtubule dynamic behavior is needed.

Given that individual microtubules switch between two distinct phases, growth and shrinkage, a natural way to characterize the dynamic behavior is using the following four parameters: microtubule growth rate (v_g), microtubule shrinkage rate (v_s), rate of switching from the growing to the shrinking phase (catastrophe frequency, f_c), and the rate of switching from shrinking back to the growing phase (rescue frequency, f_r).

Microtubule growth rate is a linear function of tubulin concentration, and, to a first approximation, can be regarded as constant over time. It is commonly reported in μm/min, although a more appropriate (SI unit) is nm/s. Typical growth rates achieved in vitro with tubulin alone are in the range of 10–50 nm/s (0.6–3 μm/min) [5], however, with the addition of MAPs [6] and crowding agents [7], they can reach 300–600 nm/s (20–40 μm/min), similar to the highest rates observed in cells [8]. Importantly, some MAPs have been shown to affect microtubule growth rates in a microtubule length-dependent fashion [9]. Furthermore, when investigated with higher spatial resolution, microtubule growth rates in general exhibit strongly dispersive behavior [10–13]. Nevertheless, on the micrometer scales characteristic for the switching transitions between growing and shrinking phases, parameterization with a constant growth rate remains useful in most cases.

Catastrophe is the switching event from microtubule growth to shrinkage, and is commonly parameterized by "catastrophe frequency." Catastrophe frequency is traditionally calculated by dividing the total number of observed catastrophe events by the total time all observed microtubules spent in the growth phase. This measurement assumes that microtubule catastrophe is a single random event, and that the probability of undergoing catastrophe is the same for any microtubule at any time. However, recent reports established that microtubule catastrophe is a multistep process, and its likelihood increases with the microtubule age [9]. Thus, a single, constant catastrophe rate is not a suitable parameter that can be used to describe the catastrophe switch. Instead, individual lifetimes of microtubule growth events should be measured, and their distributions characterized by the gamma distribution function.

Gamma distribution is a two-parameter probability distribution commonly used to describe a multistep process with n equal random steps that occur at a rate r. When $n = 1$, gamma distribution describes a single-step process, and reduces to the exponential distribution. Since MAPs regulate microtubule catastrophe by modulating the rate and step parameters of the gamma distribution [9], for a full characterization of the effects of a MAP on microtubule catastrophe, these parameters should be measured and reported.

Microtubule shrinkage rates measured in vitro with tubulin alone are typically an order of magnitude greater than the growth rates, with values in the range of 300–500 nm/s (20–30 μm/min) at microtubule plus ends. It is unclear to what extent shrinkage rates depend on the tubulin concentration [5]. Finally, rescues, transitions from the shrinking to the growing microtubule phase, are parameterized by "rescue frequency." Rescue frequency is obtained by counting all observed rescue events and dividing by the time all observed microtubules spent shrinking. Rescues in vitro with tubulin alone are rare events [5].

Given these four parameters, microtubule length distributions can be determined. A simple mathematical model [14] predicts the existence of two microtubule growth phases: "bounded" and "unbounded" growth phase. It is only when rescues are rare enough (i.e., $f_r < \frac{v_s}{v_g} f_c$), that the microtubule length distributions (and the total amount of microtubule polymer) reach a steady state (bounded phase). If rescues are more common, microtubules on average keep growing (unbounded phase), until the tubulin in solution becomes depleted. This model assumes all four parameters to be constant as a function of microtubule age; a more general description of microtubule length distributions can be obtained assuming microtubule catastrophe to be a multistep process, for both constant growth rate, as well as for a growth rate that linearly depends on microtubule length (see SI in Ref. [9]).

In this manuscript, I describe the procedure used for measuring the effects of MAPs on microtubule dynamics using the in vitro reconstitution approach. I begin by outlining the TIRF imaging assay that uses stabilized microtubule seeds bound to the surface of a flow cell as nucleation points for dynamic microtubule assembly (similar to [15]). I then continue with a detailed description of the experimental design necessary to characterize the dynamic parameters. Finally, I dedicate a section to methods and considerations to be used in data analysis, in order to obtain relevant measurements of the effects of individual or groups of MAPs on microtubule dynamics.

2 Materials

2.1 Proteins and Reagents

1. Tubulin seed mix: 40 μM porcine or bovine tubulin in BRB80, 25 % TAMRA-labeled.
2. Tubulin extension mix: 100 μM porcine or bovine tubulin in BRB80, 10 % Alexa-Fluor-488-labeled (*see* **Note 1**).
3. MAPs of interest.
4. Anti-tetramethylrhodamine, Rabbit IgG Antibody Fraction, 10 μg/ml in BRB80.
5. BRB80: 80 mM PIPES/KOH, pH 6.9, 1 mM $MgCl_2$, 1 mM EGTA (ethylene glycol tetraacetic acid).
6. Pluronic F127, 1 % in BRB80.
7. Glucose-oxidase, 2 mg/ml in BRB80.
8. Catalase, 0.8 mg/ml in BRB80.
9. DL-Dithiothreitol (DTT), 1 M in ddH_2O.
10. Casein from bovine milk, 8 mg/ml in BRB80.
11. d-(+)-Glucose, 2 M in ddH_2O.
12. $MgCl_2$, 100 mM.
13. KCl, 1 M.
14. GTP disodium salt, 25 mM in n ddH_2O.
15. Guanosine-5′-[(alpha,beta)-methyleno]triphosphate (GMPCPP), 10 mM.

2.2 Specialist Equipment, Materials, and Software

1. Airfuge Air Driven Ultracentrifuge with A95 rotor (Beckman Coulter).
2. TIRF microscope with 100× high NA objective, objective lens heater, and appropriate filter cubes (e.g., Nikon Inverted Microscope Eclipse Ti-E, CFI Apo 100× Oil Immersion Objective Lens TIRF NA 1.49, Tokai Hit Lens Heater with Velcro tape).
3. Piranha-cleaned and silanized 18 mm × 18 mm and 22 mm × 22 mm coverslips (*see* **Note 2**).
4. Fiji/ImageJ [16, 17].
5. MATLAB, The MathWorks, Inc., Natick, MA.

3 Methods

3.1 Preparation

3.1.1 Prepare Microtubule Seeds

Microtubules grown with GMPCPP, a slowly hydrolyzable GTP analogue [18], are stable against depolymerization for several hours (typical depolymerization rates of 0.2 nm/s), and are thus used as "seeds" for nucleation of microtubule extensions specifically from these GMPCPP-grown microtubules.

1. In a 600 μl centrifuge tube, mix together 2.5 μl tubulin seed mix (final tubulin concentration 2 μM), 5 μl GMPCPP (final concentration 1 mM), 0.5 μl $MgCl_2$ (at 100 mM, final concentration 1 mM), and 42 μl BRB80.
2. Incubate on ice for 5 min (*see* **Note 3**).
3. Incubate at 37 °C for 30 min (*see* **Note 4**).
4. Dilute with 350 μl RT (room-temperature) BRB80 and spin down the entire volume (400 μl) in Airfuge at 22 psi (126,000 × *g*) for 5 min (*see* **Note 5**).
5. Discard the supernatant. Resuspend the pellet (usually not visible) in 100 μl RT BRB80 to generate concentrated microtubule seed solution (*see* **Note 6**).

3.1.2 Prepare the Flow Cell

For controlled imaging of microtubule dynamics, GMPCPP-microtubule seeds are bound to the surface of a flow cell using antibodies. The surface is passivated using Pluronic F127 polymer to block nonspecific binding. The flow cell setup allows introduction and exchange of reaction solutions as needed.

1. Form two 2–3 mm wide flow cell channels by adhering two Piranha-cleaned silanized coverslips (one 18 × 18 mm, one 22 × 22 mm) using strips of Parafilm or double-sided sticky tape (*see* **Note 7**).
2. Using vacuum, perfuse 20 μl BRB80 into the flow cell (*see* **Note 8**).
3. Perfuse 20 μl anti-tetramethylrhodamine antibody (at 10 μg/ml in BRB80) into the flow cell and incubate for 5 min.
4. Wash the flow cell two times with 20 μl BRB80.
5. Perfuse 20 μl Pluronic F127 (1 % in BRB80) into the flow cell and incubate for at least 15 min (*see* **Note 9**).
6. Wash the flow cell five times with 20 μl BRB80.
7. Make 100 μl solution of 1:50 dilution of GMPCPP-microtubule seeds in RT BRB80. Perfuse 20 μl of the diluted solution into the flow cell and incubate for 2 min. Quickly check the density on the microscope, and wash out the microtubule seed solution with 20 μl BRB80 when optimal density is reached (*see* **Note 10**).

3.2 Experimental Design and Imaging

The choice of imaging conditions, including temperature, spatial and temporal resolution, will dictate what parameters of dynamic instability can be successfully measured. Controlling the reaction mix components, and performing the experiments over a range of protein concentrations is essential for the full characterization of protein effects on microtubule dynamics.

3.2.1 Choose the Imaging Conditions

1. Using an objective heater, set the desired temperature for the experiment. Typically, a temperature of 35 °C is used when working with mammalian proteins only, whereas a temperature of 28 °C is used when working with yeast proteins (*see* **Note 11**).
2. Choose the spatial resolution. In combination with the 100× TIRF objective, additional magnification can be used (e.g., Optovar lens) to reduce the pixel size (pixel size of 100 nm or less is desirable).
3. Set the time-lapse interval appropriately. Optimal acquisition is achieved when the growth/shrinkage rates are 1 spatial pixel per 1 temporal pixel (e.g., for a typical growth rate of 1 μm/min = 100 nm/ 6 s, optimal interval is 6 s for 100 nm pixel size; for shrinkage rate of 30 μm/min = 100 nm/ 0.2 s, optimal interval would be 200 ms). Due to the potential effects of photo-damage, time-lapse interval should be kept the same for a set of experiments.
4. Choose the time-lapse duration appropriately. For measuring microtubule lifetime distributions, recommended duration is at least 5× the mean lifetime (typically 15–45 min) (*see* **Note 12**).

3.2.2 Perform the Experiment

1. In a 600 μl tube mix together the polymerization mix (PM): 132 μl BRB80, 20 μl of 1 M KCl (final concentration 100 mM), 8 μl d-glucose (final concentration 80 mM), 8 μl glucose-oxidase (final concentration 80 μg/ml), 8 μl catalase (final concentration 32 μg/ml), 4 μl casein (final concentration 0.16 mg/ml), 4 μl DTT (20 mM), and 16 μl GTP (final concentration 2 mM) (*see* **Note 13**).
2. Prepare 20 μl control reaction mix by combining 10 μl PM, x μl Alexa-488 tubulin and 10 − x μl BRB80. Final tubulin concentration between 10 and 15 μM yields robust growth and dynamics at temperatures of 28–35 °C.
3. For each of the MAPs, reaction mix (RM) is prepared in the following way: 10 μl PM, x μl Alexa-488 tubulin, y μl MAP, 8 − x − y μl BRB80 (*see* **Note 14**).
4. To characterize effects of a MAP on tubulin dynamics, estimate a concentration of protein expected to be at about four times its K_m (I denote this concentration C_0). (This can be based on physiological protein concentrations, when known.) Perform a titration with increasing amounts of the MAP to investigate its effects over a range of concentrations (following dilutions are recommended: 1×, 2×, 3×, 4×, 8×, 16×, 32× diluted C_0).
5. Equilibrate the flow cell by perfusing a room temperature mix of 10 μl PM combined with 10 μl BRB80.
6. Flow in 20 μl prewarmed RM containing the proteins of interest.

7. Acquire one TRITC image using epifluorescence (to capture the microtubule seeds position), followed by a 488 nm laser line TIRF time-lapse, using the conditions chosen as outlined in Subheading 3.2.1. At the end of the time-lapse, acquire another epifluorescence TRITC image (in case there were any unexpected changes in the seeds channel, or a drift in position) (*see* **Note 15**).

3.3 Data Analysis

The first step in data analysis of time-lapse images is generating space-time plots of individual microtubules, called kymographs. Values of dynamic instability parameters and their distributions are then determined from the kymograph data.

3.3.1 Make Kymographs

1. Generate a color combined time-lapse image file. This can be accomplished in Fiji/ImageJ by creating a stack from a single snapshot of GMPCPP-microtubule seeds (e.g., using the *ConvertImageToStack* macro; make sure to create the same number of images as available in the green time-lapse) and color-combining this stack with the stack containing the dynamic TIRF time-lapse information.
2. Generate individual kymographs for all dynamic microtubules in the field of view (Fig. 1). This can be done in ImageJ/Fiji using the *Reslice* tool. Kymographs should be created for all seeds with both ends in the field of view. Those microtubule seeds that only have one end within the field of view should be discarded, as the polarity of the seed may not be reliably determined (*see* **Note 16**).

3.3.2 Measure Growth and Shrinkage Rates

1. In most cases, microtubule growth can be well approximated by a single constant growth rate.
   The rate for each growth event can be simply determined from two measurements only: the position and time at the onset of growth (x and y coordinates on a kymograph), and the position of the microtubule tip after ~30 frames (*see* **Note 17**).
2. Shrinkage rates should be measured analogous to the growth rates. However, as shrinkage rates are typically an order of magnitude faster in microtubule reconstitution assays, time-lapse intervals should be adjusted appropriately. To estimate shrinkage rates when conditions are such that microtubules shrink over less than 30 frames, when calculating mean shrinkage rate, each measurement should be weighted by the time interval spent in shrinkage.

3.3.3 Measure Catastrophe and Rescue Rates

1. To characterize microtubule catastrophe, all full growth events should be analyzed. Lifetime of an event should be measured as the time from the onset of growth from the seed, until the time when catastrophe occurs. Spatially cut off (growing out of

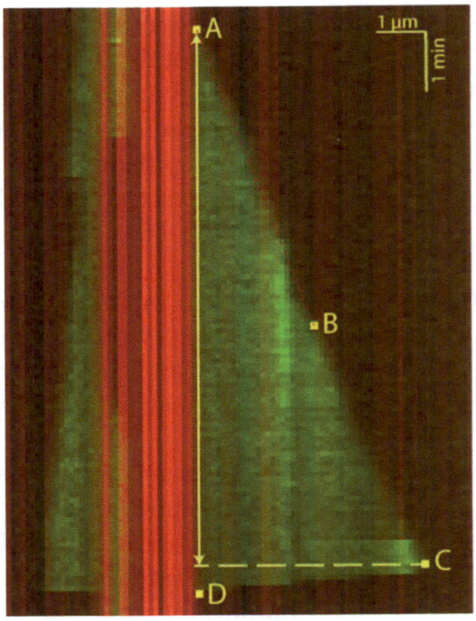

Fig. 1 Example of a single full microtubule growth/catastrophe/shrinkage event on a kymograph. The microtubule starts growing at point A, undergoes catastrophe at point C, and shrinks back to zero length at point D. Only the plus microtubule end is considered. Growth rate is determined from the slope between points A and B (temporally separated by at least 30 pixels). Event lifetime is defined by the time coordinate difference between points C and A. Shrinkage rate can be determined from the slope between points C and D (note that in this case the shrinkage rate measurement will have a large systematic error due to the limited number of frames over which the shrinkage took place)

the field of view) and temporally cut off (starting to grow prior to the onset of the time-lapse, or still growing at the end of the time-lapse) events should be disregarded. To ensure that the measurement is not favoring shorter events, the duration of the time-lapse should be at least five times the mean event lifetime. Furthermore, a smaller central region of the full field of view should be specified and all events initiated from the seeds in this region should be analyzed.

2. Distribution of microtubule lifetimes can be well described by a gamma distribution (Fig. 2). To do so, a minimum of 100 events should be analyzed, lifetimes should be plotted as a cumulative distribution plot (to avoid binning artifacts commonly resulting from histogram plots), and data fitted to the gamma distribution using maximum likelihood estimates (MLE) for the step (n) and rate (r) parameters of the gamma distribution (*see* **Note 18**). Catastrophe frequency is not a constant; rather it is an increasing function of microtubule age. It saturates at the value of the rate parameter, r. The mean

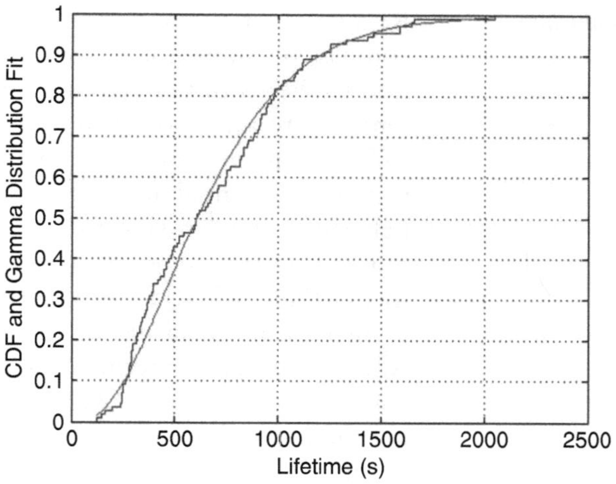

Fig. 2 Parameterization of microtubule lifetimes by gamma distribution function. Individual lifetimes are plotted as a cumulative distribution function (CDF) and fit to a gamma distribution. In this example, a total of 110 events from a single time-lapse experiment were analyzed. The maximum likelihood estimates for gamma fit parameters were: $n = 3.08\,(2.40, 3.96;\, 95\%\,\mathrm{CI})$; $r = 221.0\,\mathrm{s}\,(168.2\,\mathrm{s}, 290.3\,\mathrm{s}; 95\%\,\mathrm{CI})$. The mean lifetime was $\langle T \rangle = (670.3 \pm 3.6)\,\mathrm{s}\,(\mathrm{SE}, N = 110)$

lifetime corresponds to the ratio of step and rate parameters of the gamma distribution. Mean catastrophe rate can be reported as the inverse of the mean lifetime.

3. Rescue rates are traditionally reported as the ratio of the total number of rescues observed over the total amount of time microtubules spent in shrinkage. Alternatively, rescue lengths can be reported as the total length over which microtubules shrunk, divided by the total number of rescues observed.

3.3.4 Measure Microtubule Length Distributions

1. Microtubule length distribution can be determined by measuring the lengths of all (plus-end) microtubule outgrowths in a single time frame, making sure that the distribution has reached steady state (*see* **Note 19**).

3.3.5 Analyze Titration Data

1. Effects of a MAP on the microtubule dynamics parameters will strongly depend on the concentration of the MAP used. For a complete characterization, fit all measured values obtained from the titration in the MAP's concentration to the Michaelis–Menten model to determine the saturating rate of a dynamic parameter, as well as the concentration of the MAP needed for the half-maximum rate (*see* **Note 20**).
2. Once the effects of individual MAPs are characterized, combined effects of MAPs are investigated by performing an analogous titration in one MAP's concentration (*see* **step 4** in Subheading 3.2.2), while using a saturating concentration of another MAP.

3.4 Conclusions

Regulation of microtubule dynamics by MAPs is an important and complex phenomenon that yields itself to biochemical reconstitution studies with purified protein components. The methods presented here are used for in vitro characterization of the effects of MAPs on microtubule dynamic instability. Consistent parameterization is essential for characterizing the dynamic microtubule behavior; I thus provide a set of guidelines on how to measure and report the parameters of dynamic instability. It is my hope that this chapter will serve as a starting point for the analysis of the regulatory mechanisms of many so far uncharacterized MAPs. Adding multiple components to this in vitro reconstitution system has the potential to uncover novel collective effects of MAPs, paving the way to our understanding of the complex behaviors at play in the cell.

4 Notes

1. Tubulin can be purified and labeled according to standard protocols [19–23].
2. A common method for glass cleaning and silanization can be found in Ref. [15].
3. This step ensures efficient nucleotide exchange, as well as depolymerization of any potential oligomers in tubulin solution.
4. This procedure results in GMPCPP-microtubules of several micrometers in length, suitable to be used as seeds in the dynamic assay. Alternatively, microtubules in this step can be grown for up to 2 h, resulting in longer individual lengths and overall greater polymer mass. In that case, microtubules can be sheared by passing the solution ten times through a 27-gauge needle to achieve shorter lengths appropriate for seeds.
5. Microtubules depolymerize when exposed to cold, and thus, care should be taken that all solutions are brought to room temperature prior to contact with microtubules.
   When using Beckman Airfuge, make sure to balance the solution within 1 mg. When Airfuge is not available, appropriate ultracentrifuge substitutions can be made.
6. This solution can be kept at RT and used for several hours. I recommend growing fresh GMPCPP-microtubule seeds daily.
7. When using Parafilm, adhesion is achieved by briefly heating up the coverslip sandwich on a heat plate. Up to three channels can be prepared on the same coverslip sandwich; for longer experiments make bigger, i.e., wider and thicker, flow cells. Flow cell thickness can be increased by using two layers of Parafilm or double-sided sticky tape.

If sample holder for 22 × 22 mm coverslip size is not available, a coverslip can be similarly adhered to a piranha-cleaned and silanized microscope slide that fits most standard microscope stages. After introducing the final experimental reaction, the flow cell can then be sealed with VALAP (1:1:1 Vaseline–lanolin–paraffin) just prior to imaging. A disadvantage of such setup is that the reaction solution cannot be exchanged once the chamber is sealed.

8. This volume is adequate for a typical flow cell of volume 5–10 μl; if using bigger channels, scale all the perfusion solution volumes accordingly. Silanized glass coverslips are hydrophobic, and thus vacuum line is needed for initial introduction of solution into the flow cell. After the first step, perfusion can be performed using filter paper alone.
9. Pluronic F127 is a tri-block copolymer used to block the surface against nonspecific binding of proteins. Incubation times of up to 45 min can be used.
10. Optimal density of seeds is 20–50 per field of view. If the density of microtubules is too low, a lower dilution of GMPCPP seeds and/or longer incubation times can be used.
    Light exposure at this step should be minimized to avoid photo-damage, as no antifade solution has been introduced into the channel yet.
11. Dynamic microtubule behavior is greatly influenced by temperature [24], and thus, temperature control is essential for experimental reproducibility. Desired sample temperature can be efficiently achieved and controlled with a lens objective heater. Note that the temperature inside the flow cell will not be the same as on the objective lens; to accurately determine the sample temperature, a thin probe can be inserted into the flow cell [7].
12. When performing time-lapse imaging for longer than 15 min, particular care should be taken that the channel does not start to dry out. This can be prevented by: using larger flow-cell volume (wider channels with double layer of Parafilm or double-sided sticky tape); leaving an ample amount of reaction solution on both sides of the flow-cell; controlling the humidity (simple addition of wet tissues on the sides and a cover for the sample are usually sufficient). Flow cell can also be sealed by VALAP; however, this prevents solution exchange, which is often desirable. Any potential effects of photobleaching and/or chamber deterioration can be simply monitored by comparing the dynamics parameters (especially growth rate) at the beginning and towards the end of the time-lapse. Depletion of tubulin from solution due to polymer formation is typically not an issue in these assays, as the concentration of microtubule

seeds is in pM range, and the total amount of tubulin polymerized off seeds in nM range; ~3 orders of magnitude below typical concentration of tubulin dimers in solution.

13. This polymerization mix contains a 2× solution of components needed in the final reaction mix. It includes an oxygen-scavenging cocktail (glucose oxidase, catalase, d-glucose and DTT) that helps against photobleaching. To achieve the best effects of the anti-fade cocktail, polymerization mix should be freshly prepared. The polymerization mix can be kept on ice and used for a maximum of 2 h.
    The following modifications/additions to the polymerization mix can be considered. The ionic strength of the buffer should be adjusted accordingly, depending on the optimal conditions for the particular MAPs used. This can be done by changing the concentration of KCl, or using BRB20 instead of BRB80. Particular care should be taken when working with a combination of two or more MAPs that might have different buffer preferences. Small amounts of detergents, such as Tween 20 (usually 0.01–0.1 %) can be added to prevent protein aggregation. Additionally, small amounts of crowding agents (such as 0.1 % methylcellulose) can be used; this helps to minimize thermal fluctuations of dynamic microtubule tips away from the imaging surface. Note that the crowding agents are known to influence the microtubule dynamics parameters [7], and thus their effects should be carefully characterized in control experiments, if they are to be used.
14. Particular attention should be given to potential changes in buffer conditions. If introducing the protein significantly affects the ionic strength of the final reaction mix, the control experiments must be performed with the equimolar amounts of the protein storage buffer (e.g., if the protein storage buffer other than BRB80 is used and the dilution factor is small, such that the changes in ionic strength would be greater than 5 mM).
15. The reaction mix perfused into the flow cell will reach the desired temperature within tens of seconds, and thus waiting for a couple of minutes to start with the imaging is sufficient to ensure stable experimental conditions. If measuring microtubule length distributions, the polymer mass should be allowed to reach steady state prior to the onset of imaging. Although this will depend on the specific conditions of the assay, in many cases this is achieved within 10 min.
    A single flow cell is typically used for only one imaging condition. It is possible to wash the flow cell with 2× 20 μl BRB80, followed by the 20 μl RT 1:1 mixture of PM and BRB80, and then reuse it. Prior to introducing a new reaction mix, the

channel should be imaged to make sure all the microtubules are depolymerized and there is no leftover tubulin in the chamber.

16. When measuring any of the parameters, special attention should be given to the polarity of each microtubule seed. Microtubule plus ends can be distinguished from the minus ends by their faster (typically 2× with tubulin alone) growth rates. However, MAPs can affect this behavior. If experimental conditions do not allow for unambiguous determination of polarity, after the completion of the dynamics experiment, the reaction mix should be exchanged by one containing a labeled unidirectional motor protein (e.g., GFP-tagged kinesin-1) while keeping the same field of view.

17. Although the measurement error will be smaller the later the time-point is, care should be taken as microtubules tend to fluctuate and bend, sometimes out of the TIRF field of view, making the determination of the tip position more challenging. On the other hand, measurement of the tip position after at least 30 frames will result in less than 5 % systematic measurement error, if conditions are optimized.

18. Gamma distribution fitting using maximum likelihood estimates in MATLAB can be accomplished with the command:

    [PHAT, PCI] = mle(alltimes, 'distribution', 'gamma')

    where input "*alltimes*" denotes a column vector containing all of the measured lifetimes. This results in values of the step and rate parameters, as well as their respective 95 % confidence intervals.

19. Microtubule length distributions can also be calculated from the four parameters characterizing dynamic instability. For negligible rescue rates, length distribution density is given by:

$$p_+(x) = \frac{\Gamma(n, rt(x))}{tv_+(x)\Gamma(n)}$$

    where $t = n / r$, and $\Gamma(n, rt)$ is the incomplete gamma function (see supplementary info in Ref. [9]).

20. If potential cooperativity among molecules is suspected, a fit to the more general Hill equation is more appropriate. For example, for a given MAP X, growth rate can be fit by:

$$v([X]) = v_0 + \frac{(v_{\max} - v_0)[X]^h}{K^h + [X]^h}$$

    where h is the Hill coefficient, v_0 growth rate in the absence of the MAP, $v_{\max}$ the saturating growth rate, and K the concentration of X at half-maximum rate.

Acknowledgements

I thank Anika Rahman for help with the data analysis. I am grateful to Justin Bois, Gary Brouhard, Melissa Gardner, Anneke Hibbel, Jonathon Howard, Elizabeth Lawrence, Chloe Snider, Michal Wieczorek, and especially Marija Podolski for helpful discussions and comments on the manuscript.

References

1. Mitchison TJ, Kirschner MW (1984) Dynamic instability of microtubule growth. Nature 312: 237–242
2. Lyle K, Kumar P, Wittmann T (2009) SnapShot: microtubule regulators I. Cell 136:380–380. e1. doi:10.1016/j.cell.2009.01.010
3. Akhmanova A, Steinmetz MO (2008) Tracking the ends: a dynamic protein network controls the fate of microtubule tips. Nat Rev Mol Cell Biol 9:309–322. doi:10.1038/nrm2369
4. Howard J, Hyman AA (2007) Microtubule polymerases and depolymerases. Curr Opin Cell Biol 19:31–35. doi:10.1016/j.ceb.2006. 12.009
5. Walker RA, O'Brien ET, Pryer NK et al (1988) Dynamic instability of individual microtubules analyzed by video light microscopy: rate constants and transition frequencies. J Cell Biol 107:1437–1448
6. Zanic M, Widlund PO, Hyman AA, Howard J (2013) Synergy between XMAP215 and EB1 increases microtubule growth rates to physiological levels. Nat Cell Biol 15(6):688–693. doi:10.1038/ncb2744
7. Wieczorek M, Chaaban S, Brouhard GJ (2013) Macromolecular crowding pushes catalyzed microtubule growth to near the theoretical limit. Cell Mol Bioeng. doi:10.1007/s12195-013-0292-9
8. Srayko M, Kaya A, Stamford J, Hyman AA (2005) Identification and characterization of factors required for microtubule growth and nucleation in the early C. elegans embryo. Dev Cell 9:223–236. doi:10.1016/j.devcel. 2005.07.003
9. Gardner MK, Zanic M, Gell C et al (2011) Depolymerizing kinesins Kip3 and MCAK shape cellular microtubule architecture by differential control of catastrophe. Cell 147:1092–1103. doi:10.1016/j.cell.2011.10.037
10. Kerssemakers JWJ, Munteanu EL, Laan L et al (2006) Assembly dynamics of microtubules at molecular resolution. Nature 442:709–712. doi:10.1038/nature04928
11. Schek H, Gardner MK, Cheng J et al (2007) Microtubule assembly dynamics at the nanoscale. Curr Biol 17:1445–1455. doi:10.1016/j.cub. 2007.07.011
12. Gardner MK, Charlebois BD, Jánosi IM et al (2011) Rapid microtubule self-assembly kinetics. Cell 146:582–592. doi:10.1016/j.cell. 2011.06.053
13. Howard J, Hyman AA (2009) Growth, fluctuation and switching at microtubule plus ends. Nat Rev Mol Cell Biol 10:569–574. doi:10.1038/nrm2713
14. Verde F, Dogterom M, Stelzer E et al (1992) Control of microtubule dynamics and length by cyclin A- and cyclin B-dependent kinases in Xenopus egg extracts. J Cell Biol 118: 1097–1108
15. Gell C, Bormuth V, Brouhard GJ et al (2010) Microtubule dynamics reconstituted in vitro and imaged by single-molecule fluorescence microscopy. Methods Cell Biol 95:221–245. doi:10.1016/S0091-679X(10)95013-9
16. Schindelin J, Arganda-Carreras I, Frise E et al (2012) Fiji: an open-source platform for biological-image analysis. Nat Methods 9:676–682. doi:10.1038/nmeth.2019
17. Abràmoff MD, Magalhães PJ (2004) Image processing with ImageJ. Biophotonics Int 11: 36–42
18. Hyman AA, Salser S, Drechsel DN et al (1992) Role of GTP hydrolysis in microtubule dynamics: information from a slowly hydrolyzable analogue, GMPCPP. Mol Biol Cell 3: 1155–1167
19. Ashford AJ, Andersen S, Hyman AA (1998) Preparation of tubulin from bovine brain. In:

Celis J (ed) Cell biology: a laboratory handbook. Academic Press, New York

20. Gell C, Friel CT, Borgonovo B et al (2011) Purification of tubulin from porcine brain. Methods Mol Biol 777:15–28. doi:10.1007/978-1-61779-252-6_2
21. Castoldi M, Popov AV (2003) Purification of brain tubulin through two cycles of polymerization–depolymerization in a high-molarity buffer. Protein Expr Purif 32:83–88. doi:10.1016/s1046-5928(03)00218-3
22. Hyman AA, Drechsel DN, Kellogg D et al (1991) Preparation of modified tubulins. Meth Enzymol 196:478–485
23. Widlund PO, Podolski M, Reber SB et al (2012) One-step purification of assembly-competent tubulin from diverse eukaryotic sources. Mol Biol Cell 23:4393–4401. doi:10.1091/mbc.E12-06-0444
24. Fygenson D, Braun E, Libchaber A (1994) Phase diagram of microtubules. Phys Rev E 50:1579–1588

Chapter 5

Imaging and Quantifying the Dynamics of γ-Tubulin at Microtubule Minus Ends in Mitotic Spindles

Nicolas Lecland and Jens Lüders

Abstract

Understanding the organization of complex microtubule arrays such as the mitotic spindle requires information about the position and dynamics of microtubule plus and minus ends. Whereas plus end dynamics have been widely studied using markers such as EB1-GFP, much less is known about the dynamic properties of minus ends, in part because a suitable marker has only recently become available. Here we describe the use of photoactivatable γ-tubulin-paGFP to image and quantify the dynamics of microtubule minus ends in mitotic spindles.

Key words Spindle, Microtubule, Minus end dynamics, γ-Tubulin, Photoactivatable GFP

1 Introduction

The bipolar mitotic spindle is assembled from microtubules that are nucleated by the γ-tubulin ring complex (γTuRC). γTuRC is a multi-subunit complex composed of gamma complex proteins (GCPs) and γ-tubulin. In the γTuRC ~13 γ-tubulin molecules are arranged such that they resemble the end of a microtubule and thus provide a template for the addition of α-β-tubulin heterodimers, the building blocks of the microtubule polymer [1, 2]. The head-to-tail arrangement of the α-β-tubulin heterodimers in the polymer confers an intrinsic polarity to microtubules. The end at which nucleation occurs, the so-called minus end, is relatively stable, whereas the plus end is more dynamic and constantly undergoes transitions between growth, pause and shrinkage [3]. After nucleation γTuRC can remain associated with minus ends and can also bind to minus ends of preexisting microtubules. This "capping" of minus ends is thought to have a stabilizing function [4]. During mitosis nucleation of microtubules takes place at the centrosomes, around the chromatin, and from preexisting microtubules, by an augmin-dependent branching mechanism [5]. At metaphase two populations of microtubules have their minus ends

Paul Chang and Ryoma Ohi (eds.), *The Mitotic Spindle: Methods and Protocols*, Methods in Molecular Biology, vol. 1413, DOI 10.1007/978-1-4939-3542-0_5, © Springer Science+Business Media New York 2016

attached at centrosomes: astral microtubules that grow with their plus ends into the spindle body and towards the cell cortex, and k-fiber microtubules that form thick bundles of uniform polarity and connect with their plus ends to chromosomes. Interpolar microtubules are not directly attached at centrosomes but have their minus ends positioned within the spindle body, near the poles [6]. Their plus ends can extend far into the spindle body forming antiparallel overlaps with interpolar microtubules growing from the opposite half spindle. Even though spindles at metaphase maintain a specific size and morphology, spindle microtubules are highly dynamic. Apart from plus end dynamics, microtubule lattices constantly move towards the centrosomes, a phenomenon known as flux [7, 8]. Understanding the complex dynamic architecture of the mitotic spindle is a major challenge and requires information about position and dynamics of both microtubule plus and minus ends [9]. Plus ends can be visualized by markers that specifically bind to growing plus ends such as EB1-GFP [10].

Here we describe the methodology to study the dynamics of microtubule minus ends using a fusion of γ-tubulin with photoactivatable GFP (paGFP) as marker. We have recently used this approach to reveal poleward sorting of the minus ends of non-k-fiber microtubules nucleated within the spindle [11]. paGFP is an engineered GFP variant that does not emit significant fluorescence unless activated by irradiation with ~400 nm light [12]. The use of a laser pulse that can be directed at a specific time at a confined cellular space allows for the photoactivation and study of a subpopulation of the protein of interest. Compared to using fusions with conventional GFP, which can generate high cytoplasmic background signal, spatially restricted photoactivation improves the signal-to-noise ratio during imaging. Here we provide a step-by-step guide detailing the generation of cell lines expressing γ-tubulin-paGFP, the photoactivation of γ-tubulin-paGFP in specific cellular regions, the time-lapse imaging of activated γ-tubulin-paGFP, and the image processing required to obtain quantitative information. While our protocols were developed for studying minus end dynamics in mitotic cells, modification of the procedure may also allow imaging of minus ends in microtubule arrays of other cell types.

2 Materials

2.1 Generation and Culture of Stable Cell Lines

1. Wild-type cell line (*see* **Note 1**).
2. Plasmids encoding γ-tubulin-paGFP and mCherry-α-tubulin (*see* **Note 2**).
3. Dulbecco's modified Eagle's medium (DMEM).
4. Fetal bovine serum (FBS).
5. 0.05 % trypsin–EDTA (Life Technologies).

6. 1 % penicillin–streptomycin (PenStrep) (Life Technologies).
7. Sterile phosphate-buffered saline (PBS).
8. Lipofectamine 2000 reagent (Life Technologies) or similar DNA transfection reagent.
9. Geneticin/G418, stock solution at 50 mg/ml in water, stored at 4 °C.
10. Puromycin, stock solution at 20 mg/ml in water, stored at -20 °C.
11. 10 and 15 cm cell culture dishes.
12. 6 and 24 multi-well cell culture dishes.
13. 3.5 cm glass-bottom cell culture dishes for live imaging (for example MatTek P35G).
14. Cloning cylinders (C3983, Sigma).

2.2 Studying the Molecular Requirements for Microtubule Flux

1. Dimethyl sulfoxide (DMSO).
2. Z-Leu-Leu-Leu-al (MG132); proteasome inhibitor, will be used to block cells in metaphase; stock solution at 5 mM in DMSO, stored at -20 °C.
3. S-Trityl-l-cysteine (STLC); will be used to inhibit the kinesin motor KIF11 (also known as Eg5); stock solution at 50 mM in DMSO, stored at -20 °C.
4. Erythro-9-(2-Hydroxy-3-nonyl)adenine hydrochloride (EHNA; will be used to inhibit the motor dynein) stock solution at 32 mM in DMSO, stored at -20 °C.
5. Ciliobrevin D (will be used to inhibit the motor dynein; inhibits ATPase of cytoplasmic dynein; more specific than EHNA) stock solution at 50 mM in DMSO, stored at -20 °C.
6. siRNA oligos targeting a protein of interest.
7. Lipofectamine RNAiMAX reagent (Life Technologies) or similar siRNA transfection reagent.

2.3 Photoactivation, Image Acquisition, Processing, and Quantification

1. Spinning disc microscope with temperature and CO_2-controlled incubation chamber. We have developed this method using an Olympus IX81 microscope equipped with a Yokogawa CSU-X1 spinning disc. Images were acquired with a 1.4 Numerical Aperture 100× oil immersion objective and an iXon EMCCD Andor DU-897 camera. However, other similar equipment can also be used.
2. 405 nm laser module for photoactivation (e.g., FRAPPA module; Andor Technology).
3. Image acquisition software (e.g., ex/IQ2 from Andor).
4. Image processing and analysis software. For image processing and quantification we recommend ImageJ software, but other software can be used as well.

3 Methods

The methodology described here uses a cell line stably expressing both mCherry-α-tubulin, to visualize microtubules, and γ-tubulin-paGFP, to label microtubule minus ends. The use of a stable cell line is recommended since it minimizes cell-to-cell variability of expression levels of the recombinant proteins and thus ensures high reproducibility. In exceptional cases, e.g., for testing γ-tubulin mutants, transient transfection may be used (*see* **Note 3**). The bright mCherry-α-tubulin spindle labeling allows identification of cells undergoing mitosis and selection of the spindle region that will be used for photoactivation of γ-tubulin-paGFP. Mitotic cells at specific mitotic stages (prophase, prometaphase, metaphase, anaphase, telophase) can be easily identified based on the geometry of the mCherry-α-tubulin labeled microtubule array. Variations of the general protocol by using cells in which specific target proteins are inhibited by siRNA transfection or by drug treatment are listed at the end of the method section.

3.1 Generating a γ-Tubulin-paGFP, mCherry-α-Tubulin Stable Cell Line

1. Plate cells in a culture dish. Cells seeded in one well of a 6-well plate or one 3.5 cm dish will typically yield several stable clones.
2. Co-transfect the cells following the protocol provided by the manufacturer with the two plasmids γ-tubulin-paGFP and mCherry-α-tubulin at a 1:1 molar ratio. After 24 h replace the medium with medium containing antibiotics to select for cells containing both plasmids (for U2OS cells: geneticin 0.4 mg/ml, puromycin 20 μg/ml) (*see* **Note 4**).
3. One week after addition of selective medium, trypsinize and resuspend cells of each well/plate in 2 ml selection medium to obtain a single cell suspension (*see* **Note 5**). Transfer cells to several 15 cm plates, using a different dilution of the cell suspension for each dish (e.g., 1/500; 1/250; 1/100). Keep the cells in fresh DMEM + 10 % FBS + 0.1 % PenStrep + antibiotics. Replace medium every 3 days.
4. When colonies become visible, select a dish that contains individual, well-separated colonies. For clone isolation confine each colony by placing a cloning cylinder on the dish. Add 40 μl of trypsin–EDTA solution to the colony inside the cylinder. Incubate for 5 min at 37 °C to allow cells to detach.
5. Add 80 μl DMEM + 10 % FBS + 0.1 % PenStrep + selective antibiotics to each trypsinized colony inside the cloning cylinder and pipet up and down a few times to separate cells. For each colony transfer the cell suspension into the well of a 24-well plate (1 colony per well). Adjust the volume in each well by adding DMEM + 10 % FBS + 0.1 % PenStrep + selective antibiotics to 1 ml final volume.

6. Keep the cells growing with media changes every 2–3 days until 80 % confluency, then trypsinize and transfer each well of cells into a well of a 6-well plate.
7. Keep the cells growing with media changes every 2–3 days until they reach 80 % confluency.
8. Observe cell clones in individual wells under a fluorescence microscope to identify clones positive for expression of mCherry-α-tubulin. Select clones with relatively homogeneous expression levels throughout the cell population (*see* **Note 6**).
9. For each clone selected in **step 8** appropriate expression of γ-tubulin-paGFP will be confirmed by standard immunofluorescence microscopy and by western blotting (*see* **Note 7**).
10. Expand the clones with appropriate expression of mCherry-α-tubulin and γ-tubulin-paGFP in 10 cm dishes with DMEM + 10 % FBS + 0.1 % PenStrep + selective antibiotics. For each clone freeze and store several aliquots of cells in liquid nitrogen as backup for future use.

3.2 Preparing Cells for Live Imaging

1. 24–48 h before imaging aspirate medium from a 90 % confluent 10 cm dish of cells stably expressing γ-tubulin-paGFP and mCherry-α-tubulin. To prepare cells that transiently express these proteins *see* **Note 3**.
2. Wash two times with 1× sterile PBS.
3. Add 1 ml 0.05 % trypsin–EDTA and incubate approximately 5 min at 37 °C until cells detach from the dish.
4. Add 2 ml of DMEM + 10 % FBS + 0.1 % PenStrep + antibiotics.
5. Make dilutions of cell suspension and transfer into glass-bottom dishes. The dilutions should be adapted to the growth rate of the cell line, so that after 24–48 h, at the day of the experiment, dishes will have a confluency of 80–90 % (*see* **Note 8**). In the example of U2OS cells use 10 % of the volume obtained from the **step 4** (300 μl cell suspension plated in 2 ml medium) 24 h before the experiment will result in the appropriate cell density.

3.3 Imaging the Dynamics of γ-Tubulin- paGFP

1. Approximately 2 h before the experiment warm up the incubation chamber to 37 °C and set the CO_2 to 5 %.
2. Find round mitotic cells using the transmitted light channel. With some training the desired mitotic stages can be identified by light microscopy based on cell shape and chromosome distribution. Switch to the mCherry-α-tubulin channel to confirm the stage of mitosis based on spindle morphology. Since paGFP is not yet activated, no (or only a very weak) signal should be visible in the green channel.

3. For each cell acquire one image using transmitted light and a stack of five images (corresponding to 2 μm in the *z* axis; *see* **Note 9**) for the red and green channel, respectively. This will document the pre-activation state of the cell.
4. Use your software to select a region of interest (ROI) where the laser will be directed for photoactivation (*see* **Note 10**). Using the mCherry-α-tubulin signal in the red channel the ROI can be selected conveniently at any position inside or outside the spindle area. To image the dynamics of γ-tubulin-paGFP at microtubule minus ends in spindles, for example, the photoactivation will be done along a line that is perpendicular to the spindle axis, at a distance of 2–4 μm from the poles (*see* Fig. 1a).
5. Photoactivate γ-tubulin-paGFP in ROI by a 405 nm laser (*see* **Note 11**).
6. Immediately acquire stacks of five images (corresponding to 2 μm in the *z* axis) for the red and green channel, respectively, at 5 s intervals for 2 min (*see* **Note 12**).
7. For each experimental condition image multiple cells. Repeat experiments at least three times.

3.4 Image Processing and Quantification of γ-Tubulin-paGFP Signals

After photoactivation, changes in the distribution of γ-tubulin-paGFP fluorescence can be quantified. The method to be used depends on the spindle region in which γ-tubulin-paGFP signals are measured. Here we describe two methods, the first measures changes in fluorescence intensity at the spindle poles/centrosomes, and the second measures movement of fluorescence marks in metaphase spindles.

3.4.1 Quantification of γ-Tubulin-paGFP Signals at Centrosomes

1. In ImageJ open the green channel of each time-lapse series and perform a maximum *z*-stack projection.
2. Draw a circle of 2 μm using the oval selection tool from ImageJ.
3. Place the ROI at the spindle pole in order to surround the centrosome (*see* **Note 13**), then measure the intensity of the region (*see* Fig. 1b).
4. Place the ROI in cytoplasmic region outside of the spindle near the centrosome and measure the intensity. This value will be used for background subtraction (*see* Fig. 1b).

Fig. 1 (continued) intensity of γ-tubulin-paGFP at centrosomes. The *red* and *green* channels of the first two frames of the experiment in (**a**) are displayed. The *yellow circle* represents the ROI used to measure centrosome fluorescence, the *red circle* is the ROI used to measure background fluorescence. (**c**-**f**) Workflow for quantification of γ-tubulin-paGFP poleward movement in metaphase spindles. Details are described in the protocol. The *arrowhead* in the kymograph in panel (**e**) indicates the position of the centrosome, the *asterisk* the position of laser activation. Scale bar, 1 μm. In the kymograph in panel (**f**) the *yellow line* represents the slope of the γ-tubulin-paGFP signal in the pole-distal region (faster movement), the *green line* represents the slope of the signal closer to the pole (slower movement). Parts of this figure are extracted from our previously published work [11]

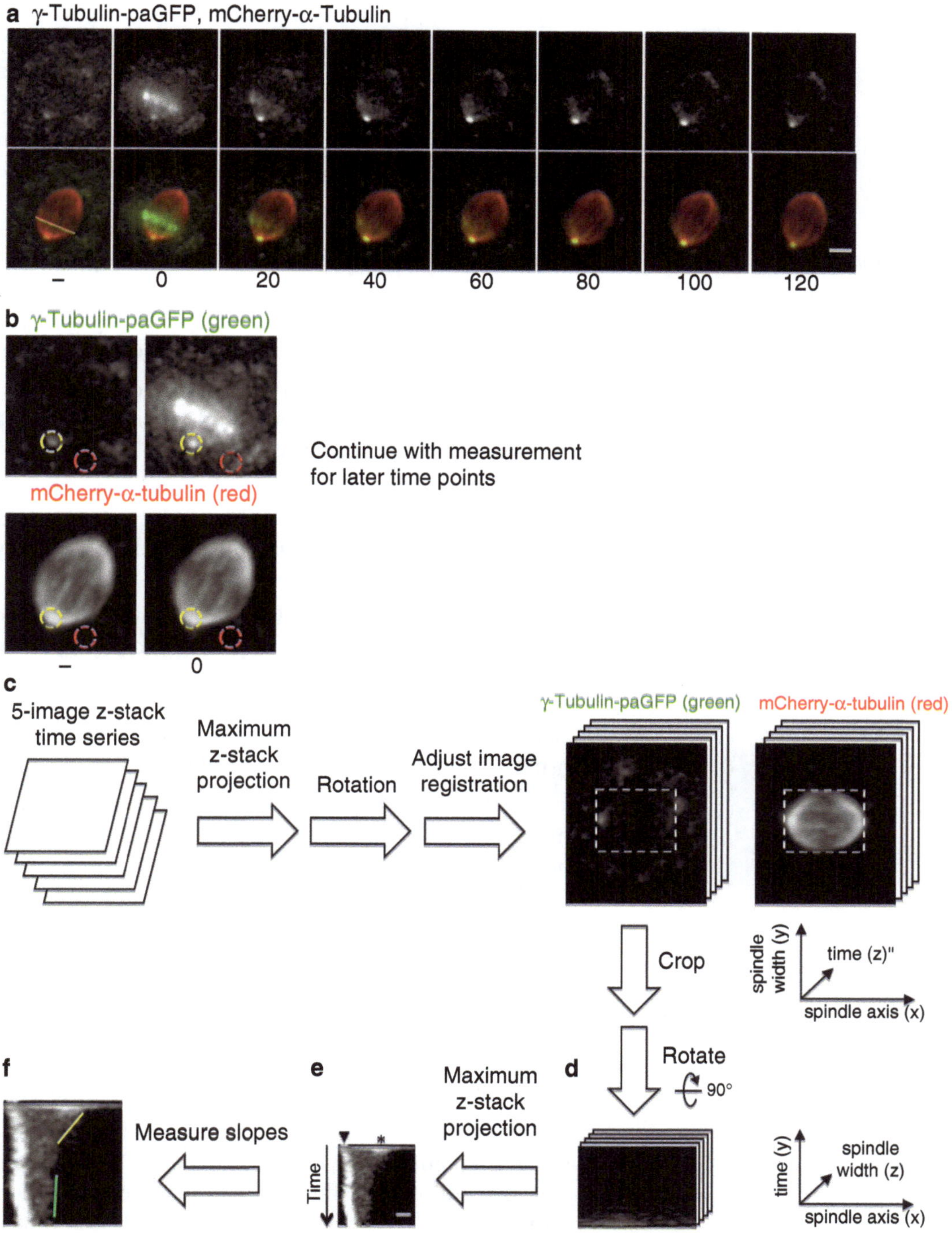

Fig. 1 Example of a photoactivation experiment and workflow for quantification of γ-tubulin-paGFP dynamics. (**a**) Metaphase U2OS cells stably expressing γ-tubulin-paGFP and mCherry-α-tubulin were subjected to laser-directed photoactivation (*yellow line*). Shown are still images of the channel for γ-tubulin-paGFP (*grey scale*) and a merge of the channels for γ-tubulin-paGFP (*green*) and mCherry-α-tubulin (*red*) before and after photoactivation. Numbers below images indicate time in seconds. Scale bar, 5 μm. (**b**) Illustration for the quantification of the

5. Copy and paste the values in an Excel spreadsheet or equivalent, then normalize the value of the centrosome intensity by subtracting the background. Repeat **steps 3–5** for each selected time point of the time-lapse.
6. For each condition, measure the mean kinetics of γ-tubulin accumulation at the centrosome for several cells in multiple independent experiments.

3.4.2 Quantification of γ-Tubulin-paGFP Signals in Metaphase Spindles

1. For quantification of the poleward movement of γ-tubulin-paGFP at the minus ends of spindle microtubules, generate kymographs in ImageJ that display positional changes of the γ-tubulin-paGFP photoactivation mark over time.
2. For each time-lapse image series (*see* Fig. 1c) open the two channels (green and red) in different windows and perform a maximum *z*-stack projection for the two files.
3. Draw a line that connects the two poles and measure the angle. Rotate each time-lapse series of images by the measured angle such that the spindle axes are positioned horizontally (*see* Fig. 1c).
4. To correct for movement of the cell and for movement of the spindle inside the cell that occurs during image acquisition, use the mCherry-α-tubulin channel to align the spindle images of each time point using the multistack registration plug-in (MultiStackReg) of ImageJ. Apply the correction to the γ-tubulin-paGFP channel (*see* Fig. 1c and **Note 14**).
5. Continue working only with the registration-corrected γ-tubulin-paGFP file. Crop the picture to contain only the spindle body using the rectangle selection tool (*see* Fig. 1c, d and **Note 15**).
6. Rotate the time-lapse image stack by 90° around the *x*-axis using the transformJ plug-in. The resulting file will display the time dimension on the *y*-axis and the spindle width on the *z*-axis (*see* Fig. 1d).
7. Create the kymograph by performing a maximum stack projection of the spindle width. The kymograph will display the spindle axis as *x*-axis and time on the *y*-axis (*see* Fig. 1e).
8. The rate of movement can be obtained from the kymograph. Draw a line that follows the fluorescence mark and adjust its position. ImageJ will display its slope. The rate in μm/min is calculated by multiplying the slope of the line by the pixel size divided by the time between frames (5 s in our case) (*see* **Note 16** and Fig. 1f).
9. To correct for potential residual spindle movement that has not been corrected for by the ImageJ plugin (*see* **step 4** above) examine the kymographs for any visible centrosome movement along the spindle axis. If this occurs, determine its rate as

described above and subtract it from the rate measured for the γ-tubulin-paGFP mark in the spindle (*see* **Note 17**).

10. For each condition, calculate the mean movement rate measured in multiple cells and in several independent experiments.

3.5 RNA Interference by Transfection of siRNA Prior to Imaging

1. To treat cells with siRNA, plate the cells in 3.5 cm dishes or in 6-well plates.
2. Transfect siRNA oligos at 50 % confluency using Lipofectamine RNAiMAX or equivalent reagent according to the manufacturer's protocol.
3. On the following day, trypsinize cells and seed on glass-bottom plates as described in Subheading 3.1. Choose a density that will allow cells to grow for 24–48 h before reaching a density of 80–90 % on the day of the imaging experiment (*see* **Note 18**).
4. Continue with imaging as described in Subheading 3.3 above.

3.6 Drug Treatments

1. Add drugs prior to imaging directly to the culture plate, rock gently to mix. The following drug concentrations and incubation times have been used in our laboratory: 32 μM EHNA or 50 μM Ciliobrevin D for 30 min to inhibit dynein (in the case of Ciliobrevin D pretreatment with MG132 is required; *see* **Note 19**); 50 μM STLC for 2 h to study Eg5 inhibition in monopolar spindles (to study Eg5 inhibition in bipolar spindles, pretreatment with MG132 is required; *see* **Note 19**). In our hands mitotic U2OS cells do not detach upon gentle rocking of the culture plate (*see* also **Note 8**). However, for other cell lines it may be better to premix the drug with medium and then replace the medium of the culture plate with this mixture by gentle pipetting to one side of the plate.
2. Continue with imaging as described in Subheading 3.3 above.

4 Notes

1. Our protocol was established using human U2OS cells. However, by adjusting the conditions other cell lines may be used.
2. A mCherry-α-tubulin expression plasmid can be obtained from Addgene (Addgene 21043). To obtain γ-tubulin (TUBG1) carrying photoactivatable GFP (paGFP) at its carboxy-terminus, full-length TUBG1 cDNA can be inserted into pEGFP-C1-based plasmids, in which EGFP had been replaced with paGFP. A γ-tubulin-paGFP plasmid can also be obtained from the authors. We recommend using C-terminally tagged γ-tubulin, which appears to be functional (based on the rescue of spindle defects caused by depletion of endogenous γ-tubulin) [11]. We have not tested N-terminal tags.

3. Generation of stable cell lines is straightforward for wild type proteins or mutants that do not affect cell proliferation upon constitutive expression. In cases where constitutive expression interferes with cell proliferation and thus generation of a stable line, the mutant protein can be expressed under the control of an inducible promoter (e.g., tetracycline-regulated; T-REx System, Life Technologies). Alternatively, transient expression can be used [11]. In this case transfect cells on plastic dishes as described in Subheadings 3.1 and 3.2 using a threefold molar excess for the γ-tubulin mutant containing plasmid over the mCherry-α-tubulin plasmid (3:1 ratio). This increases the chances that cells positive for mCherry-α-tubulin (which is used to identify transfected cells) have also received the mutant expressing plasmid. After 24 h transfer cells to glass-bottom dishes and grow cells for additional 24–48 h before imaging. This will allow shRNA-mediated depletion of endogenous γ-tubulin, if required, and reduce the initially very high γ-tubulin expression level. Cells should be grown in regular DMEM + 10 % FBS + 0.1 % PenStrep, in case of transient transfection, the use of selective antibiotics is not necessary.
4. To select for cells containing both γ-tubulin-paGFP and mCherry-α-tubulin expression plasmids with two different selection markers (antibiotic resistance) are required. We have successfully used a combination of neomycin resistance on the γ-tubulin-paGFP plasmid and puromycin resistance on the mCherry-α-tubulin plasmid. These can be selected for using geneticin/G418 and puromycin, respectively. The minimal antibiotic concentration required to kill non-resistant cells is cell line dependent and needs to be determined empirically.
5. Single cell suspension is needed at this step to obtain colonies derived from a single cell. This will ensure similar expression levels of the recombinant protein throughout the cell population. In order to obtain a single cell suspension, gently pipet the trypsinized cells up and down several times with a P1000 pipette tip.
6. Appropriate expression of mCherry-α-tubulin is determined by the visualization of spindle microtubules in live cells with only low/moderate cytoplasmic signal and the absence of cytoplasmic tubulin aggregates (sometimes observed as a result of overexpression). In case the expression level is variable between cells, a more homogeneous population can be obtained by FACS sorting based on the expression signal intensity of mCherry-α-tubulin.
7. Grow negative control cells and clones selected for stable expression of γ-tubulin-paGFP and mCherry-α-tubulin on poly-lysine-coated coverslips. To obtain good centrosome labeling we recommend fixation in cold methanol for 10 min

at -20 °C. For immunostaining we use the following method (but any standard protocol can be used): rehydrate the fixed cells in blocking buffer PBS-BT (phosphate-buffered saline (PBS), 3 % bovine serum albumin (BSA), 0.1 % Triton X-100) and incubate in the same buffer first with each of the primary and then with secondary antibodies. For each antibody incubate 30 min at room temperature and wash three times with PBS-BT between antibody incubations. Use primary antibodies produced in different species to detect paGFP tagged γ-tubulin (e.g., mouse anti-GFP, 3E6, Invitrogen, dilution: 1:500) and centrosomes (e.g., rabbit anti-CDK5RAP2, IHC-0006, Bethyl Laboratories; dilution 1:500), and secondary antibodies coupled to fluorophore (e.g., Alexa 488 anti-mouse and Alexa 568 anti-rabbit; A11029 and A11034, Life Technologies). Wash and mount the coverslips on slides with mounting medium (e.g., ProLong antifade, Life Technologies). Observe stained cells under a fluorescence microscope to confirm expression of γ-tubulin-paGFP. Ideally the expression level should be uniform across the cell population and in each cell recombinant γ-tubulin should be easily visible at centrosomes (colocalization with CDK5RAP2) and on spindle microtubules in mitotic cells. The γ-tubulin-paGFP signal in the cytoplasm should be low/moderate. In addition to CDK5RAP2 staining, some signal derived from mCherry-α-tubulin will be visible in the red channel. However, methanol fixation will reduce the fluorescence of mCherry and the residual signal will not interfere with the analysis. For western blot of proteins from negative control cells and cells stably expressing γ-tubulin-paGFP and mCherry-α-tubulin whole cell lysates are separated by SDS-PAGE, transferred to nitrocellulose or PVDF membranes, and probed with anti-γ-tubulin antibody (e.g., mouse anti-γ-tubulin, GTU-88, Sigma; dilution: 1:10,000). Endogenous γ-tubulin runs at about 50 kDa, γ-tubulin-paGFP at about 75 kDa. The expression level of tagged γ-tubulin should be similar or only moderately increased relative to endogenous γ-tubulin.

8. Round mitotic cells adhere poorly to the substrate and a relatively high cell density at the time of imaging helps to maintain attachment of these cells. In the case of U2OS cells extensive contacts with neighboring cells can prevent mitotic cells from rounding up completely, which can improve imaging.

9. The part of the spindle that is to be imaged should be positioned within the 5-image stack. One way to achieve this is to focus on the spindle pole as a reference and then select the first and last frames such that they will be focused slightly above and below (or vice versa) this position.

10. Photoactivation can result in some activation of γ-tubulin-paGFP also outside the ROI. Do not place the ROI too close to centrosomes as this will result in a very bright centrosomal photoactivation signal that may mask the relatively dim spindle-associated fluorescence. Perform the photoactivation at the center of the imaging stack selected in **step 3**.
11. Our settings were: nine repeats, dwell time 60 ns, laser power 15 %. The settings are specific to our system and need to be determined empirically for other systems.
12. Photoactivation followed by time-lapse imaging should be automated using the microscope software to ensure consistent timing. The described conditions are suitable for imaging γ-tubulin poleward movement (flux of microtubule minus ends) in the metaphase spindle of U2OS cells. Taking *z*-stacks at each time point will ensure that the spindle-associated fluorescence mark remains in focus in at least one of the imaged planes. For different cell types or mitotic stages, the parameters may have to be adjusted, for example by varying the number of images per stack (increasing or decreasing the *z*-dimension), by varying the imaging interval, and by varying the total imaging time.
13. Even though γ-tubulin-paGFP at centrosomes can display weak fluorescence even before photoactivation, the intensity might be too weak to position the ROI precisely. In this case, use the spindle structure as defined in the red channel to place the ROI at the pole. Use the ROI manager tool to apply the ROI with the same coordinates to the green channel. If required, repeat this procedure for each time point to position the ROI correctly around the centrosome.
14. Since the results of this alignment are not always perfect, a second correction can be applied as described in **step 8** of the protocol.
15. Selection of the area can be done using the red channel, it is important to keep the complete spindle within the rectangular selection at every time point of the series. Use the ROI manager tool to apply the ROI with the same coordinates to the green channel series.
16. When analyzing kymographs of γ-tubulin-paGFP poleward movement in spindles, it should be taken into account that the rate of movement slows with decreasing distance from the pole. Keeping the kymograph at a low magnification leads to a better appreciation of the slope. For better visualization of the photoactivation mark adjust the brightness and contrast settings of the display. Inverting the LUT in ImageJ might also help.
17. For this correction the position of the spindle pole serves as a reference. Any translocation along the spindle axis can be detected in kymographs shortly after photoactivation. This is

possible because centrosomes at spindle poles rapidly accumulate γ-tubulin-paGFP fluorescence, even before the spindle-associated mark reaches the pole (most likely due to a small amount of γ-tubulin-paGFP that gets activated in the cytoplasm and reaches the pole more rapidly).

18. The timing depends on and needs to be adjusted to the siRNA and the specific kinetics at which the targeted protein is depleted. In most cases maximal depletion of target proteins will occur within 48–72 h post-transfection.

19. Addition of some drugs such as Ciliobrevin D, a more specific dynein inhibitor than EHNA, or the Eg5 inhibitor STLC will not allow bipolar spindle assembly. To overcome this problem cells can be pretreated with 5 μM MG132 for at least 2 h. This treatment will block cells in metaphase and increase the percentage of cells with a bipolar spindle configuration. Following the 2 h pretreatment Ciliobrevin D or STLC can be added for 30 min to inhibit dynein and Eg5, respectively. This treatment combination will maintain most of the bipolar spindle configurations for the duration of the experiment.

References

1. Kollman JM, Merdes A, Mourey L, Agard DA (2011) Microtubule nucleation by γ-tubulin complexes. Nat Rev Mol Cell Biol 12:709–721. doi:10.1038/nrm3209
2. Teixidó-Travesa N, Roig J, Lüders J (2012) The where, when and how of microtubule nucleation – one ring to rule them all. J Cell Sci 125:4445–4456. doi:10.1242/jcs.106971
3. Akhmanova A, Steinmetz MO (2008) Tracking the ends: a dynamic protein network controls the fate of microtubule tips. Nat Rev Mol Cell Biol 9:309–322. doi:10.1038/nrm2369
4. Wiese C, Zheng Y (2000) A new function for the gamma-tubulin ring complex as a microtubule minus-end cap. Nat Cell Biol 2:358–364. doi:10.1038/35014051
5. Meunier S, Vernos I (2012) Microtubule assembly during mitosis – from distinct origins to distinct functions? J Cell Sci 125:2805–2814. doi:10.1242/jcs.092429
6. Mastronarde DN, McDonald KL, Ding R, McIntosh JR (1993) Interpolar spindle microtubules in PTK cells. J Cell Biol 123:1475–1489
7. Mitchison T, Evans L, Schulze E, Kirschner M (1986) Sites of microtubule assembly and disassembly in the mitotic spindle. Cell 45:515–527
8. Mitchison TJ (1989) Polewards microtubule flux in the mitotic spindle: evidence from photoactivation of fluorescence. J Cell Biol 109:637–652
9. Burbank KS, Mitchison TJ, Fisher DS (2007) Slide-and-cluster models for spindle assembly. Curr Biol 17:1373–1383. doi:10.1016/j.cub.2007.07.058
10. Tirnauer JS, Salmon ED, Mitchison TJ (2004) Microtubule plus-end dynamics in Xenopus egg extract spindles. Mol Biol Cell 15:1776–1784. doi:10.1091/mbc.E03-11-0824
11. Lecland N, Lüders J (2014) The dynamics of microtubule minus ends in the human mitotic spindle. Nat Cell Biol 16:770–778. doi:10.1038/ncb2996
12. Patterson GH, Lippincott-Schwartz J (2002) A photoactivatable GFP for selective photolabeling of proteins and cells. Science 297:1873–1877. doi:10.1126/science.1074952

Chapter 6

Visualizing and Analyzing Branching Microtubule Nucleation Using Meiotic Xenopus Egg Extracts and TIRF Microscopy

Matthew King and Sabine Petry

Abstract

Mitotic and meiotic spindles consist primarily of microtubules, which originate from centrosomes and within the vicinity of chromatin. Indirect evidence suggested that microtubules also originate throughout the spindle, but the high microtubule density within the spindle precludes the direct observation of this phenomenon. By using meiotic *Xenopus laevis* egg extract and employing total internal reflection (TIRF) microscopy, microtubule nucleation from preexisting microtubules could be demonstrated and analyzed. Branching microtubule nucleation is an ideal mechanism to assemble and maintain a mitotic spindle, because microtubule numbers are amplified while preserving their polarity. Here, we describe the assays that made these findings possible and the experiments that helped identify the key molecular players involved.

Key words Cell division, Mitotic spindle, Meiotic spindle, Cytoskeleton, Microtubule, Microtubule nucleation, *Xenopus laevis* egg extract, TIRF microscopy

1 Introduction

The mitotic spindle is assembled from several microtubule (MT) organizing centers, most prominently centrosomes and chromosomes, all of which require the universal MT nucleating molecule gamma-tubulin (γ-TB). More recently, MTs were also found to originate from within the body of the spindle [1, 2], and this was shown to be important for spindle assembly [3, 4]. Furthermore, a targeting factor for γ-TB to spindle MTs was identified in a whole-genome RNAi screen and termed augmin, as it increases the MT density within the spindle [5–8]. Yet the high MT density in the spindle precludes direct observation MT nucleation events, leaving the question of how MTs are generated within the body of the spindle open.

This roadblock could be circumvented by using meiotic *Xenopus* egg extract, which allowed resolving individual MTs with a low background by total internal reflection (TIRF) microscopy [9]. By simultaneously imaging MTs and their growing MT plus

Paul Chang and Ryoma Ohi (eds.), *The Mitotic Spindle: Methods and Protocols*, Methods in Molecular Biology, vol. 1413,
DOI 10.1007/978-1-4939-3542-0_6, © Springer Science+Business Media New York 2016

ends via fluorescently labeled tubulin and end-binding protein 1 (EB1), respectively, it was directly demonstrated that MTs originate from the wall of preexisting MTs. Addition of recombinant proteins and immunodepletion of endogenous proteins revealed the key molecular players to be augmin, γ-TB, RanGTP and its downstream factor "targeting factor of Xklp2" (TPX2) [9]. The goal of this chapter is to describe the assay and its variations that first characterized branching MT nucleation and its molecular players in *Xenopus* egg extract.

2 Materials

2.1 Proteins and Antibodies

2.1.1 Unlabeled and Fluorescently Labeled Tubulin

1. Purified and polymerization competent tubulin [10].
2. Biotinylated tubulin [11].
3. Alexa 568-labeled tubulin.
4. Cy5-labeled tubulin.

2.1.2 Recombinant Proteins for Addition to Extract

1. GFP-labeled H.s. End-Binding protein 1 (EB1).
2. X.l. Ran^{Q69L}.
3. X.l. TPX2.

2.1.3 Rabbit Polyclonal Antibody Production and Purification for Immunodepletion

1. Polyclonal antibody serum specific against protein of interest.
2. Affigel-10 and -15 matrix (Bio-Rad).
3. Glycerol.

2.2 Branching MT Nucleation Assay in Xenopus Egg Extract

2.2.1 Preparation of Xenopus Egg Extracts

1. Freshly prepared *Xenopus laevis* CSF extract of high quality (*see* **Note 1**).

2.2.2 Flow Chamber Assembly

1. Frosted glass slides.
2. Double-sticky tape.
3. Razorblade.
4. No. 1.5 glass coverslips (22 × 22 mm).

2.2.3 Basic Branching Reaction

1. CSF-XB buffer (10 mM K-HEPES pH 7.7, 2 mM $MgCl_2$, 0.1 mM $CaCl_2$, 100 mM KCl, 5 mM EGTA, and 50 mM sucrose).
2. Hot candle wax or nail polish.

3. Objective-based TIRF microscope equipped with a 100× 1.49 NA objective, a low noise EM-CCD camera and laser power of at least 20 mW out-of-fiber.

2.2.4 Variations of the Basic Branching Reaction

1. Sodium orthovanadate (100 mM, NEB) or purified p150-CC1 (*see* **Note 2**).

2.2.5 Immunodepletion of CSF Extract and Add-Back of Recombinant Protein

1. Dynabeads Protein A (Invitrogen).
2. DynaMag-2 Magnet (Invitrogen) for retrieving Dynabeads.
3. TBS-T (50 mM Tris–HCl, 150 mM NaCl, 0.05 % Tween 20).
4. Antibody of desired specificity and control total IgG fraction.
5. CSF-XB buffer (10 mM K-HEPES pH 7.7, 2 mM $MgCl_2$, 0.1 mM $CaCl_2$, 100 mM KCl, 5 mM EGTA, and 50 mM sucrose).
6. SDS-sample buffer, SDS-PAGE and Western blot equipment.
7. Recombinant protein that can replace the immunodepleted endogenous protein.

2.3 Branching MT Nucleation from Stabilized MT Seeds

1. Passivated coverslips [12].
2. GpCpp/GMPCPP (Jena Bioscience).
3. BRB80 buffer (80 mM K-PIPES, 1 mM $MgCl_2$, 1 mM EGTA, pH 6.8 with KOH).
4. Anti-biotin antibody (Invitrogen).
5. Kappa-casein (Sigma).
6. Oxygen scavenger system consisting of Trolox, protocatechuate 3,4-dioxygenase (PCD) and protocatechuic acid (PCA) according to Ref. [13] (all Sigma).

3 Methods

3.1 Proteins, Antibodies, and Xenopus Egg Extract

3.1.1 Unlabeled and Fluorescently Labeled Tubulin (See **Note 3**)

1. Tubulin is purified from bovine brain using two cycles of polymerization–depolymerization [10].
2. Fractions of purified tubulin are then fluorescently labeled [11] with 6((biotinoyl)amino) hexanoic acid, succinimidyl ester (Invitrogen), Alexa 568 carboxylic acid, succinimidyl ester (Invitrogen) and Cy5 NHS ester (GE). Briefly, MTs are polymerized and pelleted before being incubated with the dye at 37 °C. Labeled MTs are pelleted and depolymerized on ice. Dimeric, labeled tubulin is separated from polymerized MTs through another pelleting step at 4 °C and subsequently flash-frozen in small aliquots at 200 μM.

3.1.2 Recombinant Proteins for Extract Addition

1. GFP-labeled H.s. End-Binding protein 1 (EB1), X.l. Ran^{Q69L}, and X.l. TPX2 are purified as recently described [14–16].
2. All proteins are dialyzed into CSF-XB buffer containing 10 % sucrose, which does not disturb microtubule assembly dynamics in *Xenopus laevis* egg extract. During this process, the sample is further concentrated and introduced to sucrose, which serves as a cryopreservant.
3. Immediately after purification and dialysis, all proteins are flash frozen in small aliquots at concentrations of 20 μM (TPX2), 220 μM (Ran^{Q69L}), and 10 mg/mL (GFP-EB1).

3.1.3 Rabbit Polyclonal Antibody Production and Purification for Immunodepletion

1. Polyclonal antibodies for immunodepletion are generated at a commercial offsite facility, here Covance. The antigen is ideally the complete protein or protein domains, instead of peptides, to generate high affinity antibodies, as was performed for TPX2 and the augmin subunit Dgt4. Purified antibody against gamma-tubulin's C-terminus was a gift from Christiane Weise [17].
2. Couple antigen covalently to a matrix, such as Affigel matrix (Bio-rad) used here.
3. Perform immuno-affinity purification to isolate the antibody from the serum.
4. Supplement antibodies with 20 % glycerol before flash-freezing them in small aliquots.

3.1.4 Xenopus Egg Extract

1. Preparation of Xenopus egg extracts have previously been described in great detail with complete lists of reagents required for mitotic arrested Xenopus egg extracts [18–20]. We precisely follow all steps outlined in Ref. [19]. In order to adapt the procedure to modern equipment, the packing spin during which the eggs are concentrated without lysing is no longer conducted in a clinical centrifuge, but a table centrifuge at 150 × *g* for 60 s followed by 598 × *g* for 25 s.

3.2 Branching MT Nucleation Assay in Xenopus Egg Extract

3.2.1 Preparation of Xenopus Egg Extracts

Xenopus laevis CSF extract is prepared by the standard method (Murray [18]). Freshly made extract should be stored on ice and must not subjected to shearing forces during pipetting. Extract can either be directly used (*see* Subheading 3.2.2) or first subjected to immunodepletion (Subheading 3.2.5) before its test in the branching assay (*see* **Note 4**).

3.2.2 Flow Chamber Assembly

Assemble a flow chamber as depicted in Fig. 1 consisting of a microscope slide, two double-sticky tape strips separated by 2–4 mm, and a glass coverslip. The chamber volume should be approximately 5 μL.

3.2.3 Basic Branching Reaction (Fig. 1)

1. Pipette 7.5 μL CSF extract into a 1.5 mL tube on ice without shearing the extract via pipetting.

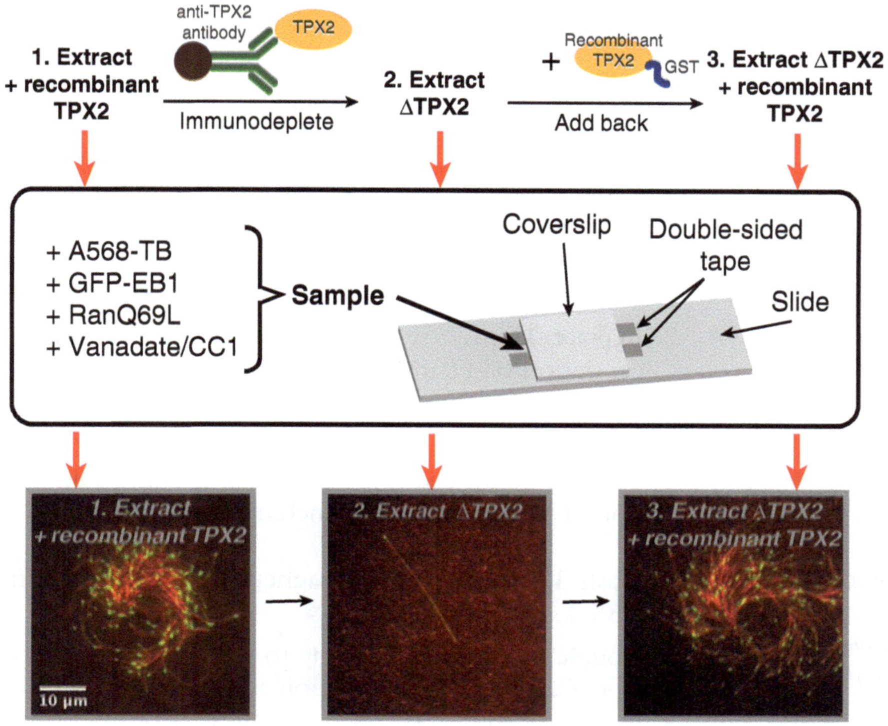

Fig. 1 Overview of the basic branching MT nucleation assay, which is performed with Xenopus egg extract. Proteins of interest, such as TPX2, can be immunodepleted to assess their phenotype and role. Recombinantly expressed and purified proteins can then be added back to the depleted extract to test whether they complement the depleted function

2. Add Alexa 568-labeled tubulin to a final concentration of 0.86 μM to visualize MTs (*see* **Note 5**).
3. Add EB1-GFP to a final concentration of 0.8 μM to visualize growing MT plus ends.
4. Add CSF-XB to obtain a final sample volume of 10 μL.
5. Mix the reaction mixture by gently pipetting it up and down once.
6. Pipette 5 μL of the reaction mixture into the flow cell to fill the channel without introducing bubbles.
7. Quickly seal the sample chamber with hot candle wax or nail polish.
8. Start imaging with an objective-based TIRF microscope equipped with a 100× 1.49 NA objective, a low noise EM-CCD camera and laser power of at least 20 mW out-of-fiber.
9. Optimize imaging parameters (laser power and exposure time) to clearly observe MTs but prevent overexposure and thus photobleaching (*see* **Note 6**).

10. Acquire frames every 2 s for 30 min.
11. Collect a stitched slide overview of fields of view using the slide explorer function in uManager [21] at equal time points (for example after 15 or 30 min; *see* **Note 7**).

3.2.4 Variations of the Basic Branching Reaction (Fig. 1)

1. Replace 0.5 μL CSF-XB buffer with 0.5 μL vanadate (10 mM) for a final concentration of 0.5 mM or p150-CC1 to a final concentration of 0.125 mg/mL to inhibit dynein and prevent MT gliding on glass.
2. Replace 0.5 μL CSF-XB buffer with Ran^{Q69L} to for a final concentration of 11 μM to release endogenous spindle assembly factors from importins, such as TPX2, and induce branching MT nucleation.
3. Replace 0.5 μL CSF-XB buffer with TPX2 to a final concentration of 1 μM to induce branching MT nucleation.

3.2.5 Immunodepletion of CSF Extract and Add-Back of Recombinant Protein (Fig. 1)

1. Wash 150 μL slurry of magnetic Dynabeads 3× in 300 μL TBS-T.
2. Couple 36 μg total antibody to washed Dynabeads overnight at 4 °C. Use total IgG fraction as a control.
3. Wash off unbound antibody with 2× 150 μL TBS-T and 2× 150 μL CSF-XB.
4. Split beads into three 50 μL fractions. Keep two on ice, remove CSF-XB from one and add 50 μL of CSF extract after setting aside an extract gel sample.
5. Incubate extract with beads for 45 min on ice and gently mix every 10 min by flicking the tube or pipetting up and down with a wide bore pipette.
6. Retrieve beads completely on a magnet for 5 min and retrieve a gel sample of extract after this first round of immunodepletion.
7. Remove CSF-XB from the second bead batch and transfer extract from first to second tube for a second round of immunodepletion.
8. Repeat one more time for a third round of immunodepletion (*see* **Note 8**).
9. Perform imaging and observe phenotype.
10. Add-back recombinant protein to immunodepleted extract at similar concentrations as the endogenous version and take a gel sample of the final reaction mixture to verify this. Perform imaging and observe phenotype.
11. Prepare SDS-PAGE gel samples by mixing 2 μL of extract with 48 μL of 1× loading buffer. Perform Western blot for immunodepleted proteins and controls, and determine extent of depletion.

3.2.6 Determining Branch Angle and Measuring MT Nucleation Kinetics (See ***Note 9****)*

1. Use ImageJ [22] for quantitating branch angle between the daughter and mother microtubules. A branch angle of 0° corresponds to parallel growth.
2. Use ImageJ [22] to duplicate the EB1 channel and apply a Gaussian Blur function with a low [1] and a high [5] radius of decay for each one.
3. Subtract the latter image sequence from the former.
4. Set a threshold to eliminate background noise before counting particles for each time frame with the Analyze Particles function in ImageJ (*see* **Note 10**).

3.3 Branching MT Nucleation from Stabilized MT Seeds

1. Coverslips are cleaned and passivated with dichlorodimethylsilane as described [12].
2. Prepare GMPCPP-stabilized MT seeds, which are biotinylated and fluorescently labeled by mixing porcine brain tubulin (1.7 mg/mL final concentration), Cy5-labeled tubulin (0.19 mg/mL final concentration, ~30 % labeled), biotin-tubulin (0.19 mg/mL final concentration), and GMPCPP (1 mM final concentration) in BRB80 buffer and incubate at 37 °C for 1 h. Store at room temperature and in the dark until use.
3. Prepare a flow channel with passivated coverslips.
4. Flow in 50 μL of 0.1 mg/mL anti-biotin antibody in BRB80 and incubate for 5 min.
5. Flow in 50 μL of BRB80 to wash the chamber.
6. Flow in 50 μL kappa-casein (1 mg/mL) and incubate for 5 min.
7. Wash flow chamber with 50 μL of BRB80.
8. Flow in 50 μL MT seeds diluted 1/500 in BRB80 and incubate for 5 min.
9. Wash in 50 μL BRB80 with an oxygen scavenger system.
10. Check MT seeds for desired density with TIRF microscopy.
11. Flow in CSF extract with desired factors to induce branching (*see* **Note 11**).
12. Image as detailed above.

4 Notes

1. Only freshly prepared *Xenopus laevis* CSF extract of high quality remains arrested in meiosis II and displays microtubule behavior necessary for spindle assembly.

2. Sodium orthovanadate and p150-CC1 are both inhibitors of the minus-end directed motor protein cytoplasmic dynein.
3. Alternatively, unlabeled and labeled tubulin can be obtained from Cytoskeleton, Inc. if necessary.
4. *Xenopus* CSF extract only remains active for short periods of time (typically 10 h for good extract). Therefore, it is critical to plan the experiments well and minimize any delays in the procedures.
5. *Xenopus* CSF extract must not be diluted. When choosing the stock concentration of a reagent to be added, maximize the concentration and balance it with the smallest volume that can be added accurately (~0.5 μL).
6. TIRF angle, laser power and exposure time need to be adjusted with the first sample in order to clearly observe dynamic MTs within a bright fluorescent background. Once that is achieved, laser power and exposure time should further minimized while keeping MTs visible to prevent overexposure of the sample, which may lead to photobleaching during the course of the reaction.
7. uManager's slide explorer function is useful for quantifying the number of fan structures on a slide and to assess local variation within a flow chamber.
8. High-affinity antibodies can deplete the protein of interest within two rounds of immunodepletion.
9. For this analysis, it is critical that the background signal is as low as possible. Usually this is only achieved with extract of highest quality.
10. Make sure that each particle detected corresponds to the EB1 signal at the growing MT ends. Each MT, even when branched, only has one EB1 comet and therefore this method directly counts MT number over time.
11. An additional step is to wash out the extract and flow in tubulin, which can branch from the attached seeds.

Acknowledgements

This work was supported by grants from the NIH/NIGMS (4R00GM100013), the Pew Scholars Program in the Biomedical Sciences, the Sidney Kimmel Foundation, and the David and Lucile Packard Foundation to S.P.

References

1. Lajoie-Mazenc I, Tollon Y, Detraves C, Julian M, Moisand A, Gueth-Hallonet C, Debec A, Salles-Passador I, Puget A, Mazarguil H et al (1994) Recruitment of antigenic gamma-tubulin during mitosis in animal cells: presence of gamma-tubulin in the mitotic spindle. J Cell Sci 107(Pt 10):2825–2837
2. Mahoney NM, Goshima G, Douglass AD, Vale RD (2006) Making microtubule sand mitotic spindles in cells without functional centrosomes. Curr Biol 16:564–569
3. Brugues J, Nuzzo V, Mazur E, Needleman DJ (2012) Nucleation and transport organize microtubules in metaphase spindles. Cell 149:554–564
4. Loughlin R, Heald R, Nedelec F (2010) A computational model predicts Xenopus meiotic spindle organization. J Cell Biol 191:1239–1249
5. Goshima G, Wollman R, Goodwin SS, Zhang N, Scholey JM, Vale RD, Stuurman N (2007) Genes required for mitotic spindle assembly in Drosophila S2 cells. Science 316:417–421
6. Goshima G, Mayer M, Zhang N, Stuurman N, Vale RD (2008) Augmin: a protein complex required for centrosome-independent microtubule generation within the spindle. J Cell Biol 181:421–429
7. Uehara R, Nozawa RS, Tomioka A, Petry S, Vale RD, Obuse C, Goshima G (2009) The augmin complex plays a critical role in spindle microtubule generation for mitotic progression and cytokinesis in human cells. Proc Natl Acad Sci U S A 106:6998–7003
8. Lawo S, Bashkurov M, Mullin M, Ferreria MG, Kittler R, Habermann B, Tagliaferro A, Poser I, Hutchins JR, Hegemann B, Pinchev D, Buchholz F, Peters JM, Hyman AA, Gingras AC, Pelletier L (2009) HAUS, the 8-subunit human Augmin complex, regulates centrosome and spindle integrity. Curr Biol 19:816–826
9. Petry S, Groen AC, Ishihara K, Mitchison TJ, Vale RD (2013) Branching microtubule nucleation in Xenopus egg extracts mediated by augmin and TPX2. Cell 152:768–777
10. Castoldi M, Popov AV (2003) Purification of brain tubulin through two cycles of polymerization-depolymerization in a high-molarity buffer. Protein Expr Purif 32:83–88
11. Hyman AA (1991) Preparation of marked microtubules for the assay of the polarity of microtubule-based motors by fluorescence. J Cell Sci Suppl 14:125–127
12. Gell C, Bormuth V, Brouhard GJ, Cohen DN, Diez S, Friel CT, Helenius J, Nitzsche B, Petzold H, Ribbe J, Schaffer E, Stear JH, Trushko A, Varga V, Widlund PO, Zanic M, Howard J (2010) Microtubule dynamics reconstituted in vitro and imaged by single-molecule fluorescence microscopy. Methods Cell Biol 95:221–245
13. Aitken CE, Marshall RA, Puglisi JD (2008) An oxygen scavenging system for improvement of dye stability in single-molecule fluorescence experiments. Biophys J 94:1826–1835
14. Petry S, Pugieux C, Nedelec FJ, Vale RD (2011) Augmin promotes meiotic spindle formation and bipolarity in Xenopus egg extracts. Proc Natl Acad Sci U S A 108:14473–14478
15. Kalab P, Weis K, Heald R (2002) Visualization of a Ran-GTP gradient in interphase and mitotic Xenopus egg extracts. Science 295:2452–2456
16. Groen AC, Maresca TJ, Gatlin JC, Salmon ED, Mitchison TJ (2009) Functional overlap of microtubule assembly factors in chromatin-promoted spindle assembly. Mol Biol Cell 20:2766–2773
17. Albee AJ, Wiese C (2008) Xenopus TACC3/maskin is not required for microtubule stability but is required for anchoring microtubules at the centrosome. Mol Biol Cell 19:3347–3356
18. Murray AW (1991) Cell cycle extracts. Methods Cell Biol 36:581–605
19. Hannak E, Heald R (2006) Investigating mitotic spindle assembly and function in vitro using Xenopus laevis egg extracts. Nat Protoc 1:2305–2314
20. Desai A, Murray A, Mitchison TJ, Walczak CE (1999) The use of Xenopus egg extracts to study mitotic spindle assembly and function in vitro. Methods Cell Biol 61:385–412
21. Edelstein AD, Tsuchida MA, Amodaj N, Pinkard H, Vale RD, Stuurman N (2014) Advanced methods of microscope control using muManager software. J Biol Methods 1(2), pii: e10
22. Rasband WS (1997/2014) ImageJ, U.S. National Institutes of Health, Bethesda, MD, USA, http://imagej.nih.gov/ij/

Chapter 7

Encapsulation of *Xenopus* Egg and Embryo Extract Spindle Assembly Reactions in Synthetic Cell-Like Compartments with Tunable Size

Matthew C. Good

Abstract

Methods are described for preparing *Xenopus laevis* egg and embryo cytoplasm and encapsulating extract spindle assembly reactions in cell-like compartments to investigate the effects of cell size on intracellular assembly. Cytoplasm prepared from the eggs or embryos of individual frogs is screened for the ability to form interphase nuclei and metaphase spindles, and subsequently packaged, along with DNA, into droplets of varying size using microfluidics. The dimensions of these cell-like droplets are specified to match the range of cell diameters present in early embryo development. The scaling relationship between droplets and spindles is quantified using live fluorescence imaging on a spinning-disk confocal microscope. By comparing the encapsulated assembly of spindles formed from cytoplasmic extracts prepared from embryos at distinct stages of *Xenopus* early development, the influence of cell composition and cell size on spindle scaling can be evaluated. Because the extract system is biochemically tractable, the function of individual proteins in spindle scaling can be evaluated by supplementing or depleting factors in the cytoplasm.

Key words Spindle assembly, Size-scaling, *Xenopus*, Egg extract, Cytoplasm, Encapsulation, Cell-like compartment, Droplet microfluidics, Emulsion, Embryogenesis

1 Introduction

The cytoplasmic extracts of *Xenopus* eggs are an established model system for investigating the molecular mechanisms of the vertebrate cell cycle in vitro [1–6]. The extracts are capable of carrying out many cellular processes, including genome and centrosome replication, and spindle and nucleus assembly. Spindle assembly is accomplished in vitro by adding DNA, in the form of either frozen demembranated sperm nuclei or freshly prepared interphase nuclei (generated by cycling extracts into interphase with calcium), to CSF-arrested extracts. Microtubules assemble around the exogenous DNA to form spindles and are imaged using labeled α/β--tubulin [7] and DNA. The extracts are especially useful because they are amenable to biochemical manipulation. Although this

Paul Chang and Ryoma Ohi (eds.), *The Mitotic Spindle: Methods and Protocols*, Methods in Molecular Biology, vol. 1413,
DOI 10.1007/978-1-4939-3542-0_7,

system has proven to be quite powerful, cytoplasmic extracts lack cell boundaries and are therefore missing much of spatial information encoded in the structure of a cell.

There is now a major thrust to understand how physical boundaries, mechanical perturbations and the spatial positioning of molecules influence the behavior of cellular processes, such as cytoskeletal or organelle assembly and signaling. For example, aspects of cell division have been reconstituted on supported lipid bilayers [8], assembly of cytoskeletal proteins has been confined to microfabricated chambers [9] or water-in-oil emulsions [10, 11], and cytoplasmic extracts have been encapsulated to create cell-like compartments [12–16].

Because the tools for microfabrication and lithography are now readily available, it is feasible to use droplet microfluidics [17, 18] to encapsulate aqueous solutions in cell-size compartments whose dimensions are precisely defined. These techniques make it possible to analyze how the geometric constraints and limited volumes of a vertebrate cell impact biochemical reactions reconstituted in vitro. For example, by combining *Xenopus* extract spindle assembly with droplet microfluidics it is possible to directly test the hypothesis that cell size regulates spindle and nucleus assembly. Importantly, with the acquisition of just a few prefabricated parts, any lab can generate microfluidic devices that can be tailored to a specific biological question.

Cell size is a largely unexplored regulatory parameter in biology. There is growing interest in the observation that cell size and organelle size are coupled. For example, during the reductive divisions of early vertebrate embryogenesis, cell size is rapidly decreased. Concomitant with the reduction in cell size, intracellular structures such as the spindle, nucleus, and centrosome also decrease in size [12, 19–22]. But how do organelles know how to reach the correct size, and how do these assembly pathways adapt to changes in cell size? One approach to directly answer these questions is to build synthetic cells whose dimensions can be tuned to match the range of cell sizes found in embryo development [12]. These non-replicative cell-like compartments contain a defined boundary and are filled with cytoplasm. In addition, because cytoplasm can be isolated from various stages of development [22, 23], it is possible to utilize these extract-filled droplets to distinguish between the relative contributions of a developmental program (e.g., composition changes) and cell size (a physical change) in organelle scaling.

Here, we describe a robust protocol for generating cytoplasm-filled compartments with tunable dimensions. Using this system, it is possible to precisely specify compartment size (from 5 to 500 μm) and to use live imaging to quantify spindle and nucleus scaling as a function of droplet size. The three methodologies described include: (1) Preparation of egg and embryo cytoplasmic extracts from individual *Xenopus laevis* frogs, (2) Fabrication of T-junction microfluidic droplet-generating devices starting from plastic replicas, and (3) Encapsulation of spindle assembly reactions inside of synthetic cell-like compartments with tunable size.

2 Materials

2.1 Egg Extract Preparation from Individual Frogs

1. Female *Xenopus laevis* frogs.
2. Pregnant mare serum gonadotropin (PMSG). 200 U/mL in sterile Milli-Q water. Store at 4 °C.
3. Human chorionic gonadotropin (HCG). 1000 U/mL in sterile Milli-Q water. Store at 4 °C.
4. 1 mL disposable syringes with 18-gauge and 27-gauge needles.
5. 4 L plastic containers with tight-fitting lids (hole-punched for air exchange).
6. 20× Marc's Modified Ringer's (MMR): 100 mM HEPES (free acid), 2 mM EDTA, 2 M NaCl, 40 mM KCl, 20 mM $MgCl_2$, 40 mM $CaCl_2$. Adjust pH to 7.8 with NaOH. Autoclave and store at room temperature.
7. 16 °C room or large incubator set to 16 °C.
8. 4 L plastic beaker and 500 mL glass beakers.
9. Plastic transfer pipettes.
10. 2 M sucrose. Filter-sterilize and store in 12.5 mL aliquots at −20 °C.
11. 1 M K-HEPES, pH 7.7, filter-sterilize and store in 5 mL aliquots at −20 °C.
12. 0.5 M K-EGTA, pH 7.7, filter-sterilize and store in 2 mL aliquots at −20 °C.
13. 2 M $MgCl_2$ solution.
14. 20× extract buffer (XB) salts: 2 M KCl, 20 mM $MgCl_2$, 2 mM $CaCl_2$. Autoclave and store at 4 °C.
15. Dejelly solution: 2 % l-cysteine base (~275 mM) in 1× XB salts, pH adjusted to 7.8 with NaOH. Prepare fresh and store at 16 °C. For 1 L solution, add 20 g l-cysteine, 50 mL 20× XB salts, and fill to 1 L. Adjust pH to 7.8 with NaOH.
16. XB: 100 mM KCl, 0.1 mM $CaCl_2$, 1 mM $MgCl_2$, 50 mM sucrose, 10 mM K-HEPES, pH 7.7. Prepare fresh and store at 16 °C. For 1 L solution, add 50 mL 20× XB salts, 25 mL 2 M sucrose, 10 mL of 1 M K-HEPES, and fill to 1 L with H_2O. Adjust pH to 7.7 with KOH.
17. CSF-XB: 1× XB + 1 mM $MgCl_2$, 5 mM K-EGTA, pH 7.7. Prepare fresh and store at 16 °C. For 400 mL solution, add 200 μL of 2 M $MgCl_2$ and 4 mL of 0.5 M K-EGTA pH 7.7 to ~400 mL of XB.

18. Leupeptin, pepstatin, chymostatin (LPC). 10 mg/mL each of leupeptin, pepstatin, chymostatin, in dimethylsulfoxide (DMSO). Store in ~55 μL aliquots at −20 °C.
19. Cytochalasin D: 10 mg/mL in DMSO. Store in ~55 μL aliquots at −20 °C.
20. CSF-XB+: to 100 mL of the previously prepared CSF-XB, add 100 μL of LPC. Prepare fresh and use immediately.
21. SW-55 ultra-clear thin-wall 5 mL centrifuge tubes (Beckman). 13×51 mm.
22. Sarstedt round-bottom 13 mL polypropylene tubes. 16.8×95 mm.
23. Refrigerated clinical centrifuge set to 16 °C.
24. High-speed refrigerated floor centrifuge containing Sorvall HB-6 rotor and rubber adapters for round-bottom centrifuge tubes (Kimble-Chase). Set to 16 °C.
25. Energy mix (50×): 190 mM creatine phosphate, 25 mM adenosine triphosphate, 25 mM $MgCl_2$, 2.5 mM K-EGTA pH 7.7. Store in 100 μL aliquots at −20 °C.

2.2 Embryo Extract Preparation

Subheading 2.2 requires reagents from Subheading 2.1, plus the following:

1. Testes freshly isolated from a male *Xenopus laevis* frog.
2. Plastic pestle for 1.5 mL microfuge tube (USA Scientific).
3. Glass petri dishes: 6 cm diameter.
4. 1/3× MMR stored at 16 °C.
5. Dissecting stereomicroscope.
6. Dejelly solution without XB salts: 2 % w/v cysteine (0.275 M final) in Milli-Q H_2O, pH to 7.8 with NaOH. Store at 16 °C.
7. 0.5 mg/mL (~13 μM) human Δ90-CyclinB1 in XB.
8. 15 mg/mL (~750 μM) human UbcH10-C114S in XB.

2.3 Fabrication of Microfluidic Droplet-Generating Devices

1. Plastic replicas: molds for casting microfluidic devices from PDMS.
2. Scotch tape.
3. Large plastic petri dishes: 150 mm diameter.
4. Slygard elastomer 184 kit: polydimethylsiloxane (PDMS) elastomer and curing agent.
5. Red solo cups.
6. 2 mL serological pipette.
7. Vacuum dessicator chamber.
8. Vacuum pump.

9. Oven set to 65 °C.
10. Transparencies.
11. 3 mm biopsy hole punch (e.g., Harris Uni-Core).
12. 0.5 mm biopsy hole punch.
13. Wire (<0.5 mm).
14. Razor.
15. Microscope slides.
16. Plasma cleaner (March Plasmod).

2.4 Extract Encapsulation and Spindle and Nucleus Assembly in Droplets

1. PDMS droplet-generating devices.
2. Aquapel rain repellant.
3. Gasket imaging chambers: laser cut 3 or 4.5 mm thick acrylic with 2–4 mm diameter circles (Fig. 4a). Affix to 18 mm glass circle using Norland 61 optical adhesive; expose to UV light for at least 2 h, or overnight.
4. Hamilton 1700 series glass syringes with removable needle: 25 μL (extract) and 50 μL (oil).
5. 28-Gauge needle for Hamilton syringes, 0.75 in. length.
6. Teflon (PTFE) tubing from Cole-Parmer: ID ~0.30 mm, OD ~0.76 mm.
7. Forceps with serrated tip.
8. Syringe pumps.
9. Stereomicroscope.
10. *Xenopus laevis* egg or embryo cytoplasmic extracts (from Subheadings 2.1 or 2.2).
11. Squalene (Sigma-Aldrich). Store at 4 °C.
12. Cithrol DPHS (Croda). Store at room temperature. Seal container with Parafilm.
13. 50 mg/mL solution of Cithrol DPHS in squalene: store 1 mL aliquots in glass vials with screw caps, topped with N_2 gas sealed with Parafilm, at 4 °C for up to 1 month.
14. Hoechst: 1 mg/mL Hoechst 33342 in Milli-Q H_2O. Store at 4 °C.
15. Demembranated sperm nuclei (a stock of 200,000 nuclei/μL). Store in 3 μL aliquots at −80 °C.
16. 50 mM $CaCl_2$ in Milli-Q H_2O. Filter-sterilize.
17. Rhodamine labeled porcine brain tubulin stock (20 mg/mL, or ~200 μM).
18. Spindle fix: 48 % glycerol, 11 % formaldehyde in a solution of 1× MMR. Add Hoechst to achieve a final concentration of 5 μg/mL Hoechst. Store at room temperature for up to 1 week.

3 Methods

Overview: The following sections describe procedures for the (1) Preparation of *Xenopus laevis* egg and embryo cytoplasmic extracts, (2) Fabrication of microfluidic droplet-generating devices from plastic molds, and (3) Encapsulation of cytoplasm and DNA in cell-size droplets to investigate spindle scaling as a function of compartment size.

3.1 Cytoplasmic Extract Preparation from Xenopus Eggs and Embryos

A key requirement of the extract experiments is the preparation of fresh egg or embryo cytoplasm capable of forming spindles with a high efficiency (percentage of sperm DNA around which bipolar spindles assemble). To ensure that at least one extract is functional, do not combine the eggs of each ovulated female frog. A significant hurdle in working with the extract system is that poor quality eggs from a single frog can inhibit the ability to (1) maintain extracts in CSF arrest, (2) cycle extracts into interphase and form nuclei, or (3) form bipolar spindles with high efficiency, after cycling nuclei back into mitosis. On the day of an experiment, my lab typically generates three separate extracts from individual frogs (yielding approximately 1 mL of pure cytoplasm from each clutch of eggs, or approximately 0.4 mL of cytoplasm from each dish of embryos) and screens these extracts for spindle and nucleus assembly in both a microfuge tube and inside cell-size droplets. Quality is scored and the best extracts are then systematically encapsulated and imaged in microfluidics droplets to plot the scaling relationship between spindle and compartment size (described in Subheading 3.3).

3.1.1 Egg Extract Preparation from Individual Frogs

See Fig. 1.

1. At 3–7 days prior to extract preparation, prime females with a subcutaneous injection of 100 U of PMSG using a 27-gauge needle and 1 mL syringe. These frogs should be used within 1 week. Numbers vary, but a typical week requiring three extract preparations necessitates the priming of 12 females.
2. At 16–18 h prior to egg collection subcutaneously inject four females with 500 U of HCG and store them separately in individual plastic containers filled with 2 L of 1× MMR (prepared with deionized water) at 16 °C. Assume that three out of four females will lay high-quality eggs, allowing the preparation of three separate single-frog extracts.
3. Analyze egg quality to determine how many extracts can be prepared. Good eggs should have clearly delineated animal (dark) and vegetal (light) poles. Clutches of eggs that are activated or lysed (white and puffy appearance) or stringy should not be used. A few lysed eggs (up to 10 in a container of ~2500 eggs) can be tolerated. These lysed eggs should be removed using a transfer pipette (*see* **Note 1**).

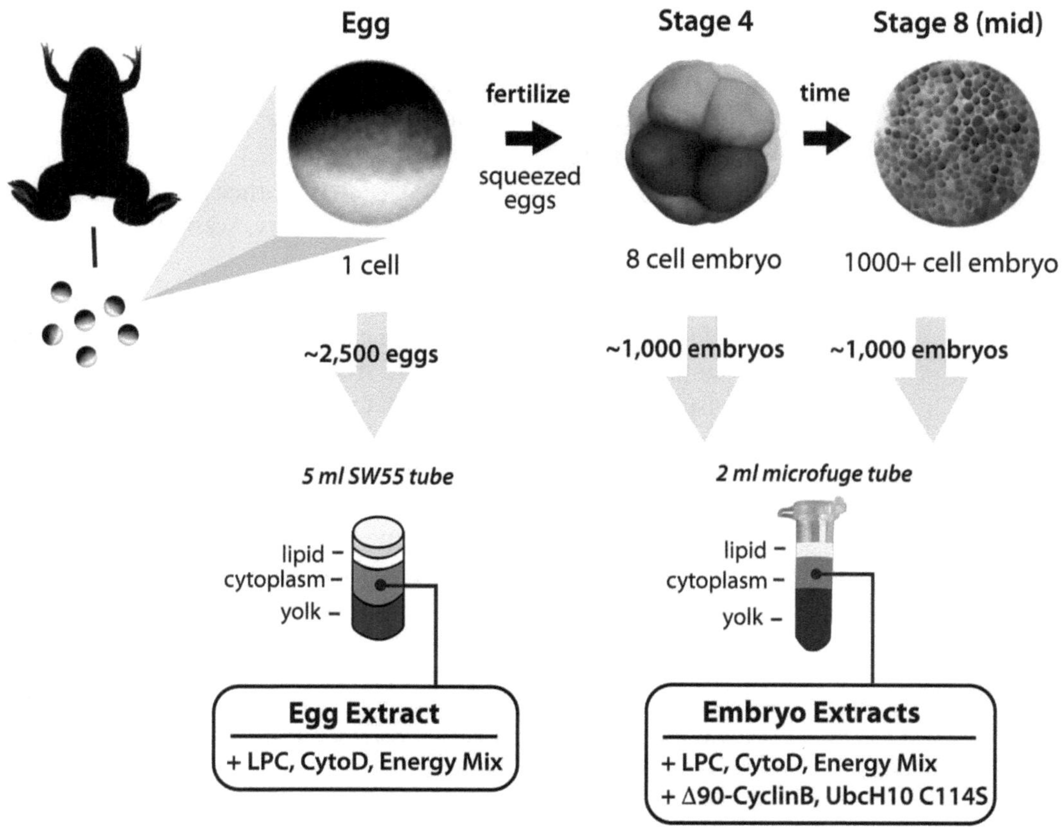

Fig. 1 Schematic of *Xenopus laevis* egg and embryo extract preparation. Starting from ovulating females, laid eggs can be collected and used to generate an "egg extract," or freshly squeezed eggs can be fertilized and synchronized embryos can be used to generate "embryo extracts" at various stages of early development

4. Separate frogs from eggs, keeping eggs in their containers at 16 °C. Frogs should be placed in a bucket of deionized water and returned to the colony by the end of the day.
5. At 2 h prior to extract preparation, begin preparing buffers, supplies and equipment (*see* **Note 2**). For three single-frog extracts: Prepare 1 L dejelly, 1 L XB, and 0.4 L CSF-XB. Ten minutes prior to the start of experiments, prepare 0.1 L CSF-XB+, and fill each of 4 Beckman SW55 tubes with 1 mL CSF-XB+ and 15 μL of 10 mg/mL cytochalasin D (final concentration of 150 μg/mL). For supplies: cut tips from 4 transfer pipettes, and leave 4 uncut. Collect one 4 L plastic beaker, 3 × 500 mL glass beakers and all reagents and supplies and use a cart to transport into the 16 °C room. Make sure that all centrifuges are set to 16 °C.

   In the next steps [6–8], gently pour buffers down the side of the beaker, and lightly swirl eggs. Between steps, pour off as much buffer as possible without exposing the eggs to air, which will shear them. Manipulate eggs carefully to prevent activation and lysing. Additionally, once eggs are dejellied, make sure to carry out subsequent washing and centrifugation steps in rapid succession.

6. Clean off the eggs: Pour off additional water inside plastic container to remove, urine, feces and other particulate matter. For containers that are especially dirty, wash the eggs once with 500 mL of 1× MMR.
7. Dejelly eggs: Pour off remaining MMR and add approximately 150 mL of dejelly solution to each of three beakers, each containing eggs from a separate frog. The egg jelly coat can be visualized by eye: the jelly coat will prevent eggs from packing directly on top of one another (there will be a clear gap between the eggs, which are dark). Swirl the eggs gently, and after 1–2 min pour off and replenish dejelly solution. Total time of dejelly will vary depending on eggs. Swirl the eggs every 20 s until they pack together tightly (up to ~5 total minutes), without gaps, when beaker it tipped to the side. Take care to not let the eggs dejelly for too long. Once ready, pour off the dejelly solution and immediately add 150 mL of XB. Begin staggering the timing of exchanging buffers in each of the 3 beakers. Keep the order the same throughout the remaining steps.
8. Wash eggs: Pour off the first addition of XB and add an additional 150 mL to each beaker. Pour off the XB and add 125 mL of CSF-XB. Pour off the CSF-XB and add 30 mL of CSF-XB+. At this point the eggs should be clean and largely pole-rotated (dark animal pole will be facing up). Do one final quick-sort to remove any lysed or misshapen eggs.
9. Transfer eggs very carefully to SW55 tube (prefilled with 1 mL of CSF-XB+ and 15 μL of 10 mg/mL cytochalasin D). Use trimmed transfer pipette, and avoid exposing the eggs to air by first drawing up 0.5 mL of CSF-XB+ buffer and then gently drawing up the eggs (Note: each transfer pipette holds a max of about 2 mL of eggs). Typically 2–3 fillings of the transfer pipette are required to deposit all the eggs from beaker containing CSF-XB+ into SW55 tube. Do not overfill the tube, exposing eggs to air. The eggs from a female which ovulates robustly will fill ~75–100 % of the volume of the SW55 tube (*see* **Note 3**).
10. Packing spin: Place SW55 tube inside 13 mL Sarstedt tube using a forceps. Transfer tubes to a refrigerated clinical centrifuge, set to 16 °C. Spin at 250×*g* for 1 min, and then at 500×*g* for an additional 30 s. After spinning, aspirate excess buffer from the top of the eggs using a transfer pipette. This step prevents buffer from diluting the cytoplasm.

11. Crushing spin: Transfer tubes with packed eggs to Sorvall HB-6 rotor containing adapters, and centrifuge at 18,000 × *g* for 15 min at 16 °C.
12. Collecting cytoplasm: Immediately after crushing the eggs remove SW55 tubes from Sarstedt tubes and place them on ice. The visibly colored layers correspond to components of the egg, top: lipid (white-yellow), middle: cytoplasm (gold), bottom: yolk granules and pigment (black/brown). Using a 18-gauge needle attached to a 1 mL syringe, puncture the thin-wall tube at the bottom of the cytoplasmic layer. Carefully remove the cytoplasmic layer without taking up any of the surrounding yolk or lipid layers. Be conservative. Expect about 0.9–1.2 mL of pure cytoplasm from a SW55 tube that was completely filled with eggs. Remove the needle and transfer the cytoplasm to a prechilled, labeled 1.5 mL microfuge tube on ice (*see* **Note 4**).
13. Supplement the extract with a final concentration of: 10 μg/mL or ~16.67 μM LPC (1:1000 of 10 mg/mL stock), 10 μg/mL or ~20 μM cytochalasin D (1:1000 of 10 mg/mL stock). Add 1:50 of energy mix to achieve a final concentration of ~38 mM creatine phosphate, 0.5 mM ATP, 0.5 mM $MgCl_2$, and 0.05 mM EGTA. Mix by inverting the microfuge tube. The extract can be used for the next 6–8 h.

3.1.2 Embryo Extract Preparation

Note: We have had the best results preparing high-quality Stage 4 (8 cell) and Stage 8 (1000s of cells) embryo extracts by fertilizing and incubating freshly squeezed eggs at 16 °C (Fig. 1). The extracts are less functional when the steps are carried out at 22 °C. Freshly squeezed eggs are necessary to achieve >90 % fertilization efficiency. Do not collect loose eggs in the container that houses the ovulating female. The methods here have been revised from the procedure described by Wilbur and Heald [23].

1. As in Subheading 3.1.1, prime females with a subcutaneous injection of 100 U of PMSG 3–7 days prior to in vitro fertilization (IVF).
2. At 14–15 h prior to squeezing, subcutaneously inject four females with 500 U of HCG and store them separately in individual plastic containers filled with 1× MMR (prepared with deionized water) at 16 °C. Assume that three out of four females will lay high-quality eggs, allowing the preparation of three separate single-frog embryo extracts.
3. 14 h after HCG injection, analyze egg quality. Avoid using females that have already laid lysed or stringy eggs, or that have laid no eggs at all.
4. Fresh egg collection: For selected frogs, promote egg-laying by squeezing females in a manner that mimics amplexus, and

collect freshly laid eggs from single females in individual 6 cm glass dishes. Repeat every 15 min (up to 4 total squeezes) or until the dish is full. Place females in deionized water in between rounds of egg collection.

5. Prepare sperm slurry: For each IVF reaction, use 1/8 of a testis recently isolated from a *Xenopus* male. Crush a piece of gonad in 1 mL of Milli-Q H_2O inside a 1.5 mL microfuge tube, using an appropriate plastic pestle, for 1–2 min.
6. Fertilize eggs: For each IVF reaction (each dish), pipette 1 mL of sperm slurry onto the eggs and incubate for 5 min at 16 °C. Wash once and replace the solution with 1/3× MMR. Let embryo development progress at 16 °C for either 4.5 h to reach mid-Stage 4, or 13.5 h to reach mid-Stage 8. To confirm fertilization, wait for 10–20 min and observe whether embryos pole-rotate such that the animal cap (dark) is facing up (*see* **Note 5**).
7. Prepare 100 mL of XB-free dejelly solution from 3 g cysteine in Milli-Q H_2O, and add 600 μL of 10 N NaOH to reach an approximate pH of 7.8.
8. Dejelly eggs at 2–3 h post-fertilization, after the first cleavage has occurred (at 16 °C). Pour off 1/3× MMR and add 20 mL of XB-free dejelly solution to each dish. Swirl gently for 2–3 min, until embryos pack tightly together when dish is tilted. Pour off dejelly and wash 3 times with 1/3× MMR. Add 15 mL of 1/3× MMR and continue incubation at 16 °C.
9. Begin sorting embryos: Using a dissecting stereomicroscope, remove embryos that have not divided, that have enlarged, or that appear misshapen. Continue to monitor developmental progression by recording the timing of the second and third divisions. Aspirate and remove embryos that are lagging or beginning to lyse.
10. Prepare remaining buffers and carry out final sort 30 min prior to embryo collection (4 h for Stage 4 and 13 h for Stage 8). Following the materials in Subheading 3.1.1 and instructions in Subheading 3.1.1, **step 5**, prepare 0.2 L of XB solution, and 100 mL of CSF-XB+ (plus 100 μL of 10 mg/mL cytochalasin D). Perform final sort to ensure that the remaining ~1000 embryos in each dish are homogeneous.
11. Wash embryos: Pour off 1/3× MMR and add in succession to each dish: 2 × 25 mL of XB, 2 × 15 mL of CSF-XB+ (plus cytochalasin D). Gently swirl and pour off buffers between each addition. Tilt dish so that surface tension helps to carefully pull the solution over the lip of the dish. This ensures that embryos will not be poured off during buffer exchanges.
12. Pack embryos: After pouring off remaining buffer, and without exposing embryos to air, carefully transfer the embryos

from a single dish into 2 mL microfuge tube using a cut transfer pipette. A full dish of embryos, unpacked, will fill the tube to a level of ~1.7 mL. Centrifuge at 200 × *g* for 1 min and then at 500 × *g* for 30 s in a benchtop microcentrifuge. Aspirate remaining buffer: first using P1000 and finally using a P200.

13. Crush embryos: Transfer 2 mL tubes containing packed eggs to a Sorvall HB-6 rotor containing microfuge tube adapters, in a superspeed centrifuge set to 16 °C. Spin at 18,000 × *g* for 12 min.
14. Collect embryo cytoplasm (*see* **Note 6**): After crushing the embryos into stratified layers, place the 2 mL microfuge on ice. The layers correspond to, top: lipid (white-yellow), middle: cytoplasm (gold), bottom: yolk granules and pigment (black/brown). Using a bent P200 pipette tip, insert the tip from above, through the lipid layer and begin carefully withdrawing cytoplasm, making sure to avoid surrounding layers. Dispense the cytoplasm into a prechilled 1.5 mL microfuge tube. Expect a yield of ~350–400 μL of pure cytoplasm for each full plate of embryos.
15. Supplement the embryo extract with to reach a final concentration of: 10 μg/mL or ~16.67 μM LPC (1:1000 of 10 mg/mL stock), 10 μg/mL or ~20 μM cytochalasin D (1:1000 of 10 mg/mL stock). Add 1:50 of energy mix to achieve a final concentration of ~38 mM creatine phosphate, 0.5 mM ATP, 0.5 mM $MgCl_2$, and 0.05 mM EGTA. Mix by inverting the microfuge tube. Additionally, to ensure that the extracts are synchronized and arrested in mitosis add a nondegradable form of cyclin B along with a dominant-negative form of UbcH10. Achieve a final concentration of: ~0.35 μM Δ90-CyclinB1 (1:40 v:v of ~13 μM stock), and ~18 μM UbcH10-C114S (1:30 v:v of ~750 μM stock). Mix by inverting tube. Embryo extracts are functional for up to 6 h (*see* **Note 7**).

3.2 Fabrication of Microfluidic Droplet-Generating Devices

Using the procedures described below, droplet-generating devices can be fabricated in a standard molecular biology lab, without the need for a clean room (Fig. 2). The key requirement is a plastic mold from which a microfluidic device can be cast using PDMS. For an overview on the process of creating a plastic mold, *see* **Note 8**. Alternatively, a number of universities and private companies have microfluidic foundries that will fabricate devices or molds on a fee-for-service basis (*see* **Note 9**).

1. Wash plastic replicas with deionized water and blow dry with pressurized air.
2. Attach clear tape around the perimeter of plastic replica to hold in elastomer poured onto the mold. About 2 mm of tape should be affixed to the wall (depth) of plastic replica and another 10 mm of tape should be visible above the feature-side

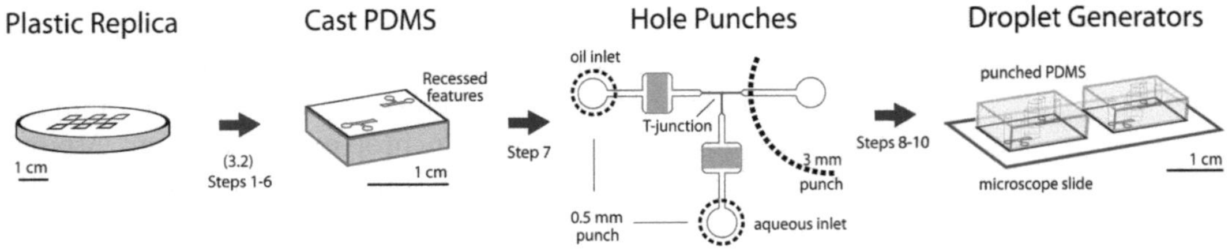

Fig. 2 Schematic of PDMS droplet-generator fabrication from a plastic mold. Starting from a plastic replica of multiple T-junction microfluidic droplet-generating device, PDMS is cast, degassed, and cured. Holes are punched in the PDMS to generate a collection chamber (3 mm hole) at the outlet, and to later connect tubing to 0.5 mm holes over inlets. Plasma-cleaned PDMS devices are then bonded to glass microscope slides. Each slide contains a total of four T-junction droplet generators. Scale bar: 1 cm

of the replica. Place 4–5 taped replicas in a large 15 cm petri dish. Features should face up.

3. Mix 10:1 (by weight) elastomer and curing agent, in a red solo cup. Stir with 2 mL pipette for 1–2 min to ensure thorough mixing. Assume approximately 20 g is required per plastic replica. Because it takes >30 min for PDMS to begin to harden, this can be done at a modest pace.
4. Slowly pour a layer of approximately 4 mm in thickness of PDMS onto taped plastic replica. Pouring a thicker layer of PDMS will make it more challenging to punch holes and connect tubing in later steps. Transfer everything to a dessicator chamber.
5. Degas the uncured PDMS. First use a house vacuum for 30 min to remove most of the bubbles, and follow with suction from a vacuum pump (~5 min). Monitor the dessicator to ensure that PDMS does not boil over under the pressure of the vacuum pump. If it starts bubbling, close down the valve on the dessicator chamber lid.
6. Cure PDMS overnight in a 65 °C oven. Allow a minimum of 2 h to thoroughly set.
7. Punch holes: Peel off tape and remove the PDMS cast carefully from the plastic replica. Place on a transparency with feature side facing down. Punch a large 3 mm hole over the outlet, just beyond the T-junction, and punch two small (0.5 mm) holes over the oil and aqueous inlets of the device. Use <0.5 mm diameter wire to remove any small plugs left in the smaller of the punched holes.
8. Cut the larger PDMS cast into individual droplet-generating devices (~1–1.5 cm rectangles) using a flat razor. Place feature side down on a second, unused transparency. The procedure can be paused here and devices can be sealed with clear tape and stored at room temperature.

9. Wash devices with deionized water to clean off and remove any residual polymer or dust in the punched holes. Dry with pressurized air.
10. Use oxygen plasma to generate hydrophilic surfaces on the PDMS casts and glass microscope slides. Place two PDMS casts and one microscope slide in Plasmod and expose to oxygen plasma for approximately 20 s. Immediately attach two PDMS casts to the microscope slide, feature side down. Using medium pressure by hand, ensure complete bonding of the PDMS and glass.
11. Place completed devices in a large 15 cm petri dish. Tape over the punched holes to ensure the devices remain dust-free. Cover and store at room temperature.

3.3 Extract Encapsulation and Spindle and Nucleus Assembly in Droplets

The successful encapsulation of *Xenopus* cytoplasmic egg and embryo extracts requires the experimenter to have prepared droplet-generating devices in Subheading 3.2, prepared cytoplasm in Subheadings 3.1.1 or 3.1.2, and to screen single-frog extracts for spindle and nucleus assembly activities. The following section describes procedures for: (1) Preparing the required equipment, supplies and reagents, (2) Evaluating extract quality, and (3, 4) Generating and imaging extract-filled droplets. All equipment, reagents and extracts should be placed in a 4 °C cold room. This strategy ensures that that microtubule polymerization is restricted, thereby preventing spindle assembly until the droplets are shifted to 22 °C (a temperature at which spindles are fully formed in 40 min) (*see* **Note 10**).

3.3.1 Preparation

Day Before Experiment

1. Treat microfluidic channels with Aquapel to ensure that they are hydrophobic. This will prevent droplets from wetting onto PDMS after being formed. Pipette 2 μL of Aquapel rain repellant into the oil inlet. The channels will lose contrast as the Aquapel is wicked into them. Blow out excess Aquapel using a N_2 stream. Cure and dry devices at 50 °C overnight (*see* **Note 11**).
2. Either place prepared imaging chambers at 4 °C, or laser cut new chambers from 3 to 4.5 mm thick acrylic and glue to 18 or 22 mm glass circle cover slips using Norland 61 optical adhesive (Expose to UV light overnight at room temperature)
3. Submerge teflon tubing in Milli-Q H_2O in a large petri dish overnight, at 4 °C. Additionally, place large items in the cold room, including syringe pumps and a stereomicroscope. Also store forceps, Hamilton syringes, pipettes and tips, and a microfuge tube rack in cold room.

Morning of Experiment

4. Remove Aquapel-treated PDMS devices from the oven, tape over the holes with Scotch tape, and move them to 4 °C. Place gasket imaging chambers in the cold room.

Immediately Before Experiment

5. Clean out Hamilton syringes and tips with appropriate reagent. Wash extract syringe with water, blow out, add 50 μL of extract (any extract), blow out, and then let it incubate with extract (any extract) until the start of experiments. Pipette 100 μL of 50 mg/mL mixture of oil and surfactant (squalene and Cithrol DPHS) into the oil syringe, blow out, add then let it incubate with the oil/surfactant mix.
6. Assemble PTFE tubing onto syringe needles using forceps. Cut tubing to an appropriate length (~6 in.).
7. Fill oil syringe and tubing with 100 μL oil/surfactant mix by pipetting into the plunger end of the syringe. Fill the aqueous syringe with 25 μL of extract. Withdraw it from a microfuge tube using the tubing.

3.3.2 Evaluate the Quality of Egg or Embryo Extracts

The goal is to identify the single-frog extract that is most capable of forming nuclei and spindles at high efficiency, both in microfuge tubes and inside microfluidic droplets. The following extract qualities are scored to identify the best extract.

1. *Evaluate mitotic character*: If demembranated sperm nuclei are added to CSF-arrested egg extracts, do spindles form with high efficiency in a 25 μL reaction (~24 μL egg extract + 0.25 μL of 50,000 sperm/μL stock + 0.25 μL of 20 mg/mL rhodamine-labeled tubulin + 0.25 μL of 20 μg/mL Hoechst) in a microfuge tube? Flick tube every 5–10 min to ensure that spindles do not clump. After 40 min at room temperature, mix 1 μL of extract reaction with 4 μL of spindle fix and image (*see* **Note 12**).
2. *Evaluate interphase character*: If 0.5 mM $CaCl_2$ and 1000/μL demembranated sperm nuclei are added to egg extracts, does the reaction (~24 μL egg extract + 0.25 μL of 200,000 sperm/μL stock + 0.25 μL of 50 mM $CaCl_2$ + 0.25 μL of 20 mg/mL rhodamine-labeled tubulin) generate large, round nuclei within 60 min at room temperate? Flick tube every 5–10 min to ensure that nuclei do not clump. Monitor progression at 45 and 60 min after calcium addition by imaging a mixture of 1 μL of extract reaction with 4 μL of spindle fix (also *see* **Note 12**).
3. *Evaluate formation of cycled spindles* from interphase nuclei + fresh CSF extract (50:50 mix). The ideal extract will form spindles at high efficiency within 40 min at 22 °C. A typical reaction is made from ~12.5 μL interphase extract containing 1000/μL fully formed sperm nuclei + 12.5 μL of CSF extract containing

0.4 mg/mL rhodamine-labeled tubulin and 0.4 μg/mL Hoechst. The 25 μL spindle assembly reaction has a final concentration of 500 nuclei/μL along with 0.2 mg/mL rhodamine-labeled tubulin and 0.2 μg/mL Hoechst. Flick tube every 5–10 min to ensure that spindles do not clump. After 40 min, mix 1 μL of extract reaction with 4 μL of spindle fix and image.

4. *Evaluate cycled spindle assembly inside hand-formed extract-in-oil emulsions.* Add 2 μL of freshly prepared cycled spindle reaction (from Subheading 3.3.2, **step 3**) to 100 μL of 50 mg/mL mixture of Cithrol DPHS in squalene in a 1.5 mL microfuge tube. Using a standard P200 tip and pipette set at 50 μL, pipette up and down 5–10 times, keeping tip pressed against the bottom of the tube. The shear forces will break the extract apart into droplets with a diameter of 20–200 μm. Transfer the droplets to a gasket imaging chamber using a pipette, and incubate for 40 min at 22 °C. Image structures formed in droplets using an inverted fluorescence microscope. This assay determines whether extract functionality is altered by the oil/surfactant mixture or the encapsulation process, and is the preferred test to determine which extract will perform at the highest level when encapsulated in microfluidic droplets.
5. *Evaluate embryo extracts*: The evaluations in the previous steps (Subheading 3.3.2, **steps 1–4**) are critical for optimizing embryo extract spindle assembly reactions. A typical embryo reaction is comprised of 85 % embryo cytoplasm, 10 % CSF-arrested egg cytoplasm, and 5 % cycled interphase nuclei (from egg extracts), along with rhodamine-tubulin and Hoechst to monitor microtubules and DNA. Choose the most mitotic egg extract for the 10 % addition (to promote mitotis), and choose the egg extract best able to make large round interphase nuclei as the source of DNA. Additionally, if time permits, test the ability of each extract to form spindles from interphase nuclei, prior to picking one embryo extract for encapsulation. The following is a typical embryo extract reaction: ~21 μL embryo extract + 1.25 μL of 6000 sperm/μL ("12×") freshly prepared nuclei in interphase egg extracts, 2.5 μL of CSF-arrested egg extract, and includes a final concentration of 0.2 mg/mL rhodamine-labeled tubulin and 0.2 μg/mL Hoechst. Monitor the reactions both in a test tube, and inside of droplet emulsions formed by hand (*see* Subheading 3.3.2, **step 4**).

Moving forward, for egg extract spindle assembly reactions, use a combination of the best CSF extract and best cycled nuclei, combining them at a 50:50 volume ratio. Generate interphase nuclei reactions at concentrations that are appropriate for the droplet sizes being generated (*see* Table 1). Additionally, it is possible to characterize the effect of the addition or depletion of factors from the extracts.

Table 1
List of common droplet diameter and volumes, and sperm concentrations necessary to generate droplets containing a single sperm nucleus

Droplet diameter (μm)	Droplet volume (L)	Ideal extract sperm conc. (number/μL)	Conventional sperm conc. (stock: 1× = 500/μL)	% Droplets loaded (assuming max of 10× sperm/reaction) (%)
1000	5.2E-07	2	0.004×	100
100	5.2E-10	1910	4×	100
80	2.7E-10	3730	7×	100
70	1.8E-10	5568	11×	91
60	1.1E-10	8842	18×	57
50	6.5E-11	15,279	31×	33
40	3.4E-11	29,842	60×	17
30	1.4E-11	70,736	141×	7
25	8.2E-12	122,231	244×	4

Spindle reactions containing the sperm concentrations highlighted in *gray* are not feasible without encapsulation. At very high sperm concentrations, spindles assembling in bulk extracts will fuse

Oil Flow Rate [μl/minute]	Extract Flow Rate [μl/hour]	Channel Width [μm]	Droplet Diameter [μm]
0.5	20.0	100	100
0.5	15.0	100	90.00
0.5	10.0	100	70-80
1.0	10.0	100	60-65
4.0	10.0	100	55-60
0.5	15.0	50	50-55
1.0	15.0	50	45-50
4.0	15.0	50	35-40
0.5	15.0	25	35
1.0	15.0	25	30
4.0	15.0	25	20-25

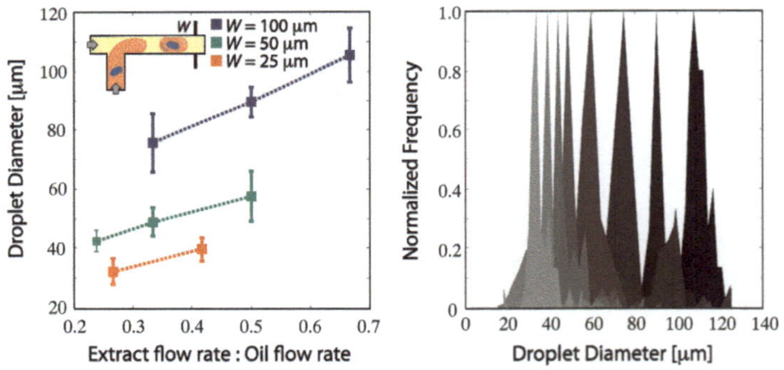

Fig. 3 Parameters for controlling droplet size in microfluidic devices. Table shows flow rates programmed for a 50 μL syringe containing a mix of oil and surfactant and 25 μL syringe filled with extracts, in devices with various channel widths. Values for generating droplets whose diameters range from 20 to 100 μm are shown. A plot of droplet diameter as a function of relative flow rate is also shown. Each peak in the histogram corresponds to a distribution of droplet sizes generated from a single experiment

3.3.3 Generate Extract-Filled Microfluidic Droplets of Specific Sizes

The channel widths and syringe flow rates described in Fig. 3 should be used as a guide to specify droplet size.

1. Switch on the syringe pump driving the oil syringe and set to rate of 0.5–1 μL/min. Insert the PTFE tubing attached to oil syringe into the punched hole over the oil inlet. Use forceps to pull the tubing near to the bottom (glass-side) of the PDMS device, as observed from the side. Place the droplet-generating

device onto the stage of stereomicroscope and watch for oil entering the channels. A reduction in contrast signals that the oil has filled the channels and the device is ready to be hooked up to the extract syringe tubing.

2. Generate spindle assembly reactions: Mix 10 μL interphase extract containing an appropriate number of sperm nuclei + 10 μL of CSF extract containing 0.2 mg/mL rhodamine-labeled tubulin and 0.2 μg/mL Hoechst in 1.5 mL microfuge tube on ice. Use immediately by transferring to the extract syringe.
3. After cleaning out the extract syringe using compressed air, fill the extract syringe by pulling back on the plunger to withdraw at least 12 μL of spindle assembly reaction through the attached PTFE tubing. Mount the syringe on the extract syringe pump. Push the plunger forward to remove any bubbles and to create a small bolus of extract. Set the appropriate flow rate (e.g., 10–15 μL/h). Use forceps to connect the tubing to the 0.5 mm punched hole over the extract inlet. Pull the tubing near to the bottom (glass-side) of the PDMS device.
4. Wait for 2 min for droplets to begin forming. Occasionally observe the device under a stereomicroscope until extract enters the T-junction. Take care not to overexpose the sample as this may increase temperature and reduce functionality.
5. As droplets begin to form, allow them to collect in the larger punched hole (collection chamber) for up to 5 min. As they accumulate, use a P20 pipette to transfer them to form a monolayer of droplets in a gasket imaging chamber.
6. Increase the oil syringe flow rate to produce slightly smaller droplets, using the same device and extract (*see* **Note 13**). Wait for 2 min, clearing out any new droplets which are likely incorrect in size. Wait for another 5 min and transfer new droplets to a second gasket imaging chamber. Allow these droplets to sit for 3 min before taking them out of cold room.
7. Repeat **steps 5–6**. Select different channel widths and flow rates to create a series of droplets that cover a range of diameters (e.g., 80/70, 60/50, 40/35, 25/20 μm). In the best-case scenario, the control and an experimental spindle assembly reaction can each be encapsulated in 8 sizes of droplets (16 total sets of droplets, each requiring separate imaging) (*see* **Notes 14** and **15**).

3.3.4 Imaging Spindle Assembly Inside of Droplets

See Fig. 4.

1. Remove droplet-filled gasket imaging chambers from the cold room and incubate at 22 °C for 40 min. Three minutes prior to the 40 min timepoint, place the imaging chamber on the stage of a spinning disk confocal microscope.
2. Identify focus using transmitted light and a 20× NA = 0.5 air objective. Set up the software to Acquire a multi-wavelength

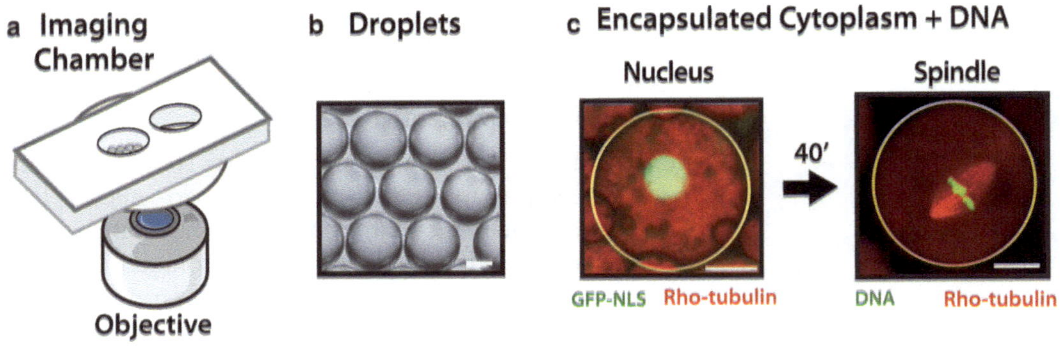

Fig. 4 Images of spindle and nucleus assembly inside of cell-size droplets. (**a**) Schematic of a monolayer of droplets being imaged inside of a gasket imaging chamber on an inverted microscope. (**b**) Representative image showing the homogeneity of 50 μm droplets generated using a T-junction microfluidic device. (**c**) Encapsulated spindle assembly reaction (a 50:50 mix of interphase nuclei : CSF-arrested cytoplasm), at time = 0. Nuclei are visualized using a GFP construct fused to a nuclear localization signal. (**d**) Encapsulated spindle assembly reaction in (**c**), after incubation at 22 °C for 40 min. Microtubules are visualized by the presence of 0.2 mg/mL rhodamine-labeled tubulin, and DNA is stained with 0.2 μg/mL Hoechst. Droplet boundaries are indicated in *yellow*. Scale bar 20 μm

(405 and 561 nm excitation) set of images, for an 8 × 8 area in XY (total of ~2 × 2 mm region of interest), across multiple slices in Z (*see* **Note 16**). Acquisition time is typically 5–7 min, depending on exposure times.

3. Repeat **steps 1–2** for each new set of droplets. If new droplets are generated every 10 min, there should be no delays in the imaging pipeline. Spindles and nuclei should form with similar efficiency inside of droplets and in unencapsulated bulk extracts (*see* **Note 17**). If spindles are not assembling properly or form with low efficiency, *see* **Note 18**.

4 Notes

1. When generating *Xenopus* cytoplasmic extracts, the most critical parameter is egg quality. Egg yield and quality are largely dependent on the age and health of the frog. It is important to allow females to rest for at least 4–6 months between ovulations and to remove frogs which repeatedly produce poor quality (lysed, stringy, or too few) eggs. Additionally, younger frogs tend to have the best quality eggs.
2. When handling frogs, eggs and embryos, be sure to use the appropriate source of water. Frogs should only be kept in deionized water or solutions made from deionized water. Industrial tap water often contains contaminants that are harmful to aquatic species. All buffers used for making extracts

should be formulated using Milli-Q H_2O, and the buffers, dejelly, XB, CSF-XB, and CSF-XB+, should always be made fresh, the day of extract preparation.

3. When working with extracts it is important to work quickly, but also carefully, to ensure that manipulations do not physically damage the eggs. For example, the smaller the window between dejellying and crushing the eggs with centrifugation, the better the extract quality. However, rapid pipetting of eggs (e.g., from beaker to SW55 tubes) will cause them to shear and lyse and can reduce overall extract activity.
4. To prevent cytoplasm from warming up when withdrawing it using a needle, only hold the SW55 tube at the very top and keep the bottom of the tube in contact with ice. It is also helpful to prechill the 1 mL syringe. In general, it is important to work quickly and keep the cytoplasm cool.
5. If fertilization efficiency is low, consider freshly preparing another clutch of eggs, dissecting a fresh set of testes and possibly using a larger portion of the testes for fertilization.
6. Prechill a box of P200 tips and make sure not to handle the 2 mL tube for an extended period of time, which will warm up the extracts.
7. Embryo extract spindle assembly reactions are much more efficient at forming spindle upon addition of 5–10 % freshly prepared CSF egg extracts. Also, the best way to coordinate timing is to prepare the egg extracts about an hour prior to generating embryo extracts (while embryos are developing).
8. To create plastic replicas for generating PDMS microfluidic devices, the following steps were performed: (I) An SU-8 mold was generated using standard soft-lithography techniques [18]. First, a 25 μm thick film of SU-8 2025 was deposited on a silicon substrate, soft-baked, and UV exposed through a custom chrome mask using a contact aligner. After a second bake, the wafers were developed in SU-8 developer and silanized under vacuum. (II) 1–2 mm of PDMS was cast on the developed and silanized wafer (from step I), degassed, and cured at 65 °C overnight. (III) A 50:50 mix of Smooth-cast 310 plastic (total 15 mL) was cast onto the PDMS and air-cured to generate the final "plastic replica" mold. From this plastic replica, PDMS microfluidic devices can be generated in any research lab (*see* Fig. 2). Because of its size, each 6 cm plastic replica contains the features of ~20–24 individual T-junction droplet-generating devices.
9. There are a number of commercial options for prototyping microfluidic devices from PDMS. Companies including Flowjem(http://www.flowjem.com/) and uFluidics (http://ufluidix.com/) will fabricate devices starting from a user-

submitted mask created in Autocad. Additionally, it is worth investigating whether a microfluidic foundry is available on your campus. Many universities have invested in microfabrication facilities for soft-lithography, and some offer custom fee-based fabrications services.

10. For those without convenient access to a cold room, or in cases in which it is preferable to keep extracts at room temperature (e.g., *Xenopus tropicalis* egg extracts), microfluidic droplets can be generated at temperatures above 4 °C. However, in these cases the process becomes time-sensitive—microtubules begin to polymerize before and during encapsulation—and therefore the experimenter must closely monitor and account for the time elapsed.
11. Residual Aquapel will inhibit spindle assembly. Make sure that it is fully removed by drying in a 50 °C oven overnight.
12. Strongly mitotic extracts are useful because they can be used to make CSF spindles in droplets or to add as a supplement to embryo extracts to improve spindle assembly efficiency. However, strongly mitotic extracts are less useful if the goal is to generate interphase nuclei from sperm DNA. Weakly mitotic or interphasic extracts are characterized by the transition into interphase even in the absence of calcium addition. Strongly interphasic extracts are useful for generating cycled nuclei although they may make it difficult to generate cycled spindles from the mixture of 50 % interphase extract + 50 % CSF-arrested extracts.
13. By staggering the removal of consecutive sets of droplets formed from the same PDMS device, a 10 min interval is created which is necessary to perform live imaging exactly 40 min after initiation of spindle assembly (via temperature shift from cold room to 22 °C).
14. Experiments in Subheading 3.3.3 are easiest to perform with two people. Ideally one person manages the droplet-generators in the cold room, and one person assembles the spindle reactions and runs the microscope. Once the best extract is selected (Subheading 3.3.2), it takes approximately 2.5–3 h to generate and image a full range of droplets sizes.
15. The major benefits of making uniformly sized microfluidic droplets rather than hand emulsions are: (1) spindles inside homogenously sized droplets are sitting in roughly the same imaging plane, making it much easier to acquire images of spindles across a large region of droplets in XY, and (2) better control over exact number droplets generated for a particular size. Although it is easy to make droplets by hand, it is difficult to generate a specified number of droplets of a given size (e.g., 1000 each for 70, 60, 50 μm droplets).

16. Because the acrylic used in making the imaging chambers is not perfectly flat, there can be 10–20 μm of tilt across the diameter of the droplet chamber. This will alter the focus from one side of the well to the other, so choose an "average" *z*-plane by finding the mean of the ideal focus in Z for spindles in frames in the upper left and lower right of the region of interest. This will ensure that the maximum number of droplets are in focus during the XY scan.
17. Extracts of the highest quality exhibit the following features: An ideal CSF extract will generate sharp-looking spindles around exogenously added sperm DNA at high efficiency, within 45 min at room temperature. Spindles should form in the absence half-spindles, or asters and spindle microtubules should look sharp, not "hairy." An ideal interphase nuclei preparation forms large, round nuclei at high efficiency within 60 min hour after addition of 0.5 mM $CaCl_2$ and sperm DNA. The best cycled spindle reactions (from 50:50 mix of interphase extracts + CSF-arrested extracts) generate bipolar spindles at extremely high efficiency. In all cases, the spindles should form easily both in bulk reactions in a test tube and within hand emulsions or 70–80 μm diameter microfluidic droplets.
18. When extracts are not performing optimally, the following issues will arise: (1) CSF extract: forms asters and nuclei rather than spindles. The extract may be interphasic and unable to hold mitotic arrest. (2) Cycled interphase extract: does not form large swollen, round nuclei within 45–60 min, but instead forms nonspherical, small nuclei. The extract may be too mitotic. (3) Encapsulated reactions: the ability to form spindles in a test tube but not inside 70–80 μm diameter droplets signals an issue with the oil/surfactant mixture. If spindles are able to form in hand emulsions but not microfluidic droplets, PDMS devices (and residual solvent) or tainted PTFE tubing are likely to blame. In this case, you often observe nuclei and asters inside microfluidic droplets (as if extracts and nuclei were stuck in interphase).

Acknowledgements

I wish to acknowledge the labs of Rebecca Heald and Daniel Fletcher for my training, and for contributions that made possible the encapsulation of *Xenopus* extract spindle assembly. I wish to thank Mike Vahey for his partnership in developing the techniques described here, and for critically reading this chapter. I also wish to acknowledge the seminal contributions of pioneers in the *Xenopus* egg extract community, including the labs of Tim Mitchison,

Andrew Murray, Marc Kirschner, Arshad Desai, Claire Walczak, and helpful advice from the labs of Tarun Kapoor, Anthony Hyman, and Thomas Surrey, which have enhanced my understanding of the extract system. This work was supported in part by a Burroughs Wellcome Fund Career Award at the Scientific Interface.

References

1. Brown KS et al (2007) Xenopus tropicalis egg extracts provide insight into scaling of the mitotic spindle. J Cell Biol 176:765–770
2. Hannak E, Heald R (2006) Investigating mitotic spindle assembly and function in vitro using Xenopus laevis egg extracts. Nat Protoc 1:2305–2314
3. Maresca TJ, Heald R (2006) Methods for studying spindle assembly and chromosome condensation in Xenopus egg extracts. Methods Mol Biol 322:459–474
4. Desai A, Murray A, Mitchison TJ, Walczak CE (1998) The use of xenopus egg extracts to study mitotic spindle assembly and function in vitro. Methods Cell Biol 61:385–412
5. Murray AW (1991) Cell cycle extracts. Methods Cell Biol 36:581–605
6. Sawin KE, Mitchison TJ (1991) Mitotic spindle assembly by two different pathways in vitro. J Cell Biol 112:925–940
7. Hyman A et al (1991) Preparation of modified tubulins. Methods Enzymol 196:478–485
8. Nguyen PA et al (2014) Spatial organization of cytokinesis signaling reconstituted in a cell-free system. Science 346:244–247
9. Laan L et al (2012) Cortical dynein controls microtubule dynamics to generate pulling forces that position microtubule asters. Cell 148:502–514
10. Baumann H, Surrey T (2014) Motor-mediated cortical versus astral microtubule organization in lipid-monolayered droplets. J Biol Chem 289:22524–22535
11. Miyazaki M, Chiba M, Eguchi H, Ohki T, Ishiwata S (2015) Cell-sized spherical confinement induces the spontaneous formation of contractile actomyosin rings in vitro. Nat Cell Biol 17:480–489
12. Good MC, Vahey MD, Skandarajah A, Fletcher DA, Heald R (2013) Cytoplasmic volume modulates spindle size during embryogenesis. Science 342:856–860
13. Hazel J et al (2013) Changes in cytoplasmic volume are sufficient to drive spindle scaling. Science 342:853–856
14. Jimenez AM et al (2011) Towards high throughput production of artificial egg oocytes using microfluidics. Lab Chip 11:429–434
15. Pinot M et al (2012) Confinement induces actin flow in a meiotic cytoplasm. Proc Natl Acad Sci U S A 109:11705–11710
16. Abu Shah E, Keren K (2014) Symmetry breaking in reconstituted actin cortices. Elife 3:e01433
17. Teh SY, Lin R, Hung LH, Lee AP (2008) Droplet microfluidics. Lab Chip 8:198–220
18. McDonald JC et al (2000) Fabrication of microfluidic systems in poly(dimethylsiloxane). Electrophoresis 21:27–40
19. Courtois A, Schuh M, Ellenberg J, Hiiragi T (2012) The transition from meiotic to mitotic spindle assembly is gradual during early mammalian development. J Cell Biol 198:357–370
20. Decker M et al (2011) Limiting amounts of centrosome material set centrosome size in C. elegans embryos. Curr Biol 21:1259–1267
21. Levy DL, Heald R (2010) Nuclear size is regulated by importin alpha and Ntf2 in Xenopus. Cell 143:288–298
22. Wuhr M et al (2008) Evidence for an upper limit to mitotic spindle length. Curr Biol 18:1256–1261
23. Wilbur JD, Heald R (2013) Mitotic spindle scaling during Xenopus development by kif2a and importin alpha. Elife 2:e00290

Part III

Methods Focused on Kinetochores and the Kinetochore-Microtubule Interface

Chapter 8

In Vitro Kinetochore Assembly

Matthew D.D. Miell and Aaron F. Straight

Abstract

The kinetochore is the primary site of interaction between chromosomes and microtubules of the mitotic spindle during chromosome segregation. Kinetochores are composed of more than 100 proteins that transiently assemble during mitosis at a single epigenetically defined region on each chromosome, known as the centromere. Kinetochore assembly and activity must be tightly regulated to ensure proper microtubule interaction and faithful chromosome segregation. Kinetochore malfunction can result in chromosome segregation defects leading to aneuploidy and cell death. As such, cell free and reconstituted systems to analyze kinetochore formation and function are invaluable in probing the biochemical activities of kinetochores. In vitro approaches to studying kinetochores have enabled the manipulation of kinetochore protein structure, function, interactions, and regulation that are not possible in cells. Here we outline a cell-free approach for the assembly of centromeres and recruitment of functional kinetochores that enables their manipulation and analysis.

Key words Kinetochore, In vitro assembly, Chromatin, Centromere, Xenopus egg extract

1 Introduction

Upon entry into mitosis more than 100 different proteins assemble at the centromeric region of chromosomes to form a kinetochore complex [1]. The proper assembly of these proteins gives rise to the activities that are essential for chromosomal attachment to the mitotic spindle, metaphase chromosome alignment, and segregation of the genome at anaphase. Once mitosis is complete, the kinetochore is then disassembled until reassembly at the next entry into mitosis.

Biochemical and genetic studies have identified many centromere and kinetochore proteins yet studying the specific functions of these proteins in cells is challenging for several reasons. First and foremost, the kinetochore complex is essential to cell viability, so that perturbing individual protein components often results in cell death without informing on the detailed function of the protein. Second, the kinetochore assembles only at the centromeric region of a chromosome during mitosis. In metazoan cells, the DNA of

Paul Chang and Ryoma Ohi (eds.), *The Mitotic Spindle: Methods and Protocols*, Methods in Molecular Biology, vol. 1413, DOI 10.1007/978-1-4939-3542-0_8, © Springer Science+Business Media New York 2016

the centromere is highly repetitive and spans many hundreds of kilobases, and thus, manipulation of centromeres and analysis of the underlying DNA presents a major challenge [2]. Finally, because cells possess a single kinetochore per chromosome and because extracting intact, functional kinetochores from the chromosome has so far proven difficult, studies of kinetochore function in vitro have been limited.

Despite these difficulties a number of successful approaches have been employed to study kinetochores. These can generally be classed into two groups, in vitro studies—where components are produced recombinantly, and ex vivo studies—where components are either extracted from cells or studied within cells but typically through tethering components to exogenous locations [3–8].

In this chapter we present a detailed method for the cell free assembly of centromeres and kinetochores that circumvents several of the major difficulties associated with studying endogenous kinetochores and enables specific perturbation and analysis of centromere and kinetochore components [9, 10]. Our system takes advantage of *Xenopus* egg extract, a highly concentrated cytoplasmic extract that can execute chromosome segregation in vitro and in which we can manipulate the cell cycle state and protein composition of the extract (Fig. 1). We couple these extracts with reconstituted chromatin substrates that resemble endogenous centromeres to build new centromeres and kinetochores on the added chromatin from the proteins present in the extract.

This system provides a major advantage in that it permits manipulation of the DNA and protein composition of the chromatin template and the protein composition of the extract to study the assembly of functional, microtubule-binding kinetochores. By labeling the input chromatin with biotin we can recover and study near-native kinetochore complexes after their assembly in the extract. A second major advantage of this system is that it is possible to release the extract from metaphase into interphase with the addition of calcium. By cycling the extract through multiple rounds of interphase and metaphase this technique supports the study of cell cycle-associated changes to centromere and kinetochore components.

The method presented in this chapter is an adapted version of a method we have published previously and some steps are identical to the previous version while others have been updated or modified [9, 10]. Here, we split the method into a number of broad steps for clarity. The first sections (Subheadings 3.1–3.9) detail the preparation of biotinylated nucleosome arrays and *Xenopus* egg extract. Four sections (Subheadings 3.10–3.13) then describe separate procedures that may be used independently for the analysis of centromere, kinetochore and related complexes in this experimental system. The final methods section (Subheading 3.14) describes the preparation of samples for analysis by immunofluorescence and subsequent data analysis.

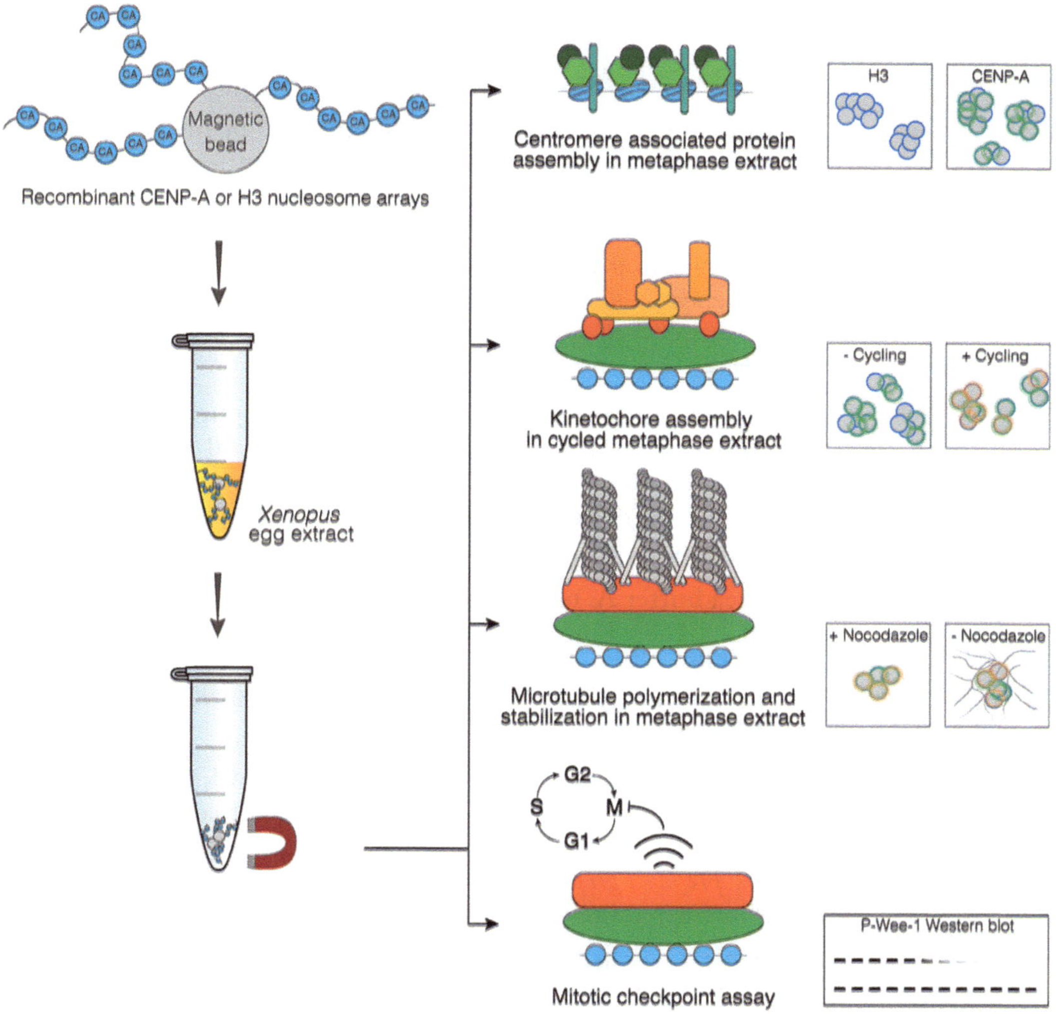

Fig. 1 Schematic overview of the in vitro kinetochore assembly process. Recombinant nucleosome arrays containing CENP-A nucleosomes can recruit functional centromere and kinetochore components when incubated in *Xenopus* egg extract. Prior attachment of the arrays to magnetic beads allow for easy reisolation of chromatin arrays and any bound complexes for further downstream analysis. Four example experiments utilizing this setup for the analysis of centromere and kinetochore complexes are summarized. These experiments are described in detail in the text

2 Materials

W buffer: 50 mM Tris–HCl (pH 7.6), 100 mM NaCl, 1 mM EDTA, 5 mM 2-mercaptoethanol, and 0.2 mM PMSF.

Freshly prepare this buffer and add 2-mercaptoethanol and PMSF right before use.

TW buffer: W buffer with 0.1 % (vol/vol) Triton X-100.

Freshly prepare this buffer and add 2-mercaptoethanol and PMSF right before use.

Unfolding buffer: 7 M guanidine-HCl, 20 mM Tris–HCl (pH 7.5), and 10 mM DTT.

Freshly prepare this buffer for each protein preparation.

Lysis buffer for CENP-A/H4 tetramer purification: 20 mM phosphate buffer (pH 6.8) (for 1 l of buffer use 9.94 ml of 1 M K_2HPO_4 and 10.06 ml of 1 M KH_2PO_4), 1 M NaCl, 5 mM 2-mercaptoethanol, 1 mM PMSF, 1 mM benzamidine, 0.05 % NP-40, and 0.2 mg/ml lysozyme.

Approximately 10 ml of lysis buffer is required per cell pellet from 1 l of bacterial cell culture. Prepare fresh and add the 2-mercaptoethanol, benzamidine, and PMSF right before use; store on ice.

Dialysis buffer for CENP-A/H4 tetramer purification (2 × 4 l): 10 mM Tris–HCl (pH 7.4), 0.75 M NaCl, 10 mM 2-mercaptoethanol and 0.5 mM EDTA.

Freshly prepare this buffer and add 2-mercaptoethanol right before use.

Urea buffer A: 6 M deionized urea, 200 mM NaCl, 10 mM Tris–HCl (pH 8), 1 mM EDTA, 5 mM 2-mercaptoethanol, and 0.1 mM PMSF.

Prepare fresh for each protein preparation and add 2-mercaptoethanol and PMSF right before use.

Urea buffer B: 6 M deionized urea, 1 M NaCl, 10 mM Tris–HCl (pH 8), 1 mM EDTA, 5 mM 2-mercaptoethanol, and 0.1 mM PMSF.

Freshly prepare for each protein preparation and add 2-mercaptoethanol and PMSF right before use.

Refolding buffer: 2 M NaCl, 10 mM Tris–HCl (pH 7.6), 1 mM EDTA, and 5 mM 2-mercaptoethanol.

Freshly prepare and add 2-mercaptoethanol and PMSF right before use.

HA buffer A: 20 mM potassium phosphate (pH 6.8) and 5 mM 2-mercaptoethanol.

This buffer can be prepared in advance and kept indefinitely at 4 °C. Add 2-mercaptoethanol right before use.

HA buffer B: 20 mM potassium phosphate (pH 6.8), 5 mM 2-mercaptoethanol, and 3.5 M NaCl.

This buffer can be prepared in advance and kept indefinitely at 4 °C. Add 2-mercaptoethanol right before use.

HA column storage buffer: 0.5 M potassium phosphate (pH 6.8).

Store at 4 °C indefinitely.

HA column cleaning buffer: 0.5 M KOH.

Store at 4 °C indefinitely.

S column (HiTrap) buffer A: 10 mM Tris–HCl (pH 7.4), 10 mM 2-mercaptoethanol, and 0.5 mM EDTA.

Freshly prepare and add 2-mercaptoethanol immediately before use.

S column (HiTrap) buffer B: 10 mM Tris–HCl (pH 7.4), 10 mM 2-mercaptoethanol, 0.5 mM EDTA, and 2 M NaCl.

Freshly prepare and add 2-mercaptoethanol immediately before use.

PEG (MW 4000 Da) 0.5 M NaCl, 20 % (wt/vol): Dissolve 20 g of PEG and 2.9 g of NaCl in water to a final volume of 100 ml.

Filter-sterilize and store at room temperature (RT) indefinitely.

TE plus 2.5 mM NaCl: 10 mM Tris–HCl (pH 8.0), 0.25 mM EDTA, and 2.5 mM NaCl.

Store at RT indefinitely.

TE, 10×: 100 mM Tris–HCl (pH 7.5) and 2.5 mM EDTA.

Filter-sterilize or autoclave, and store at RT indefinitely.

High-salt (HS) assembly buffer: 10 mM Tris–HCl, 0.25 mM EDTA, and 2 M NaCl in cold H_2O; adjust the pH to 7.5 (while cold) with HCl.

Make fresh for each assembly.

Low-salt (LS) assembly buffer: 10 mM Tris–HCl, 0.25 mM EDTA, and 2.5 mM NaCl in cold H_2O; adjust the pH to 7.5 (while cold) with HCl.

Make fresh for each assembly.

Tris-acetate, 1 M: Dissolve 12.1 g of Tris in 75 ml of water; adjust the pH to 7.9 with acetic acid. Bring the solution to a final volume of 100 ml with water.

Store at RT indefinitely.

Array digestion (AD) buffer, 10×: 500 mM potassium acetate, 200 mM Tris acetate, 10 mM DTT, and 5 mM $MgCl_2$ (pH 7.9).

Store in aliquots at −20 °C for several years.

TBE, 5×: 445 mM Tris base, 445 mM boric acid, and 10 mM EDTA (pH 8.0).

Filter-sterilize TBE and store at RT for several weeks (until substantial precipitation of the buffer has occurred).

APS, 10 % (wt/vol): Dissolve 1 g of APS in water to a final volume of 10 ml.

Store at 4 °C for several weeks.

Native acrylamide gel, 5 % (wt/vol): Mix 1.5 ml of 40 % (wt/vol) acrylamide, 4.5 ml of water, 6 ml of 1× TBE, 5–10 μl of TEMED, and 100 μl of 10 % (wt/vol) APS and pour it into a preassembled polyacrylamide gel stand. Add the comb immediately (no stack gel is required) and let the gel solidify before use.

Can be stored at 4 °C for a few days if kept moist, although typically used immediately.

Glycerol, 80 % (vol/vol): Mix 80 ml of glycerol with 20 ml of water.

Filter-sterilize and store at RT indefinitely.

DNA, 19×601 (at ~2.5 μg/μl or higher): For preparation *see* Subheading 3.5.

Store at −20 °C indefinitely

H3/CENP-A nucleosomes on biotinylated 601 array DNA, ~2 μM (calculated for 1×601 DNA): For preparation *see* Subheading 3.6.

Store at 4 °C for up to a few months.

Bead buffer, 4×: 200 mM Tris–HCl (pH 8.0), 1 mM EDTA, 300 mM NaCl, and 0.2 % (vol/vol) Triton X-100.

Store at RT for several months.

Bead buffer, 1×: 50 mM Tris–HCl (pH 8.0), 0.25 mM EDTA, 75 mM NaCl, and 0.05 % (vol/vol) Triton X-100.

Store at RT for several months.

Array buffer: 10 mM Tris–HCl (pH 7.5), 2.5 mM NaCl and 0.25 mM EDTA.

Store at RT indefinitely.

Polyvinyl alcohol, 7.5 % (wt/vol): Dissolve 7.5 g of PVA (MW 30,000–70,000 Da) in water to a final volume of 100 ml.

Filter-sterilize and store at RT indefinitely.

CaCl2, 12 mM (to release Xenopus egg extracts from CSF arrest): Mix 120 μl of 1 M $CaCl_2$ with 9.88 ml of water.

Filter-sterilize and store in small aliquots at −20 °C indefinitely.

Marc's modified Ringer's solution (MMR): 25× 150 mM Na-HEPES (pH 7.8), 2.5 mM EDTA, 2.5 M NaCl, 50 mM KCl, 25 mM $MgCl_2$, and 50 mM $CaCl_2$. For 4 l, dissolve 142.99 g of HEPES, 584.4 g of NaCl, 14.91 g of KCl, 20.33 g of $MgCl_2$, and 29.4 g of $CaCl_2$; add 20 ml of 0.5 M EDTA and adjust the pH to 7.8 with NaOH.

Store at RT for several months.

Gelatin, 5 % (wt/vol): Mix 20 g of gelatin with 80 ml of water in a glass bottle.

Autoclave and store at RT indefinitely. Before use, liquefy the gelatin by microwaving briefly.

Sucrose, 2 M: Dissolve 684.6 g of sucrose in water and adjust the final volume to 1 l.

Filter-sterilize and keep the solution at RT; in small aliquots it can be stored at −20 °C for several years.

K-HEPES, 1 M, pH 7.7: Dissolve 119.15 g of HEPES in water; adjust the pH to 7.7 with KOH and the final volume to 500 ml.

Filter-sterilize and store at RT indefinitely.

K-EGTA, 0.5 M, pH 7.7: Dissolve 95.09 g of EGTA in water; adjust the pH to 7.7 with KOH and the final volume to 500 ml.

Filter-sterilize and store at RT indefinitely.

Extract buffer (XB) salt stock, 20×: 2 M KCl, 20 mM $MgCl_2$, and 2 mM $CaCl_2$.

Filter-sterilize and keep it at RT for several months.

CSF-XB buffer, 1×: XB salt (50 ml of 20× stock), 50 mM sucrose, 10 mM K-HEPES (pH 7.7), 5 mM K-EGTA, 1 mM $MgCl_2$, and 110 μl of 10 M NaOH.

Freshly prepare this buffer before use.

Protease inhibitors, 1000×: Mix 10 mg/ml of leupeptin, pepstatin A, and chymostatin in DMSO.

Aliquot and store at −20 °C indefinitely.

CSF-XB plus protease inhibitors, 1×: Take 50 ml of freshly prepared 1× CSF-XB buffer and add 50 μl of 1000× stock protease inhibitors.

Prepare right before use.

CSF-XB plus 0.05 % (vol/vol) Triton X-100, 1×

Freshly prepare this solution.

Energy mix, 20×: 150 mM creatine phosphate, 20 mM ATP, and 20 mM $MgCl_2$.

Aliquot into 500 μl aliquots and keep at −20 °C for several months.

Cytochalasin D, 1000×: Prepare stock at a concentration of 10 mg/ml in DMSO.

Make 50 μl aliquots and keep them at −20 °C for several months.

Cytochalasin B, 100×: Prepare stock at a concentration of 10 mg/ml in DMSO.

Make 50 μl aliquots and keep them at −20 °C for several months.

Dejellying buffer: Mix 2 % (wt/vol) l-cysteine in 1× MMR at pH 7.8 (adjusted with 10 M NaOH).

Freshly prepare and dissolve l*-cysteine right before use.*

Antibody dilution buffer (AbDil): 150 mM NaCl, 20 mM Tris–HCl (pH 7.4), 0.1 % (vol/vol) Triton X-100, and 2 % (wt/vol) BSA.

Store it at 4 °C for a few weeks.

Nocodazole, 1 mg/ml: Dissolve nocodazole at a concentration of 1 mg/ml in DMSO.

Store it in small aliquots at −20 °C indefinitely.

BRB80, 5×: 400 mM PIPES (pH 6.8), 5 mM $MgCl_2$, and 5 mM K-EGTA.

Filter-sterilize and store the solution at 4 °C for several months.

Cushion: Mix 1× BRB80 and 40 % (wt/vol) glycerol.

Filter-sterilize and store at RT for several months.

Dilution buffer: 1× BRB80, 0.05 % (vol/vol) Triton X-100, and 30 % (wt/vol) glycerol.

Filter-sterilize and store at RT for several months.

CAPS transfer buffer: 10 mM CAPS (pH 11.3), 1 % (wt/vol) SDS, and 20 % (vol/vol) methanol (optional).

Freshly prepare; it is also possible to store the used buffer at 4 °C and reuse it a few times within 1 week.

Milk in PBS-Tween, 5 % (wt/vol).

Keep at 4 °C and use within 1 week.

Coverslips coated in poly-l-lysine: Wash coverslips in 1 M HCl overnight at 50–60 °C. Thoroughly rinse coverslips with water until the pH returns to neutral. Add the coverslips to a solution of 1 mg/ml poly-l-lysine in distilled water and incubate on a rocker at RT for at least 1 h. Remove the poly-l-lysine solution (this can be stored frozen and reused a few times). Rinse the coverslips thoroughly in ten changes of water. Rinse the coverslips in ethanol.

Store coverslips submerged in 70 % (vol/vol) ethanol at RT indefinitely.

Mounting medium: 20 mM Tris–HCl (pH 8.8), 90 % (wt/vol) glycerol, and 0.5 % (wt/vol) p-phenylenediamine. Prepare the solution of Tris–HCl (pH 8.8) and glycerol and bring it to the final volume with water. Add the *p*-phenylenediamine solid and dissolve it by bubbling nitrogen gas through the solution for a few hours. The solution may turn light brown. Draw into a syringe, minimizing the amount of air trapped in the syringe.

Store in the sealed syringe at −20 °C. Discard the medium once the solution turns dark brown (typically a month or two).

3 Method

3.1 H3/H4 Tetramer or H2A/H2B Dimer Expression

1. Transform histone expression plasmids (H2A, H2B, H3, and H4) and a negative control (water) into BL21(DE3) Codon Plus RIPL bacteria.
2. Plate ~20–100 μl of the transformation (including the cells-only control) onto LB/chloramphenicol/carbenicillin plates and incubate overnight at 37 °C.
3. Pick a single colony and inoculate into 5 ml of LB/chloramphenicol/carbenicillin overnight at 37 °C.
4. Dilute the starter culture to an optical density at 600 nm (OD600) of 0.005 (approximately 1:1000) in 2 l of 2×YT medium.

5. Continue growth at 37 °C, 200 r.p.m., to an OD600 of 0.5–0.6. Remove 1 ml of sample for an analytical gel in Subheading 3.3, **step 6** as a control for uninduced expression.
6. Add IPTG to a final concentration of 0.2 mM.
7. Grow the culture for a further 3 h at 37 °C.
8. Pellet cells in 1 l portions by centrifugation at 2500 × *g* at 4 °C for 15 min.
9. Discard the supernatant and resuspend each pellet in 25 ml of W buffer.
10. Combine cells for a single portion of each histone and flash-freeze in liquid nitrogen.

3.2 H3/H4 Tetramer or H2A/H2B Dimer Purification

1. Thaw cells in a 37 °C water bath with constant shaking.
2. Sonicate six times using a probe sonicator with 30 s pulses separated by 1 min intervals at 200 W. Keep the cells on ice when possible.
3. Split the lysate into two tubes and centrifuge at 25,000 × *g* for 20 min at 4 °C.
4. Discard the supernatant (save 50 μl for analysis in Subheading 3.3, **step 6**) and resuspend the pellet in each tube in 25 ml of TW buffer. For this and the following steps, ensure that the pellet is thoroughly resuspended. Use a spatula to cut and crush the insoluble material and a vortex to aid resuspension.
5. Sonicate for 30 s on maximum power.
6. Centrifuge at 25,000 × *g* for 20 min at 4 °C.
7. Discard the supernatant and resuspend the pellet in each tube in 25 ml of TW buffer.
8. Repeat **steps 6** and **7** three more times, resuspending first in TW buffer and in W buffer the second and third times.
9. Repeat **step 6** and add 175 μl of DMSO to the pellet in each tube, and then combine both portions and use a spatula to finely crush the pellet.
10. Add 13.2 ml of unfolding buffer and incubate for 1 h at RT with gentle shaking.
11. Homogenize in a Dounce homogenizer, until no more large pieces remain.
12. Centrifuge at 25,000 × *g* for 20 min at 4 °C.
13. Remove the supernatant, but save it in a fresh tube.
14. Resuspend the pellet again in 13.2 ml of unfolding buffer, centrifuge again as described in **step 12**, and combine the supernatant with the supernatant from **step 13**.

15. Dialyze this supernatant overnight into 2 l of urea buffer A, stirring at 4 °C. The following day, exchange the dialysis buffer to 2 l of fresh buffer A, and dialyze for a further 3 h at 4 °C. (A white precipitate may form during the dialysis.)
16. Centrifuge the solution at 12,000 × *g* for 20 min at 4 °C.
17. Measure the conductivity of the supernatant relative to that of the urea buffer. If necessary, dilute the supernatant with fresh urea buffer until the conductivity of the supernatant is less than 1.25 times that of the buffer alone.
18. Connect a 5 ml HiTrap Q column and a 5 ml HiTrap S column in series on an FPLC system.
19. Wash columns with three column-volumes (3 CV) of water (30 ml total), then 3 CV of urea buffer A.
20. Load the supernatant from **step 17** over the columns.
21. Wash with 2 CV of urea buffer A.
22. Remove the Q column, keeping the S column attached to the FPLC. The Q column can be reused after elution with urea buffer B.
23. Elute the S column with a linear gradient from 0 to 37.5 % urea buffer B (remainder urea buffer A) over 3 CV, and then with 100 % urea buffer B for 3 CV, collecting 1 ml fractions.
24. Run 10 μl of the peak fractions on a 20 % (wt/vol) SDS-PAGE gel. Also run the analytical gel samples taken earlier on this gel for analysis.
25. Pool the histone-containing fractions and dialyze against 4 l of water containing 0.1 mM PMSF and 5 mM 2-mercaptoethanol. Refresh the dialysis with fresh water three times to ensure complete dialysis. A white precipitate may form during the dialysis.
26. Clarify the protein preparation by centrifugation at 25,000 × *g* for 20 min at 4 °C.
27. Measure the concentration of the supernatant by Bradford assay.
28. Lyophilize the supernatant in aliquots (typically 5 mg each).

3.3 H3/H4 Tetramer or H2A/H2B Dimer Assembly

1. Resuspend lyophilized histone directly into unfolding buffer at a concentration of 10 mg/ml.
2. Mix ~10 mg of H3 and an equimolar amount of H4, or ~10 mg of H2A and an equimolar amount of H2B.
3. Dialyze against refolding buffer, with three changes of 4 l.
4. If the volume increased to more than 5 ml during dialysis, concentrate in an Amicon conical concentrator, 10,000 MWCO, to obtain a final volume of ≤5 ml.

5. Run over a Superdex 200 16/60 column (equilibrated to refolding buffer) at 1 ml/min; collect 1 ml fractions.
6. Run fractions on a 20 % (wt/vol) denaturing polyacrylamide gel, pool the fractions that have approximately equal amounts of H3 and H4 (or H2A and H2B).
7. Centrifuge the pooled histone solution for 10 min at 10,000 ×*g* at 4 °C.
8. Mix 5 μl of the supernatant with 5 μl of 8 M guanidine-HCl to denature the protein, and use a spectrophotometer to obtain the 280 nm absorption.
9. Use the theoretical extinction coefficient ε and the calculated molecular weight of each histone to precisely determine the protein concentration in milligrams per milliliter. To calculate protein concentration in mg/ml divide the protein absorbance at 280 nm by the extinction coefficient and multiply by the protein's molecular weight.
10. Aliquot the remaining pooled H3/H4 tetramer or H2A/H2B dimer in ~100 pmol aliquots, flash-freeze in liquid nitrogen, and store at −80 °C.

3.4 CENP-A/H4 Tetramer Expression and Purification

1. Repeat **steps 1** and **2** from Subheading 3.1 using a CENP-A/H4 pST39 bicistronic expression plasmid.
2. Pick a single bacterial colony, inoculate into 5 ml of LB/chloramphenicol/carbenicillin, and grow at 37 °C for 8–10 h.
3. Dilute the culture into 100 ml in LB/chloramphenicol/carbenicillin and continue growth overnight.
4. Dilute cultures 1:50–1:100 in 6 l of 2× YT medium.
5. Grow at 37 °C, with shaking, to an OD600 of 0.2 and then grow at RT until an OD600 of 0.5–0.6.
6. Add IPTG to a final concentration of 0.2 mM.
7. Continue growth at RT for ~6 h with shaking.
8. Pellet cells in 1 l portions by centrifugation at 2500 ×*g* at 4 °C for 15 min.
9. Freeze cell pellets in liquid nitrogen and store at −80 °C.
10. Equilibrate a 60 ml HA column by washing with 1 CV of 20 mM potassium phosphate (pH 6.8).
11. Clean with ~1 CV (60–100 ml) 0.5 M KOH.
12. Equilibrate with 2 CV (~120 ml) of 20 mM potassium phosphate (pH 6.8).
13. Prepare HA column buffers A and B without 2-mercaptoethanol; filter and store at 4 °C.
14. Thaw cell pellets from **step 9** by resuspending them in cold lysis buffer. Use 10 ml of lysis buffer for each liter of bacterial culture.

15. Sonicate 6×30 s at 200 W with 1 min intervals. Keep on ice between the cycles. (Optional: take an ~1 μl sample for an analytical gel in **step 21**.)
16. Centrifuge at 110,000×*g* for 1 h. (Optional: take an ~2 μl sample for an analytical gel in **step 21**.)
17. Load the supernatant onto the pre-equilibrated HA column at a flow rate of 1–2 ml/min using a peristaltic pump or FPLC. Save the flow-through. (Optional: take an ~2 μl sample for an analytical gel in **step 21**.)
18. Attach the HA column to a FPLC system if not already attached.
19. Wash the column with 6 CV (~400 ml) of 28 % HA column buffer B (corresponds to 1 M NaCl). Retain the wash. (Optional: take an ~4 μl sample for an analytical gel in **step 21**.)
20. Run a linear elution gradient from 28 % HA column buffer B (remainder HA buffer A) to 100 % buffer B over 2 CV (120 ml) and collect 6 ml fractions.
21. Run fractions of this eluate (~20 μl) and other expression and purification fractions on a denaturing 20 % polyacrylamide gel). Relevant fractions will contain visible bands at ~18 kDa (hsCENP-A) and 13 kDa (H4).
22. Pool fractions containing histones CENP-A and H4 and dialyze overnight against 4 l of dialysis buffer.
23. Change the dialysis buffer and dialyze for another 2–4 h.
24. Load the sample onto 1 ml of HiTrap SP FF column at a flow rate of 1 ml/min. Save the flow-through. (Optional: take a ~20 μl sample for an analytical gel in **step 27**.)
25. Wash with 10 CV of 37 % S column buffer B.
26. Run a linear gradient from 37 % S column buffer B (remainder S column buffer A) to 100 % S column buffer B over 20 CV to elute soluble CENP-A/H4 tetramer. Collect 0.5 ml fractions.
27. Run fractions (~20 μl) on a denaturing 20 % (wt/vol) polyacrylamide gel to determine which fractions contain CENP-A and H4 at high purity. Pool fractions containing a roughly equal ratio of CENP-A and H4 that are also free of other contaminating proteins.
28. Centrifuge the pooled histone solution for 10 min at 10,000×*g* at 4 °C.
29. Mix 5 μl of the supernatant with 5 μl of 8 M guanidine-HCl to denature the protein, and use a spectrophotometer to obtain the 280 nm absorption.
30. Use the theoretical extinction coefficient ε and the calculated molecular weight of each histone to precisely determine the protein concentration in milligrams per milliliter. To calculate

protein concentration in mg/ml divide the protein absorbance at 280 nm by the extinction coefficient and multiply by the protein's molecular weight.

31. Aliquot the remaining pooled CENP-A/H4 tetramer into ~100 pmol aliquots, flash-freeze in liquid nitrogen, and store at −80 °C.

3.5 Preparation of Biotinylated DNA

1. Transform the 19 × 601 plasmid into SURE2 bacteria (also use SURE2 cells by themselves as a control). Plate 20–100 μl of the transformation or the cells-only control, onto a LB/chloramphenicol/carbenicillin plate and incubate at 37 °C overnight. The following day, pick a single colony and use to inoculate a 5 ml starter culture in LB medium with chloramphenicol/carbenicillin. Grow for 8 h at 37 °C. Dilute the starter culture 1:1000 into 2 l of LB medium supplemented with chloramphenicol and carbenicillin and allow it to grow overnight at 37 °C.
2. Purify plasmid DNA with a Qiagen plasma giga kit tip-10000 (or other comparable method). Typically yields are 7–10 mg of purified plasmid DNA.
3. Digest 2 × 500–750 μg of plasmid DNA with EcoRI, XbaI, DraI, and HaeII (200 U of each enzyme) in a final volume of 1 ml (per reaction) at 37 °C overnight. Check digestion by running 1 μl of the digests on a 0.7 % (wt/vol) agarose gel. The 19 × 601 fragment runs at 3.8 kb, the digested backbone appears as four bands <1 kb.
4. Add 5 M NaCl to the digest to a final concentration of 0.5 M NaCl.
5. Add 20 % (wt/vol) PEG and 0.5 M NaCl to a final concentration of 5 % (wt/vol) PEG, 0.5 M NaCl.
6. Centrifuge at 5000 × *g* for 10 min at RT.
7. Transfer the supernatant into a fresh tube (take 1 μl of sample for analysis on an agarose gel in **step 10**) and keep the original tube in a rack at RT.
8. Add 20 % (wt/vol) PEG and 0.5 M NaCl to the transferred supernatant to obtain a final concentration of 5.5 % (wt/vol) PEG, 0.5 M NaCl.
9. Repeat **steps 6–8**, adding additional PEG (in 0.5 % (wt/vol) final concentration increments) and centrifuging until the final PEG concentration reaches 10 %.
10. Load each 1 μl sample onto a 0.7 % (wt/vol) agarose gel to identify fractions containing the 19 × 601 fragment. Typically, the 19 × 601 DNA fragment precipitates between 7 and 9 % final PEG concentration.

11. Resuspend precipitated DNA (from appropriate tubes stored in **step** 7) in TE (pH 8.0) and pool fractions. The volume of TE for resuspension must be determined empirically with the aim of resuspending precipitated DNA at 2–3 μg/μl.
12. Dialyze resuspended DNA in 50–200 μl dialysis buttons into TE (pH 8.0) overnight.
13. Refresh the dialysis buffer and continue dialysis for another 3–4 h.
14. Determine the concentration of the DNA using a spectrophotometer to measure the OD260. Typically, the digestion of 1 mg of 19×601 plasmid DNA yields 300–400 μg digested and purified 19×601 DNA fragment.
15. Analyze 1 μl of a 1:10 dilution of the DNA on a 0.7 % (wt/vol) agarose gel to check the purity of the sample.
16. A concentration of ~2.5 μg/μl or higher is ideal. If required, sodium acetate/ethanol-precipitate DNA and resuspend in a smaller volume of TE (pH 8.0).
17. Mix 500 μg of digested and purified 19×601 DNA in a final volume of 400 μl with biotin-14-dATP, α-thio-dGTP, α-thio-dTTP, and dCTP at a final concentration of 35 μM for each nucleotide, as well as 24 μl of Klenow fragment. Also add an appropriate amount of the reaction buffer concentrate provided with the Klenow fragment.
18. Incubate for 3 h at 37 °C.
19. Purify biotinylated 19×601 DNA using Qiagen PCR purification columns, according to the manufacturer's instructions. Elute with 30 μl of EB buffer (Qiagen) or water per column, pool the eluted fractions and precipitate using sodium acetate/ethanol precipitation. Alternatively, purify large reactions using G-50 gel filtration resin to remove the unincorporated nucleotides.
20. Resuspend in 100 μl of TE (pH 8.0) and measure the concentration at OD260 using a spectrophotometer. A final concentration of 3–5 μg/μl is ideal. Typically, the input of 500 μg of digested 19×601 DNA into the biotinylation reaction yields ~300 μg of biotinylated 19×601 DNA.
21. To analyze the biotinylation efficiency, incubate 500 ng of biotinlyated 19×601 DNA with and without 0.5 μl of Streptavidin-FITC and non-biotinylated 19×601 DNA with 0.5 μl of streptavidin-FITC in TE (pH 8.0) and 2.5 mM NaCl for 20 min at RT.
22. Run reactions on a 0.7 % (wt/vol) agarose gel. Analyze the DNA by imaging with ethidium bromide stain and the biotinylation by FITC fluorescence. Biotinylation causes the 19×601 DNA to migrate in a single band more slowly in the presence of streptavidin than in its absence.
23. Store biotinylated DNA at −20 °C.

3.6 Chromatin Array Assembly

1. Estimate the required ratios of H3/H4 and CENP-A/H4 tetramers to nucleosome positioning sites on the DNA arrays for equal stoichiometry, and use a slight excess of H2A/H2B dimers (2.2 times the molar DNA amount) (*see* **Note 1**).
2. Prepare 500 ml of HS assembly buffer and 2 l of LS assembly buffer; adjust the pH to 7.5 and store at 4 °C.
3. Set up chromatin array assembly reactions in 1.5 ml tubes on ice. Typically, we use 50 μl reactions or 200 μl reactions at a final concentration of nucleosome positioning sites of 2 μM, and make both individual H3 and CENP-A nucleosome arrays. Mix appropriate volumes of H_2O, 10× TE and 5 M NaCl to achieve a final concentration of 1× TE and 2 M NaCl. Add, in the following order, biotinylated DNA, H2A/H2B dimers, and CENP-A/H4 or H3/H4 tetramers.
4. Moisten the dialysis membrane of a small, fixed volume dialysis unit in HS assembly buffer. Pipette the assembly mixture into the dialysis unit.
5. Submerge dialysis units in the HS assembly buffer.
6. Transfer the dialysis beaker to a magnetic stir plate at 4 °C. Set the stirring speed to very low.
7. Connect a 1 ml plastic pipette to one end of two pieces of a peristaltic pump tubing, clip both pieces of the tubing with the plastic pipettes at the same end into the peristaltic pump. Set the flow speed to 0.5 ml/min. Put one pipette into the 2-l beaker containing LS assembly buffer and attach the loose end of the tubing with a binder clip to the rim of the dialysis beaker. Attach the second pipette to the rim of the dialysis beaker and put the loose end into a liquid waste container or the sink.
8. When the buffer exchange is complete (~67 h later), exchange dialysis buffer one more time with fresh LS assembly buffer and dialyze for an additional 4 h.
9. Transfer the assembly reaction into a 1.5 ml tube on ice. Determine the volume of each assembly and adjust the expected concentration accordingly.
10. Assembled arrays can be stored for up to ~3 months at 4 °C, although the nucleosome saturation should be tested within a few days of use.

3.7 Testing Nucleosome Array Saturation

1. Combine 500 ng of assembled array DNA with 1 μl of AvaI in AD buffer supplemented with 1 mM DTT and 1× BSA in a final volume of 20 μl at RT overnight. Use 500 ng of biotinylated DNA (no nucleosomes) as a control (*see* **Note 2**).
2. Add Glycerol (80 %, vol/vol) to obtain a final concentration of 20 % (6.7 μl per 20 μl reaction).

3. Run the whole reaction (~27 μl) of digested arrays and DNA on a 5 % (wt/vol) native acrylamide/bis-acrylamide/0.5 × TBE gel in 0.5× TBE as a running buffer. Also run a small volume (0.5–1 μl) of DNA ladder on this gel.
4. Mix 200 ml of 0.5× TBE with 15 μl of SYBR Gold and pour it into a gel-staining container.
5. Using a spatula, carefully detach the gel from the glass plate so that it rolls slowly into the container. Stain the gel while gently rocking or shaking for 20 min at RT.
6. Wash the gel three times with H_2O to remove residual stain.
7. Transfer the gel to a UV-transmittable glass plate.
8. Analyze the gel by fluorescence imaging. Assembly of a nucleosome onto a nucleosome positioning sequence within the array causes a band shift from ~200 bp (DNA control) to ~700 bp (mononucleosomes) (*see* **Note 2**).

3.8 Preparation of Chromatin Beads

1. Wash 45 μl of Dynabeads in ~300 μl of 1× bead buffer three times, using a magnetic tube rack to retain beads between washes (*see* **Note 3**). Split the beads into three tubes (15 μl per tube: one for CENP-A arrays, one for H3 arrays and one for the bead-only control), place them onto magnetic tube rack and aspirate the supernatant.
2. To each tube, add 7.5 μl of ~2 μM biotinylated nucleosome arrays (CENP-A or H3 saturated chromatin) or 10 μl of array buffer (for the bead only control) in a final volume of 50 μl bead buffer containing 2.5 % (wt/vol) PVA. The volume of this reaction should be scaled according to the amount of beads used. Typically, we couple 1.5 μl of chromatin arrays (~2 μM) to 3 μl of Dynabeads for each sample.
3. Attach chromatin arrays to beads by incubating at RT for at least 30 min with gentle agitation. Ensure that the shaking speed is enough to keep the beads in solution without spitting all over the inside of the tube (*see* **Note 4**).
4. Wash the chromatin-bound beads in 200–300 μl of bead buffer and store at 4 °C.

3.9 Xenopus CSF Extract Preparation

1. Prime female frogs with 0.5 ml of PMSG using a 1-ml syringe and a 25-gauge needle. Primed frogs can be used for egg laying for 2 weeks. For optimal egg laying, prime frogs again with 0.25 ml of PMSG 2 days after the first priming.
2. To induce egg laying, inject three frogs with 0.5 ml of hCG (500 U) 16–18 h before the extract will be prepared.
3. After injection, separate frogs by placing each into an individual plastic container containing at least 2 l of 1× MMR, made with deionized water. Keep at 16 °C overnight.

4. Rinse all glassware with deionized water and prepare the required buffers (2 l of 1× MMR, 200 ml of dejellying buffer with the cysteine added immediately before use, 500 ml of CSF-XB buffer). Store buffers at 16 °C.
5. Pour ~10–20 ml hot liquid gelatin solution into a water-filled glass dish and 100-ml glass beaker, immediately swirl the containers to coat the sides with gelatin. Keep the water/gelatin solution in the containers and store at RT until further use.
6. Remove frogs from their containers, rinse frog eggs twice with ~500 ml of 1× MMR and sort eggs using a glass Pasteur pipette to remove strings of eggs, puffy eggs and activated eggs (*see* **Note 5**).
7. Prepare 50 ml of CSF-XB. Transfer 2 ml into a separate tube, add cytochalasin B and transfer 1 ml into two SW-55 centrifuge tubes.
8. Pour out the water/gelatin solution from the Pyrex glass dish and rinse with ~200 ml of 1× MMR. Carefully pour the eggs into the Pyrex dish and wash twice with 200–300 ml of 1× MMR. Add the cysteine to the dejellying buffer and mix until dissolved. Pour out the 1× MMR from the Pyrex dish containing the eggs and add ~50 ml of the dejellying buffer. Swirl gently, remove the dejellying buffer and add in the remaining 150 ml of the dejellying buffer. Gently swirl again and remove any eggs that show contraction of the dark pigments. The dejellying process is complete when the eggs become closely packed when the Pyrex dish is tilted to one side. This typically takes 5–10 min and should be carefully monitored, as eggs left in dejellying solution for too long will lyse.
9. Pour off the dejellying buffer and wash the eggs three or four times with 200–300 ml of CSF-XB.
10. Empty the water/gelatin solution from the 100 ml beaker, add a few milliliters of CSF-XB buffer then carefully add the eggs.
11. Pour off the CSF-XB buffer and add ~10 ml of CSF-XB with protease inhibitors. Swirl gently, then remove the buffer and pour in the remaining 40 ml of CSF-XB with protease inhibitors.
12. Transfer the eggs into the SW-55 centrifuge tubes from **step 7** above using a fire-polished pipette.
13. Aspirate excess buffer and gently slide the tubes into 13 ml Falcon tubes without caps.
14. Centrifuge at RT for ~30 s at ~300 ×*g*, and then for ~30 s at ~500 ×*g*.
15. Remove the tubes, aspirate any excess liquid and centrifuge for 15 min at 16 °C at 9480 ×*g* in a SW-55 swinging bucket rotor.
16. Remove the tubes from the centrifuge buckets and store on ice.

17. Clean the outside of the tubes with ethanol and pierce with a 1-ml syringe and a 16-gauge needle to collect the straw-colored cytoplasmic fraction (middle layer).
18. Measure the volume of extract collected and store in a fresh tube on ice.
19. Add the appropriate amount of protease inhibitors (from 1000× stock), cytochalsin D (from 1000× stock), cold sucrose (1:40), and energy mix (1:20) to the extract and mix well. Use extract within 1–2 h as quality declines over time.

3.10 Centromere Associated Protein Assembly in Metaphase Extracts

1. Wash chromatin beads (prepared above) twice with 200–300 μl of CSF-XB supplemented with 0.05 % (vol/vol) Triton X-100.
2. Use a magnetic tube rack to collect the beads, discard the supernatant and add 100 μl of fresh CSF extract (prepared above) on ice. Mix the extract–bead mixture by pipetting gently and transfer to an 18–20 °C water bath.
3. Keep the tubes at 18–20 °C for 60 min. Flick them every 15 min to keep chromatin beads in solution.
4. Use a magnetic tube rack to collect the beads. This may take a few minutes if the extract is particularly viscous.
5. Remove the extract and wash the chromatin beads three times in 200–300 μl of cold CSF-XB buffer supplemented with 0.05 % (vol/vol) Triton X-100.
6. Resuspend chromatin beads in 100 μl of CSF-XB plus 0.05 % (vol/vol) Triton X-100 and add formaldehyde to a final concentration of 2 %. Incubate for 5 min at RT.
7. Wash three times with 200–300 μl of AbDil and adjust the concentration to 1 μl chromatin beads per 10 μl AbDil. Fixed samples can be stored in AbDil for 1–2 days at 4 °C.

3.11 Kinetochore Assembly in Cycled Extracts

1. Wash chromatin beads (prepared above) twice with approximately 200–300 μl of CSF-XB supplemented with 0.05 % (vol/vol) Triton X-100.
2. Discard the supernatant and mix chromatin beads with 50 μl of fresh CSF extract (prepared above) on ice. Pipette the extract–bead mixture up and down two or three times and transfer to a 18–20 °C water bath.
3. After 5–10 min, add $CaCl_2$ to a final concentration of 0.6 mM and incubate for 80 min at 18–20 °C. Flick the tubes every 15 min to keep chromatin beads in solution. (Optional) Take a sample for a western blot to monitor the cell cycle stage (*see* **Note 6**).
4. Add one volume (50 μl per tube) of fresh CSF extract and incubate for 90 min to drive the extract into mitosis. Flick the tubes every 15 min to keep chromatin beads in solution. If

analyzing the assembly of spindle assembly checkpoint proteins, use CSF extract that has been premixed with nocodazole (or DMSO as a control) at a final concentration of 10 μg/ml.

5. Resuspend chromatin beads in 100 μl of CSF-XB plus 0.05 % (vol/vol) Triton X-100 and add formaldehyde to a final concentration of 2 %. Incubate for 5 min at RT.
6. Wash three times with 200–300 μl of AbDil and adjust the concentration to 1 μl chromatin beads per 10 μl AbDil.

3.12 Microtubule Polymerization and Stabilization in Metaphase Extracts

1. Wash chromatin beads (prepared above) twice with 200–300 μl of CSF-XB supplemented with 0.05 % (vol/vol) Triton X-100.
2. Use a magnetic tube rack to pellet the chromatin beads, discard the supernatant and mix 2 μl of beads with 20–50 μl of freshly prepared CSF extract per spindle assembly reaction. Mix reactions by pipetting up and down two or three times, transfer to a 0.5-ml Eppendorf tube and incubate on ice (*see* **Note** 7).
3. Transfer samples to an 18–20 °C water bath and incubate for 75–90 min. Flick the tubes gently every 15 min to keep the chromatin beads in solution. (Optional) Add 100 ng/μl of nocodazole to the reaction or transfer the reaction to 4 °C for 10 min to induce microtubule depolymerization.
4. Pipette the reaction into 1 ml of dilution buffer supplemented with 2.5 % (wt/vol) formaldehyde and fix at 18–20 °C for 10 min. Use a wide tip and pipette gently to avoid breaking the assembled spindles. Mix samples by inverting the tubes.
5. Place custom-made adaptors into Corex tubes to provide a flat surface. Place a poly-l-lysine-coated coverslip on top and add 5 ml of cushion.
6. Layer fixed samples on top of the cushion. Use a wide tip and pipette gently to avoid shearing the assembled spindles.
7. Centrifuge for 20 min at 3290 × *g*.
8. Remove the coverslips and postfix in ice-cold methanol for 5 min.
9. Wash the coverslips 3 × 5 min in AbDil.
10. Block in AbDil for 30 min at RT.
11. Incubate with tubulin-specific antibodies (DM1α) at 1:2000 for 1 h at RT or at 4 °C overnight.
12. Wash coverslips 3 × 5 min in AbDil.
13. Stain using appropriate fluorophore-conjugated secondary antibodies for 30 min at RT.
14. Wash coverslips 3 × 5 min in AbDil.
15. Stain the coverslip with Hoechst stain (10 μg/ml in AbDil) for 10 min at RT.

For mounting and subsequent microscopic analysis, proceed to the immunofluorescence and data analysis section below.

3.13 Mitotic Checkpoint Assay

1. For each sample, wash 2 µl of chromatin beads (prepared above) twice with 200–300 µl of CSF-XB supplemented with 0.05 % (vol/vol) Triton X-100.
2. Take 5 µl of CSF extract, mix with 45 µl of SDS-PAGE sample buffer, and store at −20 °C for subsequent western blot analysis of the cell cycle stage.
3. Add 20 µl of fresh CSF extract to each sample from **step 1** in a 0.7 ml Eppendorf tube. Pipette up and down to mix and incubate for 5–10 min at 18–20 °C in a water bath.
4. Add $CaCl_2$ to a final concentration of 0.6 mM and mix by flicking the tube (*see* **Note 6**).
5. Incubate for 80 min at 18–20 °C, flicking the tubes every 15 min.
6. Take 5 µl of sample and mix it with 45 µl of SDS-PAGE sample buffer for western blotting, store at −20 °C. To prevent taking beads, do not flick the tubes for 15–20 min before taking the sample so that the chromatin beads sink.
7. To the remaining 15 µl of sample, add an equal volume of fresh CSF extract supplemented with either nocodazole at 10 µg/ml or DMSO as a control. Mix by pipetting up and down and avoid the addition of air bubbles.
8. Incubate for 90 min at 18–20 °C, flicking the tubes every 15 min.
9. Take 5 µl of sample ($t = 0'$) and mix it with 45 µl of SDS-PAGE sample buffer for western blotting, store at −20 °C. Prevent taking beads, as described in **step 6**.
10. Add $CaCl_2$ to the remaining 25 µl of sample to a final concentration of 0.6 mM and mix by flicking the tubes.
11. Incubate for 10 min at 18–20 °C, take 5 µl of sample ($t = 10'$) and mix with 45 µl of SDS-PAGE sample buffer for western blotting and leave at −20 °C. Again, avoid taking beads, as described in **step 6**.
12. Repeat **step 11** for $t = 20'$, $t = 30'$ and $t = 40'$ samples. Store samples in SDS sample buffer at −20 °C for a few weeks.
13. Heat western blotting samples at 95 °C for 10 min.
14. Load 20 µl of each sample (equivalent to ~2 µl of extract) on a 12.5 % (wt/vol) SDS-PAGE system.
15. Transfer separated proteins onto PVDF membrane in CAPS transfer buffer for 1 h 45 min at 400 mA.
16. Block the membrane for 30 min in 5 % (wt/vol) milk in PBS-T.

17. Incubate with primary antibodies (e.g., P-Wee-1–specific, 1:1000; tubulin-specific (DM1α), 1:2000) at 4 °C overnight.
18. Wash 3 × 5 min in PBS-T.
19. Incubate for 30 min at RT with fluorescently labeled secondary antibodies at a dilution of ~1:500 in 5 % (wt/vol) milk in PBS-T.
20. Wash 3 × 5 min in PBS-T.
21. Wash two times with PBS and analyze the western blot by fluorescence detection (*see* **Note 8**).

3.14 Immunofluorescence and Data Analysis

1. Dry poly-l-lysine-coated coverslips in a humidified chamber.
2. Pipette 20 μl of chromatin beads in AbDil (equivalent to ~2 μl of chromatin beads) onto each coverslip and let the beads attach to the coverslip for 15 min at RT, without letting them dry out.
3. Block the coverslips in AbDil by pipetting 30–50 μl on top of the attached beads and incubating for 30 min at RT.
4. Aspirate the AbDil and incubate each coverslip with antibodies to proteins of interest (diluted in AbDil) for 1 h at RT or overnight at 4 °C (*see* **Note 9**).
5. Wash coverslips for 3 × 5 min in AbDil.
6. Stain the coverslips using appropriate fluorophore-conjugated secondary antibodies for 30 min at RT.
7. Wash coverslips 3 × 5 min in AbDil.
8. When using two antibodies raised in the same animal, one must be directly conjugated to a fluorophore. Block with whole IgG of that species for 45 min before incubation with a directly labeled antibody for 1 h at RT.
9. Wash coverslips 3 × 5 min in AbDil.
10. Incubate with 30–40 μl of propidium iodide (1:1000 in AbDil) for 10 min at RT to stain DNA.
11. Wash coverslips 3 × 5 min in AbDil.
12. Wash one or two times in PBS.
13. Add ~5 μl of mounting medium to a labeled microscope slide, aspirate the coverslips, and place upside down onto the slide. Carefully remove excess mounting medium and seal the sides of the coverslip to the slide with clear nail polish. Store slides at −20 °C until analysis.
14. Image at least three fields of beads per slide, with 0.2 μm section intervals through the beads. Image a few hundred beads in total for each sample, imaging each relevant wavelength at each section. Determine exposure times for each wavelength to ensure that the camera does not saturate. Keep exposure times the same for different samples stained with the same marker.

15. Ensure images are in a format readable by the "imread" function in MATLAB (e.g., TIFF). Do not convert images to a format that uses loss-prone compression (e.g., JPEG-compressed image formats) or alters the bit depth of the images.
16. Quantify the fluorescence signal of beads in each image (*see* **Note 8**). We have provided a MATLAB script that is freely available at http://straightlab.stanford.edu/software. This script reads TIFF images where each color channel is in a separate file, finds beads using the DNA-stained images and quantifies the intensity of each bead in each color channel (*see* **Note 10**).

4 Notes

1. Arrays are easily oversaturated or undersaturated if the amount of histones used for nucleosome assembly was not quite correct. We recommend a careful titration of both tetramer–dimer ratio and the histone–DNA ratio to achieve 90–100 % array saturation. Typically, we will test DNA–tetramer ratios of 1:0.8, 1:0.9, 1:1.0, 1:1.1, and 1:1.2 to determine the ideal ratio for high nucleosome saturation. The ratio of H2A/B dimers to DNA is usually more forgiving and is usually kept at 2.2:1 in nucleosome assembly reactions.
2. In the 19 × 601 array DNA, each positioning sequence is separated by an AvaI restriction site; thus, digestion with AvaI generates 19–200-bp fragments. Migration of the mono-nucleosome fragment is dependent on the precise positioning of the core histone complex with respect to the DNA.
3. A volume of 45 μl of Dynabeads is sufficient to analyze the binding of four different centromere/kinetochore proteins to CENP-A arrays versus H3 arrays. It will also ensure enough for the H4 staining and the bead-only control (no chromatin arrays coupled) coverslips, both of which are required for subsequent quantification by fluorescence microscopy.
4. Coupling of nucleosome arrays to beads often causes the beads to clump slightly. This effect appears to be histone-dependent and is not observed with naked DNA.
5. Frogs that lay particularly low quality eggs (where strings of eggs, puffy eggs or activated eggs comprise more than 10 % of the whole batch) typically suffer from three major complaints. They may have been ovulated too frequently and must be rested for longer time periods between ovulations. They may be aged, in which case they must be replaced. Else they may be stressed, and treatment would involve removing potential stressors (such as exposure to noise or poor water quality).
6. Extract that is of poor quality may not exit from mitosis upon calcium addition. Equally, insufficient calcium addition may

also prevent mitotic exit. In either case, repeating the experiment with fresh extract and slightly more calcium usually circumvents these problems. Making sure the water temperature does not exceed 16–18 °C during incubations has also been found to help. Extract that does not arrest in mitosis as expected may occur for two reasons. Firstly, the water temperature may be too high, in which case make sure the temperature stays between 16 and 18 °C throughout the incubation. The second possibility is that kinetochore assembly was inefficient, in which case it is necessary to retry using fresh extract and adding more calcium.

7. As a control for spindle assembly, it is possible to prepare a sample with 250 *Xenopus* sperm nuclei per 1 μl of extract. Demembranated sperm is prepared exactly as described by Murray [11].
8. If no kinetochore or spindle checkpoint proteins are observed, this may be for two reasons. The extract quality may have been poor, or kinetochore assembly was inefficient due to the quality of the nucleosome array. Repeating with fresh extract and new arrays typically solves such issues.
9. In our hands, efficient histone H4 staining at 1:100 can only be obtained with incubation at 4 °C overnight.
10. For a more detailed description of this MATLAB script *see* Guse et al. [10].

References

1. Westhorpe FG, Straight AF (2013) Functions of the centromere and kinetochore in chromosome segregation. Curr Opin Cell Biol 25:334–340
2. Fukagawa T, Earnshaw WC (2014) The centromere: chromatin foundation for the kinetochore machinery. Dev Cell 30:496–508
3. Mitchison TJ, Kirschner MW (1985) Properties of the kinetochore in vitro. I. Microtubule nucleation and tubulin binding. J Cell Biol 101:755–765
4. Mitchison TJ, Kirschner MW (1985) Properties of the kinetochore in vitro. II. Microtubule capture and ATP-dependent translocation. J Cell Biol 101:766–777
5. Kingsbury J, Koshland D (1991) Centromere-dependent binding of yeast minichromosomes to microtubules in vitro. Cell 66:483–495
6. Akiyoshi B, Nelson CR, Ranish JA, Biggins S (2009) Quantitative proteomic analysis of purified yeast kinetochores identifies a PP1 regulatory subunit. Genes Dev 23:2887–2899
7. Lechner J, Carbon J (1991) A 240 kd multisubunit protein complex, CBF3, is a major component of the budding yeast centromere. Cell 64:717–725
8. Sorger PK, Severin FF, Hyman AA (1994) Factors required for the binding of reassembled yeast kinetochores to microtubules in vitro. J Cell Biol 127:995–1008
9. Guse A, Carroll CW, Moree B et al (2011) In vitro centromere and kinetochore assembly on defined chromatin templates. Nature 477:1–7
10. Guse A, Fuller CJ, Straight AF (2012) A cell-free system for functional centromere and kinetochore assembly. Nat Protoc 7:1847–1869
11. Murray AW (1991) Cell cycle extracts. In: Kay BK, Peng HB (eds) Xenopus laevis: practical uses in cell and molecular biology, vol 36, Methods in cell biology. Academic, New York, pp 581–605

Chapter 9

Biochemical and Structural Analysis of Kinetochore Histone-Fold Complexes

Tatsuya Nishino and Tatsuo Fukagawa

Abstract

The kinetochore connects chromosomes to microtubules during mitosis and therefore plays an essential role in faithful chromosome segregation. It is built at the centromeric region of the chromosome and is comprised of many protein complexes. CENP-S, -T, -W, and -X are kinetochore components with histone-folds. These proteins play important roles in establishment of kinetochore chromatin. Similar to canonical histones, these kinetochore histone-fold proteins form heteromeric complexes (CENP-S/CENP-X complex and CENP-T/CENP-W complex) and bind DNA in sequence independent manner. In addition, they form a CENP-T-W-S-X heterotetrameric complex and bind DNA in a manner that is different from both CENP-S-X and CENP-T-W. To understand how kinetochores form and function it is necessary to characterize the components in detail. Here, we describe our approaches in purification and characterization of the kinetochore histone-fold complexes.

Key words Chromosome segregation, Kinetochore, Centromere, Histone-fold, Protein complex, Protein expression, Protein purification, Protein–protein interaction, Protein–DNA interaction, Gel shift assay

1 Introduction

The kinetochore is built at the centromeric region of the chromosome. It connects chromosomes to microtubules during mitosis. In vertebrates, more than 100 proteins have been identified and their functions vary from participating in protein–protein interactions, to DNA binding, chromosome binding, and microtubule binding [1]. Many of the components form specific sub-complexes, providing the framework for an intricate protein interaction network.

Kinetochore assembly is initiated by the centromere-specific nucleosome, which contains the histone H3 variant CENP-A. Once the centromere is specified by the CENP-A nucleosome, other kinetochore proteins assemble on the centromeric nucleosomes [2]. Kinetochore components can be classified into two categories

Paul Chang and Ryoma Ohi (eds.), *The Mitotic Spindle: Methods and Protocols*, Methods in Molecular Biology, vol. 1413,
DOI 10.1007/978-1-4939-3542-0_9, © Springer Science+Business Media New York 2016

based on the timing of their localization. One group localizes constitutively to the kinetochore (constitutive complexes) whereas the other localizes to the kinetochore only during mitosis (mitotic complexes). Constitutive complexes bind chromosomes and associate with centromeres throughout the cell cycle. On the other hand, mitotic complexes are recruited to the centromere only during mitosis and are released after its completion. Constitutive complexes are termed CCAN (constitutive centromere associated network) and mitotic complexes fall into the KMN (Knl1, Mis12, Ndc80) network [3].

CCAN forms the basis of kinetochore chromatin by association with the CENP-A nucleosome. The CCAN is comprised of 16 proteins (CENP-C, -H, -I, -K, -L, -M, -N, -O, -P, -Q, -R, -S, -T, -U, -W, -X). Of these CCAN components, CENP-C and CENP-N have been reported to bind directly to the CENP-A nucleosome [4]. However, how other proteins assemble and participate in the kinetochore formation or chromosome/microtubule interaction remains largely elusive.

We have been working to identify and characterize CCAN kinetochore complexes [5–9]. We are especially focused on CENP-S, -T, -W, -X, which contain a histone fold based on sequence comparison. Biochemical analyses have revealed that CENP-S and CENP-X form a complex and CENP-T and CENP-W form a complex [7, 9]. As with other histones, both complexes bind DNA in a sequence independent manner. Crystal structure revealed that CENP-S-X and CENP-T-W complex form tetramer and dimer, respectively [10]. Careful structural comparison identified similarity between two complexes. Further biochemical and structural analyses demonstrated that they form CENP-T-W-S-X heterotetramer. Biochemical and structural characterization of these complexes is an important step to understand how kinetochore proteins assemble. In this section we introduce our strategies for preparing and characterizing these kinetochore histone-fold complexes.

2 Materials

2.1 General Equipment for Biochemical Experiments

1. Orbital shaker for bacterial culture.
2. Sonifier.
3. Centrifuge.
4. HisTrap crude column.
5. Mono S column.
6. Superdex 200 gel filtration column.
7. Superose 6 gel filtration column.

8. Amicon Ultra.
9. FPLC (Akta Purifier, GE).

2.2 Kinetochore Histone Fold Expression

1. pET and RSFduet vector (Merck) cloned with kinetochore histone folds. Chemically transformation competent *E. coli* strains, BL21Codonplus(DE3)RIL(Merck) and BL21Star(DE3) pRARE2Lys (Stratagene).
2. LB medium: 1 % Bacto-tryptone, 0.5 % Bacto yeast extract, 1 % NaCl supplemented with indicated antibiotics (kanamycin 25 μg/ml, 100 μg/ml ampicillin, 20 μg/ml chloramphenicol).
3. IPTG: 0.22 μm filtered 50 μg/ml isopropyl-b-d-thiogalactopyranoside in water.

2.3 Kinetochore Histone Fold Complex Purification Under Denaturing Condition and Refolding

1. Buffer A: 50 mM Tris–HCl (pH 8.0), 500 mM NaCl, 1 mM PMSF, 5 % glycerol.
2. Buffer B: 50 mM Tris–HCl (pH 8.0), 7 M guanidine hydrochloride, 5 % glycerol.
3. Buffer C: 50 mM Tris–HCl (pH 8.0), 500 mM NaCl, 5 % glycerol, 6 M urea, 5 mM imidazole.
4. Buffer D: 50 mM Tris–HCl (pH 8.0), 500 mM NaCl, 5 % glycerol, 6 M urea, 500 mM imidazole.
5. Buffer E: 20 mM CH_3COONa (pH 5.2), 200 mM NaCl, 5 mM 2-mercaptoethanol, 1 mM EDTA, 6 M Urea.
6. Buffer F: 20 mM CH_3COONa (pH 5.2), 900 mM NaCl, 5 mM 2-mercaptoethanol, 1 mM EDTA, 6 M urea.

2.4 Kinetochore Histone Fold Complex Co-expression and Protein–Protein Interaction

1. Buffer G: 10 mM Tris–HCl (pH 7.5), 500 mM NaCl, 1 mM dithiothreitol.
2. Buffer H: 10 mM Tris–HCl (pH 7.5), 500 mM NaCl, 1 mM dithiothreitol, 500 mM imidazole.

2.5 Kinetochore Histone Fold Complex Protein–DNA Interaction

1. 1× Native PAGE buffer: 25 mM Tris base, 192 mM glycine.
2. 2× Sample loading buffer: 10 mM Tris–HCl pH 8.0, 1 mM EDTA, 30 % sucrose.

3 Methods

3.1 Expression/ Purification

3.1.1 Preparation of Kinetochore Histone-Fold Complex by Refolding

Compared to other proteins, the amount of kinetochore proteins within a cell is limited. Thus, biochemical analyses using these native complexes are challenging. Many biochemical analyses of kinetochore complexes have been done using recombinant complexes [10–17]. Bacterial recombinant protein expression and purification were applied for preparation of the kinetochore histone-fold complexes. To prepare the complexes, two strategies can be taken: individual/refolding method or co-expression/soluble preparations. Individual expression and purification has been widely used for the preparation of eukaryotic histone complexes [18]. This method allows high expression of a single protein (*see* **Note 1**). In many cases, proteins expressed in inclusion body are less prone to proteolytic cleavage. Once purified, they are stored in lyophilized form and the combinatorial process can be used to prepare different complexes.

1. Transform BL21-CodonPlus(DE3)-RIL with pET28a-ggCENPT, ggCENP-W, ggCENP-S, or ggCENP-X. Grow *E. coli* cells in 250 ml LB with kanamycin at 37 °C to an OD600 of 1.0.
2. Induce protein expression by the addition of IPTG to final concentration of 1 mM.
3. Continue incubation at 37 °C for 3 h and harvest cells by centrifugation.
4. Resuspend bacterial pellet in 25 ml of buffer A.
5. Sonicate bacterial suspension by sonifier.
6. Centrifuge lysate and discard supernatant.
7. Resuspend pellet in buffer B and homogenize the pellet by sonication.
8. Rotate on the rolling wheel at 4 °C overnight.
9. Centrifuge suspension and collect supernatant.
10. Apply supernatant to HisTrap crude column.
11. Wash column with 50 column volume of buffer C.
12. Elute by 50 column volume linear gradient of buffer C and buffer D.
13. Check fractions by SDS-PAGE.
14. Collect peak fractions and dialyze in 5 mM Tris–HCl (pH 7.5), 2 mM 2-mercaptoethanol for overnight.
15. Check protein concentration by Bradford assay or UV absorbance at 280 nm.
16. Add 2 U of thrombin per mg of protein and incubate for 3 h at room temperature.
17. Check cleavage by SDS-PAGE.
18. Load samples onto MonoS column.
19. Elute by 50 column volume linear gradient of buffer E and buffer F.
20. Check purity of proteins by SDS-PAGE.
21. Collect peak fractions and dialyze against water overnight.

22. Recover dialyzed samples and check protein concentration.
23. Freeze samples by liquid nitrogen and evaporate water by freeze-dry centrifuge.
24. Store powdered materials at 4 °C or −20 °C.

3.1.2 Refolding of Kinetochore Histone-Fold Complexes

To refold the kinetochore histone fold complexes, we utilized similar approach to that of canonical histone octamer with modifications. Bacterially expressed insoluble proteins were affinity purified under denaturing condition. Purified complexes were refolded by stepwise removal of denaturing agent. This procedure worked well for domains that only contain histone fold. However, full length CENP-T containing CENP-T-W complex precipitated during the process. CENP-T is relatively large among the histone fold containing proteins. Histone-fold is located at the extreme C-terminus and the N terminal 500 residues are predicted to be unstructured [19]. The presence of such a long unstructured region might have interfered with the refolding process. To minimize the nonspecific interaction and aggregation, we included arginine in the final steps of refolding. Arginine facilitated refold of the full length CENP-T-W complex (Fig. 1).

1. Resuspend equimolar amount of CENP-T and -W or CENP-S and -X in buffer C.
2. Dialyze samples over night at 4 °C.
3. Transfer the dialysis tubing to 3 M urea containing buffer C and dialyze overnight.
4. Repeat the transfer/dialysis cycle and reduce urea concentration to 2, 1, 0.5, and 0 M. For preparation of full length CENP-T containing CENP-T-W complex, 0.4 M arginine was included from the 0.5 M urea containing buffer C.
5. Centrifuge the dialysis sample and remove the pellet.
6. Apply the supernatant to a Superdex 200 column.
7. Check proteins of the peak fractions by SDS-PAGE.
8. Concentrate the fractions containing CENP-T-W or CENP-S-X by Amicon Ultra filtration tubes (Millipore).
9. Measure the protein concentration by UV absorbance at 280 nm.

3.1.3 Preparation of Kinetochore Histone Fold Complexes by Bacterial Co-expression and Purification

Coexpression of protein complexes, if successful, allows simple and fast preparation of the recombinant complexes (*see* **Note 2**). Several bacterial co-expression vectors are commercially available and we have utilized pRSFduet co-expression vector (*see* **Note 3**) to express CENP-T-W or CENP-S-X complexes in *E. coli*. Both complexes were expressed as soluble materials and they were purified using nickel columns (Fig. 2).

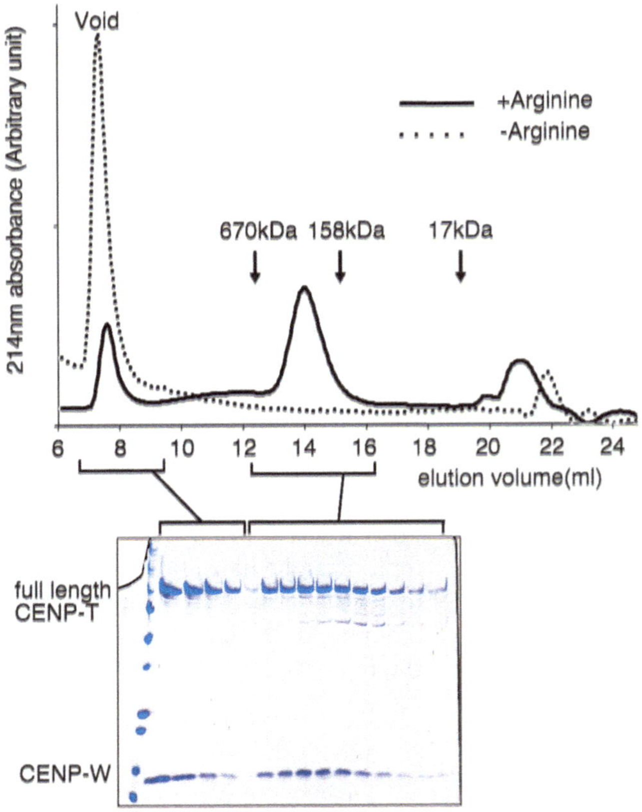

Fig. 1 Gel filtration analysis of refolded full length CENP-T-W complex. *Top panel*: Chromatogram of Superose 6 gel filtration column. Refolding was performed in the presence (*Bold*) or absence (*Dotted*) of arginine hydrochloride. *Bottom panel*: SDS-PAGE gel of the CENP-T-W peak fraction. Refolded CENP-T-W complex was loaded onto Superose 6 10/300 column in 10 mM Tris–HCl pH 7.5, 2 M NaCl, 2 mM 2-mercaptoethanol. Note that in the absence of arginine, most of the refolded material eluted in the void fraction. Arginine promoted CENP-T-W complex formation

1. Transform BL21-Star(DE3)pRARE2Lys with pRSFduet-his6tev-ggCENP-T-strepII-tev-ggCENP-W,pRSFduet-his6tev-ggCENP-S-strepII-tev-ggCENP-X. Grow *E. coli* cells in 250 ml LB containing 10ug/ml kanamycin at 37 °C to an OD600 of 1.0.
2. Induce protein expression by the addition of IPTG to a final concentration of 0.1 mM and continue incubation for overnight at 16 °C.
3. Harvest cells by 4000 × *g* centrifugation for 10 min at 4 °C. At this point, cells can be frozen at −20 °C until purification.
4. Thaw and resuspend the bacterial cells in 40 ml of buffer G.
5. Sonicate resuspension using Branson sonicator.
6. Centrifuge lysate at 20,000 × *g*.

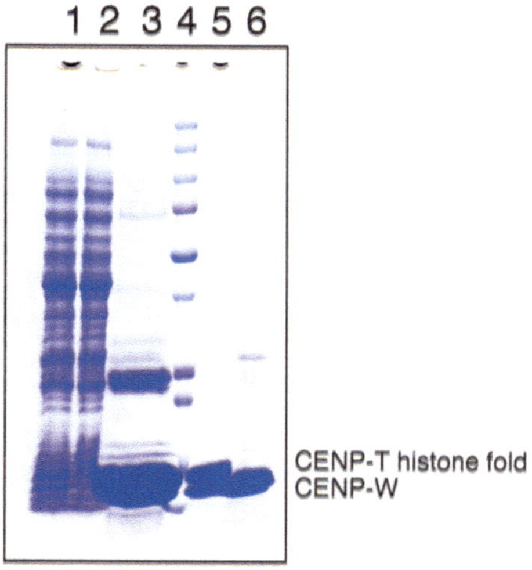

Fig. 2 Co-expression and purification of CENP-T-W histone fold domain. SDS-PAGE analysis of CENP-T-W purification steps is shown. *Lane 1*: Cell lysis supernatant *lane 2*: HisTrap flow through fraction, *lane 3*: HisTrap elute fraction; *lane 4*: protein molecular weight marker; *lane 5*: gel filtration peak fraction; *lane 6*: TEV protease cleaved fraction

7. Collect supernatant and apply them to HisTrap crude (5 ml column size).
8. Wash column with 50 column volume of Buffer G containing 20 mM imidazole.
9. Elute by 50 column volume linear gradient of Buffer G and Buffer H.
10. Check fractions by SDS-PAGE.
11. Apply peak fractions to Superdex 200 16/60 column equilibrated with Buffer G.
12. Check fractions by SDS-PAGE.
13. Concentrate peak fractions by Amicon Ultra filtration tubes (10 kDa cutoff) using fixed angle centrifuge at 5000 × *g*
14. Cleave affinity tag by recombinant TEV protease at 30 °C overnight.
15. Check cleavage of proteins by SDS-PAGE.
16. Apply cleaved protein complex through HisTrap crude (5 ml column size).
17. Collect flow through fraction and concentrate the fraction by Amicon Ultra filtration tubes (10 kDa cutoff).
18. Apply concentrated sample onto Superdex 200 16/60 column equilibrated with Buffer G.
19. Measure the protein concentration by UV absorbance at 280 nm.

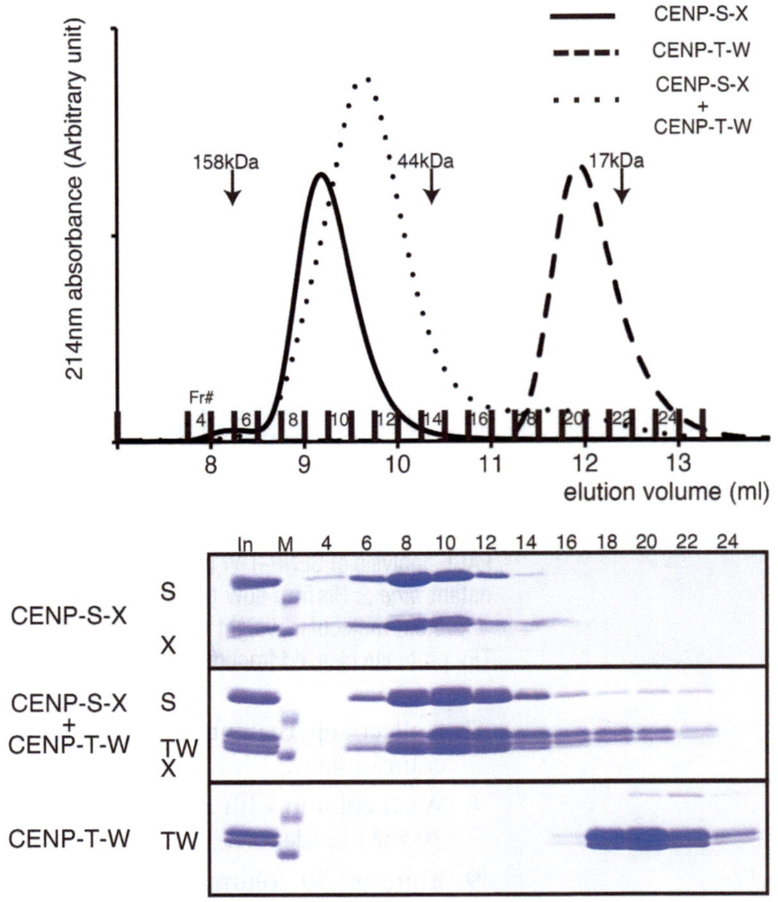

Fig. 3 CENP-S-X and CENP-T-W interaction analysis by gel filtration. *Top panel*: Gel filtration chromatogram of CENP-S-X (*bold line*), CENP-T-W (*broken line*), and CENP-T-W/CENP-S-X (*dotted line*). Protein sample (100 μl of 25 μM) was loaded onto Superdex 10/300 gel filtration column equilibrated with Buffer G. *Bottom panel*: Indicated elute fractions were analyzed by SDS-PAGE and Coomassie staining

3.2 Protein–Protein Interaction Analysis of Kinetochore Histone Fold Complex

Gel filtration is a convenient and reliable method to detect protein–protein interaction. Complex formation can be monitored by UV absorbance and protein compositions of eluate fractions visualize by SDS-PAGE (Fig. 3). Stoichiometry of the complex can be analyzed by mixing different ratio of the components. Sample elution volume depends on Stoke's radii of the molecule. If gel filtration is coupled with multi-angle light scattering, then molecular weight of the complex can be measured as well. The interaction can be further characterized by surface plasmon resonance, isothermal titration calorimetry, and fluorescence anisotrophy. Through these analyses, stoichiometry, affinity constants (association and dissociation constants), and thermodynamics can be measured.

1. Equilibrate Superdex 200 10/300 column with buffer G.
2. Prepare 200 μl solution of CENP-T-W and CENP-S-X at final concentration of 25 μM each in buffer G. Incubate on ice for 10 min. For control, prepare two 25 μM solutions of both complexes.
3. Load 100 μl of solutions onto Superdex 200 10/300 column using 100 μl sample loop. Measure 280 nm absorbance to detect protein peak (optional: measure 254 and 214 nm to detect nucleic acid and peptide bonds, respectively. Measurement at 214 nm can detect dilute protein samples). Collect 200 μl elute fractions starting from void volume.
4. Check the elute fractions by SDS-PAGE.

3.3 Protein–DNA Interaction Analysis of Kinetochore Histone Fold Complex

Gel shift assay or EMSA (electrophoresis mobility shift assay) is a commonly used technique to detect protein–DNA interaction. Protein–DNA mixture is run through native polyacrylamide gel. Protein-DNA complex usually migrates slower than the free DNA due to change in size and/or charge. For sequence specific DNA binding proteins, protein-DNA complex can be detected when a particular DNA is used. On the other hand, histone and related proteins bind DNA nonspecifically. CENP-S-X and CENP-T-W complexes similarly bound to Widom 601, α-satellite repeat (human centromereDNA), chicken centromere DNA or linear dsDNA from pUC119 plasmid without any sequence preferences [10]. Properties of the complex can be estimated from the band pattern of the protein-DNA complex. CENP-S-X forms a regular ladder pattern, suggesting that multiple molecules of complexes bind to single DNA molecule, whereas CENP-T-W forms an aggregated complex that does not enter the gel (Fig. 4). In addition, CENP-T-W-S-X forms a distinct complex. These properties may be attributed to the differences in protein charge or in DNA binding properties. Chicken CENP-S-X complex has overall pI of 6.93 and chicken CENP-T-W histone fold domain has pI of 10.17. As for CENP-S-X, the number of ladder changed when different DNA length was used whereas band pattern of CENP-T-W did not change.

1. Prepare 10 μl of solution containing 2.5 μM 145 bp Widom601 dsDNA and CENP-T-W (10, 20, or 40 μM); CENP-S-X (10, 20, or 40 μM); an equimolar mixture of CENP-T-W and CENP-S-X (10, 20, or 40 μM each); histone octamer (2.5 or 5 μM).
2. Incubate for 15 min at room temperature.
3. Add 10 μl of 2× sample buffer
4. Load 5 μl of sample to 5–15 % gradient polyacrylamide gel (Wako Chemicals).

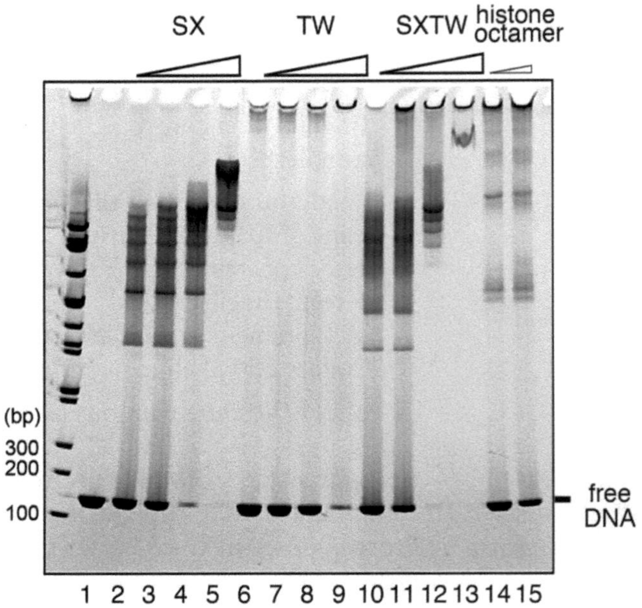

Fig. 4 Histone-fold and DNA interaction analysis by gel shift assay. Widom601 145 bp double stranded DNA (*lane 1*) was mixed with increasing amount of CENP-S-X (*lanes 2, 3, 4, 5*), CENP-T-W (*lanes 6, 7, 8, 9*), CENP-T-W/CENP-S-X mixture (*lanes 10, 11, 12, 13*), or histone octamer (*lanes 14 and 15*). Details of the specific condition are as given in the text. After incubation for 15 min at room temperature, sample was loaded onto native PAGE gel and run for 1 h at 20 mA. DNA was stained with ethidium bromide and image was obtained by UV illuminator

5. Run at 20 mA for 1 h at 4 °C.
6. Stain the gel with 10 μg/ml ethidium bromide in water for 5 min.
7. Visualize ethidium bromide stained DNA using UV illuminator.

4 Notes

1. For expression of individual proteins, fusion with MBP or GST sometimes enhances expression levels and/or solubility. These fusion proteins also allow affinity purification. We fuse these fusion proteins at the N-terminus of the target protein. In between the MBP/GST and the target, TEV or Prescission protease recognition sites are introduced. After affinity purification step, fusion proteins are cleaved by TEV/Prescission protease. MBP/GST protein is removed by affinity columns and the target is further purified by gel filtration.
2. Two or more duet vectors with different antibiotics can be introduced in the same bacteria enabling expression/purification of

trimeric, tetrameric, pentameric, and hexameric complexes. Expression level from different vectors may vary so we test different combinations.

3. For co-expression of two proteins by pETduet or RSFduet vectors, gene order sometimes affect level of protein expression. We normally test both ways whenever possible.

Acknowledgements

We thank members of the Fukagawa lab in particular Kozo Takeuchi for his help with sample preparation and characterization. We also acknowledge Kiyomi Kita and Mayumi Takahashi for their contributions.

References

1. Santaguida S, Musacchio A (2009) The life and miracles of kinetochores. EMBO J 28(17):2511–2531. doi:10.1038/emboj.2009.173
2. Perpelescu M, Fukagawa T (2011) The ABCs of CENPs. Chromosoma 120(5):425–446. doi:10.1007/s00412-011-0330-0
3. Takeuchi K, Fukagawa T (2012) Molecular architecture of vertebrate kinetochores. Exp Cell Res 318(12):1367–1374. doi:10.1016/j.yexcr.2012.02.016
4. Carroll CW, Milks KJ, Straight AF (2010) Dual recognition of CENP-A nucleosomes is required for centromere assembly. J Cell Biol 189(7):1143–1155. doi:10.1083/jcb.201001013
5. Okada M, Cheeseman IM, Hori T, Okawa K, McLeod IX, Yates JR 3rd, Desai A, Fukagawa T (2006) The CENP-H-I complex is required for the efficient incorporation of newly synthesized CENP-A into centromeres. Nat Cell Biol 8(5):446–457. doi:10.1038/ncb1396
6. Kwon MS, Hori T, Okada M, Fukagawa T (2007) CENP-C is involved in chromosome segregation, mitotic checkpoint function, and kinetochore assembly. Mol Biol Cell 18(6):2155–2168. doi:10.1091/mbc.E07-01-0045
7. Hori T, Amano M, Suzuki A, Backer CB, Welburn JP, Dong Y, McEwen BF, Shang WH, Suzuki E, Okawa K, Cheeseman IM, Fukagawa T (2008) CCAN makes multiple contacts with centromeric DNA to provide distinct pathways to the outer kinetochore. Cell 135(6):1039–1052. doi:10.1016/j.cell.2008.10.019
8. Hori T, Okada M, Maenaka K, Fukagawa T (2008) CENP-O class proteins form a stable complex and are required for proper kinetochore function. Mol Biol Cell 19(3):843–854. doi:10.1091/mbc.E07-06-0556
9. Amano M, Suzuki A, Hori T, Backer C, Okawa K, Cheeseman IM, Fukagawa T (2009) The CENP-S complex is essential for the stable assembly of outer kinetochore structure. J Cell Biol 186(2):173–182. doi:10.1083/jcb.200903100
10. Nishino T, Takeuchi K, Gascoigne KE, Suzuki A, Hori T, Oyama T, Morikawa K, Cheeseman IM, Fukagawa T (2012) CENP-T-W-S-X forms a unique centromeric chromatin structure with a histone-like fold. Cell 148(3):487–501. doi:10.1016/j.cell.2011.11.061
11. Petrovic A, Pasqualato S, Dube P, Krenn V, Santaguida S, Cittaro D, Monzani S, Massimiliano L, Keller J, Tarricone A, Maiolica A, Stark H, Musacchio A (2010) The MIS12 complex is a protein interaction hub for outer kinetochore assembly. J Cell Biol 190(5):835–852. doi:10.1083/jcb.201002070
12. Screpanti E, De Antoni A, Alushin GM, Petrovic A, Melis T, Nogales E, Musacchio A (2011) Direct binding of Cenp-C to the Mis12 complex joins the inner and outer kinetochore. Curr Biol 21(5):391–398. doi:10.1016/j.cub.2010.12.039
13. Schleiffer A, Maier M, Litos G, Lampert F, Hornung P, Mechtler K, Westermann S (2012) CENP-T proteins are conserved centromere receptors of the Ndc80 complex. Nat Cell Biol 14(6):604–613. doi:10.1038/ncb2493
14. Malvezzi F, Litos G, Schleiffer A, Heuck A, Mechtler K, Clausen T, Westermann S (2013) A structural basis for kinetochore recruitment of the Ndc80 complex via two distinct centromere receptors. EMBO J 32(3):409–423. doi:10.1038/emboj.2012.356
15. Nishino T, Rago F, Hori T, Tomii K, Cheeseman IM, Fukagawa T (2013) CENP-T

provides a structural platform for outer kinetochore assembly. EMBO J 32(3):424–436. doi:10.1038/emboj.2012.348
16. Hornung P, Troc P, Malvezzi F, Maier M, Demianova Z, Zimniak T, Litos G, Lampert F, Schleiffer A, Brunner M, Mechtler K, Herzog F, Marlovits TC, Westermann S (2014) A cooperative mechanism drives budding yeast kinetochore assembly downstream of CENP-A. J Cell Biol 206(4):509–524. doi:10.1083/jcb.201403081
17. Takeuchi K, Nishino T, Mayanagi K, Horikoshi N, Osakabe A, Tachiwana H, Hori T, Kurumizaka H, Fukagawa T (2014) The centromeric nucleosome-like CENP-T-W-S-X complex induces positive supercoils into DNA. Nucleic Acids Res 42(3):1644–1655. doi:10.1093/nar/gkt1124
18. Luger K, Rechsteiner TJ, Richmond TJ (1999) Expression and purification of recombinant histones and nucleosome reconstitution. Methods Mol Biol 119:1–16. doi:10.1385/1-59259-681-9:1
19. Suzuki A, Hori T, Nishino T, Usukura J, Miyagi A, Morikawa K, Fukagawa T (2011) Spindle microtubules generate tension-dependent changes in the distribution of inner kinetochore proteins. J Cell Biol 193(1):125–140. doi:10.1083/jcb.201012050

Chapter 10

Measuring Kinetochore–Microtubule Attachment Stability in Cultured Cells

Keith F. DeLuca, Jacob A. Herman, and Jennifer G. DeLuca

Abstract

Duplicated sister chromatids connect to the mitotic spindle through kinetochores, large proteinaceous structures built at sites of centromeric heterochromatin. Kinetochores are responsible for harnessing the forces generated by microtubule polymerization and depolymerization to power chromosome movements. The fidelity of chromosome segregation relies on proper kinetochore function, as precise regulation of the attachment between kinetochores and microtubules is essential to prevent mitotic errors, which are linked to the initiation and progression of cancer and the formation of birth defects (Godek et al., Nat Rev Mol Cell Biol 16(1):57–64, 2014; Ricke and van Deursen, Semin Cell Dev Biol 22(6):559–565, 2011; Holland and Cleveland, EMBO Rep 13(6):501–514, 2012). Here we describe assays to quantitatively measure kinetochore–microtubule attachment stability in cultured cells.

Key words Kinetochore, Microtubule, Spindle, Mitosis, Chromosome

1 Introduction

Kinetochores are considered the "orchestrators" of mitosis because they physically connect chromosomes to spindle microtubules and transduce forces through these connections to facilitate chromosome congression and segregation [1, 2, 4–6]. In addition, kinetochores are responsible for regulating the attachment strength to spindle microtubules, ensuring faithful and timely chromosome segregation. In early mitosis, attachments between chromosomes and the spindle are labile and kinetochore–microtubule turnover is high, which results in a cycle of continuous binding and release that prevents the accumulation of incorrect kinetochore–microtubule attachments. Conversely, in late mitosis attachments are stabilized and kinetochore–microtubule turnover is low so that forces can be generated to drive chromosome congression and segregation [1, 5–7].

Over 100 unique proteins make up the vertebrate kinetochore, and many of these are known to contribute to the generation and regulation of kinetochore–microtubule attachments [8–10].

Paul Chang and Ryoma Ohi (eds.), *The Mitotic Spindle: Methods and Protocols*, Methods in Molecular Biology, vol. 1413,
DOI 10.1007/978-1-4939-3542-0_10, © Springer Science+Business Media New York 2016

How these kinetochore proteins work together to build, maintain, and regulate this interface is not yet clear but is an active area of study. Not unexpectedly, the quantitative assessment of kinetochore–microtubule attachment stability is a critical aspect of understanding the formation and regulation of the interface and also in understanding certain cellular pathologies, including those that arise upon oncogenic transformation [2, 3, 11–13]. This chapter describes methods to quantitatively assess kinetochore–microtubule attachment stability in cultured vertebrate cells.

2 Materials

2.1 Equipment

1. Incubator for growing cells (37 °C, 5 % CO_2).
2. Laminar flow hood.
3. Water bath set to 37 °C.
4. Olympus CK2 tissue culture microscope with a 10× phase objective.
5. Hemocytometer.
6. DeltaVision Personal DV microscope (or similar imaging system) equipped with: a temperature-controlled environmental chamber, motorized *X*, *Y*, *Z* stage, independent excitation and emission filters, a quad bandpass dichroic mirror for imaging 350, 488, 568, and 647 nm fluorophores, a CoolSnap HQ2 CCD camera, and an Olympus 60×/1.42NA Planapochromat DIC oil immersion lens.

2.2 Cell Culture and Transfection

1. Cultured cells. HeLa, RPE1 (immortalized human retinal pigment epithelial cells) and PtK1 (immortalized rat kangaroo kidney epithelial cells) (*see* **Note 1**).
2. 35 mm and 6-well polystyrene tissue culture dishes (Falcon).
3. Sterile, acid-washed coverslips. Heat coverslips in a loosely covered glass beaker in 1 M HCl at 50–60 °C for 8 h. Cool to room temperature, and rinse briefly with doubly distilled water (ddH_2O). Fill beaker with ddH_2O and sonicate in water bath for 15 min. Repeat rinse and sonication step two more times. Pour off liquid, fill beaker with 50 % EtOH in ddH_2O, and sonicate in water bath for 15 min. Pour off liquid and repeat wash and sonication step with 70 % EtOH and then with 95 % EtOH. Store coverslips in 95 % EtOH. When ready to use, remove single coverslips with forceps and hold briefly over a flame before proceeding. We prepare both round #1.5 25 mm diameter coverslips (Warner) and square, #1.5 22×22 mm coverslips (VWR).
4. Glass-bottomed 35 mm dishes for live-cell imaging. Using a drill press and a #14 step drill bit, drill out a ¼ inch hole into

the bottom of a 35 mm culture dish. Use a deburring tool to smooth out the cut surface and remove shavings by blowing with pressurized air. Affix an acid-washed, round, #1.5 25 mm diameter glass coverslip to the bottom of the dish using 100 % clear silicone rubber (Aquarium sealant). Apply a small amount of silicone around the entire circumference of the drilled hole. Place the coverslip atop the silicone, invert the dish and press firmly on a flat surface. Ensure the silicone spreads out evenly, sealing up all gaps along the circumference of the hole. Allow 24 h for the silicone to cure. Fill the dish with 70 % EtOH and allow to sit for 10 min. Remove the EtOH and place under UVC light for 10 min prior to use.

5. Growth media for cells. HeLa: Dulbecco's Modified Eagle Medium (DMEM) (Gibco); RPE1: 50 % DMEM/50 % Ham's F-12 Medium (Gibco); PtK1: Ham's F-12 Medium (Gibco). Supplement each with 10 % fetal bovine serum (FBS) (Seradigm) and 1 % antibiotic/antimycotic solution (Gibco).
6. 0.25 % trypsin, stored in 2 mL aliquots at −20 °C.
7. Oligofectamine (Life Technologies). Store at 4 °C and keep the lid tightly closed when not in use.
8. Transfection media: Opti-MEM (Life Technologies) + 10 % FBS made just prior to use.

2.3 Immunofluorescence and Live-Cell Imaging

1. PBS buffer: 137 mM NaCl, 2.7 mM KCl, 10 mM Na_2HPO_4, 2 mM KH_2PO_4 (pH 7.2).
2. PHEM buffer: 60 mM PIPES, 25 mM HEPES, 10 mM EGTA, 4 mM $MgSO_4$ (pH 7.0).
3. 2× PHEM buffer: 120 mM PIPES, 50 mM HEPES, 20 mM EGTA, 8 mM $MgSO_4$ (pH 7.0).
4. PHEM-T wash buffer: Prepare PHEM + 0.1 % Triton X-100 and sonicate in water bath for 5 min. Make solution fresh on day of use.
5. Lysis solution: PBS or PHEM + 0.5 % Triton X-100 (*see* **Note 2**).
6. Paraformaldehyde (PFA) fixation solution: Prepare a 4 % PFA solution in PHEM buffer diluted from a 16 % stock solution (Electron Microscopy Science) (*see* **Note 2**).
7. Methanol fixation solution: 95 % MeOH + 5 mM EGTA. Store at −20 °C.
8. Primary antibodies: mouse monoclonal tubulin antibody (Sigma), human anti-centromere antibody (ACA) (Antibodies Inc.), mouse monoclonal Hec1 9G3 antibody (Genetex), rabbit polyclonal phospho-specific antibodies to Hec1 Ser44 and Hec1 Ser55 [30].
9. Microcystin (VWR). Store in 100 mM single-use aliquots in water at −20 °C.

10. Secondary antibodies: donkey anti-mouse, anti-human, and anti-rabbit IgG, conjugated to Alexa 488, Alexa 568, or Alexa 647 (Jackson ImmunoResearch).
11. Blocking solution: 10 % boiled donkey serum (BDS). To prepare, add 10 mL lyophilized BDS (Jackson ImmunoResearch) to 25 mL Milli-Q water and boil for 10 min. Bring the total volume to 50 mL using 2× PHEM. Add sodium azide to a final concentration of 0.05 %. Centrifuge solution in a JA-20 rotor (Beckman) for 1 h at 40,000×g. Filter supernatant through a 0.22 μm filter and store at 4 °C.
12. Thymidine (Sigma). Prepare a 10 mM stock solution in PBS and store at −20 °C.
13. Nocodazole (Tocris). Prepare a 10 mM stock solution in DMSO and store at −20 °C.
14. ZM447439 (Tocris). Prepare a 10 mM stock solution in DMSO and store at −20 °C.
15. MG132 (Tocris). Prepare a 10 mM stock solution in DMSO and store at −20 °C.
16. STLC (Tocris). Prepare a 10 mM stock solution in DMSO and store at −20 °C.
17. Microtubule destabilizing buffer: For HeLa and RPE1 cells, use standard ice-cold growth media to depolymerize microtubules. For PtK1 cells, prepare the following buffer: PHEM + 1 mM MgATP, 1 mM $CaCl_2$, 0.5 % Triton X-100 (made fresh and cooled to 4 °C on day of experiment).
18. Mounting media: 20 mM Tris (pH 8.0), 0.5 % *N*-propyl gallate, 90 % glycerol. Vortex until completely dissolved, and store protected from light at 4 °C.
19. DAPI: 4′,6-diamidino-2-phenylindole. Make a 0.5 μg/mL stock solution in PBS. Store at 4 °C protected from light for up to 1 month.
20. Humid chambers made with 150 mm diameter polystyrene tissue culture dishes lined with damp Kimwipes and Parafilm.
21. Coverslip Staining Jars (Ted Pella).

3 Methods

We typically measure kinetochore–microtubule attachment stability in response to some cellular perturbation such as depletion of a protein, depletion of a protein followed by a "rescue" (exogenous expression of a wild-type, fluorescently tagged, and/or mutant version of the protein), treatment with a pharmacological agent, or cellular transformation. Here we describe our standard lab technique for siRNA (small, interfering RNA)-mediated protein depletion paired with rescue by transient exogenous protein expression. However, in some cases, it is advantageous to carry out siRNA-mediated protein depletion in cells stably transfected with an inducible "rescue" protein of interest (*see* **Note 3**).

For analyzing kinetochore–microtubule attachment stability phenotypes, it is useful to synchronize cultures to enrich for mitotic cells. While both HeLa and RPE1 cells are responsive to a double thymidine-mediated synchronization, PtK1 cells are not, which may be an important point to consider when choosing a cell line of interest (*see* **Note 1**). Below we describe a method for carrying out silence and rescue experiments in synchronized cultures followed by multiple methods to quantitatively measure kinetochore–microtubule attachment stability.

3.1 siRNA-Mediated Silence and Rescue in Synchronized Cells

RNA interference (RNAi) is currently the most widely used method used for depletion of a protein of interest from cultured cells (*see* **Note 4**), and this method is often combined with exogenous expression of (1) a mutant version of the depleted protein to precisely dissect its function and (2) a wild-type version of the protein to determine if any observed depletion phenotypes are specific and not due to off-target effects. In both cases, it is useful to tag the exogenously expressed protein with a fluorescent protein to confirm expression and to determine subcellular localization. For these experiments, it is important to first test the exogenously expressed wild-type fusion protein in rescue assays to ensure that the tag does not impair protein function. If deleterious phenotypes are observed in depleted cells rescued with fluorescently tagged, wild-type protein, the tag can be moved to the opposite end of the protein (N- vs. C-terminus) or alternatively, a small peptide tag (e.g., HA, FLAG, myc) can be used instead and detected using an appropriate antibody.

1. Seed cells in culture dishes containing 22×22 mm, #1.5 acid-washed coverslips or live-cell dishes. Seed cells such that they are approximately 40–50 % confluent the following day. Typically, 70,000–90,000 cells per 9 cm^2 will achieve this level of confluency. Maintain cells in an incubator set to 37 °C and 5 % CO_2 for 24 h.
2. Remove media and add fresh media containing 2 mM thymidine. Incubate cells for 16 h.
3. Release the cells from the first thymidine arrest by removing media and washing with 5 mL PBS warmed to 37 °C. Repeat the PBS wash two more times and replace final wash with 900 μL Opti-MEM warmed to 37 °C.
4. For each coverslip, prepare two solutions in separate tubes. In the first, mix 50 μL Opti-MEM and 6 μL Oligofectamine (*see* **Note 5**). In the second, mix 50 μL Opti-MEM and 8 μL of a 20 μM stock solution of siRNA. If performing a rescue experiment, also add 1 μg of plasmid DNA encoding the rescue protein to the tube (*see* **Note 5**). Mix by inverting multiple

times, and incubate at room temperature for 5 min. For each test siRNA, prepare a sample containing a negative control siRNA (e.g., siRNA to luciferase, Qiagen).

5. Combine the contents of the two tubes, mix by inverting multiple times, and incubate for 20 min at room temperature. Mix by flicking every 5 min during the incubation time.
6. Add the contents of the tube (114 μL) to the culture dishes containing 900 μL of Opti-MEM.
7. Incubate cells at 37 °C, 5 % CO_2 for 2 h. After the incubation, add 100 μL FBS to each dish and incubate for 6 h.
8. Prepare 1 mL Opti-MEM + 10 % FBS containing 4 mM thymidine. Add this to the media in the culture dish to achieve a final concentration of 2 mM thymidine. Incubate at 37 °C, 5 % CO_2 for 16 h.
9. Release the cells from the second thymidine arrest by removing the media and washing with 5 mL PBS warmed to 37 °C. Repeat the PBS wash two more times, replace with 2 mL standard growth media, and incubate 9–10 h. HeLa cells begin to enter mitosis ~9 h after release, while RPE1 cells enter mitosis ~7 h after release (*see* **Note 6**). To generate a population of cells that enter mitosis in the absence of kinetochore–microtubule attachments, add nocodazole to the cell media (final concentration 10 μM) 2 h prior to anticipated mitotic entry time. To generate a population of metaphase-arrested cells, add MG132 to the cell media (final concentration 10 μM) 2 h prior to anticipated mitotic entry time.

3.2 Kinetochore-Fiber Intensity Analysis

Human kinetochores typically accumulate 12–22 "end-on" attached microtubules during mitosis to form a stable kinetochore-fiber (k-fiber) [14–16]. Electron microscopy can be used to accurately count the number of attached microtubules; however, this method is time-consuming and not typically feasible for analyzing large numbers of kinetochores from cells subjected to multiple experimental perturbations. An alternative approach for quantifying stably attached kinetochore microtubules is to measure the fluorescence intensity of k-fibers. We recently demonstrated that this technique can reliably measure very small, but significant changes in kinetochore–microtubule attachment stability [17].

1. Seed cells into wells of a 6-well dish containing acid-washed coverslips and treat as desired (e.g., silence and rescue procedure as described above).
2. On the day of the experiment, prepare and warm lysis solution (PHEM + 0.5 % Triton X-100) to 37 °C. Remove growth media from cells and gently wash with PBS + 5 mM EGTA warmed to 37 °C. Remove wash and replace with lysis solution. Incubate

for 4 min at 37 °C. Remove lysis solution and gently wash with PBS + 5 mM EGTA warmed to 37 °C.

3. Remove wash and add ice-cold 95 % MeOH + 5 mM EGTA. Incubate for 5 min at room temperature. Move cells to a −20 °C freezer and incubate for an additional 20 min. Remove cells from freezer and wash 3 × 5 min with PBS at room temperature.
4. Using forceps, remove coverslips from wells and place cell-side-up in Parafilm-lined humid chambers. Add 200 μL of blocking solution (10 % BDS) to each coverslip, cover the chamber, and incubate for 1 h at room temperature.
5. Dilute primary antibodies in 5 % BDS (in PHEM) to the following concentrations: alpha-tubulin, 1:300 and ACA (anti-centromere antibody), 1:300. Add 200 μL of solution to each coverslip and incubate for 1 h at room temperature or 10–14 h at 4 °C.
6. Remove primary antibody and place coverslips into Coverslip Staining Jars (Ted Pella). Wash 3 × 5 min by filling jar with freshly made PHEM-T wash buffer. Carry out one final rinse with 1 × PHEM.
7. Dilute secondary antibodies to 1:300 in 5 % BDS (in PHEM). For these experiments, we use Alexa 568-conjugated donkey anti-mouse and Alexa 647-conjugated donkey anti-human secondary antibodies. Remove coverslips from staining jars and place cell-side-up onto parafilm. Add secondary antibodies and incubate for 1 h at room temperature, protected from light.
8. Remove secondary antibodies and wash 5 × 3 min with PHEM-T wash buffer in Coverslip Staining Jars. Keep coverslips protected from light.
9. Wash coverslips once with PHEM, add 0.5 μg/mL DAPI solution, and incubate 2 min at room temperature. Remove DAPI solution by aspiration and rinse 5 × 2 min in PHEM-T wash buffer. Remove last wash solution and replace with PHEM. Keep coverslips protected from light.
10. Mount coverslips. Prewash standard microscope slides with 70 % EtOH and dry thoroughly using Kimwipes. Add 20 μL mounting media to each slide. Using forceps, pick up coverslips by a corner, dab off excess PHEM onto a Kimwipe and gently place the coverslip cell-side-down atop the mounting media. Use a folded Kimwipe to gently draw off excess mounting media by applying slight pressure along the edges of the coverslip. Secure the coverslip to the slide by applying nail polish along the edges.
11. Image cells using a fluorescence microscope capable of collecting z-stacks. Cells should be chosen which are not tilted (pole-to-pole axis is parallel to the coverslip surface and both spindle poles can be visualized in a single plane), and are of the appropriate phase (i.e., late prometaphase or metaphase). Acquire

z-stacks from the bottom to the top of the spindle at 0.2 μm intervals using a 60× objective with a 1.6× magnification lens inserted into the light path.

12. Open non-deconvolved, uncompressed z-stacks of imaged kinetochores and microtubules in an appropriate image analysis software program (we use SoftWorX, Applied Precision). Create a 5×5 pixel region centered on the kinetochore–microtubule contact point (Fig. 1) as well as two 1×1 pixel regions adjacent to the microtubule contact point in regions void of microtubules (these will be used to measure background fluorescence). Log the total integrated intensities of all three regions. To calculate the background fluorescence intensity, average the intensities of the two 1×1 pixel regions and multiply by 25. To calculate the corrected intensity of the k-fiber, subtract this total background fluorescence intensity from the total integrated intensity of the 5×5 k-fiber pixel region.

13. If a kinetochore protein is exogenously expressed for a specific experiment, it is important to include only cells expressing similar levels of expressed protein in final analyses. This can be done by calculating the kinetochore fluorescence intensity of the exogenously expressed proteins. For this, we use a method described by Hoffman et al. [18]. Draw a 5×5 and 7×7-pixel region (inner region and outer region, respectively) around a kinetochore and collect the fluorescence intensity of each. For

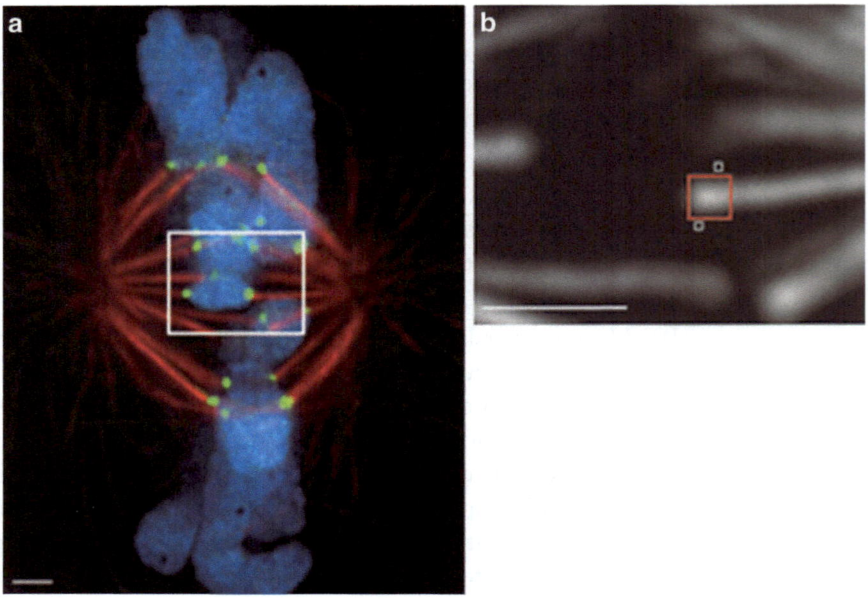

Fig. 1 Kinetochore-fiber intensity measurements. (**a**) Metaphase PtK1 cell. Shown is an overlay of tubulin (*red*), Hec1 (*green*), and DAPI (*blue*) fluorescence images from a single z-stack plane. (**b**) Zoomed view of the *boxed region* in panel (**a**) showing the tubulin channel only. The fluorescence intensity of the k-fiber at the kinetochore–microtubule interface (*red boxed region* in **b**) is recorded as well as the small 1×1 pixel-region boxes, which are used to calculate the background fluorescence. Scale bar is 1 μm

each kinetochore, make sure to use the image plane where the fluorescence intensity of the signal of interest is highest (do this by scrolling through the z-stack and identify the plane with the peak intensity). Log the integrated intensities of each region. Calculate the background fluorescence intensity by first subtracting the integrated intensity of the inner region from that of the outer region. Then scale this number appropriately by multiplying it by the area of the inner region divided by the difference between the area of the outer region and the area of the inner region. Determine the final background-corrected kinetochore fluorescence intensity by subtracting the scaled background value from the integrated fluorescence intensity of the inner region. Finally, to account for immuno-staining variabilities, measure the kinetochore fluorescence intensity of the ACA signal and calculate the ratio of the ACA signal to the test kinetochore protein signal when reporting final measurements.

3.3 Inter-Kinetochore Distance Analysis

When stable microtubule attachments are generated at each of the two opposing sister kinetochores of a kinetochore pair, pulling forces result in centromere stretching and an increase in the distance between the two kinetochores. Consequently, inter-kinetochore distances increase as cells transition from prophase to metaphase, or from completely unattached to fully attached. Thus, inter-kinetochore distance measurements are commonly used to assess kinetochore–microtubule attachment stability [19, 20] (*see* **Notes** 7 and **8**). Recently, we demonstrated that inter-kinetochore distances of bi-oriented sister kinetochore pairs decrease coordinately in response to experimentally induced weakening of kinetochore–microtubule attachment stability, and furthermore, that small decreases in attachment stability (~10 %) can be reliably detected by measuring inter-kinetochore distances [17].

1. Seed cells in wells of a 6-well plate containing acid-washed coverslips and treat as desired (e.g., silence and rescue procedure as described above).
2. On the day of the experiment, prepare and warm lysis solution (PHEM + 0.5 % Triton X-100) and fixation solution (4 % PFA in PHEM) to 37 °C.
3. Remove culture media from cells and gently wash with PHEM warmed to 37 °C. Remove PHEM by aspiration, replace with fixation solution, and incubate at 37 °C for 20 min.
4. Remove fixation solution by aspiration, and replace with lysis solution. Incubate for 4 min at 37 °C. Remove lysis solution by aspiration and wash 3 × 5 min with PHEM at room temperature.
5. Using forceps, remove coverslips from wells and place cell-side-up in Parafilm-lined humid chambers. Add 200 μL of blocking solution to each coverslip, cover the chamber, and incubate for 1 h at room temperature.

6. Aspirate off serum and add primary antibodies diluted in 5 % BDS. We typically use the 9G3 Hec1 antibody to mark kinetochores (used at a 1:3000 dilution); however, antibodies to other kinetochore proteins can also be used (*see* **Note 9**). Incubate for 1 h at room temperature or 10–14 h at 4 °C. For each 22 × 22 mm coverslip, add 200 μL primary antibody solution.
7. Complete immunostaining procedure by following **steps 6–10** in Subheading 3.2 above.
8. Image cells using a fluorescence microscope capable of collecting z-stacks. Cells should be chosen which are not tilted and are of the appropriate phase. Acquire z-stacks from the bottom to the top of the spindle at 0.2 μm intervals using a 60× objective with a 1.6× magnification lens inserted into the light path.
9. For data analysis, open non-deconvolved, uncompressed z-stacks in an appropriate image analysis software package. Identify sister kinetochore pairs that are not tilted with respect to the pole-to-pole axis and whose maximal kinetochore fluorescence intensities share a single plane. Mark the centroids of each kinetochore (*see* **Note 10**) and use the "distance" tool to measure the distance between the two centroids (Fig. 2). For a single experiment, we typically analyze at least 15 cells and measure at least ten kinetochore pairs per cell.

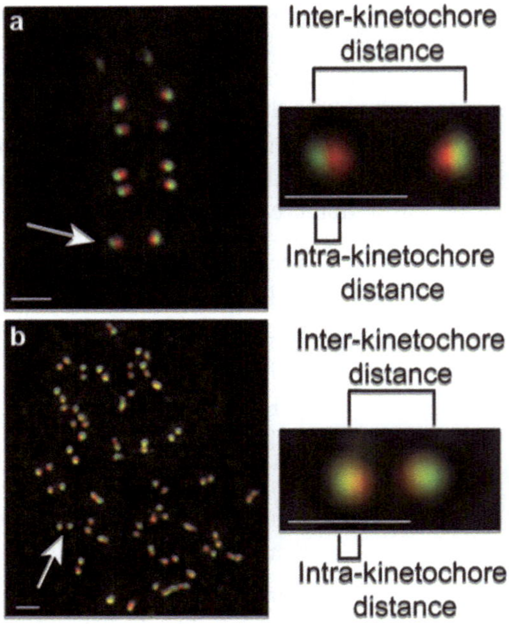

Fig. 2 Inter- and intra-kinetochore distance measurements. (**a**) Metaphase and (**b**) prophase RPE1 cell. Shown for each are overlays of Hec1 (9G3 antibody) (*green*) and CENP-A (*red*) fluorescence images from a single z-stack plane. The *arrows* point to the kinetochore pair shown in the *zoomed images* on the *right*. Scale bar in all images is 1 μm

10. To determine the "rest length" of sister kinetochores, which is the inter-kinetochore distance of completely unattached kinetochores, we carry out these measurements in prophase cells, just prior to nuclear envelope breakdown (Fig. 2), or from cells treated with 10 μM nocodazole for 2 h prior to fixation to depolymerize all microtubules.

3.4 Whole Spindle Intensity Measurements After Cold-Induced Depolymerization

Microtubule plus-ends embedded in kinetochores are more stable than free microtubule plus-ends, and therefore kinetochore microtubules are more resistant to depolymerization induced by cold temperatures, increased hydrostatic pressure, microtubule poisons, and exposure to calcium than their unattached counterparts [21–24]. Given this differential response, the stability of kinetochore microtubules can be readily assessed through a cold-induced microtubule depolymerization assay, in which cells are incubated in ice-cold media or buffer to depolymerize the majority of non-kinetochore microtubules (Fig. 3). The remaining cold-stable polymer is quantified to determine, on a whole-cell basis, the level of kinetochore–microtubule stability.

1. Seed HeLa or RPE1 cells in wells of a 6-well dish containing acid-washed coverslips and treat as desired (e.g., silence and rescue procedure as described above). *See* **Note 11** for PtK1-cell specific protocol modifications.
2. On the day of the experiment, prepare lysis solution (PHEM + 0.5 % Triton X-100) and fixation solution (4 % PFA in PHEM) and keep at room temperature.
3. Remove culture media from cells by aspiration, replace with culture media cooled to 4 °C, and place dishes in a shallow ice-bath. Incubate for 10 min (*see* **Note 12**).
4. Remove cells from ice-bath, aspirate off cold media, and perform a quick wash with PHEM at room temperature.

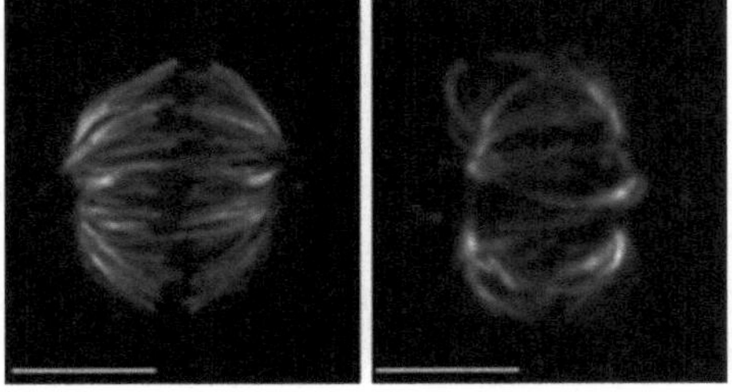

Fig. 3 Cold-induced microtubule depolymerization assay. An RPE1 cell (*left*) and an RPE1 cell transformed with oncogenic RAS (*right*) were subjected to a cold-induced depolymerization assay. The kinetochore–microtubule attachments in the transformed cell are weaker, and therefore less microtubule polymer remains after the procedure. Scale bar is 5 μm

5. Remove PHEM, replace with lysis buffer, and incubate for 4 min at room temperature.
6. Remove lysis solution, replace with 4 % PFA, and incubate for 20 min at room temperature. Wash 3×5 min using PHEM, remove coverslips from wells, and place in Parafilm-lined humid chambers. Add 200 μL of blocking solution to each coverslip, cover the chamber, and incubate for 1 h at room temperature.
7. Remove serum by aspiration and add 200 μL of primary antibody solution (α-tubulin antibody diluted 1:300 in 5 % BDS + PHEM). Incubate 10–14 h at 4 °C.
8. Complete staining procedure by following **steps 6–10** in Subheading 3.2.
9. Image cells using a fluorescence microscope capable of collecting z-stacks. Ideally, cells should be chosen which are not tilted and are of the appropriate phase. For each cell, acquire a z-stack containing a total of 20 image planes, 0.2 μm apart. Make sure to include ten images below and ten images above the middle of the cell, at which point both spindle poles are in focus.
10. Prior to data analysis, ensure that the cold-induced microtubule depolymerization was successful. Typically, we image multiple interphase cells (which lack kinetochore microtubules) to confirm that the majority of microtubules are depolymerized (*see* **Note 12**). For data analysis, using SoftWorx imaging software, open a z-stack of non-deconvolved images, compress the stack, and sum the total tubulin fluorescence. Draw a rectangle (rectangle 1) around the entire spindle, making sure to include the spindle poles. Log the total integrated intensity and area of the rectangle. To calculate background fluorescence, draw multiple 10×10 pixel regions around the outside of the spindle, log the total integrated intensities of each, and calculate the average integrated intensity per pixel. Multiply the average integrated intensity per pixel value by the area of rectangle 1. To obtain the background-corrected spindle intensity, subtract the calculated background value from the total integrated intensity of rectangle 1.

3.5 Kinetochore Fluorescence Intensity Quantification of Hec1 Phosphorylation

Kinetochore–microtubule attachments are regulated in large part by the essential mitotic kinase, Aurora B [25–29]. Arguably the most critical Aurora B substrate involved in controlling kinetochore–microtubule attachment stability is Hec1/Ndc80, whose tail domain is phosphorylated at multiple sites by Aurora B kinase during mitosis [30]. We generated antibodies to phosphorylated serine residues within the Hec1 tail and demonstrated that during normal mitotic progression, Hec1 phosphorylation on Ser44 and Ser55 is high on unattached kinetochore and decreases as stable, end-on kinetochore–microtubule attachments are formed [30]. The kinetochore fluorescence intensity of Hec1 phosphorylation can be used as a readout for kinetochore Aurora B activity and thus kinetochore–MT attachment stability [31].

1. Seed cells in wells of a 6-well plate containing acid-washed coverslips and treat as desired.
2. On the day of the experiment, prepare lysis solution: PHEM + 1 % Triton X-100 + 100 nM microcystin (phosphatase inhibitor) and fixation solution: 4 % PFA in PHEM. Warm each to 37 °C.
3. Remove culture media from cells and gently wash with PHEM warmed to 37 °C. Remove PHEM by aspiration and replace with lysis solution. Incubate for 4 min at 37 °C (*see* **Note 13**).
4. Replace lysis solution with fixation solution and incubate at 37 °C for 20 min. Remove fixation solution by aspiration and wash 5 × 3 min with PHEM at room temperature.
5. Follow the procedure described above in Subheading 3.2, **steps 4–10** for immunostaining except use the following primary antibodies: phospho-Hec1 (pSer44 or pSer55) at 1:3000, Hec1 9G3 at 1:3000, and ACA at 1:300.
6. Image cells as described in Subheading 3.2, **step 11**.
7. Measure the kinetochore fluorescence intensities of phospho-Hec1 (pSer44 or pSer55 antibody channel) and total Hec1 (9G3 antibody channel) as described in Subheading 3.2, **step 13**. Calculate the ratio of phosphoHec1 to total Hec1 to determine if an experimental perturbation affects the phosphorylation status of Hec1, total Hec1 levels at kinetochores, or both.

3.6 Kinetochore–Microtubule Error Correction Efficiency Assay

In early mitosis, cells typically generate multiple incorrect kinetochore–microtubule attachments including those that are syntelic (both sister kinetochores attached to microtubules from a single pole) and merotelic (one kinetochore attached to microtubules from both poles) [2, 32, 33]. To ensure that erroneous attachments do not accumulate during mitosis and persist through anaphase, kinetochore–microtubule turnover in early mitosis is high, which results in release of incorrect attachments and allows kinetochores to "reset" and form new, correct attachments. This phenomenon, referred to as error correction, is controlled by Aurora B kinase, which phosphorylates kinetochore substrates to promote kinetochore–microtubule release and prevent the premature stabilization of erroneous attachments The stability of kinetochore–microtubule attachments affects the efficiency of this error correction process such that hyper-stable attachments antagonize the correction process and weak attachments promote it [34–36]. To quantify error correction efficiency, we use the small molecule STLC to inhibit the kinesin-5 microtubule motor, Eg5, which is required for spindle pole separation in early mitosis. Inhibition of Eg5 activity leads to the formation of monopolar spindles and results in the generation and accumulation of many syntelically attached kinetochore pairs [37–39]. Upon STLC wash-out, the motor activity of Eg5 is reactivated, and the two spindle poles separate at a rate limited by the kinetics of kinetochore–microtubule release [35, 36]. Thus, we use the rate of spindle pole separation in living cells to determine the efficiency of kinetochore–microtubule error correction [17]. *See* **Note 14** for an alternative fixed-cell protocol.

1. Seed cells in 35 mm glass-bottomed, live-cell dishes and treat as desired. Use cells that are either stably or transiently expressing a fluorescently labeled spindle pole marker (e.g., γ-tubulin) (*See* **Note 15**).
2. On the day of the experiment, remove growth media from cells and replace with media containing 10 μM STLC warmed to 37 °C (*See* **Note 16**). Incubate cells for 2 h at 37 °C.
3. Remove STLC-containing media and perform four quick washes with standard growth media warmed to 37 °C. Replace final wash with pre-warmed filming media.
4. Quickly place culture dish in the heated chamber (or heated slide holder) of a microscope with a high precision *x*, *y*, *z* stage. Identify and mark the positions of multiple monopolar cells, identified by DIC imaging of chromosomes and fluorescence imaging of the pole maker.
5. Time-lapse image cells every 60 s for a total of 80 min, using a 60×/1.42NA Planapochromat DIC oil immersion lens. Collect both DIC and fluorescence images.
6. Using an appropriate image analysis software, measure the distance between the spindle poles at each time point. Plot the distance and time in *x* and *y* to generate a curve for each cell measured. The distance between the two spindle poles typically increases linearly and then reaches a plateau (Fig. 4). Using the linear portion of each graph, calculate the slope of the line to determine the rate of spindle pole separation (for control cells, this includes the first 6–10 time points). For each experimental condition tested, we average the separation rates of at least 15 cells.
7. In addition to the test samples, perform a control experiment in which cells are released from the STLC incubation into media containing the Aurora B kinase inhibitor ZM447439, which stabilizes kinetochore–microtubule attachments, prevents efficient error correction, and results in decreased rates of spindle pole separation [17, 30, 34, 36].

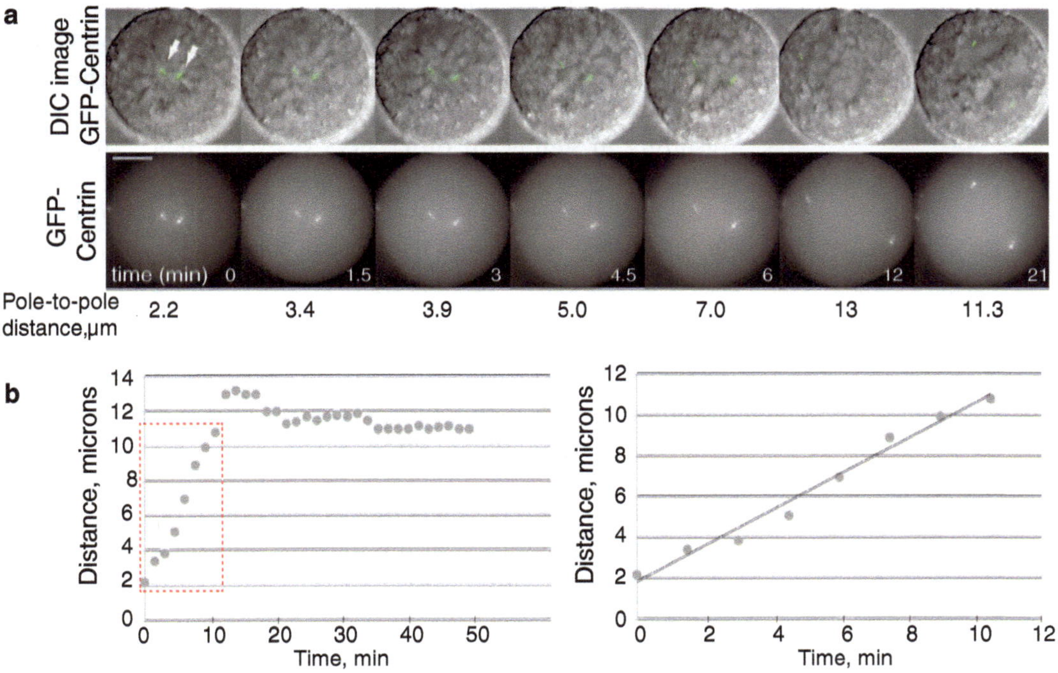

Fig. 4 Kinetochore–microtubule error correction assay. (**a**) Time-lapse image series of a HeLa cell treated with STLC for 2 h and released into fresh, drug-free media at time 0. The *top row* shows DIC images overlaid with the centrin-GFP images. White arrows point to the two centrin spots. The *bottom row* shows only the centrin-GFP channel. Time is in minutes. (**b**) Quantification of spindle pole separation rate. The *boxed region* in the graph on the *left* is expanded in the graph on the *right*

4 Concluding Remarks

In this Chapter, we have described basic methods used to quantify kinetochore–microtubule attachment stability in cultured cells. Typically, k-fiber turnover increases with decreasing kinetochore–microtubule attachment stability [40–42], and similarly, kinetochore oscillations increase with decreasing attachment stability [17, 30]. However, it is important to note that changes in both k-fiber turnover and kinetochore oscillations can result from changes in kinetochore–microtubule dynamics as well as kinetochore–microtubule stability.

5 Notes

1. There are many issues to consider when choosing cell lines for study. HeLa cells (a human cell line derived from cancerous cervical tissue) are widely used and have several advantages. HeLa cells are immortalized, they grow rapidly even under sub-optimal culture conditions, and they are easily transfected.

In addition, HeLa cells are historically the most common cell line used for studying mitosis and kinetochores [43–45]. One significant disadvantage to using HeLa cells is that they are a cancer cell line and have undergone a host of genetic changes resulting in significant differences between their genome and the genome of normal, non-cancerous human cells [46]. Not unexpectedly, cancer cells differ phenotypically from healthy, non-transformed cells in many aspects [47–51], and therefore results obtained in HeLa cells may not be analogous to results obtained in non-cancer cells. Because of this issue, the use of non-transformed cells in experiments either in place of, or in addition to, HeLa cells is advised. A non-transformed cell line commonly used by those in the mitosis field is RPE1, which is a human retinal pigment epithelial cell line immortalized with telomerase [11, 51–54].

Most human cells, whether transformed or not, tend to round up significantly during mitosis. This property, together with their large number of small chromosomes (46 chromosomes in RPE1 cells and on average between 70 and 100 in most HeLa cell lines), renders high resolution imaging of the kinetochore–microtubule attachment interface difficult. Because of this, we commonly use PtK1 cells, which are immortalized rat kangaroo kidney epithelial cells [55–57], to study the kinetochore–microtubule interface. In addition to only having 12 large chromosomes, these cells remain relatively flat during mitosis, which is advantageous for microscopic imaging. Two downsides of using PtK1 cells for kinetochore studies are (1) the genome has not been sequenced, and therefore to study a PtK1 protein of interest, the encoding gene must be identified, sequenced, and cloned [58, 59], and (2) PtK1 cell cultures are not readily synchronized.

2. The lysis solution, fixation solution, and incubation conditions should be optimized for each new antibody (or combination of antibodies) tested. Typically we test at least three concentrations of Triton X-100 for the lysis solution: 0.5, 1, and 2 %. For formaldehyde-based fixation solutions, we begin by testing at least two concentrations of PFA: 2 and 4 %. In addition, for each new antibody tested, we determine if optimal staining is achieved by lysing prior to fixing, or vice versa.

3. This chapter describes a "rescue" method in which cells are simultaneously transfected with siRNA to deplete a protein of interest and a DNA plasmid encoding a modified version of the depleted protein. This method is highly effective for many mitotic proteins; however, in some instances we have found it advantageous to generate stable cell lines that can be induced to express a gene of interest from a specific genomic location. For instance, in our hands the kinetochore protein KNL1 has

proven difficult to reproducibly transfect and express in cultured cells. We therefore generated cell lines stably expressing the wild-type protein fused to GFP as well as a variety of GFP-tagged mutant versions whose expression can be induced with the addition of doxycycline using the HeLa Flp-In T-REx system (Life Technologies) [31].

4. Currently, the most common method for protein depletion from cultured cells is RNA interference (RNAi). One disadvantage to this technique is that the level of protein depletion is variable, and in most cases, a complete knock-out is not achieved. There are technologies now available that offer complete gene knock-out through genome editing, and current tools include Zinc-finger nucleases (ZFNs), transcription activator-like effector nucleases (TALENs), and the clustered, regularly interspaced, short palindromic repeat (CRISPR)/Cas endonuclease system [60–64]. The growing commercialization of these tools will undoubtedly provide increased access to a wide array of research groups, and the use of these approaches will likely become routine for the study of mitotic proteins.

5. We find that Oligofectamine works well for siRNA-mediated depletion of many mitotic components, but not all, and therefore it is sometimes necessary to troubleshoot this aspect of the protocol. Other transfection reagents that we use in our lab include Effectene (Qiagen) and Lipofectamine 2000 (Invitrogen). It is also important to consider the turnover of the targeted protein of interest when carrying out RNAi. Many mitotic factors are destroyed at the end of the mitosis and resynthesized during the following cell cycle, which typically results in efficient depletion after 24 h. However, if a protein is long-lived and turnover is low, longer siRNA incubation times may be necessary. Finally, it may be necessary to troubleshoot the amount of plasmid DNA added for rescue experiments. When working with a new plasmid, we initially test a range between 0.5 and 2 μg in order to identify the amount of DNA required to rescue a given depletion phenotype.

6. For these assays, it is important to establish mitotic entry and exit times after the second thymidine release. When studying a new cell line, we initially fix cells at various time points post-release (e.g., $t=6$, 7, 8, 9, and 10 h), stain the chromosomes with DAPI, and image the cells to determine average mitotic entry and exit times so that the appropriate time-point for fixation is used for a given experiment.

7. While inter-kinetochore distances increase from prophase to metaphase as stable attachments are formed, it is important to note that inter-kinetochore distances also increase in response to loss of sister chromatid cohesion function [65–67]. If defects in sister chromatid cohesion are suspected, it is

important to confirm this by carrying out additional assays, such as preparing chromosome spreads and scoring sister chromatid separation [68, 69].

8. Upon formation of stable kinetochore–microtubule attachments, individual kinetochores undergo architectural changes, as evidenced by an increased distance between the inner and outer kinetochore. This increase in *intra*-kinetochore distance is possibly correlated to spindle assembly checkpoint silencing and can also be used as a measure of kinetochore–microtubule attachment stability [70, 71]. To measure intra-kinetochore distances, we typically immuno-stain cells with antibodies to CENP-A (an inner kinetochore component) and Hec1 (an outer kinetochore component) (Fig. 2). We then measure the distance between the two CENP-A puncta in a sister kinetochore pair, subtract this from the distance between the two Hec1 puncta and divide this value by 2 to obtain the *intra*-kinetochore distance [72].
9. Almost any kinetochore protein can be used to measure inter-kinetochore distances, but it is important to consider that the distances will vary depending on the kinetochore marker used. For example the distance between Hec1 puncta at metaphase is ~150–200 nm greater than distance between CENP-A puncta [72, 73].
10. When kinetochore centroids are determined manually ("by eye"), we enlarge the image to 800× and use pseudo-color display to localize the centroid by determining the pixel(s) with the highest intensity. Alternatively, various software packages in which kinetochore centroids are determined from Gaussian distributions can also be used.
11. For PtK1 cells, we use the following microtubule depolymerization buffer: PHEM + 1 mM MgATP, 1 mM $CaCl_2$, 0.5 % Triton X-100, pH 7.0. Cells are incubated in ice-cold depolymerization buffer in a shallow ice-bath for 30 min.
12. The degree of cold-stable microtubule polymer remaining after incubation in ice-cold depolymerization buffer depends on the incubation time. The procedure is intended to depolymerize all non-kinetochore microtubules, and therefore the incubation time can be estimated by determining the time required to depolymerize all microtubules in interphase cells, which lack kinetochores. It is important to note that even after prolonged incubation in cold depolymerization buffer, several thick bundles of cold-stable fibers typically remain in most interphase cells.
13. Most cultured cells tend to round up during mitosis which can result in significant cell loss from coverslips. If cell loss due to rounding is a problem, a "quick-fix" can be carried out, in

which the growth media is removed and a 4 % PFA fixation solution warmed to 37 °C is added to the cells for 5–10 s prior to the first PHEM wash.

14. As an alternative to the live-cell protocol described, a fixed-cell method for assessing kinetochore–microtubule error correction efficiency can be carried out [30, 35]. In this case, cells are treated with STLC as described and fixed at various time-points post-STLC washout. Cells are immunostained using tubulin and ACA antibodies, imaged, and scored based on spindle phenotype: monopolar, bipolar prometaphase, or bipolar metaphase. Control, untreated cells typically form bipolar spindles and metaphase plates within 60 min of STLC washout.
15. Cells transfected with fluorescently labeled α or β tubulin can also be used to score spindle pole separation. In this case, the pole-to-pole distance is measured using the center of each tubulin aster. Additionally, cells expressing labeled H2B can be helpful in discerning chromosome alignment.
16. Some cell types fail to release efficiently from a 10 μM STLC arrest and persist with monopolar spindles (e.g., RPE1 cells). In this case, 100 μM monastrol can be used instead.

References

1. Godek KM, Kabeche L, Compton DA (2014) Regulation of kinetochore-microtubule attachments through homeostatic control during mitosis. Nat Rev Mol Cell Biol 16(1):57–64
2. Ricke RM, van Deursen JM (2011) Correction of microtubule-kinetochore attachment errors: mechanisms and role in tumor suppression. Semin Cell Dev Biol 22(6):559–565
3. Holland AJ, Cleveland DW (2012) Losing balance: the origin and impact of aneuploidy in cancer. EMBO Rep 13(6):501–514
4. Sarangapani KK, Asbury CL (2014) Catch and release: how do kinetochores hook the right microtubules during mitosis? Trends Genet 30(4):150–159
5. Foley EA, Kapoor TM (2013) Microtubule attachment and spindle assembly checkpoint signalling at the kinetochore. Nat Rev Mol Cell Biol 14(1):25–37
6. Caldas GV, DeLuca JG (2014) KNL1: bringing order to the kinetochore. Chromosoma 23:169–181
7. Sacristan C, Kops GJ (2014) Joined at the hip: kinetochores, microtubules, and spindle assembly checkpoint signaling. Trends Cell Biol 25(1):21–28
8. Santaguida S, Musacchio A (2009) The life and miracles of kinetochores. EMBO J 28(17):2511–2531
9. Fukagawa T, Earnshaw WC (2014) The centromere: chromatin foundation for the kinetochore machinery. Dev Cell 30(5):496–508
10. Hori T, Fukagawa T (2012) Establishment of the vertebrate kinetochores. Chromosome Res 20(5):547–561
11. Ding Y, Hubert CG, Herman J, Corrin P, Toledo CM, Skutt-Kakaria K, Vazquez J, Basom R, Zhang B, Risler JK, Pollard SM, Nam DH, Delrow JJ, Zhu J, Lee J, DeLuca J, Olson JM, Paddison PJ (2012) Cancer specific requirement for BUB1B/BubR1 in human brain tumor isolates and genetically transformed cells. Cancer Discov 3(2):198–211
12. Bakhoum SF, Thompson SL, Manning AL, Compton DA (2009) Genome stability is ensured by temporal control of kinetochore-microtubule dynamics. Nat Cell Biol 11(1):27–35
13. Ertych N, Stolz A, Stenzinger A, Weichert W, Kaulfuß S, Burfeind P, Aigner A, Wordeman L, Bastians H (2014) Increased microtubule assembly rates influence chromosomal instability in colorectal cancer cells. Nat Cell Biol 16(8):779–791
14. Wendell KL, Wilson L, Jordan MA (1993) Mitotic block in HeLa cells by vinblastine: ultrastructural changes in kinetochore-microtubule attachment and in centrosomes. J Cell Sci 104(Pt 2):261–274

15. McEwen BF, Chan GK, Zubrowski B, Savoian MS, Sauer MT, Yen TJ (2001) CENP-E is essential for reliable bioriented spindle attachment, but chromosome alignment can be achieved via redundant mechanisms in mammalian cells. Mol Biol Cell 12(9):2776–2789
16. DeLuca JG, Dong Y, Hergert P, Strauss J, Hickey JM, Salmon ED, McEwen BF (2005) Hec1 and nuf2 are core components of the kinetochore outer plate essential for organizing microtubule attachment sites. Mol Biol Cell 16(2):519–531
17. Zaytsev AV, Sundin LJR, DeLuca KF, Grishchuk EL, DeLuca JG (2014) Accurate phosphoregulation of kinetochore-microtubule affinity requires unconstrained molecular interactions. J Cell Biol 206:45–59
18. Hoffman DB, Pearson CG, Yen TJ, Howell BJ, Salmon ED (2001) Microtubule-dependent changes in assembly of microtubule motor proteins and mitotic spindle checkpoint proteins at PtK1 kinetochores. Mol Biol Cell 12(7):1995–2009
19. Waters JC, Chen RH, Murray AW, Salmon ED (1998) Localization of Mad2 to kinetochores depends on microtubule attachment, not tension. J Cell Biol 141(5):1181–1191
20. DeLuca JG, Moree B, Hickey JM, Kilmartin JV, Salmon ED (2002) hNuf2 inhibition blocks stable kinetochore-microtubule attachment and induces mitotic cell death in HeLa cells. J Cell Biol 159(4):549–555
21. Brinkley BR, Cartwright J Jr (1975) Cold-labile and cold-stable microtubules in the mitotic spindle of mammalian cells. Ann N Y Acad Sci 253:428–439
22. Salmon ED, Goode D, Maugel TK, Bonar DB (1976) Pressure-induced depolymerization of spindle microtubules. III. Differential stability in HeLa cells. J Cell Biol 69(2):443–454
23. Mitchison T, Evans L, Schulze E, Kirschner M (1986) Sites of microtubule assembly and disassembly in the mitotic spindle. Cell 45(4):515–527
24. Cassimeris L, Rieder CL, Rupp G, Salmon ED (1990) Stability of microtubule attachment to metaphase kinetochores in PtK1 cells. J Cell Sci 96(Pt 1):9–15
25. Biggins S, Severin FF, Bhalla N, Sassoon I, Hyman AA, Murray AW (1999) The conserved protein kinase Ipl1 regulates microtubule binding to kinetochores in budding yeast. Genes Dev 13(5):532–544
26. Lampson MA, Cheeseman IM (2011) Sensing centromere tension: aurora B and the regulation of kinetochore function. Trends Cell Biol 21(3):133–140
27. Pinsky BA, Kung C, Shokat KM, Biggins S (2006) The Ipl1-Aurora protein kinase activates the spindle checkpoint by creating unattached kinetochores. Nat Cell Biol 8(1):78–83
28. Carmena M, Wheelock M, Funabiki H, Earnshaw WC (2012) The chromosomal passenger complex (CPC): from easy rider to the godfather of mitosis. Nat Rev Mol Cell Biol 13(12):789–803
29. Funabiki H, Wynne DJ (2013) Making an effective switch at the kinetochore by phosphorylation and dephosphorylation. Chromosoma 122(3):135–158
30. DeLuca KF, Lens SM, DeLuca JG (2011) Temporal changes in Hec1 phosphorylation control kinetochore-microtubule attachment stability during mitosis. J Cell Sci 124(Pt 4):622–634
31. Caldas GV, DeLuca KF, DeLuca JG (2014) KNL1 facilitates phosphorylation of outer kinetochore proteins by promoting Aurora B kinase activity. J Cell Biol 203(6):957–969
32. Salmon ED, Cimini D, Cameron LA, DeLuca JG (2005) Merotelic kinetochores in mammalian tissue cells. Philos Trans R Soc Lond B Biol Sci 360(1455):553–568
33. Gregan J, Polakova S, Zhang L, Tolić-Nørrelykke IM, Cimini D (2011) Merotelic kinetochore attachment: causes and effects. Trends Cell Biol 21(6):374–381
34. Lampson MA, Renduchitala K, Khodjakov A, Kapoor TM (2004) Correcting improper chromosome-spindle attachments during cell division. Nat Cell Biol 6(3):232–237
35. Vader G, Cruijsen CW, van Harn T, Vromans MJ, Medema RH, Lens SM (2007) The chromosomal passenger complex controls spindle checkpoint function independent from its role in correcting microtubule kinetochore interactions. Mol Biol Cell 18(11):4553–4564
36. Sundin LJ, Guimaraes GJ, Deluca JG (2011) The NDC80 complex proteins Nuf2 and Hec1 make distinct contributions to kinetochore-microtubule attachment in mitosis. Mol Biol Cell 22(6):759–768
37. Mayer TU, Kapoor TM, Haggarty SJ, King RW, Schreiber SL, Mitchison TJ (1999) Small molecule inhibitor of mitotic spindle bipolarity identified in a phenotype-based screen. Science 286(5441):971–974
38. Kapoor TM, Mayer TU, Coughlin ML, Mitchison TJ (2000) Probing spindle assembly mechanisms with monastrol, a small molecule inhibitor of the mitotic kinesin, Eg5. J Cell Biol 150(5):975–988
39. Skoufias DA, DeBonis S, Saoudi Y, Lebeau L, Crevel I, Cross R, Wade RH, Hackney D, Kozielski F (2006) S-trityl-L-cysteine is a reversible, tight binding inhibitor of the human kinesin Eg5 that specifically blocks mitotic progression. J Biol Chem 281(26):17559–17569
40. Kabeche L, Compton DA (2013) Cyclin A regulates kinetochore microtubules to promote faithful chromosome segregation. Nature 502(7469):110–113

41. DeLuca JG, Gall WE, Ciferri C, Cimini D, Musacchio A, Salmon ED (2006) Kinetochore microtubule dynamics and attachment stability are regulated by Hec1. Cell 127(5):969–982
42. Cimini D, Wan X, Hirel CB, Salmon ED (2006) Aurora kinase promotes turnover of kinetochore microtubules to reduce chromosome segregation errors. Curr Biol 16(17): 1711–1718
43. Hsu TC (1954) Cytological studies on HeLa, a strain of human cervical carcinoma, I. Observations on mitosis and chromosomes. Tex Rep Biol Med 12(4):833–846
44. Moses MJ, Counce SJ (1974) Electron microscopy of kinetochores in whole mount spreads of mitotic chromosomes from hela cells. J Exp Zool 189(1):115–120
45. Brinkley BR, Rao PN (1973) Nitrous oxide: effects on the mitotic apparatus and chromosome movement in HeLa cells. J Cell Biol 58(1):96–106
46. Landry JJ, Pyl PT, Rausch T, Zichner T, Tekkedil MM, Stütz AM, Jauch A, Aiyar RS, Pau G, Delhomme N, Gagneur J, Korbel JO, Huber W, Steinmetz LM (2013) The genomic and transcriptomic landscape of a HeLa cell line. G3 (Bethesda) 3(8):1213–1224
47. Orr B, Compton DA (2013) A double-edged sword: how oncogenes and tumor suppressor genes can contribute to chromosomal instability. Front Oncol 3:164
48. Luo J, Emanuele MJ, Li D, Creighton CJ, Schlabach MR, Westbrook TF, Wong KK, Elledge SJ (2009) A genome-wide RNAi screen identifies multiple synthetic lethal interactions with the Ras oncogene. Cell 137(5):835–848
49. Herman JA, Toledo CM, Olson JM, DeLuca JG, Paddison PJ (2015) Molecular pathways: regulation and targeting of kinetochore-microtubule attachment in cancer. Clin Cancer Res 21:233–239
50. Thompson SL, Bakhoum SF, Compton DA (2010) Mechanisms of chromosomal instability. Curr Biol 20(6):R285–R295
51. Salimian KJ, Ballister ER, Smoak EM, Wood S, Panchenko T, Lampson MA, Black BE (2011) Feedback control in sensing chromosome biorientation by the Aurora B kinase. Curr Biol 21(13):1158–1165
52. Jiang XR, Jimenez G, Chang E, Frolkis M, Kusler B, Sage M, Beeche M, Bodnar AG, Wahl GM, Tlsty TD, Chiu CP (1999) Telomerase expression in human somatic cells does not induce changes associated with a transformed phenotype. Nat Genet 21(1):111–114
53. Mikhailov A, Cole RW, Rieder CL (2002) DNA damage during mitosis in human cells delays the metaphase/anaphase transition via the spindle-assembly checkpoint. Curr Biol 12(21):1797–1806
54. Uetake Y, Sluder G (2004) ll cycle progression after cleavage failure: mammalian somatic cells do not possess a "tetraploidy checkpoint". J Cell Biol 165(5):609–615
55. Walen KH, Brown SW (1962) Chromosomes in a marsupial (Potorous tridactylis) tissue culture. Nature 194:406
56. Levan A, Nichols WW, Peluse M, Coriell LL (1966) The stemline chromosomes of three cell lines representing different vertebrate classes. Chromosoma 18(2):343–358
57. Levan G (1970) Contributions to the chromosomal characterization of the PTK 1 rat-kangaroo cell line. Hereditas 64(1):85–96
58. Stout JR, Rizk RS, Kline SL, Walczak CE (2006) Deciphering protein function during mitosis in PtK cells using RNAi. BMC Cell Biol 23:7–26
59. Guimaraes GJ, Dong Y, McEwen BF, Deluca JG (2008) Kinetochore-microtubule attachment relies on the disordered N-terminal tail domain of Hec1. Curr Biol 18(22):1778–1784
60. Gaj T, Gersbach CA, Barbas CF 3rd (2013) ZFN, TALEN, and CRISPR/Cas-based methods for genome engineering. Trends Biotechnol 31(7):397–405
61. Carroll D (2011) Genome engineering with zinc-finger nucleases. Genetics 188(4):773–782
62. Joung JK, Sander JD (2013) TALENs: a widely applicable technology for targeted genome editing. Nat Rev Mol Cell Biol 14(1):49–55
63. Doudna JA, Charpentier E (2014) Genome editing. The new frontier of genome engineering withCRISPR-Cas9.Science346(6213):1258096
64. Song M, Kim YH, Kim JS, Kim H (2014) Genome engineering in human cells. Methods Enzymol 546:93–118
65. Kitajima TS, Hauf S, Ohsugi M, Yamamoto T, Watanabe Y (2005) Human Bub1 defines the persistent cohesion site along the mitotic chromosome by affecting Shugoshin localization. Curr Biol 15(4):353–359
66. Ritchie K, Seah C, Moulin J, Isaac C, Dick F, Bérubé NG (2008) Loss of ATRX leads to chromosome cohesion and congression defects. J Cell Biol 180(2):315–324
67. Daum JR, Potapova TA, Sivakumar S, Daniel JJ, Flynn JN, Rankin S, Gorbsky GJ (2011) Cohesion fatigue induces chromatid separation in cells delayed at metaphase. Curr Biol 21(12):1018–1024
68. Sonoda E, Matsusaka T, Morrison C, Vagnarelli P, Hoshi O, Ushiki T, Nojima K, Fukagawa T, Waizenegger IC, Peters JM, Earnshaw WC, Takeda S (2001) Scc1/Rad21/Mcd1 is required for sister chromatid cohesion and kinetochore function in vertebrate cells. Dev Cell 1(6):759–770
69. Tang Z, Sun Y, Harley SE, Zou H, Yu H (2004) Human Bub1 protects centromeric sister-chromatid cohesion through Shugoshin

during mitosis. Proc Natl Acad Sci U S A 101(52):18012–18017

70. Maresca TJ, Salmon ED (2009) Intrakinetochore stretch is associated with changes in kinetochore phosphorylation and spindle assembly checkpoint activity. J Cell Biol 184(3):373–381
71. Uchida KS, Takagaki K, Kumada K, Hirayama Y, Noda T, Hirota T (2009) Kinetochore stretching inactivates the spindle assembly checkpoint. J Cell Biol 184(3):383–390
72. Suzuki A, Badger BL, Wan X, DeLuca JG, Salmon ED (2014) The architecture of CCAN proteins creates a structural integrity to resist spindle forces and achieve proper Intrakinetochore stretch. Dev Cell 30(6):717–730
73. Wan X, O'Quinn RP, Pierce HL, Joglekar AP, Gall WE, DeLuca JG, Carroll CW, Liu ST, Yen TJ, McEwen BF, Stukenberg PT, Desai A, Salmon ED (2009) Protein architecture of the human kinetochore microtubule attachment site. Cell 137(4):672–684

Chapter 11

Studying Kinetochores In Vivo Using FLIM-FRET

Tae Yeon Yoo and Daniel J. Needleman

Abstract

Kinetochores play essential roles in coordinating mitosis, as a mechanical connector between chromosome and microtubule and as a source of numerous biochemical signals. These mechanical and biochemical behaviors of kinetochores change dynamically in cells during mitosis. Therefore, understanding kinetochore function requires an imaging tool that quantifies the protein–protein interactions or biochemical changes with high spatiotemporal resolution. FRET has previously been used in combination with biosensors to probe protein–protein interactions and biochemical activity. In this chapter, we introduce FLIM-FRET, a lifetime-based method that quantifies FRET, and describe the use of FLIM-FRET as a method for studying dynamic kinetochore behavior in cells with high spatiotemporal resolution.

Key words FLIM-FRET, Kinetochore, Bayesian analysis, Live cell imaging

1 Introduction

During mitosis, normally functioning eukaryotic cells rapidly and accurately segregate their chromosomes. Inaccurate chromosome segregation gives rise to aneuploidy, which can lead to cancer and death [1, 2]. The proper attachment of microtubules to chromosomes is necessary for accurate chromosome segregation [3, 4]. The interaction between chromosomes and microtubules is predominately mediated by the kinetochore, a highly complex protein structure located at the centromeric region of each chromatid. The main functions of kinetochores can be divided into two parts: attachment and signaling. The kinetochore can attach to microtubules even as the microtubules dynamically switch between growth and shrinking phases. These attachments can withstand significant tension, and therefore provide the physical linkage and coupling between microtubules and chromosomes [5, 6]. The kinetochore is also a source and target of a myriad of regulatory biochemical signals required for the correction of erroneous kinetochore–microtubule attachments and the spindle assembly checkpoint [7–9]. Though both of these main functions of kinetochores are very

Paul Chang and Ryoma Ohi (eds.), *The Mitotic Spindle: Methods and Protocols*, Methods in Molecular Biology, vol. 1413, DOI 10.1007/978-1-4939-3542-0_11,

important and have been studied extensively, many aspects of their function and behavior remain poorly understood. One difficulty is a lack of methods for measuring biochemical activities at kinetochore with high temporal and spatial resolution.

Förster Resonance Energy Transfer (FRET) has been widely used as a probe for protein–protein interactions and protein activity in vivo. FRET is the process in which a donor fluorophore transfer the energy from its excited state to an acceptor fluorophore a few nanometers in proximity to the donor. In order to utilize FRET as a probe for protein–protein interaction, one protein of interest is labeled with a donor fluorophore, and the other with an acceptor fluorophore (Fig. 1). FRET can also be used to measure enzyme activity, by adopting FRET biosensors engineered to undergo a conformational change upon the modification by the enzyme of interest, which leads to a change in intramolecular FRET (Fig. 1). Many types of FRET biosensors have been developed and used in a variety of cell biology studies [10–15].

There are two widely used approaches to measure FRET: intensity-based and lifetime-based approaches. Intensity-based approaches include sensitized emission measurements and acceptor photobleaching methods. Though the sensitized emission measurement is the simplest assay to perform, a major limitation is

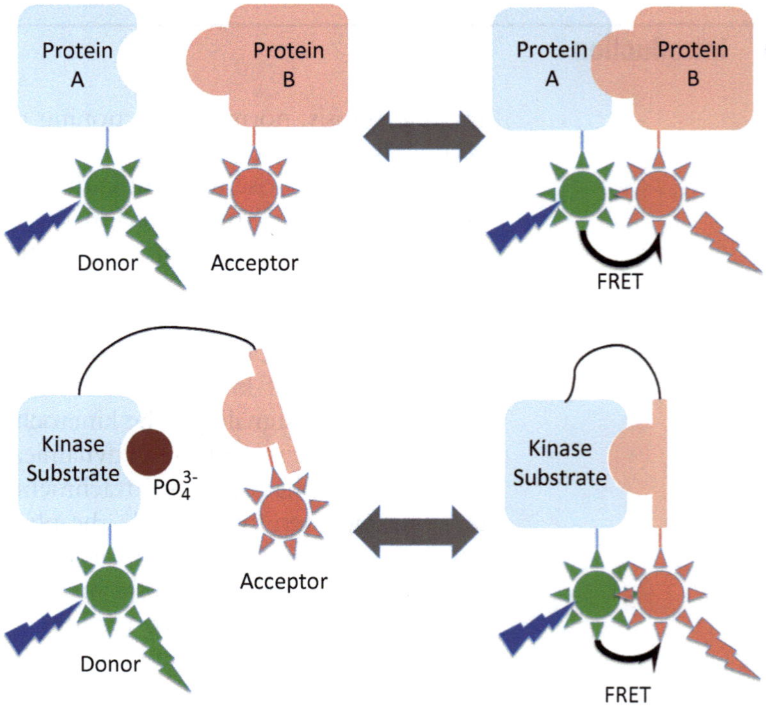

Fig. 1 The usages of FRET for studying (*top*) protein–protein interaction using fluorophore-labeled proteins, and (*bottom*) kinase activity using engineered FRET bio-sensor

the requirement of complicated correction and normalization procedures and the dependence on the relative concentration of donor and acceptor fluorophores, which is hard to control in practice. The acceptor photobleaching method, on the other hand, does not require the complicated correction procedures, but is difficult to use for time-varying measurements, and is also prone to artifacts due to the motion of acceptors and photodamage from the use of high intensities required for photobleaching. Therefore, this method is not suited for the study of dynamic kinetochores.

Fluorescence Lifetime Imaging (FLIM) is a lifetime-based approach to measure FRET. The fluorescence lifetime refers to the average time that a fluorophore spends in the excited state before relaxing and emitting a photon. The lifetime depends on both its intrinsic photo-physical properties and on its microenvironment, including the local pH, temperature, viscosity, and the presence of quenching, such as by FRET. FLIM-FRET is an imaging and analysis technique that probes fluorophores' lifetimes and quantifies FRET using the measured lifetimes. FLIM-FRET provides a highly quantitative measure of FRET, which is difficult to achieve with intensity-based measurements, because FLIM-FRET uncouples two factors that contribute to the overall FRET signal: the FRET efficiency of an individual FRET pair and the number of FRET pairs. Moreover, FLIM-FRET does not require irreversible reaction, such as acceptor photobleaching, and combined with Bayesian analysis, it is capable of measure FRET with high spatiotemporal resolution. These advantages make FLIM-FRET suitable for the study of dynamic kinetochores in live cell.

Here we describe the Bayesian analysis of FLIM-FRET as a tool to study kinetochore function in live cell at up to sub-millisecond time scales with single kinetochore resolution.

2 Materials

2.1 Fluorescence Lifetime Imaging (FLIM)

There are several variants of FLIM, which differ in how the fluorescence lifetime is measured and fluorophores are excited. We use time-domain FLIM (vs. frequency-domain) for more straightforward analysis, and two-photon excitation (vs. single-photon excitation) for the flexibility and low photon-induced damage.

(a) Femtosecond Ti:sapphire pulse laser (Mai Tai DeepSee, Spectral Physics)

In time-domain FLIM, an ultrashort pulsed laser is used as an excitation light source. The desired period of the laser is ~10 ns so that the nanosecond-scale fluorescence decay is well captured in a single laser period. Ti:sapphire lasers are particularly convenient when used for two-photon excitation because they are tunable between 690 and 1040 nm, allowing for the use of many different fluorophores.

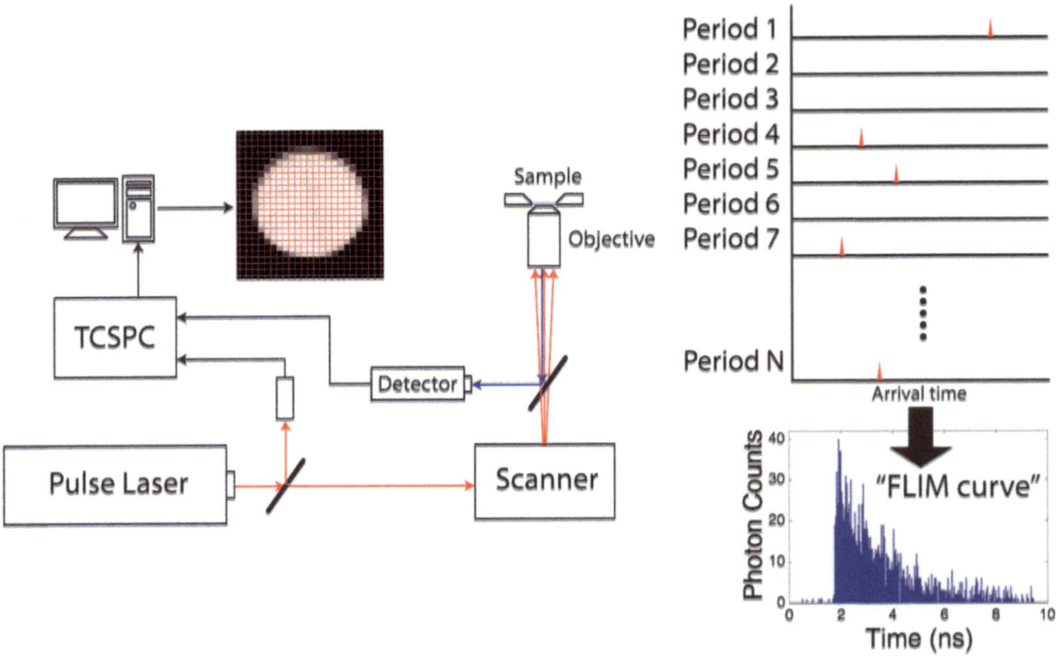

Fig. 2 (*left*) Arrangement and connections of major components in FLIM system. *Red* and *blue arrows* indicate the path of the excitation laser beam and that of emission photon, respectively. (*right*) Construction of FLIM curve by TCSPC

(b) Time-Correlated Single Photon Counting module (SPC-150, Becker & Hickl GmbH)

Time-Correlated Single Photon Counting (TCSPC) is a technique that detects a single photon and measures the timing of the photon arrival with picosecond time resolution [16]. TCSPC determines the time between the laser pulse and the arrival of the emitted photon at the detector, for each photon detected. After accumulating a number of photons, their arrival times can be used to construct a histogram, called the "fluorescence decay" or "FLIM curve" (Fig. 2)

(c) Laser scanning system (DCS-120, Becker & Hickl GmbH)

(d) Detector (HPM-100-50, Becker & Hickl GmbH)

Time-domain FLIM requires detectors with single-photon sensitivity and fast response.

(e) Optics

All the optics in the excitation light path, including the objective, lenses, and mirrors should be suitable for near-infrared wavelength pulsed laser. For the study of kinetochores, an objective with high Numerical Aperture (NA) is highly preferred because it allows a more compact point spread function (PSF), which results in better spatial resolution as well as more efficient two-photon excitation.

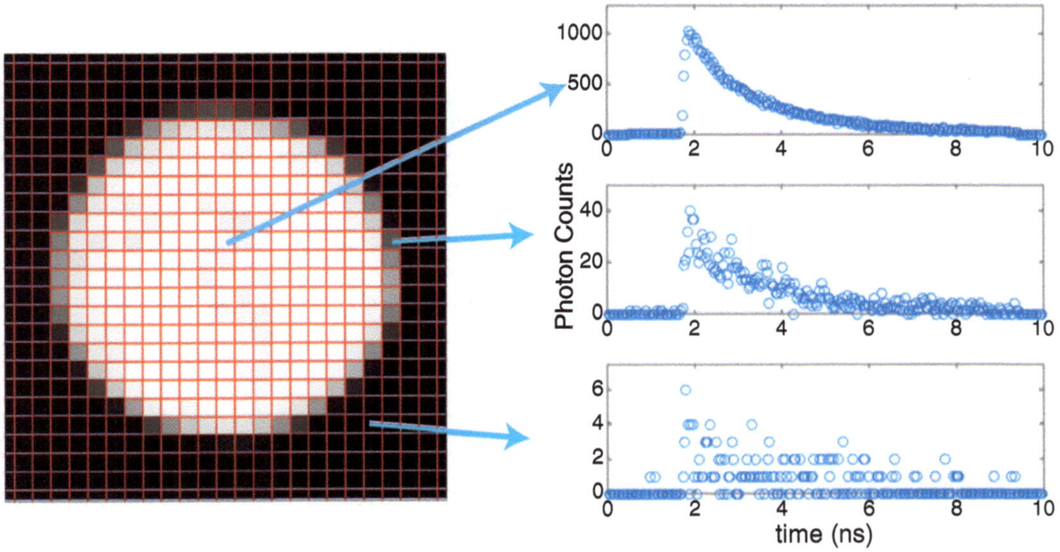

Fig. 3 Example of FLIM image. (*left*) Intensity image obtained by integrating FLIM curve at each pixel over the arrival time. (*right*) FLIM curves at three different pixels with different number of photons

It is also necessary to have appropriate optics to modulate the excitation intensity. This can be achieved by the combination of polarization optics, neutral density filters, or opto-electronic elements such as a Pockels cell.

(f) FLIM software (SPCM, Becker & Hickl GmbH)
To generate a FLIM image, FLIM software, either custom built or purchased, controls the scanning of the excitation laser over the region of interest and records the pixel location and the arrival time of each emitted photon using TCSPC. A FLIM curve is then generated for each pixel in the image (Fig. 3). These FLIM curves could be analyzed separately or photons from multiple pixels, such as all the pixels from a given kinetochore, could be grouped and analyzed together.

2.2 Mitotic Live Cell Imaging

We use U2OS tissue culture cells, but the protocol can be generalized to any type of cells.

(a) Imaging media (FluoroBrite™ DMEM, Life Technologies)
Imaging media should be carefully selected such that it does not contain significant autofluorescence, while still ensuring cell viability during imaging. The amount of serum should be reduced to a minimal amount, and vitamins such as riboflavin and pyridoxal can be removed completely [17]. Imaging media suited to human tissue culture cells is commercially available. If CO_2 control is unavailable, the imaging media should be supplemented with 10 mM HEPES to buffer the cells against changes in the pH.

(b) Sample cover glass (GG-25-pdl, Neuvitro)
Poly-d-lysine coating enhances the attachment and viability of adherent tissue culture cell.

(c) Temperature-controlled microscope sample chamber
We use an open microscope chamber that has electrical temperature control and an aperture where 25 mm round cover glass can be mounted.

(d) Objective heater (Bioptechs)

(e) Objective Z-stage (PIFOC, Physik Instrumente), if 3D imaging is needed.

3 Methods

We briefly describe the instrumentation, protocol, and analysis for FLIM-FRET measurement of kinetochores in vivo. More detail on the instrumentation and theory of FLIM-FRET is available in several other publications [16, 18–21].

3.1 Two-Photon FLIM Instrumentation

The essential components of a time-domain FLIM system are described in Fig. 2. The optics between the pulse laser and scanning system should be carefully examined and aligned before each experiment, as they influence the spatial resolution and the uniformity of the excitation intensity over the scanning area. The two most important optical components to optimize are the telescope, the combination of lenses that expands or reduces the size of the excitation beam, and the periscope assembly, the combination of mirrors that translates the beam. In order to fully utilize the numerical aperture of the objective, the telescope needs to be adjusted such that the excitation beam is well collimated when entering the objective and large enough to overfill the back aperture of the objective. The periscope assembly should be adjusted such that the beam is perpendicular and centered when entering the objective (*see* **Note 1**).

3.2 Live Cell Kinetochore FLIM-FRET Imaging

This section provides step-by-step instruction on live cell kinetochore FLIM-FRET imaging. We present an example using a FRET biosensor for Aurora kinase B activity fused to Hec1, an outer kinetochore protein, in U2OS cells. See Fuller et al. [15] for more information on the Aurora kinase B FRET biosensor.

1. Choice of fluorescent proteins (*see* **Note 2**).

   We chose mTurquiose2 (Cyan) and YPet (Yellow) to be the donor and acceptor fluorescent proteins in the Aurora kinase B FRET reporter. mTurquiose2 is well suited for FLIM-FRET, because of its monoexponential fluorescence decay, long fluorescence lifetime (~4 ns), and excellent brightness with the quantum yield of 93 % [22].

2. Make a plasmid that carries the DNA sequence of the desired kinetochore protein tethered to the FRET probe using standard cloning techniques.
3. Transfect cells.

   Since the brightness of the donor fluorescent protein at kinetochores has influence on the precision of FLIM-FRET measurement, high expression of the labeled protein is preferred as long as overexpression does not affect its behavior. Transient transfection results in some cells having multiple plasmids, which therefore express the construct at very high level. If transient transfection is difficult or undesirable, stable cell lines can be constructed using viruses or genome-editing techniques such as CRISPR.
4. Prepare a sample for imaging by culturing cells on a sample cover glass. Incubate in complete media for about 2 days, at 37 °C, 5 % CO_2.
5. Switch the complete growth media to the imaging media 30 min before imaging.
6. Prepare the FLIM setup: turn on and warm up the Ti:sapphire laser, pre-warm the microscope sample chamber and objective lens heater to 37 °C. Check the alignment of the optical system (*see* **Note 1**), and choose a proper emission filter. Calibrate the device that modulates the excitation intensity (*see* **Note 3**).

   The Ti:sapphire laser needs to be set to the optimal two-photon excitation of the donor fluorescent protein, which is 865 nm for mTurquoise2. The optical alignment should always be checked before imaging, since a defective PSF worsens image quality and two-photon excitation efficiency.

   The emission filter must be carefully chosen such to minimize signal from the acceptor, reduce the detected cellular background autofluorescence, and maximize the emission collected from the donor. For the mTurquiose2/YPet pair, we use ET470/40 m filter (Chroma).

   The alignment and other imaging setting should be verified by imaging a standard sample before the experiment (*see* **Note 4**).
7. Mount the sample on the temperature-controlled microscope chamber

   Layer the coverslip with the imaging media that is equilibrated at 37 °C, 5 % CO_2, and then cover with mineral oil to prevent evaporation. Clean the bottom of the coverslip with a Kimwipes and ethanol before mounting the chamber to the microscope stage.
8. Acquire FLIM images with the appropriate imaging settings.

   The following is the general guideline of choosing imaging settings: determine the scanning region such that the region of interest just fits into it; using a larger scanning area will lower the frame rate. The pixel size should be chosen such that each diffraction-limited spot, such as a kinetochore, is composed of about 10 pixels. The integration time should be short enough that moving kinetochores are not blurred, and long enough that more than a few hundreds photons per kinetochore are accumulated and kinetochores are clearly distinguished against the background noise.

For kinetochore imaging of U2OS cells, high quality data can be obtained with an NA 1.25, 40× water-immersion objective, a 14 μm × 14 μm scanning region, 128 × 128 pixels, and a 3–5 s integration time. Smaller regions can be scanned at higher speeds. The repeat time—the time between two consecutive images—will be determined by the aim of the study: studying kinetochore oscillations in metaphase may require a repeat time of 10 s, while for studying long-term changes, a repeat time of a few minutes may be preferable. The excitation beam power is typically set to 1–2 mW. Higher power may induce photodamage, with symptoms including metaphase arrest, prolonged metaphase, shrinkage of the spindle, and an increase in autofluorescence (*see* **Note 5**).

9. Record the instrument response function (IRF) of the FLIM system by measuring second harmonic generation (SHG).

   The IRF plays a very important role in FLIM analysis, and therefore an accurate measurement of the IRF is crucial. The IRF can easily be recorded by measuring second harmonic generation (SHG) signal from a urea crystal. SHG is a nonlinear optical process in which a material that lacks inversion symmetry converts incident photons into photons with half the wavelength of the incident photons. SHG is an ultrafast scattering process so the FLIM curve of SHG provides an accurate measure of the IRF of the FLIM system.

   To make a urea crystal sample for IRF measurements, dissolve urea in water and drop ~10 μl of the urea solution on the coverslip. As the water evaporates, it leaves crystallized urea on the coverslip. Put the coverslip on a microscope slide and seal it. This urea crystal IRF standard can be used repeatedly.

   To record the IRF, set up the emission bandpass filter such that it transmits the photon of the wavelength half of the excitation wavelength. For example, 425/30 can be used for 865 nm excitation. Since the IRF is specific to the imaging settings, every imaging setting other than the emission filter must remain the same. Place the urea crystal on the objective, and park the excitation laser where the crystal generates significant SHG. Make sure that the number of time bins is set to the maximum value. Record SHG for 10–30 s, allowing a smooth IRF curve to be generated. For more information about recording the IRF, refer to the application note published by Becker & Hickl [23] (*see* **Note 6**).

3.3 Analysis of Kinetochore FLIM-FRET Imaging

3.3.1 FLIM-FRET Models

If there is no donor engaged in FRET, then the number of photons $n(t)$ decays monoexponentially with a long lifetime τ_D:

$$n(t) \propto A\exp\left(-\frac{t}{\tau_D}\right) + (1-A) \tag{1}$$

where the parameter A, which ranges from 0 to 1, determines the fraction of detect photons emitted by the donor. In most applications, the dark noise is low, and thus A is usually very close to 1.

On the other hand, if there is a mixture of donors, with some donors engaged in FRET (all with the same FRET efficiency) and the others not engaging in FRET, then the decay of the number of detected photons is biexponential:

$$n(t) \propto A\left\{f\exp\left(-\frac{t}{\tau_D}\right) + (1-f)\exp\left(-\frac{t}{\tau_{FRET}}\right)\right\} + (1-A) \tag{2}$$

where f is the fraction of donors that are not engaged in FRET, and τ_D and τ_{FRET} are the lifetimes of donor not in FRET state and in FRET state, respectively. τ_D and τ_{FRET} can be translated into the FRET efficiency by:

$$E_{FRET} = 1 - \frac{\tau_{FRET}}{\tau_D} = \frac{1}{1+\left(\frac{r}{R_0}\right)^6} \tag{3}$$

where r is the separation between donor and acceptor fluorophores and R_0 is the Förster radius [24] (*see* **Note 7**).

Figure 4 shows experimentally measured FLIM curves from two samples, where no donors are engaging in FRET and where some donors are in a FRET state. It is apparent that the FLIM curve decays faster when there is FRET, as predicted. Upon closer inspection, it is evident that the data significantly departs from the simple exponential model: the real data appears to be shifted and broadened relative to the model. This discrepancy comes from the limited temporal resolution of the FLIM system and from the finite travel time of photons and information in the system. Fortunately, these discrepancies are determined by the instrument configuration, and can be measured and incorporated into the model. These effects are captured by the instrument response function (IRF), which is what the FLIM system would record from a sample with an infinitely short lifetime. As mentioned above, the IRF can be experimentally measured and usually appears to be a sharp, skewed peak, shifted from $t = 0$ by the same extent as the exponential decay in FLIM curve (Fig. 5).

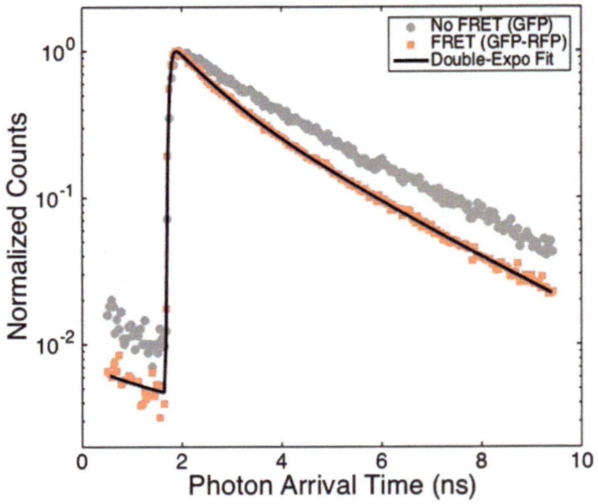

Fig. 4 Normalized FLIM curve on log-scale of donor when there is no donor engages in FRET (*gray*) and when some fraction of donors are engages in FRET (*orange*)

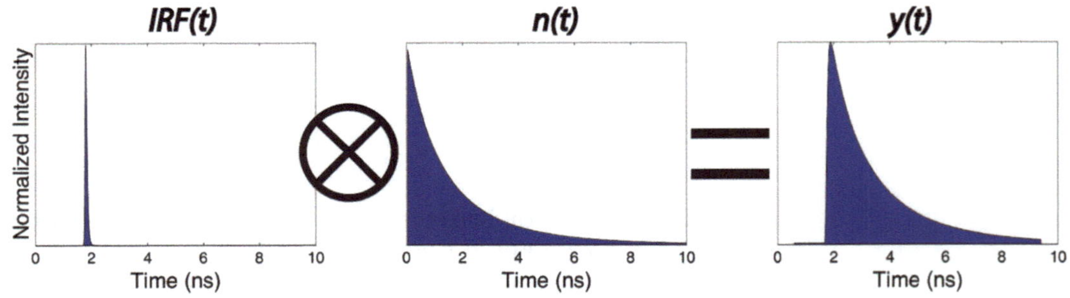

Fig. 5 The illustration of building a complete FLIM model by convolving IRF and exponential decay model

As illustrated in Fig. 5, a complete FLIM model can be constructed by convolving the exponential decay model (Eqs. 1 or 2) with the measured IRF:

$$y = C_{\text{norm}} \cdot \text{IRF}\left(t - t_{\text{s}}\right) \otimes n\left(t\right) \quad (4)$$

where C_{norm} is a normalization constant, IRF is the measured instrument response function, t_{s} is the shift of the measured IRF relative to the real IRF, and $n(t)$ is the exponential decay model given by Eqs. 1 or 2. The shift t_{s} is necessary for precise fitting because the location of the measured IRF may be different by very small extent, usually within 100 ps, from the real IRF. The constant C_{norm} is determined by normalizing the area under the curve to the total photon counts. Thus, Eqs. 1 or 2 describes the behavior of the donor fluorophore, while the IRF describes the behavior of the FLIM instrument.

The more realistic model of Eq. 4 is sufficient to describe experimental data. Measured FLIM curves can thus be fit to Eq. 4 to determine the five free parameters: t_s, A, τ_D, f, and τ_{FRET}.

3.3.2 Identification and Tracking of Kinetochores

We developed a MATLAB graphical user interface (GUI) that semi-automatically analyzes kinetochore FLIM-FRET data, which allows high-throughput kinetochore analysis.

First, load FLIM images (Fig. 6, *see* **Note 8**). Then identify and track kinetochores using standard particle tracking techniques. We modified the particle tracking MATLAB codes developed and used in Pelletier et al. [25], and incorporated it into the GUI. As shown in Fig. 6b, each kinetochore trajectory is automatically labeled with a number, and the GUI enables us to manually modify mislabelled kinetochore features, or remove features that are incorrectly identified as kinetochores. By fitting 2D Gaussian to each kinetochore, the location of the kinetochore can be determined with sub-pixel precision, and can be used for distance and velocity measurements.

3.3.3 FLIM-FRET Measurement of Kinetochore with Bayesian Analysis

Assuming that the fluorescence lifetimes of FRET and non-FRET states are constant throughout the movie, we can precisely determine the lifetimes by grouping all the FLIM curves from the kinetochores in every image to obtain a FLIM curve with a large number of photons (Fig. 6c), and make inference using the biexponential FLIM model to accurately estimate the fluorescence lifetimes (Fig. 6d). In the analysis of FLIM curves in each frame or each kinetochore with a fewer number of photons, we can then fix the lifetimes and IRF shift to the values obtained from the sum of many FLIM curves, and perform the biexponential FLIM-FRET Bayesian analysis with a reduced number of free parameters to precisely measure the fraction of the donors FRETing at each timepoint or at each kinetochore (Fig. 6e, f).

Getting accurate and precise estimation of parameters in a biexponential FLIM-FRET model is not a simple task, especially when the number of photons composing the FLIM curve is small. When the number of photons is of the order of 10^5 or more, commonly used nonlinear least-square fitting of the Eq. 4 may provide the estimates of the FLIM parameters with reasonable accuracy and small bias. However, when the number of photons is very limted, which is the case for the study of dynamic kinetochores, a Bayesian approach with a complete probabilistic model must be employeed, in order to minimize the bias resulting from the simplified assumption inherent to least-square fitting.

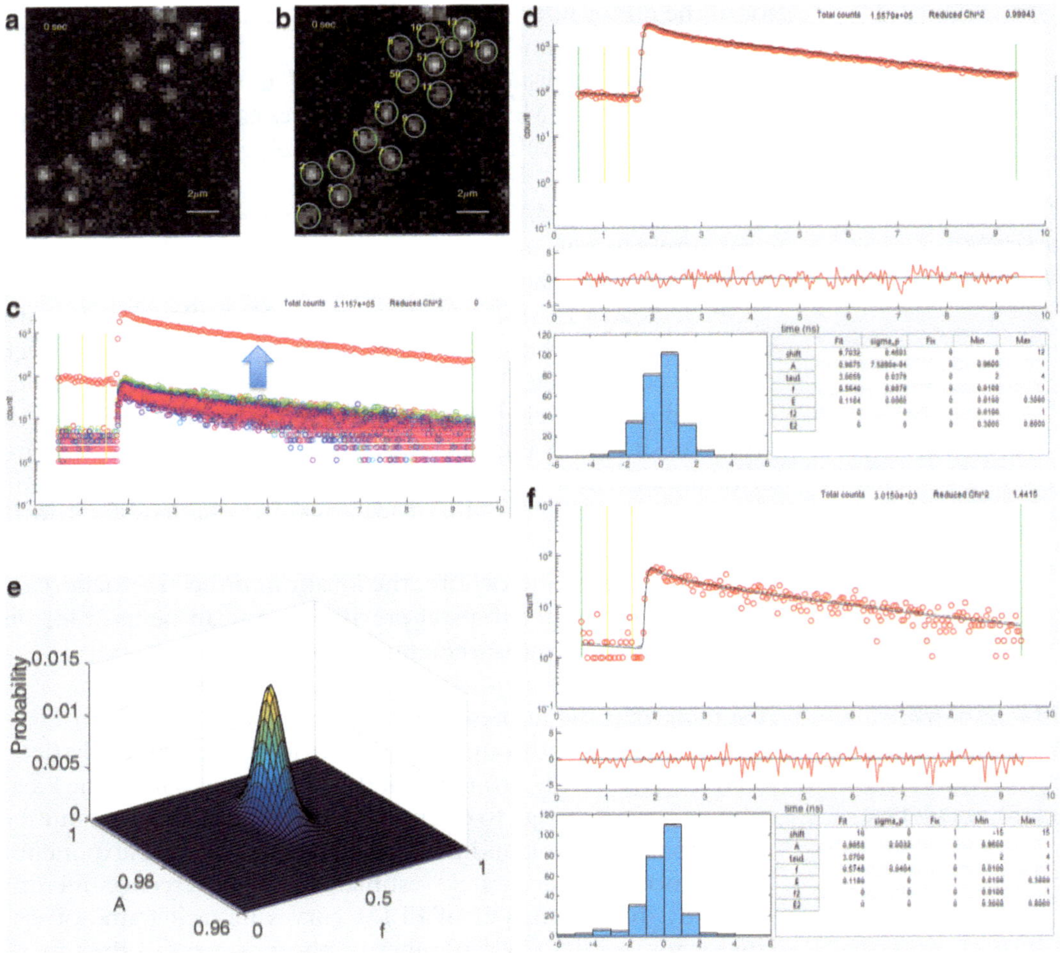

Fig. 6 Kinetochore FLIM-FRET measurement procedures. (a) Loading FLIM images saved in a special data format. (b) Identification of kinetochores. (c) Summing the FLIM curves obtained from every kinetochore, and (d) determination of the fluorescence lifetimes. (e, f) Bayesian FLIM analysis on the FLIM curve from one frame

Let θ be the set of parameters that governs the FLIM model, and $y = \{yi\}$ be the observed FLIM data, where yi is the number of photons detected in i-th time bin of the FLIM curve. The aim of Bayesian inference is to obtain the posterior probability distribution of the parameters of interest, $p(\theta \mid y)$, through the Bayes rule:

$$p(\theta \mid y) \propto p(y \mid \theta) p(\theta) \tag{5}$$

The common name of the first term in the product $p(y \mid \theta)$ is the "likelihood function," which is the probability of the observed data y given the parameter values θ. The likelihood function is based on the data generation process. Let Δt be the width of the time bin with which the FLIM data is acquired. Then the likelihood function can be written as

$$p(y|\theta) \propto \prod_{i=1}^{N} P\left(t_{\text{ar}} \in \left[(i-1)\Delta t, i\Delta t\right] | \theta\right)^{y_i} \quad (6)$$

where t_{ar} is the photon arrival time, and N is the number of time bins. Assuming that the size of the time bin is very small compared to the time scale of fluorescence decay, the probability that the arrival time t_{ar} falls in the i-th time bin is approximated by a Riemann sum:

$$P\left(t_{\text{ar}} \in \left[(i-1)\Delta t, i\Delta t\right] | \theta\right) \cong \sum_{k=(i-1)K+1}^{k=iK} h_\theta\left(k\Delta t\right)\Delta t \quad (7)$$

where $\tilde{\Delta} t$ is the smallest time bins with which IRF is measured, and the ratio $K = \frac{\Delta t}{\Delta \tilde{t}}$ is a positive integer. $h_\theta\left(k\Delta\tilde{t}\right)$ in the Eq. 7 can be written as:

$$h_\theta\left(k\tilde{\Delta} t\right) = \left(\text{IRF} \otimes (Ag_\theta + (1-A)\right)\left(k\tilde{\Delta} t\right) \cong \sum_l \text{mIRF}\left[l - b_{\text{shift}}\right]\left(Ag_\theta\left((k-l)\tilde{\Delta} t\right) + (1-A)\right) \quad (8)$$

where mIRF is the IRF measured in the finest time bins with size $\Delta\tilde{t}$, and b_{shift} is an integer parameter that determines the approximate shift of measured IRF from the theoretical IRF. The function $g\theta$ depends on the number of exponentials in FLIM model. For example, when there are two species of donors, one of which are engaged in FRET with the same efficiency and the other not, $g\theta$ is expressed as:

$$g_\theta\left(t_{\text{d}}\right) = f\mathrm{e}^{-\frac{t_{\text{d}}}{\tau_{\text{D}}}} + (1-f)\mathrm{e}^{-\frac{t_{\text{d}}}{\tau_{\text{FRET}}}} \quad (9)$$

We ignored the "pile-up" effect that results from photons that are emitted from the fluorophores excited in the previous pulses and appear in the current laser period, as typically this effect is very small for commonly used fluorophores with lifetime shorter than 4 ns [16]. Equation 8 can be efficiently evaluated by using a fast convolution algorithm combined with fast Fourier transformation, rather than directly carrying out the summation.

The Bayes rule (Eq. 5) results in the posterior distribution:

$$p(\theta | y) \propto p(\theta) \prod_{i=1}^{N} P\left(t_{\text{ar}} \in \left[(i-1)\Delta t, i\Delta t\right] | \theta\right)^{y_i} \quad (10)$$

The prior distribution $p(\theta)$ is often chosen to be flat between reasonable upper and lower limits of the parameters, i.e., $p(\theta) = 1$ (*see* **Note 9**).

Equation 10 needs to be numerically evaluated, which can be done using a number of different methods. The grid search method is conceptually simple to implement. In this method, a grid is formed by dividing the range of values for each of parameter in $\theta = \{b_{\text{shift}}, A, f, \tau_D, \tau_{\text{FRET}}\}$ into a number of discrete levels, and the posterior distribution is computed at each grid point. The computation time grows exponentially with the number of free parameters, and therefore the use of this method is restricted to low-dimensional searches. For example, if the values of b_{shift}, τ_D, and τ_{FRET} are known, and a search is performed over the other two parameters, this method works well with a reasonably fine grid and a computation time of few seconds. On the other hand, if there are many free parameters, Markov Chain Monte Carlo (MCMC) is a faster method. The most popular variants of MCMC are Gibbs sampling and Metropolis–Hastings algorithm [26–29]. We do not discuss the details of these algorithms here. Figure 7 shows the marginal posterior distributions constructed by the grid search and Gibbs sampling, when only b_{shift} is fixed and the other four parameters are free. The two methods produced almost identical result, while the Gibbs sampling is roughly ten times faster than the grid search, taking ~10 min vs. ~100 min to run.

Once the posterior distribution is calculated, a variety of point and interval estimations of the parameters can readily be made. If the marginalized posterior distribution of a parameter is symmetric, its mean is a natural point estimate for the parameter. If the marginalized posterior distribution is truncated or asymmetric, the posterior mode may be a better point estimate. The Bayesian probability interval, or credible interval, can be also found from the marginal posterior distribution and be used as an interval estimate of parameter.

One essential advantage of using Bayesian analysis is its explicit use of probability for quantifying uncertainty in inferences. This makes the propagation of uncertainty perspicuous, especially when combining estimates from multiple datasets. Another important advantage of Bayesian FLIM analysis is its superior accuracy and precision compared to other fitting methods. It has been shown that Bayesian analysis can be used to determine fluorescence lifetimes of a monoexponential fluorescence decay with no bias with as little as 50 photons, while LS and MLE cause significant bias at such low photon counts [30]. We also confirmed that when applied to double-exponential fluorescence decay data Bayesian FLIM analysis is capable of estimating the fraction of one exponential component relative to the other with no bias with only 200 photons. Using standard labeling levels and illumination intensities, one photon can be collected from a single kinetochore every microsecond. Thus, with Bayesian FLIM analysis it should be possible to measure the fraction of fluorophores engaged in FRET at a single kinetochore every 200 μs, enabling the study of submillisecond dynamics of protein behaviors at kinetochores.

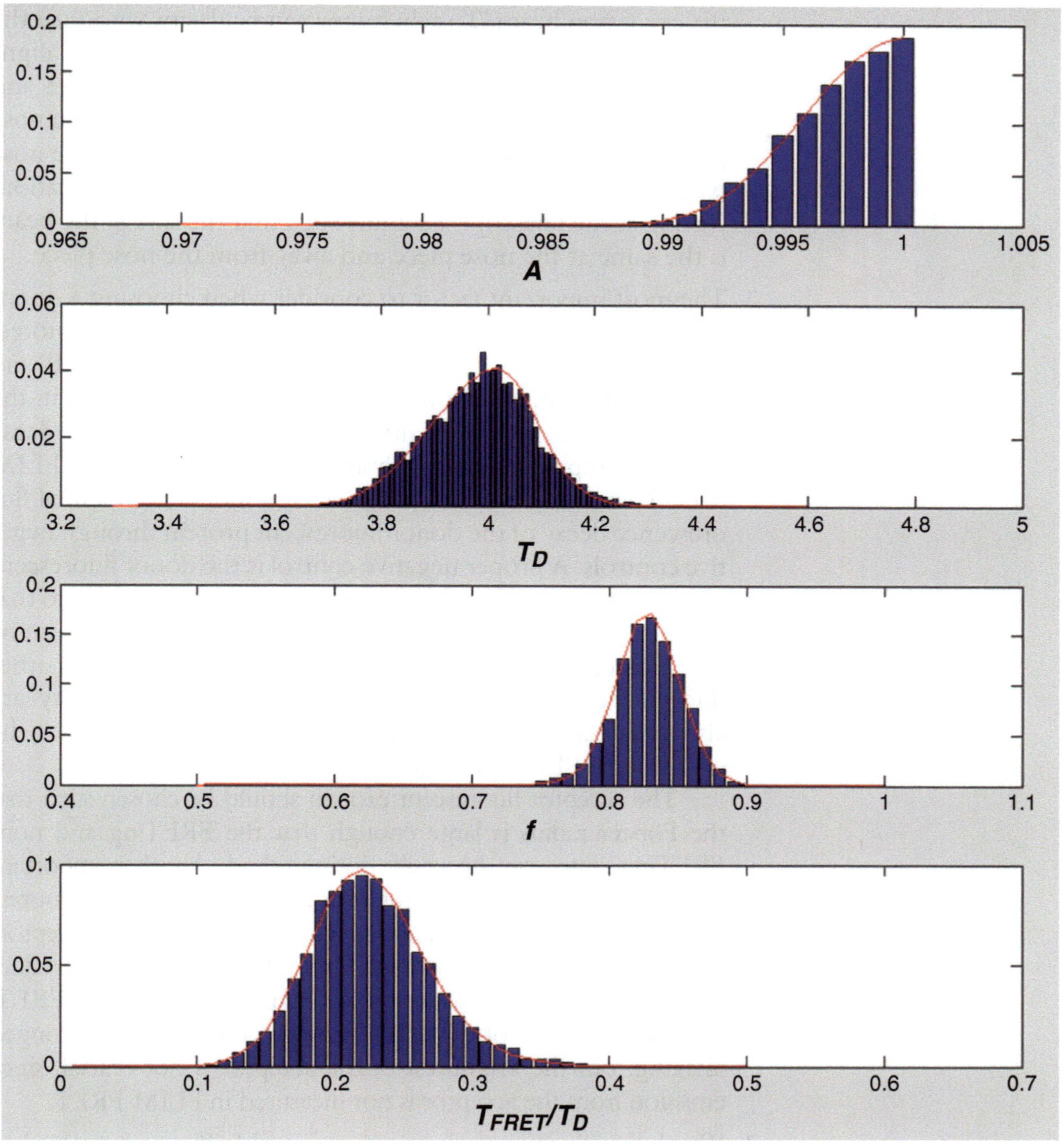

Fig. 7 Marginal posterior distribution constructed by the grid search (*red line*) and Gibbs sampling (*blue bar*). The sizes of the grid are 0.001 for *A*, and 0.01 for the other three parameters. 5000 values were stimulated by Gibbs sampling

4 Notes

1. Some alignment tools can facilitate the adjustment and alignment of the periscope and telescope assembly. Fluorescing alignment disks (such as VRC2RMS, Thorlabs), which can be mounted on the nose piece of microscope, is particularly useful as it can be used to visualize the size and position of the beam entering the objective. We also use a fluorescing alignment disk in combination with a long lens tube (such as SM1E60, Thorlabs) to ensure the perpendicularity and collimation of

the excitation beam. To achieve perpendicularity, we adjust the periscope assembly such that the beam is centered on the alignment disk when the disk is mounted on the nose piece and when it is mounted on a lens tube that is installed on the nose piece (i.e., the alignment disk is positioned away from the nose piece by the length of the lens tube). To achieve collimation, we adjust the telescope assembly such that the size of the beam is the same at the nose piece and away from the nose piece.

2. The most important factor to consider when choosing a donor fluorescent protein is the complexity of the donor's fluorescence decay. Some fluorescent proteins, such as ECFP and EYFP, exhibit multiexponential fluorescence decay even in the non-FRETing state. Multiexponential fluorescence decay greatly complicates the analysis and interpretation of FLIM data. Therefore, it is essential to verify a mono-exponential fluorescence decay of the donor fluorescent protein through negative controls. A proper negative control is the donor fluorescent protein targeted to the same protein as the FRET probe, so that the local environmental effects on the fluorescence decay can be investigated in advance of the FRET measurements. As in other fluorescence microscopies, brightness and photostability are also important, since the number of photons determines the precision of FLIM-FRET measurement.

   The acceptor fluorescent protein should be chosen such that the Förster radius is large enough that the FRETing and non-FRETing states can be easily distinguished. Another important factor to consider is the maturation time of the acceptor fluorescent protein. Slow maturation leads to a large fraction of acceptors being unfolded or immature, and therefore results in the reduction of the number of donor molecules that can engage in FRET. The photostability of the acceptor is also essential for prolonged imaging, but the brightness of the acceptor is not crucial, since emission from the acceptor is not measured in FLIM-FRET.

3. We place a linear polarizer, a motorized half-wavelength plate and another linear polarizer in series in the laser path to modulate the excitation intensity; the transmission of the laser changes as the angle of the half-wavelength plate changes. Before every experiment, we create a mapping between the angle of the half-wavelength plate and the excitation laser intensity, by measuring the laser power at the objective lens.

4. We normally use Alexa dye on a coverslip to check the uniformity of the excitation, and 200 nm fluorescent microsphere (T14792, Life Technologies) to check the quality of the point spread function. When the system is aligned properly, the full width at half maximum of the point spread function should be ~300 nm laterally and ~700 nm axially.

5. In order to prevent the emergence of artifacts related to photodamage and photobleaching, one should carefully choose

appropriate imaging settings by performing negative control experiments. The best negative control sample is cells expressing the protein of interest labeled with only the donor. To test an imaging setting, take images of the negative control sample with the imaging setting, and check whether the observed fluorescence lifetime and/or brightness of the donors change over time. Photodamage is known to result in the increase in autofluorescence [31], which contaminates the fluorescence signal from donor and may change the observed fluorescence lifetime. Photobleaching also may change the observed fluorescence lifetime as it increases the relative contribution of the autofluorescence. Photodamage and photobleaching are highly nonlinear with excitation intensity, following a power law with the exponent greater than 2, while the two-photon excitation is proportional to intensity squared [32, 33]. Therefore, photodamage and bleaching can be significantly suppressed by lowering the excitation intensity, without sacrificing the fluorescence signal from the donor as much. For this reason, as a way to enhance the signal-to-noise ratio, using longer integration time is preferred over using higher excitation intensity.

6. When acquiring the IRF using an SHG sample, remove the condenser lens because it may reflect the SHG and cause additional peak in the measured IRF curve.
7. It is not recommended to use Eq. 3 to attempt to measure the distance between donor and acceptor fluorophores. One major difficulty is that FRET efficiency is highly dependent on the relative orientation of the donor and acceptor fluorophores. The distance-FRET efficiency predicted by Eq. 3 holds only for FRET efficiency averaged over all orientations, assuming that the orientations are isotropic. In practice, it is difficult to know the relative orientations of the donor and acceptor.
8. The MATLAB code that imports FLIM data generated by Becker & Hickl SPCM software is available from Becker & Hickl upon request.
9. Any prior knowledge can be incorporated into the posterior distribution through the prior distribution. For example, some studies impose an exponential prior on the lifetime [30], i.e., $p(\theta)=\alpha e^{-\alpha\tau}$.

References

1. Gordon DJ, Resio B, Pellman D (2012) Causes and consequences of aneuploidy in cancer. Nat Rev Genet 13:189. doi:10.1038/nrg3123
2. Rajagopalan H, Lengauer C (2004) Aneuploidy and cancer. Nature 432:338–341
3. Bakhoum SF, Thompson SL, Manning AL (2009) Genome stability is ensured by temporal control of kinetochore–microtubule dynamics. Nat Cell Biol 11:27. doi:10.1038/ncb1809
4. Cimini D, Degrassi F (2005) Aneuploidy: a matter of bad connections. Trends Cell Biol 15:442, http://www.sciencedirect.com/science/article/pii/S0962892405001601
5. Nicklas RB (1983) Measurements of the force produced by the mitotic spindle in anaphase. J Cell Biol 97:542. doi:10.1083/jcb.97.2.542
6. Inoué S, Salmon ED (1995) Force generation by microtubule assembly/disassembly in mito-

sis and related movements. Mol Biol Cell 6:1619. doi:10.1091/mbc.6.12.1619
7. Li R, Murray AW (1991) Feedback control of mitosis in budding yeast. Cell 66:519–531
8. Musacchio A, Salmon E (2007) The spindle-assembly checkpoint in space and time. Nat Rev Mol Cell Biol 8:379–393
9. Godek KM, Kabeche L, Compton DA (2015) Regulation of kinetochore-microtubule atta chments through homeostatic control during mitosis. Nat Rev Mol Cell Biol 16:57. doi:10.1038/nrm3916
10. Violin JD, Zhang J, Tsien RY, Newton AC (2003) A genetically encoded fluorescent reporter reveals oscillatory phosphorylation by protein kinase C. J Cell Biol 161:899–909
11. Chen Y, Saulnier J, Yellen G, Sabatini B (2014) A PKA activity sensor for quantitative analysis of endogenous GPCR signaling via 2-photon FRET-FLIM imaging. Front Pharmacol 5:56
12. Espenel C, Acharya BR, Kreitzer G (2013) A biosensor of local kinesin activity reveals roles of PKC and EB1 in KIF17 activation. J Cell Biol 203:445–455
13. Pereira AM, Tudor C, Kanger JS, Subramaniam V, Martin-Blanco E (2011) Integrin-dependent activation of the JNK signaling pathway by mechanical stress. PLoS One 6:e26182
14. Gavet O, Pines J (2010) Progressive activation of CyclinB1-Cdk1 coordinates entry to mitosis. Dev Cell 18:533–543
15. Fuller B et al (2008) Midzone activation of aurora B in anaphase produces an intracellular phosphorylation gradient. Nature 453:1132 –1136
16. Becker W (2012) The bh TCSPC handbook. Becker & Hickl GmbH, Berlin
17. Bogdanov A, Kudryavtseva E, Lukyanov K (2012) Anti-fading media for live cell GFP imaging. PLoS One 7:e53004
18. Lleres D, Swift S, Lamond AI (2007) Detecting protein-protein interactions in vivo with FRET using multiphoton fluorescence lifetime imaging microscopy (FLIM). Curr Protoc Cytom. doi: 10.1002/0471142956.cy1210s42
19. Berezin M, Achilefu S (2010) Fluorescence lifetime measurements and biological imaging. Chem Rev 110:2641–2684
20. Festy F, Ameer-Beg S, Ng T, Suhling K (2007) Imaging proteins in vivo using fluorescence lifetime microscopy. Mol Biosyst 3:381–391
21. Ebrecht R, Paul C, Wouters F (2014) Fluorescence lifetime imaging microscopy in the medical sciences. Protoplasma 251:293305
22. Goedhart J et al (2011) Structure-guided evolution of cyan fluorescent proteins towards a quantum yield of 93%. Nat Commun 3:751
23. Becker & Hickl GmbH (2008) Recording the Instrument Response Function of a Multiphoton FLIM System
24. Lakowicz JR (2011) Principles of fluorescence spectroscopy, vol 954. Springer, New York, NY
25. Pelletier V, Gal N, Fournier P, Kilfoil M (2009) Microrheology of microtubule solutions and actin-microtubule composite networks. Phys Rev Lett 102:188303
26. Casella G, George EI (1992) Explaining the Gibbs sampler. Am Stat 46:167. doi:10.1080/00031305.1992.10475878
27. Geman S, Geman D (1984) Stochastic relaxation, gibbs distributions, and the Bayesian restoration of images. IEEE Trans Pattern Anal Mach Intell PAMI-6:721–741
28. Gelman A et al (2013) Bayesian data analysis, vol 675. Chapman and Hall/CRC, Boca Raton, FL
29. Lawrence CE et al (1993) Detecting subtle sequence signals: a gibbs sampling strategy for multiple alignment. Science 262:208–214
30. Rowley M, Barber P, Coolen ACC, Vojnovic B (2011) Bayesian analysis of fluorescence lifetime imaging data. Proc SPIE 7903:790325-1
31. Galli R et al (2014) Intrinsic indicator of photodamage during label-free multiphoton microscopy of cells and tissues. PLoS One 9:e110295
32. Hopt A, Neher E (2001) Highly nonlinear photodamage in two-photon fluorescence microscopy. Biophys J 80:2029
33. Patterson G, Piston D (2000) Photobleaching in two-photon excitation microscopy. Biophys J 78(4):2159

Part IV

Methods Focused on the Spindle Pole

Chapter 12

Purification of Fluorescently Labeled *Saccharomyces cerevisiae* Spindle Pole Bodies

Kimberly K. Fong, Beth Graczyk, and Trisha N. Davis

Abstract

Centrosomes are components of the mitotic spindle responsible for organizing microtubules and establishing a bipolar spindle for accurate chromosome segregation. In budding yeast, *Saccharomyces cerevisiae*, the centrosome is called the spindle pole body, a highly organized trilaminar structure embedded in the nuclear envelope. Here we describe a detailed protocol for the purification of fluorescently labeled spindle pole bodes from *S. cerevisiae*. Spindle pole bodies are purified from yeast using a TAP-tag purification followed by velocity sedimentation.

This highly reproducible TAP-tag purification method improves upon previous techniques and expands the scope of in vitro characterization of yeast spindle pole bodies. The genetic flexibility of this technique allows for the study of spindle pole body mutants as well as the study of spindle pole bodies during different stages of the cell cycle. The ease and reproducibility of the technique make it possible to study spindle pole bodies using a variety of biochemical, biophysical, and microscopic techniques.

Key words Spindle pole body (SPB), Purification, Mitosis, Centrosome, TAP-tag

1 Introduction

The mitotic spindle ensures accurate segregation of genetic material in a dividing cell. The centrosome is a crucial component of this macromolecular machine, nucleating and organizing microtubules. Because components of the mitotic spindle are highly conserved throughout eukaryotes, the yeast spindle serves as an excellent model.

In yeast, the centrosome equivalent is the spindle pole body. The spindle pole body is a highly organized laminar structure, consisting of three plaques. The core spindle pole body components form the central plaque, which is embedded in the nuclear envelope throughout the cell cycle [1, 2]. The inner plaque sits on the nuclear side of the central plaque and is the site of nuclear microtubule nucleation. The inner plaque contains the γ-tubulin small complex, which is essential for microtubule nucleation, and

Paul Chang and Ryoma Ohi (eds.), *The Mitotic Spindle: Methods and Protocols*, Methods in Molecular Biology, vol. 1413, DOI 10.1007/978-1-4939-3542-0_12,

Spc110, a linker protein that binds the γ-tubulin small complex to the core of the spindle pole body [3–5]. The outer plaque is on the cytoplasmic side of the central plaque. This structure also contains the γ-tubulin small complex, which is bound to the spindle pole body via interactions with Spc72 [6].

Yeast spindle pole bodies (SPBs) have historically been studied by thin section electron microscopy of yeast cells [1, 2, 7] or purified to identify components, determine structure, and examine function of the SPB in vitro [3, 8–11]. A method to purify SPBs out of the lesser characterized yeast *S. uvarum* generated high yields of purified SPBs, but was limited in the ability to perform genetic analyses and the SPBs could only be visualized by electron microscopy [3]. More recently, SPBs were co-purified from *S. cerevisiae* with a TAP-tagged nuclear pore component, Mlp2 [12]. Expanding on this method, we discovered that TAP-tagging Spc97, a component of the spindle pole body, increases yield. Adding fluorescent tags to spindle pole body components helps to both visualize the SPBs and quantify yields. Here we offer a detailed purification protocol for growing, harvesting, and lysing cells to purify spindle pole bodies by TAP-tag purification followed by velocity sedimentation.

2 Materials

2.1 Growing, Harvesting, and Lysing Cells

1. *S. cerevisiae* strain expressing *SPC97-TAP::kanMX* with *SPC42-mCherry::hphMX or S. cerevisiae* strain expressing *SPC97-TAP::kanMX* with *SPC42-mCherry::hphMX* and *TUB4-GFP::kanMX*.
2. YPD media: 1 % yeast extract, 2 % peptone, 2 % glucose.
3. Resuspension buffer: 20 mM sodium-HEPES buffer (pH 7.4), 1.2 % polyvinylpyrrolidone (average MW 40,000), 1 mM dithiothreitol, 1 mM phenylmethanesulfonyl fluoride, 4 μg/ml aprotinin, 4 μg/ml chymostatin, 4 μg/ml leupeptin, 4 μg/ml pepstatin, 10 mM sodium fluoride, 1 mM sodium pyrophosphate, 1 mM β-glycerophosphate.
4. PM 100 ball mill grinder (Retsch).
5. Stainless steel 125 ml planetary ball mill grinding jar with 20 mm diameter mill grinding balls (Retsch).
6. Liquid nitrogen.
7. 18 G needle.

2.2 Spindle Pole Body Purification

1. Extraction Buffer with 300 mM NaCl (EB1 w/ 300 mM NaCl): 20 mM sodium-HEPES buffer (pH 7.4), 300 mM NaCl, 0.5 % Triton X-100, 2 mM $MgCl_2$, 100 μM GTP, 1 mM ATP, 1 mM dithiothreitol, 1 mM phenylmethanesulfonyl fluoride, 4 μg/ml

aprotinin, 4 μg/ml chymostatin, 4 μg/ml leupeptin, 4 μg/ml pepstatin, 10 mM sodium fluoride, 1 mM sodium pyrophosphate, 1 mM β-glycerophosphate, 5 % glycerol.

2. Extraction Buffer with 200 mM NaCl (EB1 w/ 200 mM NaCl): 20 mM sodium-HEPES buffer (pH 7.4), 200 mM NaCl 0.5 % Triton X-100, 2 mM $MgCl_2$, 100 μM GTP, 1 mM ATP, 1 mM dithiothreitol, 1 mM phenylmethanesulfonyl fluoride, 4 μg/ml aprotinin, 4 μg/ml chymostatin, 4 μg/ml leupeptin, 4 μg/ml pepstatin, 10 mM sodium fluoride, 1 mM sodium pyrophosphate, 1 mM β-glycerophosphate, 5 % glycerol.
3. TEV cleavage buffer: 40 mM sodium-HEPES buffer (pH 7.4), 200 mM NaCl, 2 mM $MgCl_2$, 1 mM GTP, 1 mM ATP, 1 mM EDTA (pH 8), 1 mM dithiothreitol, 5 % glycerol.
4. PCU-2-110 Homogenizer (Kinematica).
5. M-270 Epoxy Dynabeads (Invitrogen) conjugated to rabbit IgG (MP Biomedicals) according to manufacturer's protocol.
6. Magnetic stands for 50 ml Falcon tube and 1.7 ml microcentrifuge tubes.
7. TEV protease (stored in 50 mM Tris buffer (pH 7.5), 1 mM EDTA, 5 mM DTT, 50 % glycerol, 0.1 % Triton X-100).

2.3 Velocity Sedimentation

1. Sucrose solutions: 10, 20, 30, 40 %, and 2.5 M sucrose in 10 mM Bis-Tris buffer (adjust to pH 6.5 with HCl), 0.1 mM $MgCl_2$.
2. Thick wall polycarbonate tubes: 11 × 34 mm (Beckman Coulter).
3. Ultracentrifuge Rotor TLS-55 (Beckman Coulter).

3 Methods

3.1 Growing, Harvesting, and Lysing Cells

This protocol has been adapted from Michael Rout's lab (http://lab.rockefeller.edu/rout/protocols). The lysis technique with the Retsch PM 100 ball mill grinder increases the percent of cells lysed and decreases protein degradation, resulting in higher yields.

1. Grow 10 ml YPD culture of cells overnight and maintain in log phase growth to inoculate 2 l culture.
2. Grow 2 l YPD culture of cells overnight to 4.5×10^7 cells/ml (still in log phase).
3. Pellet cells at 4000 × *g* for 10 min at 4 °C in 1 l bottles.
4. Resuspend cell pellet in 25 ml cold dH_2O on ice. Transfer cells to 50 ml Falcon tubes and pellet cells at 2600 × *g* for 5 min at 4 °C.
5. Combine two cell pellets and resuspend in 15 ml cold dH_2O. Spin down at 2600 × *g* for 5 min at 4 °C.

6. Resuspend pellet in 15 ml cold resuspension buffer. Pellet at 2600 × *g* for 15 min at 4 °C. Decant the supernatant.
7. Pellet again at 2600 × *g* for 15 min at 4 °C to remove excess buffer. At this point, the pellet should be a relatively dry, thick paste.
8. Cool a 50 ml Falcon tube and fill with liquid nitrogen. Using an 18 G needle, poke holes in the tube lid to allow liquid nitrogen vapor to escape.
9. With a spatula, scoop out the cell paste and transfer to a 20 ml syringe. Press the cell paste into the liquid nitrogen in the Falcon tube, creating noodles.
10. When all cells have been frozen, loosely cap tube to allow liquid nitrogen to escape. Store at −80 °C.
11. To lyse cells, cool a 125 ml stainless steel grinding jar and stainless steel grinding balls in liquid nitrogen.
12. Empty the contents of one 50 ml Falcon tube of frozen cell noodles into the grinding jar and fill the jar with grinding balls.
13. Ensure the counterweight is properly set.
14. Start grinding program: 3 × 1 min grinding cycles at 400 rpm, reversing directions with each cycle.
15. At the end of the grinding program, cool the grinding jar in liquid nitrogen.
16. Repeat the grinding program and cooling seven more times, or until you get satisfactory lysis.
17. When lysis is complete, scoop out the lysed cell dust and transfer to the cooled 50 ml Falcon tube (*see* **Note 1**). Store lysed cells at −80 °C.

3.2 Spindle Pole Body Purification

Previous purification protocols were modified to increase the yield and stability of purified spindle pole bodies. Prior techniques were laborious and time-intensive, whereas this protocol has reduced the purification to a few hours.

1. Resuspend 4 g of lysed cells in 20 ml of cold lysis buffer.
2. Homogenize for 30 s at speed 5 using PCU-2-110 Homogenizer (Kinematica).
3. Clear lysate at 2000 × *g* for 10 min at 4 °C.
4. While clearing the lysate, prepare IgG magnetic beads. In a 1.7 ml microcentrifuge tube, take 250 μl of Dynabeads conjugated to rabbit IgG, and place in a magnetic stand to magnetize and collect the beads. Remove the supernatant and resuspend in 250 μl EB1 w/ 300 mM NaCl. Magnetize and

repeat wash twice. (We do our own conjugations by following the manufacturer's instructions.)

5. Transfer cleared cell lysate to a clean 50 ml Falcon tube. Add 250 μl of IgG conjugated Dynabeads.
6. Incubate on a nutator (Clay Adams) to prevent beads from settling for 30 min at 4 °C.
7. Magnetize beads and resuspend in 100 μl EB1 w/ 200 mM NaCl and transfer to 1.7 ml low retention microcentrifuge tube.
8. Incubate 2 min on a nutator. Magnetize beads and repeat wash with EB1 w/ 200 mM NaCl twice.
9. Wash once with 100 μl TEV cleavage buffer. Magnetize beads and resuspend in 50 μl TEV cleavage buffer. Add 1 μg TEV.
10. Incubate on a rotator for 2 h at 4 °C (*see* **Note 2**).
11. Magnetize beads and take supernatant. Store supernatant on ice until velocity sedimentation (*see* **Note 3**).

3.3 Velocity Sedimentation

The soluble pool of γ-tubulin small complex and other spindle pole body components exist in the higher fractions of the sucrose gradient (fractions 1–7). Spindle pole bodies, as determined by western blot analysis and verified by mass spectrometry, migrate near the 2.0 M–2.5 M sucrose interface. Intact spindle pole bodies reproducibly fall in fractions 9–11 (out of 12). The velocity sedimentation effectively separates the soluble pool of spindle pole body components from the intact spindle pole bodies, resulting in a more homogeneous sample.

1. Using a wide-bore tip, load 200 μl of 2.5 M sucrose into a 1 ml thick wall polycarbonate tube for the TLS-55 rotor.
2. Carefully, to avoid mixing, use wide-bore tips to layer 40, 30, 20, and 10 % sucrose solutions in tube.
3. Let sit at 4 °C for 2 h to allow the gradient to equilibrate.
4. Load purified sample onto the sucrose gradient with a wide-bore tip and centrifuge at 50,000 rpm for 5 h at 4 °C in a TLS-55 swinging bucket rotor.
5. Stepwise remove 90 μl fractions from the top of the gradient using a wide-bore tip.
6. To identify which fractions of the sucrose gradient contain spindle pole bodies, spindle pole body components Spc110 and Spc97 can be probed via western blot analysis (*see* Fig. 1).
7. Flash freeze 5 μl aliquots of the spindle pole body-containing fractions in liquid nitrogen and store at −80 °C.

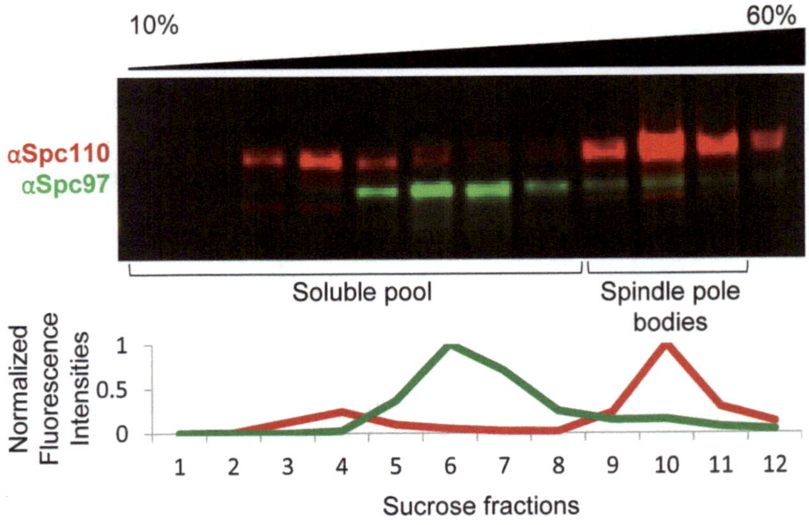

Fig. 1 Western blot analysis of sucrose gradient fractions. Twelve fractions (90 μl each) were taken from the top of the sucrose gradient, with sucrose concentration increasing from *left* (10 %) to *right* (60 %). Sucrose fractions were probed for Spc110 (in *red*) and Spc97 (in *green*) by western blot analysis. Integrated intensities were normalized to the peak intensity. The soluble pool of spindle pole body components migrate in fractions 1–7 while fully formed spindle pole bodies migrate in fractions 9–11. The presence of all spindle pole body components in fractions 9–11 was further confirmed by mass spectrometry. Note that the antibody against Spc110 is much more sensitive than the antibody against Spc97

4 Notes

1. The ease by which the lysed cells are removed from the grinding jar varies greatly. It helps to fill the jar entirely, with a full 50 ml Falcon tube of cell noodles, as well as thoroughly cool the jar between grinding programs. This will result in a loose powder that can easily be scooped out and transferred to a 50 ml Falcon tube for storage. Insufficient cooling between grinding programs can cause the cell dust to cake on the grinding jar walls, requiring substantial effort to scrape off.
2. At this step, incubation on a rotator is preferred over a nutator. With the smaller volume during TEV cleavage, the nutator is insufficient for preventing the Dynabeads from settling to the bottom of the tube.
3. At this step in the purification, you have spindle pole bodies at a high concentration, but contaminated with nucleoli, fragments of the nucleus, and soluble γ-tubulin small complex. If spindle pole body concentration is of more importance than purity, the TEV eluate from this step can be used in downstream experiments. If purity is of high importance, the TEV eluate can be further purified by velocity sedimentation; however, the concentration of spindle pole bodies will decrease.

References

1. Moens PB, Rapport E (1971) Spindles, spindle plaques, and meiosis in the yeast *Saccharomyces cerevisiae*. J Cell Biol 50:344–361
2. Byers B, Goetsch L (1975) Behavior of spindles and spindle plaques in the cell cycle and conjugation of *Saccharomyces cerevisiae*. J Bacteriol 124(1):511–523
3. Rout MP, Kilmartin JV (1990) Components of the yeast spindle and spindle pole body. J Cell Biol 111:1913–1927
4. Knop M, Schiebel E (1997) Spc98p and Spc97p of the yeast gamma-tubulin complex mediate binding to the spindle pole body via their interaction with Spc110p. EMBO J 16(7):1550–1564
5. Nguyen T, Vinh DB, Crawford DK, Davis TN (1998) A genetic analysis of interactions with Spc110p reveals distinct functions of Spc97p and Spc98p, components of the yeast gamma-tubulin complex. Mol Biol Cell 9(8):2201–2216
6. Knop M, Schiebel E (1998) Receptors determine the cellular localization of a gamma-tubulin complex and thereby the site of microtubule formation. EMBO J 17(14):3952–3967
7. Winey M, Mamay CL, O'Toole ET, Mastronarde DN, Giddings TH Jr, McDonald KL, McIntosh JR (1995) Three-dimensional ultrastructural analysis of the *Saccharomyces cerevisiae* mitotic spindle. J Cell Biol 129(6):1601–1615
8. Peterson JB, Ris H (1976) Electron-microscopic study of the spindle and chromosome movement in the yeast *Saccharomyces cerevisiae*. J Cell Sci 22:219–242
9. Byers B, Shriver K, Goetsch L (1978) The role of spindle pole bodies and modified microtubule ends in the initiation of microtubule assembly in *Saccharomyces cerevisiae*. J Cell Sci 30:331–352
10. Hyams JS, Borisy GG (1978) Nucleation of microtubules in vitro by isolated spindle pole bodies of the yeast *Saccharomyces cerevisiae*. J Cell Biol 78(2):401–414
11. Bullitt E, Rout MP, Kilmartin JV, Akey CW (1997) The yeast spindle pole body is assembled around a central crystal of Spc42p. Cell 89(7):1077–1986
12. Niepel M, Strambio-de-Castillia C, Fasolo J, Chait BT, Rout MP (2005) The nuclear pore complex-associated protein, Mpl2p, binds to the yeast spindle pole body and promotes its efficient assembly. J Cell Biol 170(2):225–235

Chapter 13

A Cell-Free System for Real-Time Analyses of Centriole Disengagement and Centriole-to-Centrosome Conversion

Rajesh Kumar Soni and Meng-Fu Bryan Tsou

Abstract

Centriole or centrosome number in cycling cells is strictly maintained through coordinated duplication and segregation. Duplication is limited to once only per cell cycle by separating the assembly event that occurs in S/G2 phase from the two licensing events, centriole disengagement and centriole-to-centrosome conversion, both of which occurs in mitosis. In addition to duplication licensing, centriole-to-centrosome conversion also enables centrioles to associate with spindle poles and thereby to segregate equally during cell division. Centriole disengagement and centriole-to-centrosome conversion thus constitute the major regulatory module ensuring centrosome homeostasis in cycling cells. Using *Xenopus* egg extracts and purified engaged centrioles, we here describe an in vitro assay allowing us to synchronously induce the initiation of centriole disengagement and centrosome formation, pause the reaction anytime during the process, and more importantly, preserve "reaction intermediates" for further analyses.

Key words Centriole, Centrosome, Disengagement, Duplication licensing, Centriole-to-centrosome conversion, MTOC, PCM, *Xenopus* egg extract

1 Introduction

Animal centrioles form the core of the centrosome or microtubule-organizing center (MTOC), and have profound effects on nearly all microtubule-based processes including spindle formation and cell division. Centriole or centrosome numbers in cycling cells are carefully maintained [1]. Vertebrate cells begin the cell cycle with two centrioles, each of which duplicates in S phase, generating two pairs of mother–daughter centrioles that are tightly attached or engaged to each other. In mitosis, the two pairs of engaged centrioles segregate equally to two daughter cells through association with spindle poles, and subsequently lose their tight engagement at late mitosis, restoring the normal copy number of centrioles for the next cell cycle.

Centrioles can duplicate and segregate (through association with spindle poles) only when they have been converted to cen-

Paul Chang and Ryoma Ohi (eds.), *The Mitotic Spindle: Methods and Protocols*, Methods in Molecular Biology, vol. 1413, DOI 10.1007/978-1-4939-3542-0_13,

trosomes [2]. The conversion is a process in which newborn centrioles acquire the competence to recruit the pericentriolar material (PCM) and thereby function as the MTOC or centrosome. Centriole-to-centrosome conversion starts in early mitosis depending on Plk1 and CEP295 [2, 3], and completes at the end of the cell cycle together with centriole disengagement, giving rise to one previously converted and one newly converted centrioles that are both MTOC-competent and disengaged. These disengaged, converted centrioles serve as mother centrioles that duplicate in the following S phase and carries the newborn, MTOC-non-competent daughter centriole to which it is engaged through the segregation process, ensuring centriole homeostasis. Importantly, only centrioles that are both converted and disengaged can duplicate, whereas non-converted centrioles, engaged or not, are "infertile" [2, 3]. These two requirements, centriole-to-centrosome conversion and centriole disengagement, fully exclude unlimited duplication in one cell cycle.

Aberration in centriole/centrosome numbers is known to negatively impact spindle formation, chromosome segregation and cell division [4, 5], but mechanisms underlying centriole-to-centrosome conversion and centriole disengagement remain elusive. A useful approach is to develop an in vitro assay recapitulating both processes. Here, we describe a cell-free system in *Xenopus* egg extracts [6] allowing real-time monitoring and characterization of centriole disengagement and centrosome formation as they occur in late mitosis [7].

2 Materials

2.1 Isolation of Human Engaged Centrioles

1. HeLa cell expressing centrin2-GFP [7].
2. Dulbecco's modified Eagle's medium (DMEM) supplemented with 10 % fetal bovine serum (FBS) and 1 % penicillin–streptomycin (all Invitrogen).
3. Aphidicolin (Sigma): 2 mg/ml in DMSO as the stock solution (1000×), stored in −20 °C.
4. Nocodazole (Sigma).
5. Hypotonic buffer (20 mM K-HEPES pH 7.8, 5 mM K-acetate, 0.5 mM $MgCl_2$, 0.5 mM DTT make fresh).
6. Cell Scraper (Nunc).
7. Falcon Cell Strainers (Fisher).
8. DNase I (Stratagene): 10 U/μl.
9. Sucrose gradient buffer (SGB): 10 mM HEPES pH 7.8, 0.1 % betamercaptoethanol, 0.1 % Triton X-100.

10. Dounce tissue grinder.
11. Discontinuous sucrose gradient. From the bottom of centrifuge tubes, sequentially load 5 ml of 70 % (W/W) sucrose, 3 ml of 50 % (W/W) sucrose, and 3 ml of 40 % (W/W) sucrose solutions. Sucrose solutions are made in sucrose gradient buffer.
12. Benchtop centrifuge (Fisher).
13. Ultracentrifuge with a SW 32 Ti or SW 28 rotor (Beckman).
14. Ultra-Clear™ centrifuge tube (Beckman; 344058).
15. Microscope (Zeiss; Axio Imager A1).

2.2 CSF Extract Preparation from Xenopus laevis

1. Pregnant mare serum gonadotropin (PMSG) (Sigma): 100 U/ml in sterile water.
2. Human chorionic gonadotropin (HCG) (Sigma): 100 U/ml in sterile water.
3. MMR buffer: 100 mM NaCl, 2 mM KCl, 1 mM $MgCl_2$, 2 mM $CaCl_2$, 5 mM HEPES, pH 7.8.
4. Extract buffer (XB): 10 mM potassium HEPES pH 7.7, 1 mM $MgCl_2$, 0.1 mM $CaCl_2$, 100 mM KCl, 50 mM sucrose—make fresh.
5. CSF-XB: 10 mM HEPES pH 7.7, 2 mM $MgCl_2$, 0.1 mM $CaCl_2$, 100 mM KCl, 5 mM EGTA, 50 mM sucrose—make fresh.
6. 2 M sucrose.
7. Dejellying solution: 2 % cysteine in XB, pH 7.8—make fresh.
8. Protease inhibitors (1000×): 10 mg/ml of leupeptin, chymostatin, and pepstatin in DMSO
9. Cytochalasin D in DMSO (10 mg/ml; 1000×).
10. 18 gauge needle (BD 305195).
11. Swing bucket rotor (e.g., H-6/Sorvall or SW-55/Beckman) and centrifuge tubes (e.g., Sorvall 03124).

2.3 Analyses of Centriole Disengagement and Centriole-to-Centrosome Conversion in Egg Extracts

1. Glass slides and 12 mm round coverslips (Fisher Scientific).
2. Liquid nitrogen.
3. Razor blade (Fisher Scientific; heavy duty; S95932).
4. Slide jar (Wheaton).
5. 100 % methanol (−20 °C).
6. PBS: phosphate buffered saline.
7. PBST: 0.1 % Triton X-100 in PBS.
8. PBSB: 3 % bovine serum albumin (BSA) in PBS.
9. Antibodies: anti-GFP (BioLegend), anti-human C-Nap1 [7], anti-human SAS-6 (Santa Cruz), anti-γ-tubulin (Santa Cruz),

anti-pericentrin (Abcam), and appropriate secondary antibodies (Invitrogen), all diluted in PBSB.

10. Clearing solution [8]: mix 1 volume of benzyl alcohol to 2 volumes of benzyl benzoate (1:2; both from Sigma).
11. Nail polish.
12. Fluorescence microscope (e.g., Axio Imager A1, Zeiss).

3 Methods

3.1 Isolation of Engaged Centrioles Labeled with Centrin-GFP

1. Our centrosome isolation is based on previously published protocols [9–11], with some critical modifications.
2. Grow 7–10 large plates (150 mm) of HeLa cells expressing Centrin-GFP to ~70 % confluency.
3. Add 2 μg/ml aphidicolin and incubate for 18 h to arrest cells at S phase.
4. After 18 h, add nocodazole to 5 μg/ml, and incubate for 1 h to depolymerize microtubules, which facilitates the detachment of centrioles from nuclei.
5. Remove the medium and rinse cells twice with ice-cold hypotonic buffer (HB).
6. Add 20 ml of HB to each 150 mm plate and incubate at 4 °C for 10 min to swell cells.
7. Remove HB and scrape cells off the plate using cell scrapers.
8. Break the swollen cells in a Dounce tissue grinder using a loose pestle, and transfer the lysate to a Falcon tube. 5–10 strokes are sufficient to release centrosomes.
9. Centrifuge the lysate at $2000 \times g$ in a benchtop centrifuge with a swing bucket rotor at 4 °C for 5 min to pellet nuclei, and collect the supernatant.
10. Wash the pellet with 10 ml of HB to release trapped centrosomes and spin again at $2000 \times g$ at 4 °C for 5 min. Collect the supernatant.
11. Combine supernatants, add Triton X-100 to 0.5 %, and spin again at $2000 \times g$ at 4 °C for 5 min. Collect the supernatant.
12. Filter the supernatant through 70 μm cell strainers.
13. Incubate the supernatant with 1 μg/ml (2 U/ml) of DNase I on ice for 30 min.
14. Load the final supernatant onto discontinuous sucrose gradients in ultra-clear centrifuge tubes (25×89 mm; Beckman) and spin at $122{,}000 \times g$ with SW 28 or SW 32 Ti rotor at 4 °C for 1 h.

15. Puncture the bottom of the ultracentrifuge tube with 18 gauge needles and collect 0.5 ml fractions in 1.5 ml tubes on ice. Centrosomes are normally trapped at the interphase of 70 and 50 % sucrose gradients.
16. Spot 1 μl of each fraction on a slide and check the yield of centrioles (GFP labeled) with fluorescence microscopy. An example of engaged centrioles in a peak fraction is shown in Fig. 2a.
17. Aliquot and freeze centrosome-containing fractions in liquid nitrogen, and store at −80 °C. The concentration of centrosomes in peak fractions is around $2.5 \times 10^4/\mu l$.

3.2 CSF Extract Preparation from *Xenopus laevis*

1. Our preparation of CSF extracts is similar to what is described by Murray [6].
2. Day 1: Inject female frogs (*Xenopus laevis*) with 50 U pregnant mare serum gonadotropin (PMSG) to prime ovulation.
3. Day 5: Inject the same female frogs with 250 U human chorionic gonadotropin (HCG) to induce final ovulation. Each injected frog should be placed individually in a static tank containing 1.5 l of 1× MMR buffer at 18–20 °C for 16 h during which ovulation occurs.
4. Eggs are collected at day 6 and washed with fresh MMR buffer at 18 °C.
5. Remove MMR, treat eggs with 2 % cysteine in extract buffer (dejellying solution) to remove the jelly coat.
6. Gently wash dejellied eggs three times with extract buffer (XB), three times with CSF-XB, and two times with CSF-XB supplemented with proteinase inhibitors (leupeptin, chymostatin, and pepstatin at 10 μg/ml), and cytochalasin D (10 μg/ml).
7. Using fire polished pasteur pipettes, transfer the eggs drop by drop into centrifuge tubes (e.g., Sorvall #03124; 13 × 100 mm).
8. Remove excess CSF-XB on top of the eggs.
9. Pack the eggs by centrifuging at 200 × *g* for 1 min and then 500 × *g* for 1 min, and aspirate the excess buffer.
10. Crush the eggs by centrifugation at 10,000 rpm (16,400 × *g*) for 10 min at 16 °C in a swinging bucket rotor (e.g., H-6/ Sorvall with proper adaptors).
11. Collect the cytoplasmic fraction, which is the middle layer in straw color, with an 18-gauge needle by puncturing the side of the centrifuge tube. Immediately place the extract in cold eppendorf tubes on ice.
12. Add 1/1000 volume of protease inhibitors and cytochalasin D, and 1/40 volume of 2 M sucrose to the extract. Mix thoroughly with blunted pipette tips.
13. Fresh CSF extracts should be used within 1 or 2 h.

3.3 Setup of Centriole Disengagement and Centriole-to-Centrosome Conversion Reaction in CSF-Released Extracts

1. This assay is based on Tsou and Stearns [7].
2. Combine 40 μl of fresh CSF extract with 1 μl of isolated engaged centrioles (approximately 2.5×10^4) in a 0.5 ml eppendorf tube and incubate at RT (~23 °C) for 10 min. Make sure centrioles are mixed thoroughly in the extract (with blunted pipette tips).
3. Add 0.4 μl of 40 mM $CaCl_2$ to trigger centriole disengagement and centriole-to-centrosome conversion, both of which occur robustly in the following 30–40 min during which the CSF extract exits mitosis (CSF-released extract).
4. Reactions can be stop and preserved at different time points for future analyses by adding 1 μl of 0.5 M EDTA followed by instant freezing in liquid nitrogen. Alternatively, the behavior of centrioles or centrosomes can be examined in real time as the reaction proceeds (*see* Subheading 3.4 below).

3.4 Examining the Centriole Disengagement and Centrosome Assembly Reaction in Real Time

At different time points after calcium addition, aliquot 1 μl of the reaction mix onto a glass slide (Fig. 1a), and squash the extract into a thin layer with a coverslip (12 mm round) (Fig. 1b, c). The behavior of GFP-labeled centrioles in a live reaction can then be directly observed under fluorescence microscopy, as revealed in Fig. 2b for centriole disengagement.

1. Alternatively, the squashed extract can be instantly frozen by immersing the whole slide into liquid N_2 for 1 min (Fig. 1d), capturing the reaction intermediates for further analyses.
2. To immunostain centrioles in the extract, take out the slide from liquid N_2 and quickly remove or pry off the coverslip with

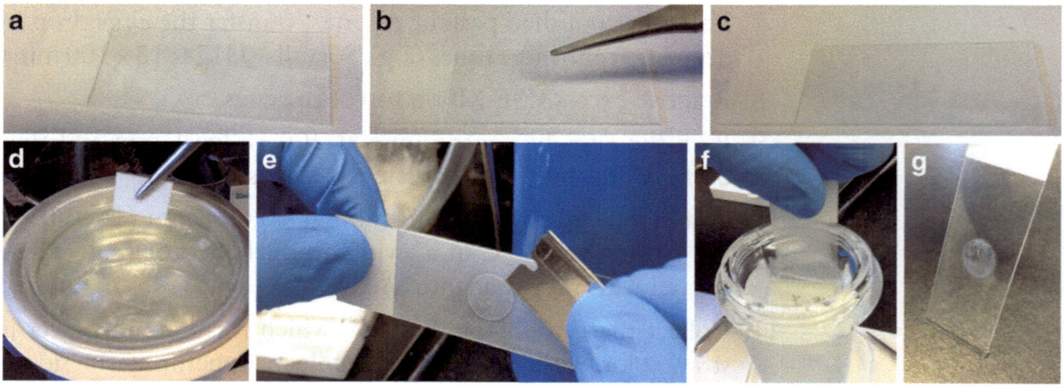

Fig. 1 Preparation of a thin layer of frozen extract for immunostaining. (**a**) Aliquot 1 μl of the reaction mix onto a glass slide. (**b**, **c**) Squash the extract into a thin layer using a 12 mm rounded coverslip. (**d**) Instantly freeze the squashed extract by immersing the whole slide into liquid N_2 for 1 min. (**e**) Take out the slide from liquid N_2 and quickly remove or pry off the coverslip with a single-edge razor blade while the extract is still frozen. (**f**) Immediately fix the frozen extract by immersing the slide into chilled 100 % methanol (−20 °C) in a slide jar for 5 min. (**g**) After fixation, a thin, turbid layer of the extract is left on the slide

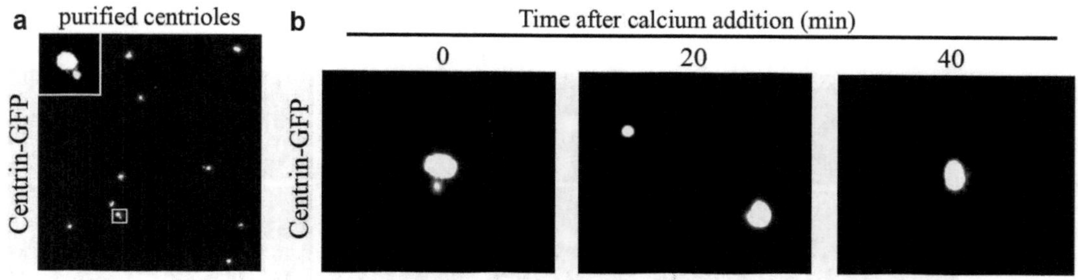

Fig. 2 Centriole disengagement in real-time reactions. (**a**) GFP-labeled, engaged centrioles purified from HeLa cells. GFP signals are visualized by fluorescence microscopy. (**b**) Monitoring of centriole disengagement. Purified engaged centrioles were incubated with CSF extracts, and calcium was added to release the arrest. Centrioles at indicated time points after calcium addition were visualized directly by fluorescence microscopy. Note the separation of daughter centrioles (weak/small GFP foci) from mothers

a single-edge razor blade (Fig. 1e), leaving a thin layer of frozen extract on the slide.

3. Immediately fix the frozen extract by immersing the slide in chilled 100 % methanol (–20 °C) in a jar for 5 min (Fig. 1f). After fixation, a thin, turbid layer of the extract can be seen stably attached to the slide (Fig. 1g).
4. Rehydrate the fixed extract with PBST for 5 min, and then with PBS for another 5 min in separate jars to wash out the detergent (Triton X-100) before staining.
5. Without drying or touching the rehydrated extract, carefully dry the rest of the slide by wiping off the PBS around the squashed sample using tissue paper, leaving the extract in a small square of hydrated area for the following immunostaining.
6. Cover the rehydrated extract with 30–40 μl blocking solution (PBSB) and keep the slide in a wet chamber for 1 h at RT (23 °C)
7. To examine centriole disengagement or centriole-to-centrosome conversion, incubate the extract with 30–40 μl of selected primary antibodies (e.g., anti-GFP, γ-tubulin, C-Nap1, or pericentrin), and keep the slide in a wet chamber for 1 h at RT, or overnight at 4 °C.
8. Wash the slide in jars twice with PBST for 5 min, and then with PBS for another 5 min.
9. Wipe off the excess PBS buffer around the sample, apply 30–40 μl of secondary antibodies onto the sample, and let them incubate for 1 h at RT in a wet chamber.
10. Wash the slide three times with PBST, and dehydrate the extract by immersing the slide in 100 % chilled methanol twice in separate jars, each for 5 min. Use freshly made 100 % methanol to ensure complete dehydration of the extract.
11. Take out the slide from methanol, and let methanol evaporate at RT for 1–2 min.

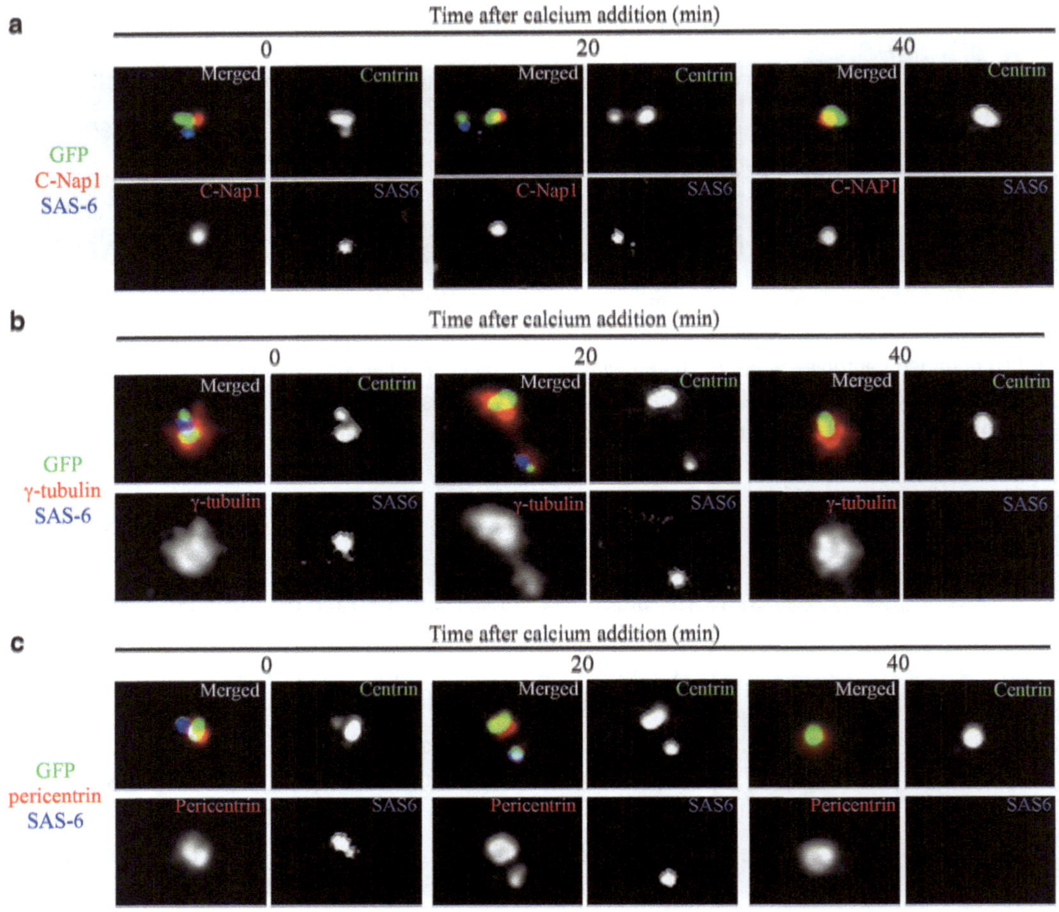

Fig. 3 Analyses of centriole disengagement and centriole to centrosome conversion in frozen extracts by immunostaining. (**a**–**c**) Centrioles and the relative position of the mother and daughter in the extract were fixed and examined with indicated antibodies at indicated time points after calcium addition. GFP-centrin marks all centrioles, C-Nap1 highlights mother centrioles (**a**), and SAS-6 labels only the daughter. Note that daughter centrioles start to recruit γ-tubulin (**b**) and pericentrin (**c**) as they are being disengaged from the mother (20 min time points). SAS-6 is stably maintained in disengaged daughter centrioles, as the egg extract naturally lacks the activity of APC^{cdh1} mediated protein degradation. At the 40 min time point, centrioles were fully disengaged, showing only the mother centriole

12. After the dehydrated extract becomes air-dry and appears turbid, apply 10 μl of clearing solution (benzyl alcohol–benzyl benzoate in 1:2 ratio) on top of the dry extract [8].
13. As the extract instantly becomes transparent, mount a 22 × 22 mm coverslip onto the clearing solution, remove the excess solution, and seal the coverslip with nail polish. Avoid trapping air bubbles in between the slide and coverslip.
14. The slide is now ready for viewing. An example of the immunostaining result that reveals centriole disengagement and centriole-to-centrosome conversion is shown in Fig. 3 (*see* also **Note 13** below).

4 Notes

1. We use HeLa cells for centriole isolation, because centrosomes can be easily detached from nuclei through the procedure described here, which is not the case for some cell types. However, after a few strokes of douncing, check the lysed cells under microscope to make sure that centrosomes are released from cells and efficiently detached from nuclei.
2. Do not use a tight pestle to break cells, as it may fragment the nuclei into small pieces.
3. Sucrose gradients should be freshly made. Do not disturb the gradient when loading the lysate.
4. Engaged centrioles normally peak at fractions 9–11.
5. The robustness of the assay critically depends on the quality of the extract and thus the quality of eggs. Energy mix is not required for this assay.
6. The coverslip needs to be removed quickly while the squashed extract remains frozen.
7. It is not a good idea to fix the squashed, frozen extract with PFA, as the extract would melt, and then quickly come off the slide during the fixation.
8. Importantly, in the frozen extract, the relative position of mother and daughter centriole in each pair of "disengaging centrioles" can be instantly captured, and permanently maintained after fixation, allowing further analyses by immunostaining (Fig. 3).
9. A PBS wash (at the end of a wash series) to remove Triton X-100 is critical, as the residual detergent on the slide would affect the surface tension of the antibody solution added later, making the immunostaining process problematic.
10. Similarly, antibodies (both primary and secondary) should be diluted in PBSB free of detergent.
11. The clearing solution (benzyl alcohol–benzyl benzoate) is not water soluble, so the extract must be completely dehydrated with 100 % methanol and air-dried before mounting.
12. The clearing solution possesses little or no anti-fade activity, so the subsequent imaging process needs to be done quickly with care.
13. Figure 3 shows the coincidence of centriole disengagement (Fig. 3a–c) and centriole-to-centrosome conversion (Fig. 3b, c) in CSF-released extracts. GFP-centrin marks all centrioles, whereas C-Nap1 highlights mother centrioles, and SAS-6 labels only the daughter. At the 20 min time point after calcium addition, daughter centrioles start to recruit γ-tubulin and pericentrin as they are disengaging from the mother (Fig. 3b, c),

marking the conversion of centrioles to centrosomes. Note that as the egg extract naturally lacks APC^{cdh1}-mediated protein degradation, SAS-6 is stably maintained in disengaged daughter centrioles, serving nicely as a daughter marker.

References

1. Nigg EA, Stearns T (2011) Nat Cell Biol 13:1154–1160
2. Wang WJ, Soni RK, Uryu K, Tsou MF (2011) J Cell Biol 193:727–739
3. Izquierdo D, Wang WJ, Uryu K, Tsou MF (2014) Cell Rep 8:957–965
4. Ganem NJ, Godinho SA, Pellman D (2009) Nature 460:278–282
5. Poulton JS, Cuningham JC, Peifer M (2014) Dev Cell 30:731–745
6. Murray AW (1991) Methods Cell Biol 36:581–605
7. Tsou MF, Stearns T (2006) Nature 442:947–951
8. Gard DL, Hafezi S, Zhang T, Doxsey SJ (1990) J Cell Biol 110:2033–2042
9. Mitchison TJ, Kirschner MW (1986) Methods Enzymol 134:261–268
10. Blomberg-Wirschell M, Doxsey SJ (1998) Methods Enzymol 298:228–238
11. Bornens M, Paintrand M, Berges J, Marty MC, Karsenti E (1987) Cell Motil Cytoskeleton 8:238–249

Chapter 14

Assays to Study Mitotic Centrosome and Spindle Pole Assembly and Regulation

Vladimir Joukov, Johannes C. Walter, and Arcangela De Nicolo

Abstract

Faithful chromosome segregation during cell division requires proper bipolar spindle assembly and critically depends on spindle pole integrity. In most animal cells, spindle poles form as the result of the concerted action of various factors operating in two independent pathways of microtubule assembly mediated by chromatin/RanGTP and by centrosomes. Mutation or deregulation of a number of spindle pole-organizing proteins has been linked to human diseases, including cancer and microcephaly. Our knowledge on how the spindle pole-organizing factors function at the molecular level and cooperate with one another is still quite limited. As the list of these factors expands, so does the need for the development of experimental approaches to study their function. Cell-free extracts from *Xenopus laevis* eggs have played an instrumental role in the dissection of the mechanisms of bipolar spindle assembly and have recently allowed the reconstitution of the key steps of the centrosome-driven microtubule nucleation pathway (Joukov et al., Mol Cell 55:578–591, 2014). Here we describe assays to study both centrosome-dependent and centrosome-independent spindle pole formation in *Xenopus* egg extracts. We also provide experimental procedures for the use of artificial centrosomes, such as microbeads coated with an anti-Aurora A antibody or a recombinant fragment of the Cep192 protein, to model and study centrosome maturation in egg extract. In addition, we detail the protocol for a microtubule regrowth assay that allows assessment of the centrosome-driven spindle microtubule assembly in mammalian cells.

Key words Mitosis, Spindle assembly, Spindle pole, Centrosome, Pericentriolar material, Microtubule-organizing center, Cep192, *Xenopus* egg extract, Cell-free system

1 Introduction

Spindle poles are inherent components of the mitotic/meiotic spindle and represent focal collections of microtubule (MT) minus ends towards which chromatids move during anaphase [1]. In addition to their role as centers of convergence of the chromosomes in nascent daughter cells, the spindle poles play two other important roles by maintaining spindle integrity in face of the forces of chromosome alignment and segregation [2, 3] and by anchoring spindle MTs, which is a prerequisite for chromosome segregation. Proper spindle pole assembly is, therefore, essential

Paul Chang and Ryoma Ohi (eds.), *The Mitotic Spindle: Methods and Protocols*, Methods in Molecular Biology, vol. 1413,
DOI 10.1007/978-1-4939-3542-0_14, © Springer Science+Business Media New York 2016

for accurate cell division and maintenance of genome stability. Indeed, supernumerary or malformed poles can lead to micronuclei formation and aneuploidy, which are the hallmarks of many cancers [2–4]. Not surprisingly, there is a growing list of spindle pole proteins implicated in cancer and other diseases [5].

In most animal cells, the position of each spindle pole is defined by the centrosome, a non-membrane-bound organelle that serves as the major MT-organizing center both in interphase and in mitosis. The centrosome consists of a pair of centrioles surrounded by pericentriolar material (PCM). Centrioles serve as templates for the assembly of new centrosomes and cilia and duplicate once per cell cycle (in S phase). The PCM nucleates and anchors MTs and participates in other centrosome functions. In addition to their role in spindle assembly, centrosomes are involved in a variety of other cellular processes, including motility, protein homeostasis, intracellular signaling, and immune response [6–8].

In the G2 phase of the cell cycle, the two centrosomes dramatically increase in size and MT-nucleating capacity owing to the recruitment of additional PCM components, such as the γ-tubulin ring complex (γ-TuRC), which serves as a MT template. This process, termed centrosome maturation, leads to the formation of two dense radial arrays of MTs (centrosomal MT asters) [6, 7]. Centrosome maturation and the consequent MT nucleation and anchoring are driven by a multistep Aurora A (AurA)-Plk1 kinase cascade locally organized by the key regulator of centrosome biogenesis, Cep192 [9, 10]. Notably, this cascade and the resulting focal MT nucleation can be recapitulated by artificial clustering of endogenous Cep192-Aur A-Plk1 complexes on the surface of protein-coated microbeads in metaphase (M-phase)-arrested *Xenopus* egg extract [10].

After the nuclear envelope breakdown, spindle MTs are also generated in the vicinity of chromatin by a completely different mechanism dependent on the small GTPase Ran. Specifically, Ran-GTP, which is produced by the chromatin-bound guanine nucleotide exchange factor for Ran, RCC1 [11] promotes bipolar spindle assembly by two cooperative mechanisms: (1) it releases multiple spindle assembly factors, such as TPX2, NuMA, HuRP, and NuSAP, from their inhibitory binding to importin α/β, thus promoting MT nucleation and growth, and (2) it enables the binding of certain other spindle assembly factors to exporting karyopherins (exportins), which is critical for the localization of these factors to specific spindle compartments [11]. The MTs generated by chromatin, centrosomes, and by two additional sources, kinetochores and MTs walls, are then organized by specialized proteins into a bipolar spindle [2, 3, 12].

Interestingly, although centrosomes are associated with and define the position of the spindle poles, they are not required for spindle assembly per se, since cells lacking these organelles (e.g., female meiotic cells) can form bipolar spindles and faithfully segregate chromosomes. Thus, the centrosome-independent pathways of MT assembly suffice to assemble a functional bipolar spindle. Both the formation of acentrosomal spindle poles and the conversion of centrosomal MT asters into spindle poles were shown to require Ran-GTP and at least three groups of proteins: (1) MT minus end-associated spindle assembly factors (such as γ-TuRC and the MT crosslinking protein NuMA); (2) motor proteins (such as cytoplasmic dynein and kinesins); and (3) factors that affect MT growth and shrinkage (such as chTOG/XMAP215, katanin, and MT-destabilizing kinesins) [2, 3, 10, 13].

Our understanding of the mechanisms of spindle pole assembly has greatly benefited from studies in cell-free *Xenopus laevis* egg extracts. Such extracts are naturally arrested in M-phase of meiosis II and, when supplemented with demembranated sperm nuclei, form bipolar spindles [14, 15]. Moreover, since after the release of the arrest, the inherently oscillating activity of Cdk1/cyclin B drives cell cycle transitions in *Xenopus* egg extracts, this experimental system recapitulates (and allows the dissection of) both mitotic and interphase processes [15, 16]. The function of a protein of interest can be investigated in egg extracts by immunodepletion/reconstitution approaches or by the use of various inhibitors or dominant negative mutants. Studies in *Xenopus* egg extracts have allowed researchers to experimentally demonstrate the existence of and to recapitulate each of the four major independent pathways of spindle MT assembly, i.e., those mediated by the chromatin [17–21], kinetochores [22, 23], MT walls [24], and centrosomes [10].

This chapter primarily focuses on the application of *Xenopus* egg extracts to the analysis of mitotic centrosome and spindle pole assembly. Detailed protocols on the preparation of *Xenopus* egg extracts and demembranated sperm nuclei and on their use as tools to study spindle assembly can be found elsewhere [14–16, 25]. We first outline an assay for the analysis of acentrosomal spindle pole assembly. We then describe assays to model and study centrosome maturation using both natural (centrosomes) and artificial (beads coated with an anti-Aur A antibody or a recombinant Cep192 protein fragment) templates. In addition, we provide a protocol for the so-called MT regrowth assay that can be used to determine MT-nucleating activity of mitotic centrosomes in cultured mammalian cells.

2 Materials

2.1 Analysis of Acentrosomal Spindle Pole Assembly in Xenopus Egg Extracts

1. Recombinant human or *Xenopus* Ran(Q69L) protein. Protocols for preparation of recombinant Ran(Q69L) can be found elsewhere [26, 27] (*see* **Note 1**).
2. 1 M HEPES–KOH (pH 7.7). Dissolve HEPES in deionized water, adjust pH with KOH, filter-sterilize, and store at 4 °C (*see* **Note 2**).

3. 20× extract buffer (XB)-salts: 2 M KCl, 20 mM $MgCl_2$, 2 mM $CaCl_2$. Filter-sterilize and store at 4 °C.
4. 1.5 M sucrose. Dissolve in deionized water, filter-sterilize, and store at 4 °C.
5. 1 M dithiothreitol (DTT). Dissolve in deionized water, filter-sterilize, and store in 100 μL aliquots at −20 °C.
6. XB: 10 mM HEPES–KOH (pH 7.7), 100 mM KCl, 1 mM $MgCl_2$, 0.1 mM $CaCl_2$, 50 mM sucrose, 1 mM DTT. Make fresh from the stock components without DTT (*see* **items 2–4**). Add DTT from a 1 M stock (*see* **item 5**) prior to use.
7. 0.5 M EDTA (pH 8.0). Add EDTA powder to deionized water and adjust pH with NaOH until the powder dissolves. Filter-sterilize and store at room temperature.
8. Guanosine 5′-triphosphate (GTP) sodium salt (Sigma-Aldrich).
9. 1 M $MgCl_2$. Dissolve in deionized water, filter-sterilize, and store at 4 °C.
10. 1.5 mL Eppendorf tubes.
11. 0.5 mL Eppendorf tubes.
12. Dialysis cassettes or dialysis tubing.
13. Protein concentrators, 0.5 mL, molecular-weight cutoff <10,000 kDa, low nonspecific binding (EMD Millipore Amicon or Thermo Scientific).
14. M-phase-arrested *Xenopus* egg extract. Detailed protocols for the preparation of the extract are described in refs. 14, 15 and 25. The extract should be quality tested before use (*see* **Notes 3** and **4**).
15. Water bath at 21 °C with a floating rack for microcentrifuge tubes (*see* **Note 5**).
16. Rhodamine-labeled tubulin (Cytoskeleton). Briefly spin the tubes and dissolve the content (20 μg of the lyophilized protein) in 2 μL of XB, to obtain a final concentration of 10 mg/mL. Snap-freeze the tubes in liquid nitrogen and store at −80 °C (*see* **Note 6**).
17. 10× MMR: 50 mM HEPES (pH 7.8), 1 M NaCl, 20 mM KCl, 10 mM $MgCl_2$, 20 mM $CaCl_2$, 1 mM EDTA (from a 0.5 M stock, *see* **item** 7). Dissolve the ingredients in deionized water, adjust pH with NaOH, filter-sterilize, and store at room temperature.
18. 16 % Formaldehyde in sealed ampoules (Ted Pella) (*see* **Note** 7).
19. 10 mg/mL Hoechst 33342/33258. Dissolve in deionized water and store at 4 °C.

20. Spindle fix solution: 50 % (v/v) glycerol, MMR (from a 10× stock, *see* **item 17**), 4 % formaldehyde (from a freshly opened 16 % stock, *see* **item 18**), 1 μg/mL Hoechst 33342/33258 (from a 10 mg/mL stock, *see* **item 19**). Make fresh prior to each experiment.
21. Microscope slides, 75 × 25 mm (Thermo Fisher Scientific) (*see* **Note 8**).
22. Coverslips, circular, 12 mm, # 1 (Thermo Fisher Scientific).
23. Straight fine-tip forceps (e.g., watchmaker forceps #5).
24. Nail polish.
25. Epifluorescent microscope.

2.2 Analysis of Centrosome Maturation in Xenopus Egg Extracts

1. 5 mg/mL Nocodazole (Sigma-Aldrich). Dissolve in DMSO and store in 50 μL aliquots at −20 °C.
2. Demembranated sperm nuclei (2×10^5/μL). Prepare as described in refs. 15 or 16.
3. 5× BRB-80: 400 mM PIPES (pH 6.8), 5 mM $MgCl_2$, 5 mM EGTA. Dissolve in deionized water and adjust pH with KOH. Filter-sterilize and store at 4 °C.
4. 20 % Triton X-100 (v/v). Dissolve in deionized water by mixing on a rotator or magnetic stirrer. Store at 4 °C.
5. Spindle dilution buffer: BRB-80 (from a 5× stock, *see* **item 3**), 30 % glycerol (v/v), 0.5 % Triton X-100 (from a 20 % stock, *see* **item 4**), 2 % formaldehyde (from a 16 % stock, *see* Subheading 2.1). Mix all ingredients except formaldehyde and store at room temperature for several weeks. Add formaldehyde just before use.
6. Cushion buffer: BRB-80 (from a 5× stock, *see* **item 3**), 40 % glycerol (v/v). Store at room temperature for several weeks.
7. BRB-80. Prepare from a 5× stock, *see* **item 3**.
8. 15 mL (18 × 102 mm) glass high strength centrifuge tubes (Kimble-Chase Kontes, Cat. #45500-15) equipped with custom-made adaptors and a metal hook or an equivalent tool for lifting the adaptors. These tools are described and schematically depicted in ref. 14.
9. Poly-d-lysine-coated coverslips, circular, 12 mm, #1 (Corning Biocoat, Cat. #354086).
10. Perforated holder with a lid for circular, 12 mm coverslips (made of an organic solvent-resistant plastic). An empty container from the Poly-d-lysine-coated circular coverslips (Corning Biocoat, Cat. #354086) ideally suits this purpose. Make a mark with a scratch on one side of the holder, to guide positioning of the inserted coverslips in the same direction (relative to the sample side).

11. Disposable plastic transfer pipets (1–2 mL).
12. High-speed refrigerated centrifuge with an HB-6 (Sorvall) or equivalent rotor.
13. Glass jar with a lid for microscope slides filled with methanol at −20 °C. Store closed in a freezer for up to 2 weeks. Replace methanol after each experiment.
14. Cell culture plates, 24 wells.
15. Parafilm.
16. 1 M Tris–HCl (pH 7.5). Dissolve Tris base in deionized water, adjust pH to 7.5 with concentrated HCl, and store at room temperature.
17. TBS-T: 25 mM Tris–HCl (pH 7.5) (from a 1 M stock, *see* **item 16**), 150 mM NaCl, 0.05 % Tween 20.
18. TBS-T/5 % goat serum: Add goat serum to TBS-T (*see* **item 17**), mix, and filter through a 2-μM filter. Prepare fresh before each experiment.
19. Pasteur pipettes.
20. Vacuum aspirator.
21. Aluminum foil.
22. Specific antibodies for proteins of interest.
23. Fluorochrome-labeled secondary antibody(ies).
24. Vectashield Mounting Medium with DAPI (Vector Laboratories).
25. Cycling egg extract. Prepare as described in detail in ref. 15 (*see* **Note 9**).

2.3 Using Anti-Aur A Antibody-Coated Beads to Model and Study Mitotic Centrosome Formation

1. Dynabeads Protein A (Life Technologies).
2. 0.5 mL siliconized Eppendorf tubes.
3. Variable speed vortex mixer.
4. Magnetic particle concentrator for microtubes referred to in the text as "magnet" (Dynal/Life Technologies).
5. Phosphate-buffered saline (PBS): 140 mM NaCl, 3 mM KCl, 8 mM Na_2HPO_4, 2 mM KH_2PO_4 (pH 7.4). Prepare by dissolving a commercially available stock solution or powder in deionized water and store at 4 °C.
6. 50 mg/mL bovine serum albumin (BSA), molecular biology grade. Dissolve in PBS (*see* **item 5**) and store in 1 mL aliquots at −20 °C. The aliquots can be repeatedly frozen/thawed.
7. PBS/2 mg/mL BSA/0.05 % Tween 20. Prepare before the experiment by mixing PBS (*see* **item 5**), BSA (from a 50 mg/mL stock, *see* **item 6**), and Tween 20.
8. Microcentrifuge.

9. Affinity purified rabbit polyclonal anti-AurA antibody (αAurA) [9]. Dilute in PBS to obtain a concentration of 1 mg/mL (*see* **Note 10**).
10. Variable speed rotator (such as Cole-Parmer Roto-Torque Rotator).
11. 0.2 M sodium borate buffer (pH 8.5)/0.02 % Tween 20. Prepare by dissolving/diluting a borate buffer stock (Thermo Scientific) and adding Tween 20.
12. Dimethyl pimelimidate (DMP), 50 mg vials individually sealed in moisture-proof packages (Thermo Scientific).
13. Crosslinking stop buffer: 100 mM Tris–HCl (pH 7.5) (from a 1 M stock, *see* Subheading 2.2), 150 mM NaCl, 0.05 % Tween 20.
14. PBS/0.1 % Tween 20. Add Tween 20 to PBS (*see* **item 5**).
15. PBS-500/0.1 % Tween 20. Add 210 mg of NaCl to 10 mL of PBS/0.1 % Tween 20 (*see* **item 14**) and mix until NaCl dissolves.
16. 20 % sodium azide. Dissolve in water and store at room temperature (highly toxic when ingested or inhaled).
17. XB/1 mg/mL BSA. Make fresh for each experiment by adding BSA (from a 50 mg/mL stock, *see* **item 6**) to XB (*see* Subheading 2.1). Keep on ice.
18. XB/1 mg/mL BSA/0.05 % Tween 20. Add Tween 20 to the XB/1 mg/mL BSA (*see* **item 17**), mix, and keep on ice.
19. XB/0.1 % Tween 20. Add Tween 20 to XB (*see* Subheading 2.1), mix, and keep on ice.

2.4 Using Cep192 Beads to Model and Study Mitotic Centrosome Formation

1. LB medium: 10 g/L tryptone, 5 g/L yeast extract, 10 g/L NaCl. Dissolve in deionized water and sterilize by autoclaving.
2. *E. coli* cells expressing Cep192$^{1-1000}$-GST fusion proteins. Transform *E. coli* cells optimized for protein expression (e.g., BL21) with a cDNA encoding Cep192$^{1-1000}$ wild type (wt) (Cep192$^{1-1000}$-wt) or its mutants of interest in a pGEX vector (GE Healthcare) [9, 10].
3. Temperature-controlled shaker incubator for flasks.
4. Laboratory flasks (100 mL–2 L).
5. UV–visible-wavelength spectrophotometer.
6. 1 M Isopropyl-β-d-thiogalactopyranoside (IPTG). Dissolve in deionized water and store at −20 °C.
7. 1 M Tris–HCl (pH 8.0). Dissolve Tris base in deionized water, adjust pH with concentrated HCl, and store at room temperature.
8. Protease inhibitor cocktail (e.g., cOmplete protease inhibitor cocktail tablets from Roche Life Science).

9. NETS buffer: 20 mM Tris–HCl (pH 8.0) (from a 1 M stock, *see* **item 7**), 100 mM NaCl, 1 mM EDTA (from a 0.5 M stock, *see* Subheading 2.1), 10 % glycerol (v/v), 1.5 % Sarkosyl (*N*-Lauroylsarcosine), 5 mM β-mercaptoethanol. Store at 4 °C. Prior to use, add protease inhibitors [e.g., 1 tablet of cOmplete (Roche Life Science) per 25 mL].
10. 15 mL Falcon tubes.
11. Lysozyme powder (Sigma-Aldrich). Store at −20 °C.
12. Sonicator (with a probe).
13. Refrigerated microcentrifuge.
14. NETN buffer: 20 mM Tris–HCl (pH 8.0) (from a 1 M stock, *see* **item 7**), 100 mM NaCl, 1 mM EDTA (from a 0.5 M stock, *see* Subheading 2.1), 10 % glycerol, 0.5 % (v/v) Nonidet P-40 (NP-40), 5 mM β-mercaptoethanol. Store at 4 °C.
15. NETN buffer/500 mM NaCl. Dissolve 232 mg of NaCl in 10 mL of NETN buffer (*see* **item 14**).

2.5 Analysis of Centrosome-Driven Spindle MT Assembly in Mammalian Cells

1. Coverslips, 22 mm × 22 mm, #1.
2. Cell culture dishes, 35-mm or 60-mm.
3. Dulbecco's Modified Eagle Medium (DMEM) containing 2 mM l-glutamine, 100 U/mL penicillin, and 100 μg/mL streptomycin. Store sterile at 4 °C.
4. DMEM/10 % fetal bovine serum (FBS). Add FBS to DMEM (*see* **item 3**). Store sterile at 4 °C.
5. Humidified incubator at 37 °C, 5 % CO_2.
6. 10 mM RO-3306 (EMD Millipore). Dissolve in DMSO and store in aliquots at −20 °C.
7. 10 mM MG-132 (Boston Biochem). Dissolve in DMSO and store in aliquots at −20 °C.
8. Ice water bath. Fill an ice bucket with ice and add water until the top layer of the ice is covered.
9. Water bath at 37 °C.
10. Methanol. Chill to −20 °C before the experiment.

3 Methods

3.1 Analysis of Acentrosomal Spindle Pole Assembly in Xenopus Egg Extracts

Accumulated evidence has revealed a key role of Ran-GTP in bipolar spindle assembly [11]. Indeed, the formation of bipolar spindles can be recapitulated in M-phase-arrested *Xenopus* egg extract independently of chromatin and centrosomes by either adding DNA-coated beads (which mimic chromatin and recruit RCC1) [11, 18, 21], or simply increasing the concentration of RanGTP [20, 21].

The addition of a constitutively active Ran mutant defective in GTP hydrolysis [Ran(Q69L)-GTP] to the extract promotes the formation of MT asters and spindles with focused poles akin to those in acentrosomal meiotic spindles [21]. Moreover, in this setting, spindle pole focusing depends on the function of proteins known to be implicated in the formation of acentrosomal spindle poles [27, 28] (Fig. 1). Thus, the RanGTP-induced MT assembly assay can be used as a simplified tool for the dissection of the

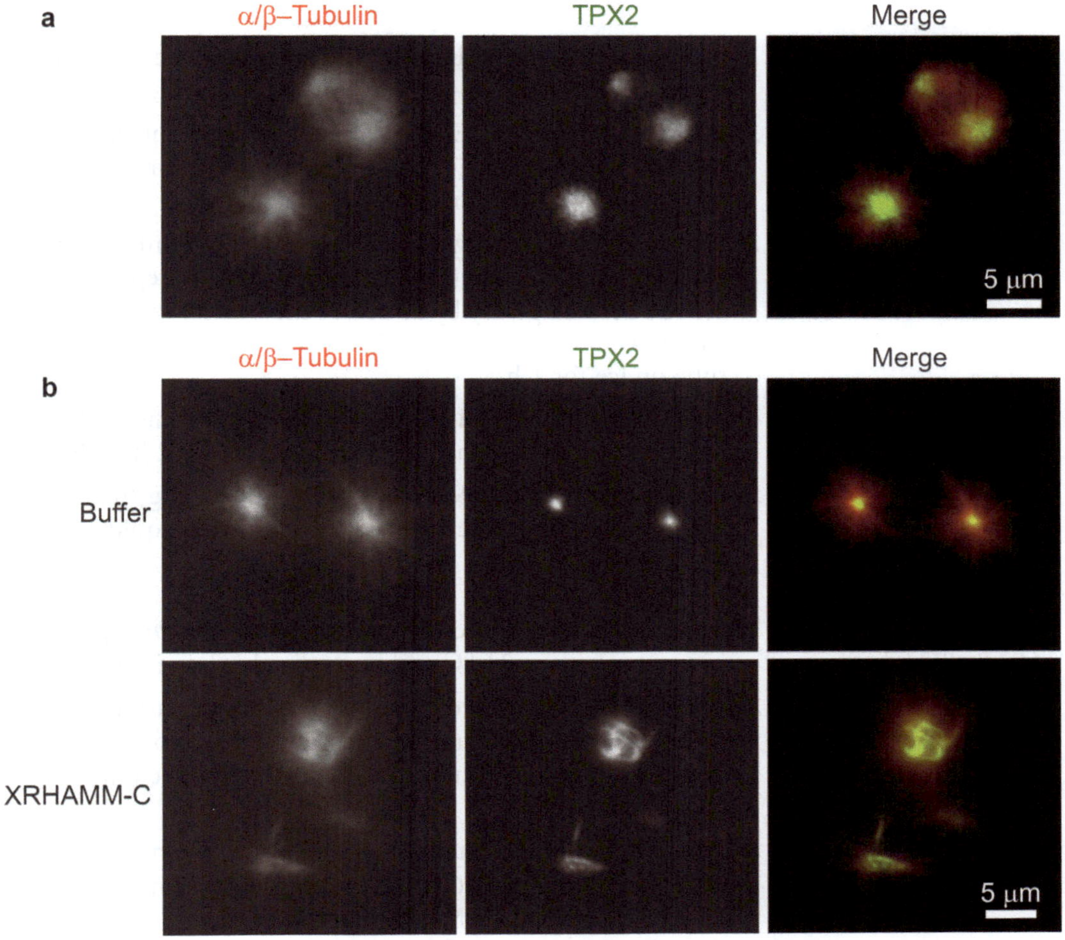

Fig. 1 RanGTP-induced MT asters and spindles formed in M-phase-arrested *Xenopus* egg extract. (**a**) The extract was supplemented with rhodamine-labeled tubulin and 15 μM of Ran(Q69L)-GTP and incubated at 21 °C for 20 min. MT structures were spun down onto a coverslip, stained with an anti-TPX2 antibody, and analyzed by immunofluorescence. A representative aster and a spindle are shown. (**b**) The extract was supplemented with rhodamine-labeled tubulin, a trace amount (5 μg/mL) of an Alexa Fluor 488-labeled anti-TPX2 antibody, and with either XB (*upper panels*) or a recombinant C-terminal fragment of XRHAMM (XRHAMM-C) [27], which is a protein partner of TPX2 in spindle assembly [28]. After the addition of 15 μM of Ran(Q69L)-GTP followed by the incubation at 21 °C for 20 min, the extracts were analyzed by fluorescence microscopy. Note the disruption of the poles of MT asters in the XRHAMM-C-supplemented extract, owing to the interference of XRHAMM-C with the function of endogenous XRHAMM [27]

mechanisms and factors involved in acentrosomal spindle pole assembly. Below, we describe protocols for the preparation of Ran(Q69L)-GTP (Subheading 3.1.1) and for the analysis of RanGTP-promoted spindle MT assembly (Subheading 3.1.2). The RanGTP-induced MT structures can be visualized by fluorescence microscopy of the extracts supplemented with rhodamine-labeled tubulin (Subheading 3.1.2) or by immunofluorescence as described below for the centrosomal MT asters (Subheading 3.2.2). A detailed protocol for the use of DNA-coated beads to study acentrosomal spindle assembly can be found in ref. 14.

3.1.1 Preparation of Ran(Q69L)GTP

1. Dialyze recombinant Ran(Q69L) against two changes of XB and concentrate to ~5–20 mg/mL (*see* **Note 1**).
2. Add 10 μL of 0.5 M EDTA to 1 mL of the Ran(Q69L) solution (yielding a final EDTA concentration of 5 mM) and mix by pipetting up and down several times.
3. Weight out 10–40 mg of GTP in a new 1.5 mL Eppendorf tube and add the Ran(Q69L) solution from **step 2** to it (*see* **Note 11**). Mix gently by inverting the tube or by pipetting up and down until the powder dissolves. Spin briefly and place the tube on ice for 1 h.
4. Add 50 μL of 1 M $MgCl_2$ to the tube, mix, and dialyze the mixture against two changes of XB at 4 °C for 2 h each.
5. Concentrate Ran(Q69L)-GTP to ~10–20 mg/mL (~0.4–0.8 mM) and snap-freeze in 3–5 μL aliquots in liquid nitrogen. Store at −80 °C.

3.1.2 RanGTP-Induced Spindle MT Assembly in Xenopus Egg Extracts

1. Thaw an aliquot of the M-phase-arrested *Xenopus* egg extract in a water bath at 21 °C and immediately place it on ice.
2. Add rhodamine-labeled tubulin to the extract to a final concentration of ~50–75 ng/μL, mix gently by pipetting up and down (*see* **Note 12**), and place the extract on ice until use (*see* **Note 13**).
3. Pipet 10 μL of each extract sample to a new 0.5 mL Eppendorf tube and add Ran(Q69L)-GTP to a final concentration of 15 μM (*see* **Note 14**). Mix by pipetting up and down and place the tube(s) in a water bath at 21 °C.
4. At the desired time intervals (~5–40 min) (*see* **Note 15**), pipet onto a microscope slide(s) 1.2 μL-drops (up to 6 drops per slide) of the spindle fix solution in a number corresponding to the number of extract samples. To each drop, add 0.8 μL of egg extract from **step 3**.
5. Using straight fine-tip forceps, gently lower a circular 12 mm coverslip on each drop of the fixed extract sample. After the fluid has evenly distributed under the coverslips, seal the edges

of each coverslip with nail polish. Avoid moving the coverslips. Air-dry and store the slides with fixed extract samples in the dark at room temperature (up to 1 month).

6. Examine the samples by fluorescence microscopy (*see* **Note 16**).

3.2 Analysis of Centrosome Maturation in Xenopus Egg Extracts

Analysis of centrosome assembly and function can be carried out in *Xenopus* egg extracts supplemented with demembranated sperm nuclei. Each sperm nucleus contains a pair of "naked" (i.e., almost devoid of PCM) centrioles attached to it. When sperm nuclei are added to the extract, the centrioles recruit PCM components and, depending on the cell cycle phase of the extract, form either a mitotic or an interphase centrosome. Centrosome maturation (which represents a transition of the centrosome from an interphase to a mitotic state) can be recapitulated and studied in the so-called cycling egg extracts, which oscillate between S and M phases due to the inherent periodic changes in the activity of the Cdk1/cyclin B complex in the egg cytoplasm [15]. The preparation and handling of cycling extracts is, however, a daunting task. Thus, it is more practical to model and study centrosome maturation in M-phase-arrested egg extracts. The experimental conclusions can then be verified in cycling extracts or cultured cells.

When a naïve M-phase-arrested egg extract is supplemented with sperm nuclei, the centrioles rapidly recruit PCM components and within minutes form a mitotic centrosome that nucleates a MT aster (Fig. 2, upper panels). Importantly, all these processes initially occur independently of the chromatin/Ran-GTP-dependent pathway of MT assembly, which becomes active only at later time points.

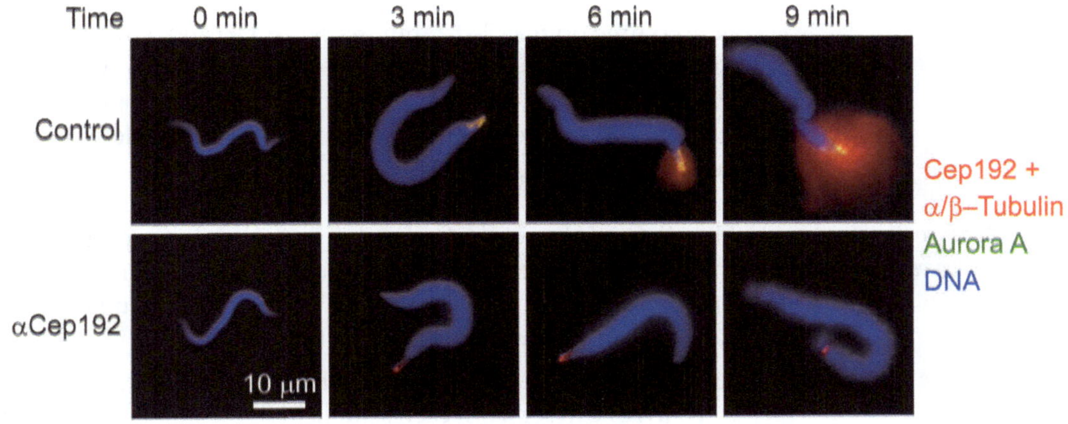

Fig. 2 Time course of the recruitment to the sperm nuclei-associated centrioles of Cep192 and AurA and of centrosome-driven MT assembly in M-phase-arrested *Xenopus* egg extract. The extract was supplemented with 100 μg/mL of non-immune IgG (*upper panels*) or αCep192 (*lower panels*), an N-terminally directed anti-Cep192 antibody that inhibits Cep192 function, in part by interfering with the binding of AurA [9]. Note that the targeting of AurA to centrosomes and the centrosome-driven MT assembly are compromised in the presence αCep192. MTs and Cep192 are in *red*, AurA is in *green*, and DNA is in *blue*. Colocalization of AurA and Cep192 produced a *yellow* signal (Reprinted with permission from Ref. 33, copyright 2011)

Below, we describe a protocol for the analysis of centrosome maturation in M-phase-arrested extract in two steps: setting up and monitoring the mitotic centrosome assembly reactions (Subheading 3.2.1) and preparing the samples for immunofluorescence (Subheading 3.2.2). We then outline the procedure for the analysis of centrosome maturation in cycling egg extracts (Subheading 3.2.3). A protocol for the preparation of such extracts is described in ref. 15 (*see* **Note 9**).

3.2.1 Setting Up and Monitoring the Mitotic Centrosome Assembly Reactions in M-Phase-Arrested Xenopus Egg Extracts

1. Thaw an aliquot of the M-phase-arrested extract (Subheading 2.1) and place it on ice.
2. Add to the extract either nocodazole to a final concentration of 15 μg/mL (for the analysis of PCM protein recruitment) (*see* **Note 17**) or rhodamine-labeled tubulin (Subheading 2.1) to a final concentration of ~50–75 ng/μL (for the analysis of centrosome-driven MT assembly) and mix by pipetting up and down.
3. Thaw an aliquot of demembranated sperm nuclei and immediately place it on ice.
4. Pipet 10–30 μL of each egg extract sample (*see* **Note 18**) to a 0.5 mL Eppendorf tube and place the tubes on ice.
5. To initiate the centrosome assembly reactions, add sperm nuclei to a final concentration of 1.5×10^3/μL to the first extract sample, mix by pipetting up and down, and place the tube in a 21 °C water bath (Subheading 2.1). Repeat this procedure for the other extract samples with the same time interval.
6. To monitor the formation of centrosomal MT asters, at the desired time points (2–12 min, *see* **Note 19**) and with the same time interval between the samples as that in **step 5**, withdraw 0.8-μL aliquots of the rhodamine-tubulin-supplemented extracts and analyze by fluorescence microscopy as described in Subheading 3.1.2, **steps 4–6**.

3.2.2 Preparing Samples for Immunofluorescence

1. Place 1.5 mL Eppendorf tubes (in a number corresponding to the number of extract samples multiplied by the number of time points to be analyzed) on a rack at room temperature. Add 0.5 μL of spindle dilution buffer to each tube.
2. At a selected time interval(s) (*see* **Notes 18** and **19**), withdraw 5–10 μL of extract from each sample in Subheading 3.2.1, **step 5**, and add to the corresponding 1.5 mL tube with spindle dilution buffer. Mix the tube immediately by inverting and incubate at room temperature for 10 min.
3. During the incubation, prepare custom-modified spin-down glass centrifuge tubes with adaptors [14] in the number corresponding to the number of the diluted fixed extract samples in

step 2. To each tube, add 4 mL of cushion buffer and a circular, 12 mm, poly-d-lysine-coated coverslip. Using a metal hook, submerge the coverslip, placing it horizontally on top of the adaptor.

4. Using a disposable plastic transfer pipet, layer each diluted fixed extract sample onto the cushion buffer.
5. Spin down the tubes at 10,000 rpm (~16,000 × *g*) in an HB-6 rotor for 15 min at 16 °C (*see* **Note 20**).
6. Place the tubes on a rack and aspirate the fluid leaving a ~1 cm layer. Using a disposable plastic transfer pipet, slowly add ~4 mL of BRB-80 to each tube along the tube's wall. Aspirate BRB-80, leaving a ~0.5 cm layer of the cushion buffer above the coverslip.
7. While wearing gloves, use a metal hook to carefully lift the adaptor towards the top of the tube. When the adaptor has emerged approximately half-length out of the tube, press it with a finger against the wall of the tube and hold. Using fine-tip forceps, lift the coverslip and insert it into a slot of the perforated holder for circular, 12 mm coverslips. Make sure that the sample side of the coverslip is directed towards the mark on the holder. Repeat this step for the other spun-down samples.
8. Close the coverslip holder with a lid and place it into a glass jar for microscope slides filled with methanol at −20 °C. Incubate for 5 min.
9. Remove the coverslip holder from methanol and place it on a paper towel. Open the holder and, using fine-tip forceps, transfer each coverslip, sample side up, to a well of a 24 well cell culture plate filled with ~1.5 mL of TBS-T. Aspirate TBS-T and refill the wells with TBS-T. Cover the 24 well cell culture plate with a lid, seal with Parafilm, wrap in aluminum foil, and store at 4 °C until immunofluorescence analysis (up to several weeks).
10. Aspirate TBS-T from the wells with coverslips and wash the coverslip twice with ~1.5 mL of TBS-T. For each wash, fill the wells with TBS-T directing the fluid stream at the wall of the well. Gently swirl the plate and leave it on the bench for 2 min. Aspirate TBS-T.
11. Add to each well 0.5 mL of TBS-T/5% goat serum and incubate at room temperature for 10 min.
12. Dilute the primary antibody to the recommended concentration in TBS-T/5 % goat serum (~50–75 μL per coverslip).
13. Lay out a sheet of Parafilm that is slightly larger in size than the 24 well plate on a clean flat surface (*see* **Note 21**).
14. For each coverslip, pipet ~50–75 μL of the diluted primary antibody on Parafilm. Make sure that the distance between the drops of antibody solution is at least 3–4 cm.

15. Using fine-tip forceps, take the first coverslip from **step 11** and remove excess fluid by gently blotting the reverse side and edges of the coverslip with a Kimwipes. Place the coverslip, sample side down, onto the antibody drop. Repeat this step for the remaining coverslips.
16. Cover the bottom of the lid of the cell culture dish with a stack of ~3–5 layers of Kimwipes and spray with distilled water from a wash bottle. Gently tap down to ensure that the Kimwipes are attached to the lid and remove the excess of water with dry Kimwipes. Tightly cover the coverslips on the Parafilm with the lid. If a fluorochrome-labeled antibody is used, cover the lid with aluminum foil. Incubate at room temperature for 1 h.
17. Squirt ~0.5 mL of TBS-T under each coverslip to lift it up. Remove the coverslips from the Parafilm and transfer them to the wells of a 24 well cell culture plate filled with TBS-T.
18. Wash the coverslips in TBS-T three times, 3 min each, at room temperature.
19. For indirect immunofluorescence, repeat **steps 13–18** using, instead of the primary antibody, a fluorochrome-labeled secondary antibody diluted to an appropriate concentration in TBS-T.
20. Apply ~20 μL drops of Vectashield Mounting Medium with DAPI onto a microscope slide (up to 6 drops per slide). Using fine-tip forceps, take each coverslip from the 24 well plate and remove excess fluid by gently blotting the reverse side and edges of the coverslip with a Kimwipes. Place the coverslip, sample side down, onto the drop of Vectashield.
21. To remove the excess of Vectashield, place a stack of 3–5 layers of Kimwipes on top of the microscope slide and cover with 3–5 paper towels. Apply gentle pressure on the paper towels avoiding sideways movements.
22. Seal the margins of the coverslips with nail polish and allow to air-dry. Gently remove residual salts from the coverslip with a wet Kimwipes and air-dry.
23. Store slides in the dark at 4 °C until imaging.

3.2.3 Analysis of Centrosome Maturation in Cycling Egg Extracts

1. Prepare cycling egg extract as described in ref. 15 and supplement it with rhodamine-labeled tubulin (Subheading 2.1) (*see* **Note 9**).
2. Pipet 50 μL of each extract sample into a 0.5 mL Eppendorf tube and add sperm nuclei (Subheading 2.2) to a final concentration of 1×10^3/μL. Mix gently by pipetting up and down and place the tube in a water bath at 21 °C (Subheading 2.1).
3. Every 10–15 min (*see* **Note 22**), withdraw two aliquots from each extract sample: a 0.5–1 μL aliquot for the analysis by SDS-PAGE/Western blotting and a 0.8 μL aliquot for moni-

toring chromatin and MT dynamics by fluorescence microscopy, as described in Subheading 3.1.2, **steps 4–6**.

4. After large nuclei have formed in the extract, watch for the signs of centrosome maturation, as evident by the appearance of MT asters, which often (but not always) neighbor the nuclei.
5. After detecting centrosomal MT asters in the control (mock) extract sample, withdraw a 10 μL aliquot from each extract sample and analyze by immunofluorescence as described in Subheading 3.2.2. If necessary, withdraw an additional aliquot(s) of each extract with a ~15–20 min interval for immunofluorescence analysis.
6. Analyze aliquots of the extracts from **step 3** by Western blotting of the PCM proteins of interest and of the markers of cell cycle progression (such as cyclin B and phosphotyrosine 15 of Cdk1/Cdc2).

3.3 Using Anti-Aur A Antibody-Coated Beads to Model and Study Mitotic Centrosome Formation

Centrosome maturation is driven by a multistep Cep192-organized signaling cascade, in which Cep192 serves both as an activating scaffold for AurA and Plk1 and as an anchoring platform for Plk1 phosphorylation-mediated γ-TuRC recruitment [10]. The cascade is initiated by AurA autophosphorylation at T288/T295 (in human/*Xenopus* AurA, respectively), which results from the recruitment and consequent accumulation and clustering of the Cep192/AurA/Plk1 complexes at centrosomes. AurA then activates Plk1 by phosphorylating it at T210/T201 (in human/*Xenopus* Plk1, respectively) in the T-loop. Plk1, in turn, phosphorylates Cep192 to generate the binding sites for γ-TuRC and, possibly, other PCM components, thus promoting MT nucleation [10]. Notably, γ-TuRC recruitment and MT nucleation can be mimicked in egg extract independently of the centrosomal milieu, by local oligomerization/clustering of endogenous Cep192 complexes on the surface of beads coated with a bivalent αAurA. In this case, Cep192 (all of which is bound to AurA in egg cytoplasm) is co-recruited with AurA to the surface of αAurA beads. The αAurA-mediated dimerization of Cep192-AurA complexes then promotes AurA autophosphorylation at T295, thus triggering the Cep192-organized AurA-Plk1 cascade [9, 10]. Hence, αAurA beads act as artificial centrosomes and can be used to model and study centrosome maturation at steps downstream of the Cep192-dependent AurA activation. The centrosome-like behavior of αAurA beads in egg extract was first reported by Tsai and Zheng [13] and the mechanism of this phenomenon was subsequently elucidated by Joukov and colleagues [9, 10]. Below, we describe the procedure for the preparation of αAurA beads (Subheading 3.3.1) followed by the protocols for the application of αAurA beads to the modeling of centrosome-driven MT assembly (Subheading 3.3.2) and PCM protein recruitment (Subheading 3.3.3).

3.3.1 Preparing αAurA Beads

1. Resuspend Dynabeads Protein A thoroughly by vortexing the stock vial and transfer 0.2 mL of the beads to a new 0.5 mL siliconized Eppendorf tube (*see* **Note 23**). Place the tube on the magnet for 1 min and pipet off the supernatant.
2. Remove the tube from the magnet, add 0.4 mL of PBS/2 mg/mL BSA/0.05 % Tween 20, and resuspend the beads by vortexing. Briefly spin the tube, place it on the magnet for 1 min, and pipet off the supernatant.
3. Wash the beads two more times, with 0.4 mL and 0.2 mL of PBS/2 mg/mL BSA/0.05 % Tween 20 (*see* **Note 24**).
4. Remove the tube from the magnet and add 0.2 mL of αAurA (1 mg/mL). Mix the beads by pipetting up and down and place the tube on a rotator at 4 °C for 2 h.
5. Remove the tube from the rotator, spin it briefly, and place on the magnet for 1 min. Transfer the supernatant to a clean 0.5 mL tube and measure the concentration of IgG in the supernatant. Add 0.5 μL of sodium azide to the αAurA supernatant and store at 4 °C (*see* **Note 25**).
6. Resuspend the beads in 0.2 mL of 0.2 M sodium borate buffer (pH 8.5)/0.02 % Tween 20 at room temperature and transfer to a clean 1.5 mL Eppendorf tube. Bring the volume to 1 mL with the same buffer. Vortex the tube, spin it briefly, and place on the magnet for 1 min. Pipet off the supernatant.
7. Add 1 mL of 0.2 M sodium borate buffer (pH 8.5)/0.02 % Tween 20 to the tube with the beads, vortex, and leave at room temperature.
8. Transfer a 20 μL aliquot of the beads to a new 0.5 mL siliconized Eppendorf tube. Place the 0.5 mL Eppendorf tube on a magnet for 1 min, remove the supernatant, and add 20 μL of SDS-PAGE sample buffer to the tube. Analyze this (pre-crosslinked) αAurA bead sample in **step 13**.
9. Bring a sealed package with a vial of DMP to room temperature. Open the package and the vial and weight out 6.5 mg of DMP on a weigh paper. Immediately add the DMP powder to the 1.5 mL Eppendorf tube with αAurA beads in **step 7**, close the tube, and invert it several times to dissolve DMP. Place the tube on a rotator and incubate at room temperature for 45 min (*see* **Note 26**).
10. Vortex the 1.5 mL tube with αAurA beads and transfer a 20 μL aliquot to a clean 0.5 mL siliconized Eppendorf tube. Place the 0.5 mL Eppendorf tube on the magnet for 1 min, remove the supernatant, and add 20 μL of SDS-PAGE sample buffer to the tube. Analyze this (post-crosslinked) αAurA bead sample in **step 13**.

11. Briefly spin the 1.5 mL tube with αAurA beads and place it on the magnet for 1 min. Remove the supernatant and wash the beads twice in 1 mL of the crosslinking stop buffer (*see* Subheading 2.3). Add 1 mL of the crosslinking stop buffer to the beads, vortex, and leave the tube at 4 °C overnight.
12. Briefly spin the 1.5 mL tube with αAurA beads and place it on the magnet for 1 min. Pipet off the supernatant and wash the beads twice with 1 mL of PBS-500/0.1 % Tween 20 and three times with 1 mL of PBS/0.1 % Tween 20. Resuspend the beads in 200 μL of PBS/0.1 % Tween 20 and transfer to a new 0.5 mL siliconized Eppendorf tube. Add 0.5 μL of 20 % sodium azide to the tube and vortex. Spin the tube briefly, seal it with Parafilm, and store at 4 °C.
13. To assess the efficiency of the crosslinking of αAurA to the Dynabeads Protein A, run the pre-crosslinked and post-crosslinked bead samples from **steps 8** and **10**, together with a protein molecular weight marker, on a SDS-PAGE gel followed by staining the gel with a protein-binding dye (e.g., Coomassie Blue).

3.3.2 Analysis of αAurA Bead-Induced MT Assembly

1. Thaw an aliquot of the M-phase-arrested extract and add to it rhodamine-labeled tubulin (Subheading 3.1.2, **steps 1** and **2**).
2. Vortex the stock tube of αAurA beads (Subheading 3.3.1) and pipet 2 μL to a 0.5 mL siliconized tube containing 200 μL of XB/1 mg/mL BSA/0.05 % Tween 20. Wash the beads twice in 200 μL of XB/1 mg/mL BSA/0.05 % Tween 20 and once in 50 μL of XB/1 mg/mL BSA. Place the tube on the magnet for 1 min, pipet off the supernatant, and resuspend the beads in 10 μL of XB/1 mg/mL BSA.
3. Pipet 15 μL of the rhodamine-tubulin-supplemented extract from **step 1** into a clean 0.5 mL siliconized Eppendorf tube and add 0.5 μL of αAurA beads from **step 2**. Mix by pipetting up and down and place the tube on ice for 30 min. Mix the extract gently every 5–10 min by flicking the tube or by swirling with a pipette tip.
4. Briefly spin the tube and add to the extract Ran(Q69L)-GTP (Subheading 3.1.1) to a final concentration of 8 μM. Mix by pipetting up and down and place the tube in a 21 °C bath (*see* **Note 27**).
5. Every 5 min of the incubation, gently flick the tube to mix the beads and take a 0.8 μL aliquot for the analysis by fluorescence microscopy as described in Subheading 3.1.2, **steps 4–6**. The αAurA bead-induced MT asters and spindles form after ~5–30 min (depending on the quality of the extract) incubation of the extract at 21 °C (Fig. 3a, b).

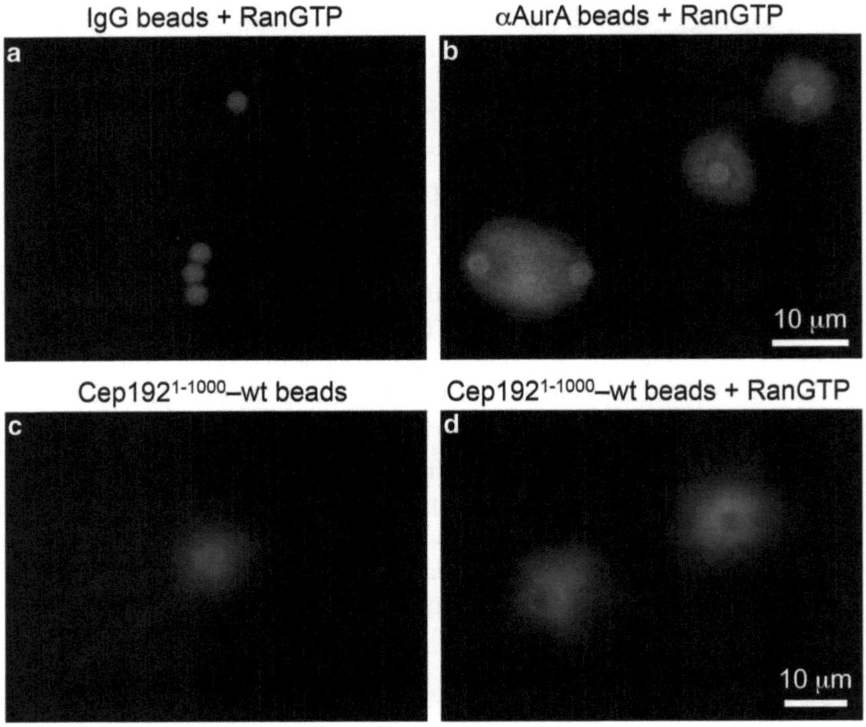

Fig. 3 MT assembly promoted by αAurA beads and Cep192$^{1-1000}$-wt beads in M-phase-arrested *Xenopus* egg extract [10]. The extract was supplemented with rhodamine-labeled tubulin and with beads coated with non-immune IgG (**a**), αAurA (**b**), or Cep192$^{1-1000}$-wt (**c**, **d**) followed by the addition of 8 μM of Ran(Q69L)-GTP (**a**, **b**, **d**) or XB (**c**). Extract aliquots were withdrawn and analyzed by fluorescence microscopy

3.3.3 Analysis of PCM Protein Recruitment Promoted by αAurA Beads

1. Add nocodazole (from a 5 mg/mL stock, Subheading 2.2) to a final concentration of 15 μg/mL to the M-phase-arrested egg extract and mix by pipetting up and down several times.
2. Vortex the stock tube of αAurA beads (Subheading 3.3.1) and transfer ($n+1$) μL of the bead suspension (where "n" is the total number of extract samples to be analyzed) to a 0.5 mL siliconized Eppendorf tube with 200 μL of XB/1 mg/mL BSA/0.05 % Tween 20. Vortex, spin briefly, and place the tube on the magnet for 1 min. Pipet off the supernatant and wash the beads twice in 200 μL of XB/1 mg/mL BSA/0.05 % Tween 20.
3. Place the tube on the magnet for 1 min, pipet off the supernatant, and add to αAurA beads $(n+1)\times 20$ μL of XB/1 mg/mL BSA/0.05 % Tween 20. Mix the bead suspension by pipetting up and down several times and transfer 20 μL-aliquots to clean 0.5 mL siliconized Eppendorf tubes. Briefly spin the tubes and place them on the magnet.
4. Pipet off all supernatant from the first tube in **step 3** and put the tube on ice. Add 25 μL of the extract from **step 1** (*see* **Note 28**) and Ran(Q69L)-GTP (Subheading 3.1.1) to a final concentration of 8 μM. Mix by pipetting up and down and place the tube

on a rotator (Subheading 2.3) at 4 °C at a low speed. Repeat this step for the other extract samples/tubes in **step 3**.

5. At the desired time interval(s) (20 min–1 h), take the first tube from the rotator, add to it 200 μL of ice-cold XB/0.1 % Tween 20, mix by pipetting up and down, and place the tube on ice. Repeat this step for the other extract samples of the same time point.
6. Spin the tubes briefly and place them on the magnet for 1 min.
7. Pipet off the supernatant from each tube and transfer the tubes to ice.
8. Wash the beads three times with 200 μL of ice-cold XB/0.1 % Tween 20. After the last wash, pipet off the supernatant and add to each tube 30 μL of SDS-PAGE sample buffer. Analyze the samples by SDS-PAGE/Western blotting with antibodies against the proteins of interest.

3.4 Using Cep192 Beads to Model and Study Mitotic Centrosome Formation

The binding sites for AurA, Plk1, γ-TuRC, and the MT polymerase XMAP215 in Cep192 all reside within the N-terminal 1000-amino acid domain of this large (2638 amino acids/290 kDa) protein [10]. Accordingly, beads coated with an oligomerized recombinant form of this domain (Cep192$^{1-1000}$-wt) recapitulate the Cep192-organized AurA-Plk1 kinase cascade and assemble MT asters similar to those formed by αAurA beads in M-phase-arrested egg extract. Moreover, the centrosome-like behavior of Cep192$^{1-1000}$-wt beads does not require the addition of exogenous Ran-GTP to the extract (Fig. 3c, d). However, unlike αAurA beads, Cep192$^{1-1000}$-wt beads do not form bipolar MT spindles, suggesting that this process requires either a region of Cep192 C-terminal to the serine-1000 residue or another AurA-interacting protein(s) [10]. Nevertheless, because Cep192$^{1-1000}$-wt beads represent a defined template, which can be easily manipulated experimentally (e.g., by mutations or other modifications of the bead-bound recombinant Cep192), they can be used as a powerful tool for dissection of the mechanisms of centrosome maturation at steps downstream of AurA activation. Below, we describe a protocol for the preparation of Cep192$^{1-1000}$-wt beads by binding of the corresponding recombinant GST-tagged Cep192 fragment to the beads coated with an anti-GST antibody (Subheading 3.4.1). We then describe how to use Cep192$^{1-1000}$-wt beads as a model to study centrosome-driven MT assembly (Subheading 3.4.2) and PCM protein recruitment (Subheading 3.4.3).

3.4.1 Preparing Cep192$^{1-1000}$ Beads

1. Inoculate 500 mL of LB in a 2 L flask with 10 mL of an overnight culture of *E. coli* expressing the Cep192$^{1-1000}$-wt-GST fusion protein (or its mutant counterpart) [9, 10] and incubate on a shaker at 37 °C. When cell density reaches OD_{600} of ~0.4–0.6 (~1–2 h), add IPTG to a final concentration of 1 mM and continue incubation on a shaker at 25 °C for 4–6 h.

2. Spin down the bacteria by centrifugation at 6000 × *g*, 4 °C for 20 min and resuspend the pellet in 5 mL of ice-cold NETS buffer (*see* **Note 29**). Transfer the bacterial lysate to a 15 mL Falcon tube and place the tube on ice.
3. Weight 5 mg of lysozyme in a 1.5 mL Eppendorf tube, dissolve it in ~200 μL of deionized water, and add to the bacterial lysate. Invert the tube several times and incubate it on ice for 30 min, inverting the tube every 10 min.
4. Sonicate the bacterial lysate on ice, at ~40 % amplitude, 4–6 times (20 s each), with a 1-min interval. Repeat the sonication cycles, if needed, until the lysate is no longer viscous (*see* **Note 30**).
5. Transfer the sample to a 10–15 mL centrifugation tube or to several 1.5 mL Eppendorf tubes and centrifuge at 16,000 × *g*, 4 °C for 30 min.
6. Transfer the supernatant to a new 15 mL Falcon tube and add Triton X-100 (from a 20 % stock, Subheading 2.2) to a final concentration of 4 %. Incubate the tube with a slow rotation for 1 h at 4 °C.
7. Place the tube on ice and snap-freeze 0.5 mL aliquots of the Cep192$^{1-1000}$-wt-GST lysate in liquid nitrogen. Store at −80 °C (*see* **Note 31**).
8. Prepare anti-GST beads according to the protocol described above for αAurA beads (Subheading 3.3.1) except that a rabbit affinity purified anti-GST antibody is used instead of αAurA (*see* **Note 32**).
9. Thaw an aliquot of the Cep192$^{1-1000}$-wt-GST lysate from **step 7** in a water bath at 21 °C and place it on ice.
10. Vortex the stock tube of the anti-GST beads from **step 8** and pipet 10 μL of the bead suspension to a new 0.5 mL siliconized Eppendorf tube with 200 μL of NETN buffer (*see* **Note 33**). Vortex, spin down briefly, and place the tube on the magnet for 1 min. Remove supernatant and wash the beads with 200 μL of NETN buffer. Place the bead suspension on the magnet for 1 min.
11. Pipet off NETN buffer and add 100 μL of the Cep192$^{1-1000}$-wt-GST lysate to the anti-GST beads. Resuspend the beads by pipetting up and down and place the tube on a rotator at 4 °C for 1 h.
12. Spin down the tube briefly and place it on ice. Wash the beads twice with 200 μL NETN buffer/500 mM NaCl and three times with 200 μL of XB/1 mg/mL BSA/0.05 % Tween 20 (Subheading 2.3, all solutions should be kept on ice). Resuspend Cep192$^{1-1000}$-wt beads in 50 μL of XB/1 mg/mL BSA/0.05 % Tween 20 and place the tube with the beads on ice until use (*see* **Note 34**).

3.4.2 Analysis of Cep192$^{1-1000}$ Bead-Induced MT Assembly

1. Thaw one aliquot of the M-phase-arrested extract and add to it rhodamine-labeled tubulin (Subheading 3.1.2, **steps 1** and **2**).
2. Pipet 15 μL of the rhodamine-tubulin-supplemented extract into a new 0.5 mL Eppendorf tube. Add 0.5 μL of Cep192$^{1-1000}$-wt beads (Subheading 3.4.1), mix by pipetting up and down several times, and immediately place the tube on a rotator at room temperature.
3. Every 5–10 min of the incubation, remove the tube from the rotator and take a 0.8 μL aliquot for the analysis by fluorescence microscopy as described in Subheading 3.1.2, **steps 4–6**. Return the tube to the rotator.

3.4.3 Analysis of PCM Protein Recruitment Promoted by Cep192$^{1-1000}$ Beads

1. Add nocodazole (from a 5 mg/mL stock, Subheading 2.2) to a final concentration of 15 μg/mL to the M-phase-arrested extract and mix by pipetting up and down several times.
2. Mix the suspension of Cep192$^{1-1000}$-wt beads (Subheading 3.4.1) and pipet 10 μL aliquots to the new 0.5 mL siliconized Eppendorf tubes (*see* **Note 35**). Spin briefly and place the tubes on the magnet.
3. Pipet off all supernatant from the first tube and transfer the tube to ice. Add 25 μL of the nocodazole-supplemented egg extract to the tube, mix by pipetting up and down several times, and place the tube on the rotator at room temperature. Repeat this step for the other tubes in **step 2**.
4. At the desired time intervals (*see* **Note 35**), take the tube(s) from the rotator and place it on ice. Add 200 μL of ice-cold XB/0.1 % Tween 20 (Subheading 2.3) to the tube and mix by pipetting up and down. Spin down briefly and place the tube on the magnet for 1 min.
5. Pipet off the supernatant and wash the beads three times with 200 μL of ice-cold XB/0.1 % Tween 20. After the last wash, pipet off the supernatant and add 30 μL of SDS-PAGE sample buffer to the tube. Analyze the samples by SDS-PAGE/Western blotting of the proteins of interest.

3.5 Analysis of Centrosome-Driven Spindle MT Assembly in Mammalian Cells

Spindle pole assembly can be assessed in mammalian cells by immunofluorescence of α/β tubulin and spindle pole-specific or enriched proteins, such as Cep192, Cep215, pericentrin, ASPM, γ-tubulin, TPX2, and NuMA. To this end, adherent cells are typically cultured on coverslips, enriched in mitosis, and fixed (usually in methanol or formaldehyde). The coverslips are then analyzed by immunofluorescence by a procedure like that described for the extract-derived spindle MT structures (Subheading 3.2.2, **steps 10–23**). However, the fact that most cultured animal cells have a centrosome makes it difficult to use this experimental system to study acentriolar spindle formation. Nevertheless, this task can be accomplished by the ablation of centrioles by laser irradiation, microsurgery or by genetic or pharmacological means [29, 30]. To study the MT-nucleating activity of centrosomes, a so-called MT regrowth assay is commonly used. In this assay, cells are first incubated on ice or in the presence of a MT-depolymerizing drug, such as nocodazole, to depolymerize MTs. Then the cells are quickly returned to MT growth-permissive conditions (by adding a warm medium or by washing out the nocodazole) and, after a short (a few seconds to several minutes) incubation at 37 °C, they are fixed and analyzed by immunofluorescence of α/β tubulin to assess MT regrowth from the centrosomes (Fig. 4). Below, we describe a MT regrowth assay optimized for the analysis of centrosome-driven spindle MT assembly in HeLa cells.

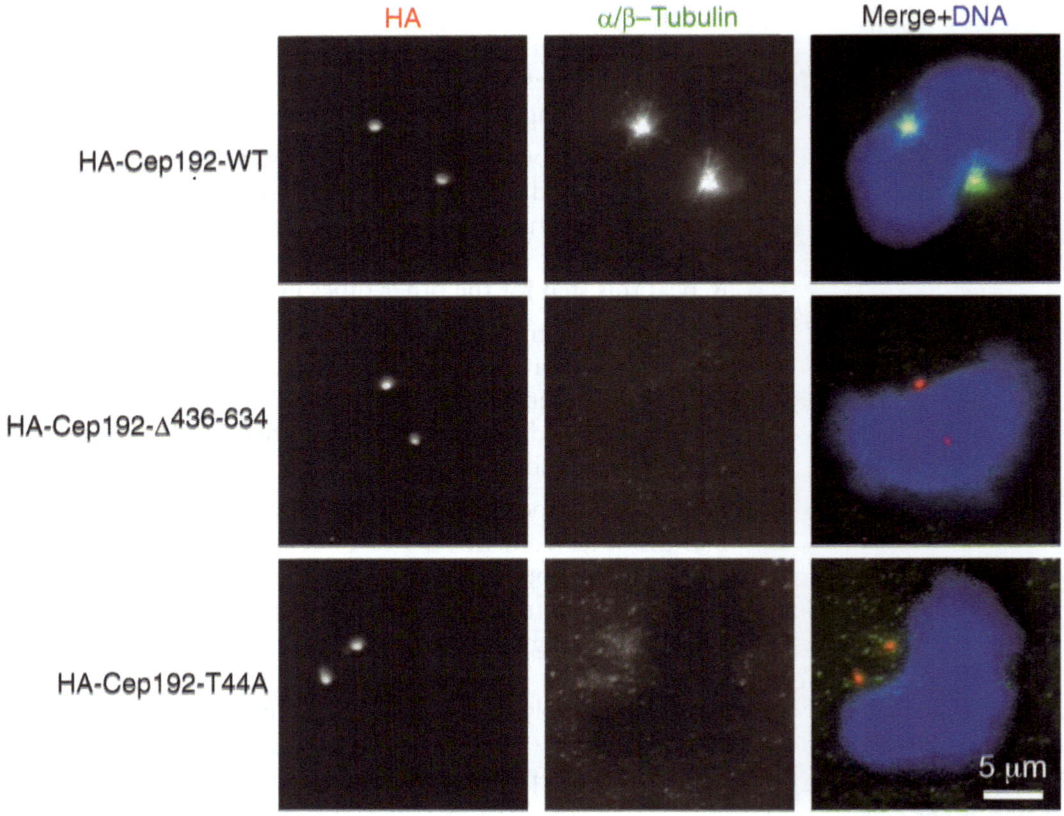

Fig. 4 MT regrowth assay in Cep192-siRNA-treated cells expressing HA-tagged human Cep192-wt or its mutant counterparts lacking the AurA-binding domain (HA-Cep192-$\Delta^{436-634}$) or the N-terminal Plk1-docking threonine 44 (HA-Cep192-T44A). The cells were subjected to a MT regrowth assay followed by immunofluorescence with anti-HA tag and anti-α/β tubulin antibodies. Note the lack of centrosomal MTs and incomplete centrosome separation due to the ablation of the Cep192-organized AurA-Plk1 cascade in cells expressing the mutant forms of Cep192 [10]

1. Using fine-tip forceps, dip each 22 mm × 22 mm #1 coverslip in 95 % ethanol in a beaker, dab off excess fluid against the wall of the beaker, and ignite the ethanol on the coverslip in a flame. After the ethanol has burned off, allow the coverslip to chill and place it in a sterile 35 or 60 mm cell culture dish.
2. Plate HeLa cells in DMEM/10 % FBS in the dishes with coverslips at a density of ~2.5×10^5 or ~5×10^5 cells (per 35 mm or 60 mm dish, respectively). Place the dishes in a humidified incubator at 37 °C, 5 % CO_2 for 24–30 h.
3. Add RO-3306 [31] at a final concentration of 9 μM to each dish, swirl gently, and continue incubating the dish in a humidified incubator at 37 °C, 5 % CO_2 for 18–19 h (*see* **Note 36**).
4. Aspirate the medium and wash the cells three times in ~4 mL or 6 mL (per 35 mm or 60 mm dish, respectively) of DMEM at 37 °C. Add 2 mL or 4 mL (per 35 mm or 60 mm dish, respectively) of DMEM/10 % FBS to the cells and incubate the dishes in a humidified incubator at 37 °C, 5 % CO_2 for 30 min (*see* **Note 37**).
5. Add MG-132 at a final concentration of 20 μM to the cell culture medium (*see* **Note 38**), swirl each dish gently, and place it on an ice water bath for 1 h to depolymerize microtubules.
6. Aspirate the medium and quickly put the dish on the water surface in a water bath at 37 °C. Immediately add 2 mL or 4 mL (per 35 mm or 60 mm dish, respectively) of preheated to 37 °C DMEM/10 %/FBS to the dish and start the timer.
7. After the desired incubation time (20 s–5 min) (*see* **Note 39**), do the following as quickly as possible: aspirate the medium from the dish, wash the cell layer with preheated (37 °C) PBS (Subheading 2.3), aspirate the fluid, and fill the dish with methanol at −20 °C.
8. Cover the dish and store at −20 °C until immunofluorescence analysis.
9. Repeat **steps 6–8** for the other cell culture dishes.
10. Aspirate methanol and wash the cells three times, 5 min each, in TBS-T (Subheading 2.2). Analyze the cells by immunofluorescence of α/β tubulin and of a centriolar/PCM marker, as described in Subheading 3.2.2, **steps 10–23** (use doubled volumes of solutions for washing/staining the coverslips).

4 Notes

1. Recombinant human Ran(Q69L) can be prepared as previously described [27]. Briefly, a cDNA encoding human Ran(Q69L) with N-terminal 6×His-tag and S-tag sequences in a pET29a(+) vector (Novagen) is expressed in BL21 Star (DE3) bacterial

cells (Life Technologies). The protein is purified using Ni-NTA Agarose (Qiagen) according to the manufacturer's instructions. A protocol for the production of an untagged recombinant human Ran(Q69L)-GTP can be found in ref. 26.

2. All solutions described in this chapter should be prepared in deionized water, unless otherwise specified.
3. Although freshly made egg extracts always perform better, all procedures described here (except the experiments in cycling extracts) can be carried out in frozen/thawed extracts. To freeze the extract, add 33 μL of 1.5 M sucrose per each mL of the extract, mix by pipetting up and down, and snap-freeze 50–100 μL aliquots in liquid nitrogen. Store the extract in liquid nitrogen or at −80 °C. Repeated freeze–thaw cycles of egg extracts are not recommended.
4. Egg extracts can be subjected to various experimental treatments, such as immunodepletion and/or addition of inhibitors or recombinant protein(s) of interest, followed by analysis using the methods described in this chapter.
5. An ice bucket filled with water at 21 °C can be used as a water bath in the experiments in egg extracts.
6. Rhodamine-labeled tubulin should preferably be added to the extract prior to the experimental procedures (e.g., immunodepletion or protein additions) to minimize the variability of the fluorescence signals between experimental samples due to pipetting errors. Freeze the remaining aliquot of rhodamine-labeled tubulin in liquid nitrogen and store at −80 °C. Up to three freeze–thaw cycles did not negatively affect the properties of rhodamine-labeled tubulin in the assays described in this chapter.
7. After opening, the ampoule with 16 % formaldehyde can be sealed with Parafilm and stored at room temperature for several weeks.
8. The microscope slides should be clean and devoid of any particles or glass powder. If necessary, clean each slide with Kimwipes soaked in ethanol prior to use.
9. For preparation of cycling egg extracts, we use a modified protocol by Murray [15], wherein eggs are activated by incubation in the presence of the calcium ionophore A23187 (Sigma) [32], instead of by electric pulse.
10. We use a rabbit affinity purified antibody generated against a bacterially expressed 6×His-tagged N-terminal (amino acids 2–199) fragment of *Xenopus* AurA [9]. Beads coated either with this antibody [9, 10] or with a rabbit polyclonal antibody against full-length *Xenopus* AurA [13] nucleate MTs in a centrosome-like fashion in M-phase-arrested *Xenopus* egg

extracts. The purified antibody is dialyzed against PBS/10 % sucrose, concentrated, snap-frozen in aliquots in liquid nitrogen, and stored at −80 °C.

11. The molar concentration of GTP in the solution should be ~100-fold higher than that of Ran(Q69L). M.w. of GTP is 523 Da. M.w. of Ran(Q69L) is ~24.4 kDa.
12. Unless otherwise specified, mixing of egg extracts should be performed by gentle pipetting up and down (5–10 times), avoiding bubbles.
13. To visualize a spindle protein of interest, the extract can also be supplemented with a trace amount of the relevant fluorochrome-labeled antibody, which does not, by itself, affect spindle MT assembly [27, 28] (Fig. 1b).
14. It is critical to minimize the volume of the components added to the egg extract because dilution of the extract negatively affects its performance. The total volume of the additives should not exceed 20 % of the volume of the extract.
15. The speed of formation and the size of the RanGTP-induced MT structures, as well as the proportion between MT asters and spindles, vary depending on the quality of the extract. Extensive experimental procedures, such as multiple rounds of immunodepletion, may inhibit MT aster formation and abolish MT spindle formation.
16. Since the quality of the fixed extract samples deteriorates with time, it is desirable to analyze the samples within a day or two after the preparation.
17. The addition of nocodazole to the extract prevents centrosome-driven MT assembly and improves the visualization of centrosomes by immunofluorescence, in part, by eliminating the background binding of antibodies to MTs. Of note, centrosome maturation precedes and occurs independently of spindle MT assembly.
18. The volume of the extract should be calculated based on the specific purpose of the experiment, particularly depending on whether and at how many time points the samples need to be analyzed by immunofluorescence (Subheading 3.2.2). In most cases, immunofluorescence of centrosomes at two time points (e.g., 5 and 10 min) is sufficient to assess PCM protein recruitment (in the nocodazole-supplemented extracts) and centrosome-driven MT assembly (in the rhodamine-tubulin-supplemented extracts).
19. The formation of centrosomal MT asters begins at ~2 min and peaks at ~10–12 min after the addition of sperm nuclei to the extract. At later time points, centrosomal MT asters may detach from sperm nuclei and the growth and architecture of

centrosomal MTs might be affected by the chromatin/RanGTP-driven pathway of spindle MT assembly.

20. If necessary, several rounds of centrifugation can be performed. After each spin-down, thoroughly rinse the spin-down glass tubes and adaptors with deionized water and dry with Kimwipes. Prolonged (up to 2 h) incubation of the extract samples in spindle dilution buffer did not affect immunofluorescence of centrosomes and MT structures.
21. To prevent wrinkling and bending of the Parafilm, place it on the surface of a bench sprayed with distilled water from a wash bottle. Using Kimwipes, gently press the Parafilm and remove the excess of water from its margins.
22. The speed of cell cycle progression depends on the quality of the extract (some extracts do not cycle) and varies from experiment to experiment. The performance of the extract is very sensitive to prolonged storage on ice, presence of impurities, dilution, and other treatments. Cycling egg extracts need to be freshly prepared and handled with extreme care. Experiments should be well planned and performed immediately after preparation of the extract, without delays.
23. Siliconized/low binding Eppendorf tubes can be used instead of the conventional Eppendorf tubes also in all other experiments described in this chapter.
24. Unless otherwise specified, in all procedures described in this chapter, wash the magnetic beads as follows: (1) add an appropriate washing buffer to the tube containing the beads; (2) resuspend the beads by vortexing for ~20 s; (3) briefly (~5 s) spin the tube with the beads in a microcentrifuge; (4) place the tube on the magnet for ~1 min to retrieve the beads; (5) accurately pipet off the supernatant from the tube on the magnet with a 200 μL or smaller (depending on the volume of the supernatant) pipette tip; (6) immediately transfer the tube with the beads to ice and proceed to the next step of the experiment.
25. Because the amount of αAurA in the binding reaction exceeds the binding capacity of the Dynabeads Protein A, a substantial amount of non-bound antibody may remain in the supernatant and can be used for other applications. The volumes of Dynabeads Protein A and αAurA can be scaled up, if necessary.
26. The purpose of the incubation of αAurA beads with the cross-linker, DMP, is to covalently bind αAurA to Protein A on the bead surface. Because of the extreme sensitivity of DMP to the moisture, the vial should be opened immediately before use and cannot be stored. It is, therefore, practical to purchase small aliquots of the DMP powder.

27. The addition of Ran(Q69L)-GTP to the reaction promotes the αAurA bead-mediated MT growth and spindle formation [13]. However, Ran(Q69L)-GTP also induces the assembly of MT asters and spindles (like those seen in Fig. 1a) independently of αAurA beads. The proportion of such "beadless" MT structures increases with the increase of the Ran(Q69L)-GTP concentration and at later time points.
28. The αAurA bead-to-extract ratio may vary to a certain extent. We recommend αAurA bead-to-extract ratios of ~1:25–1:70 for protein binding experiments and ~1:100–1:200 for the analysis of MT assembly.
29. Because of the poor solubility of the bacterially expressed Cep192$^{1-1000}$-wt, a "semi-denaturing" buffer containing Sarkosyl (NETS buffer) is used to lyse the bacteria. Sarkosyl is then neutralized by incubating the lysate with Triton X-100.
30. To improve sonication efficiency, the sonication can be carried out in a 30 mL round-bottom tube.
31. The Cep192$^{1-1000}$ lysates can be frozen/thawed several times.
32. An anti-GST antibody can be affinity isolated with a GST-resin from an antiserum generated against any GST-tagged protein. The volumes of Dynabeads Protein A and anti-GST antibody can be scaled up, if necessary.
33. The volumes of the anti-GST bead suspension and of the Cep192$^{1-1000}$-wt lysate can be scaled up accordingly if a larger amount of Cep192$^{1-1000}$-wt beads is needed for the experiments.
34. Beads preloaded with mutant forms of Cep192$^{1-1000}$ or with other fragments of Cep192 can be prepared using an analogous procedure. Cep192-coated beads can be stored on ice for several days. The density of the bead-bound Cep192 can be assessed by analyzing an aliquot of the beads by SDS/PAGE/Western blotting.
35. The number of tubes depends on the number of samples and time points to be analyzed. In a typical assay, the PCM protein recruitment is analyzed at two time points, 20 and 40 min after the addition of Cep192$^{1-1000}$ beads to the egg extract.
36. RO-3306 is a Cdk1 inhibitor, which arrests cells at the G2/M transition [31].
37. During the 30 min incubation following the wash-out of RO-3306, the cells exit the G2/M arrest and enter mitosis.
38. MG-132 prevents mitotic exit and, hence, increases the proportion of cells in mitosis.
39. Robust regrowth of MTs from mitotic centrosomes can be seen already after 20–30 s of the incubation.

References

1. Compton DA (1998) Focusing on spindle poles. J Cell Sci 111(Pt 11):1477–1481
2. Heald R, Walczak CE (2009) Mitotic spindle assembly mechanisms. In: De Wulf P, Earnshaw WC (eds) The kinetochore: from molecular discoveries to cancer therapy. Springer Science + Business Media, LLC, New York, NY, pp 231–268
3. Fant X, Merdes A, Haren L (2004) Cell and molecular biology of spindle poles and NuMA. Int Rev Cytol 238:1–57
4. Nicholson JM, Cimini D (2011) How mitotic errors contribute to karyotypic diversity in cancer. Adv Cancer Res 112:43–75
5. Manning AL, Compton DA (2008) Structural and regulatory roles of nonmotor spindle proteins. Curr Opin Cell Biol 20(1):101–106
6. Bettencourt-Dias M, Glover DM (2007) Centrosome biogenesis and function: centrosomics brings new understanding. Nat Rev Mol Cell Biol 8(6):451–463
7. Nigg EA, Raff JW (2009) Centrioles, centrosomes, and cilia in health and disease. Cell 139(4):663–678
8. Bornens M (2012) The centrosome in cells and organisms. Science 335(6067):422–426
9. Joukov V, De Nicolo A, Rodriguez A, Walter JC, Livingston DM (2010) Centrosomal protein of 192 kDa (Cep192) promotes centrosome-driven spindle assembly by engaging in organelle-specific Aurora A activation. Proc Natl Acad Sci U S A 107(49):21022–21027
10. Joukov V, Walter JC, De Nicolo A (2014) The Cep192-organized Aurora A-Plk1 cascade is essential for centrosome cycle and bipolar spindle assembly. Mol Cell 55(4):578–591
11. Kalab P, Heald R (2008) The RanGTP gradient - a GPS for the mitotic spindle. J Cell Sci 121(Pt 10):1577–1586
12. Meunier S, Vernos I (2012) Microtubule assembly during mitosis - from distinct origins to distinct functions? J Cell Sci 125(Pt 12):2805–2814
13. Tsai MY, Zheng Y (2005) Aurora A kinase-coated beads function as microtubule-organizing centers and enhance RanGTP-induced Spindle Assembly. Curr Biol 15(23):2156–2163
14. Hannak E, Heald R (2006) Investigating mitotic spindle assembly and function in vitro using Xenopus laevis egg extracts. Nat Protoc 1(5):2305–2314
15. Murray AW (1991) Cell cycle extracts. Methods Cell Biol 36:581–605
16. Lebofsky R, Takahashi T, Walter JC (2009) DNA replication in nucleus-free Xenopus egg extracts. Methods Mol Biol 521:229–252
17. Gruss OJ et al (2001) Ran induces spindle assembly by reversing the inhibitory effect of importin alpha on TPX2 activity. Cell 104(1):83–93
18. Heald R et al (1996) Self-organization of microtubules into bipolar spindles around artificial chromosomes in Xenopus egg extracts. Nature 382(6590):420–425
19. Ohba T, Nakamura M, Nishitani H, Nishimoto T (1999) Self-organization of microtubule asters induced in Xenopus egg extracts by GTP-bound Ran. Science 284(5418): 1356–1358
20. Wilde A, Zheng Y (1999) Stimulation of microtubule aster formation and spindle assembly by the small GTPase Ran. Science 284(5418):1359–1362
21. Carazo-Salas RE et al (1999) Generation of GTP-bound Ran by RCC1 is required for chromatin-induced mitotic spindle formation. Nature 400(6740):178–181
22. Guse A, Carroll CW, Moree B, Fuller CJ, Straight AF (2011) In vitro centromere and kinetochore assembly on defined chromatin templates. Nature 477(7364):354–358
23. Kelly AE et al (2007) Chromosomal enrichment and activation of the Aurora B pathway are coupled to spatially regulate spindle assembly. Dev Cell 12(1):31–43
24. Petry S, Groen AC, Ishihara K, Mitchison TJ, Vale RD (2013) Branching microtubule nucleation in Xenopus egg extracts mediated by augmin and TPX2. Cell 152(4):768–777
25. Desai A, Murray A, Mitchison TJ, Walczak CE (1999) The use of Xenopus egg extracts to study mitotic spindle assembly and function in vitro. Methods Cell Biol 61:385–412
26. Dasso M, Seki T, Azuma Y, Ohba T, Nishimoto T (1994) A mutant form of the Ran/TC4 protein disrupts nuclear function in Xenopus laevis egg extracts by inhibiting the RCC1 protein, a regulator of chromosome condensation. EMBO J 13(23):5732–5744
27. Joukov V et al (2006) The BRCA1/BARD1 heterodimer modulates Ran-dependent mitotic spindle assembly. Cell 127(3):539–552

28. Groen AC et al (2004) XRHAMM functions in Ran-dependent microtubule nucleation and pole formation during anastral spindle assembly. Curr Biol 14(20):1801–1811
29. Debec A, Sullivan W, Bettencourt-Dias M (2010) Centrioles: active players or passengers during mitosis? Cell Mol Life Sci 67(13):2173–2194
30. Mason JM et al (2014) Functional characterization of CFI-400945, a Polo-like kinase 4 inhibitor, as a potential anticancer agent. Cancer Cell 26(2):163–176
31. Vassilev LT et al (2006) Selective small-molecule inhibitor reveals critical mitotic functions of human CDK1. Proc Natl Acad Sci U S A 103(28):10660–10665
32. Iwabuchi M, Ohsumi K, Yamamoto TM, Sawada W, Kishimoto T (2000) Residual Cdc2 activity remaining at meiosis I exit is essential for meiotic M-M transition in Xenopus oocyte extracts. EMBO J 19(17):4513–4523
33. Joukov V (2011) Aurora kinases and spindle assembly: variations on a common theme? Cell Cycle 10(6):895–903

Part V

Methods Focused on the Cellular Functions of Microtubule Motor Proteins

Chapter 15

Analyzing Spindle Positioning Dynamics in Cultured Cells

Tomomi Kiyomitsu

Abstract

As spindle positioning during mitosis is a highly dynamic process, live cell imaging is a key technology when studying its underlying mechanisms. Recent advances in imaging tools and microscope systems have enabled us to simultaneously visualize several cellular components in living cells with high temporal and spatial resolution. By combining live cell imaging with functional assays such as RNAi-based depletions, drug inhibition, and micropatterned coverslips, novel and unexpected mechanisms of spindle positioning have been uncovered. In this chapter, I present methods for analyzing the dynamics of spindle positioning in cultured cells.

Key words Live cell imaging, Spindle positioning, Spindle oscillation, Cortical dynein, Membrane elongation

1 Introduction

The mitotic spindle plays multiple essential roles during cell division. First, the spindle ensures equal distribution of the replicated chromosomes to maintain genomic information [1]. Second, the mitotic spindle defines the distribution of polarized cell fate determinants by controlling spindle "orientation," which is critical to determine whether the division of a mother cell results in identical or distinct daughter cells [2]. Finally, the position of the anaphase spindle specifies cell cleavage sites, thus defining the relative size of the daughter cells [3], which is likely a critical influence on daughter cell characteristics [4].

In most mitotic cells, spindle orientation and position are regulated by cortical pulling forces exerted on astral microtubules that emanate from the spindle poles [2]. In particular, the minus-end directed microtubule-based motor, cytoplasmic dynein, plays a key role in generating cortical pulling forces on astral microtubules [5, 6]. Thus, analyzing the dynamics of cortical dynein localization is a key requirement to study spindle orientation and position. Unexpectedly, recent findings indicate that myosin-dependent

Paul Chang and Ryoma Ohi (eds.), *The Mitotic Spindle: Methods and Protocols*, Methods in Molecular Biology, vol. 1413, DOI 10.1007/978-1-4939-3542-0_15, © Springer Science+Business Media New York 2016

contractile forces at the polar cell cortex also control spindle position by expanding the polar membrane and altering cellular boundaries during anaphase [4, 7, 8]. In this chapter, I present methods to analyze mechanisms of spindle positioning in live, cultured cells.

2 Materials

2.1 Cell Culture

1. Cell line of interest (e.g., HeLa cells stably expressing DHC-GFP [9, 10]).
2. DMEM supplemented with 10 % fetal bovine serum (FBS), 1 % penicillin–streptomycin, and l-glutamine.
3. CO_2-independent medium (Life Technologies).
4. Glass bottom dish for live cell imaging (MatTek).
5. CO_2 incubator controlled at 5 % CO_2 and 37 °C.
6. Wide-field or confocal microscope with appropriate filter set or lasers, respectively.
7. Microscope incubation chamber that maintains constant temperature.

2.2 For Visualizing the Dynamics of Chromosomes, Membrane, and the Spindle During Mitosis

1. 100 μg/ml Hoechst 33342 (Sigma-Aldrich).
2. Cell lines of interest (e.g., HeLa cells stably expressing DHC-GFP and mCherry-histone H2B and Lifeact-mCherry [4]).

2.3 For Visualizing Dynamics of Spindle Oscillation During Metaphase

1. 10 mM MG132 (Sigma-Aldrich).
2. Software such as Photoshop, Metamorph, or ImageJ.

2.4 For Combining Cell-Cycle Synchronization with RNAi-Mediated Knockdown

1. Lipofectamine RNAiMAX (Life Technologies).
2. siRNAs of interest.
3. Thymidine (Sigma-Aldrich).

2.5 For Spindle Orientation Assay Using Micropattern Chips

1. Coverslip with micropatterned fibronectin (CYTOO).
2. 1-well or 4-well micropatterned chambers (CYTOO).

2.6 For Analyzing Membrane Elongation in Mitotic Cells

1. Flavopiridol (Sigma-Aldrich).
2. Nocodazole (Sigma-Aldrich).

3 Methods

Although live cell imaging has been widely used in the cell division field [11, 12], simultaneous visualization of several cellular components, paired with careful observation at high temporal and spatial resolution, has continued to reveal novel mechanisms and/or phenomena of mitosis. Live cell imaging becomes even more powerful in combination with other functional assays, such as RNAi-mediated knockdown, drug inhibition, and micropatterned coverslips [4, 13, 14]. Together, these techniques provide critical information that cannot be obtained with fixed cell approaches. Here, I present methods to analyze spindle positioning by visualizing chromosomes, the membrane, the mitotic spindle, and their associated proteins during mitosis in cultured cells. In addition, I describe methods in detail to analyze spindle oscillations, spindle orientation, and membrane elongation.

3.1 Visualizing the Dynamics of Chromosomes, Membrane, and the Mitotic Spindle During Mitosis

When cells enter mitosis, cellular structures such as chromosomes, membrane, actin cytoskeleton, and microtubule network are dynamically reorganized to achieve successful mitosis. I present methods that visualize these cellular components with chemicals or fluorescent proteins, respectively. The following methods can be applied to a wide variety of cultured cells, including HeLa, RPE1, and BHK cells.

3.1.1 Visualizing Chromosomes with Hoechst 33342

Although histone H2B has been widely used as a marker to visualize chromosomes in living cells [15], Hoechst 33342 dye can be also used to visualize chromosomes during mitosis [16]. By using Hoechst to visualize chromosomes, this enables the simultaneous observation chromosomes together with GFP- or mCherry-tagged proteins (Fig. 1). In addition, this eliminates the need to generate a stable cell line expressing the fluorescent H2B fusion. However, image quality and time resolution is limited to reduce phototoxicity during time-lapse imaging when using Hoechst (*see* **Note 1**). Other DNA stains such as CYTO11 [17] (Molecular Probes, Excitation/Emission maxima ~508/527 nm) and membrane dyes such as FM4-64 [4] (Life Technologies) and DiI [17] (Life Technologies), are also used for live cell imaging.

1. Plate cells in a glass bottom dish (e.g., MatTek P35G-1.5-20-C. *see* **Note 2**) and culture in standard tissue culture medium (e.g., DMEM supplemented with 10 % FBS and 1 % penicillin–streptomycin) for 1 day.
2. Replace medium with CO_2-independent medium (Life Technologies, supplemented with 10 % FBS and 1 % penicillin–streptomycin).

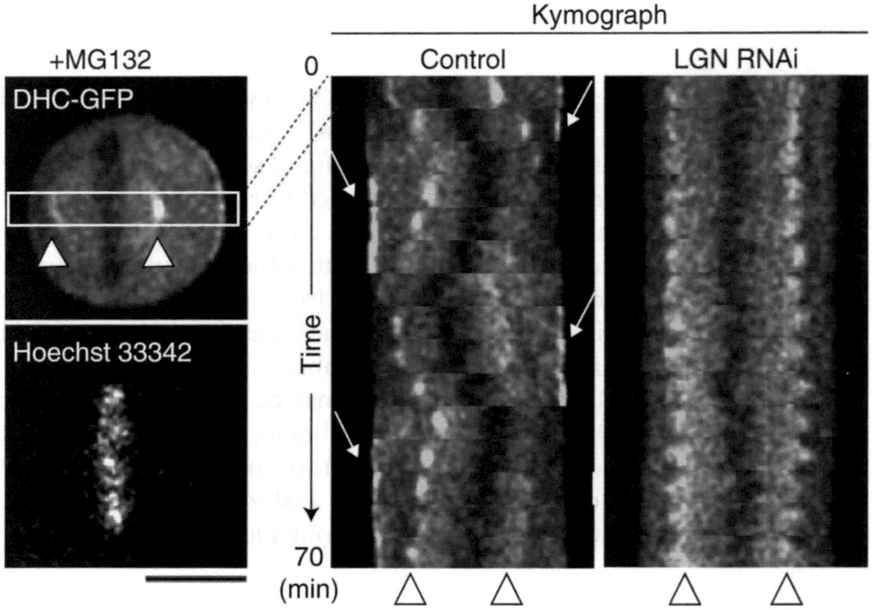

Fig. 1 Spindle oscillation in metaphase HeLa cells. *Left*: Live images of Hoechst 33342 and DHC-GFP in a MG-132 arrested metaphase cell. *Right*: Kymographs showing DHC-GFP in control and LGN-depleted cells demonstrate that the spindle moves towards the dynein-enriched side of the cell and oscillates as cortical dynein is dynamically redistributed (*arrows*). LGN depletion eliminates spindle oscillation as well as cortical dynein localization. Scale bar, 10 μm. Adapted from Ref. [10]

3. Add Hoechst 33342 (stored at 4 °C in the dark, stock concentration 100 μg/ml in sterile H_2O) to a final concentration of 50–100 ng/ml.
4. Incubate 30 min or more in a CO_2 incubator to facilitate incorporation of Hoechst 33342 into the cells (*see* **Note 3**).
5. (Optional) Wash the cells three times with standard tissue culture medium or CO_2-independent medium, and then add CO_2-independent medium. Incubate the cells for 30 min or more in a CO_2 incubator.
6. Move to the pre-warmed microscope stage or heated chamber (37 °C) and observe cells with the DAPI filter. To reduce phototoxicity, minimize exposure time and intensity of light with ND (neutral density) filter. A longer time interval between exposures is also recommended during time-lapse imaging (e.g., 100 ms exposure with DAPI filter, 32 % ND filter, 40× objective lens, 5-min intervals in Deltavision microscope with Lumencor SPECTRA light source).

3.1.2 Visualizing Chromosomes, Membrane, and the Mitotic Spindle Using Fluorescent Proteins

In addition to stains and dyes, fluorescent-tagged peptides or proteins allow the visualization of subcellular structures during mitosis. For example, instead of using Hoechst 33342, expression of fluorescent protein-tagged Histone H2B can be used to visualize chromosomes [15]. Alpha-tubulin or other microtubule binding proteins, such as EB3, can be used to label the spindle [12]. Small peptides, such as LifeAct (17-amino acids), are useful for monitoring F-actin [18], whereas the neuromodulin N-terminal sequence or the C-terminal CAAX box prenylation motif can be used to target fusion proteins to the plasma membrane [19, 20]. Standard organelle marker proteins have been previously reported [12], and additional information may be also found in other sites such as the Michael Davidson Fluorescent Protein Collection on Addgene: (https://www.addgene.org/fluorescent-proteins/davidson/). In addition, the subcellular localization, interaction profile, and RNAi-mediated knockdown phenotypes of hundreds of proteins have been analyzed in mitotic HeLa cells [11, 21] and the information is available in the MitoCheck database: (http://www.mitocheck.org/). By expressing these proteins or peptides tagged with colored fluorescent proteins such as BFP [22], GFP, mCherry [23], iRFP [24], and IFP2.0 [25], different structures and proteins can be simultaneously observed in living cells.

Whereas transient transfection is useful to express marker genes, transfected genes are typically expressed 18–24 h after transfection, and can be analyzed within 24–96 h. In addition, only sub-populations of cells express the marker proteins, and their expression levels are variable between cells. Thus, establishing monoclonal cell lines that stably and homogeneously express fluorescent-tagged marker proteins is a critical step to facilitate single-color or multicolor live cell imaging. The detailed methods for generating stable cell lines using conventional transfection and retroviral transfection are extensively documented [12, 26]. These exogenous methods tend to result in the overexpression of the tagged proteins. Thus, it is important to select a monoclonal cell line that moderately expresses the tagged proteins so as not to affect the function of these proteins or other biological processes. For this purpose, bacterial artificial chromosomes (BACs) carrying mouse genes are useful to express proteins of interest under the control of their native promoter [9]. Recently, several genome-editing techniques have been developed to knock-in cDNA of fluorescent proteins into specific genomic loci [27, 28]. These techniques are well suited to establish mammalian cell lines to visualize and analyze the dynamics of spindle positioning during mitosis.

3.2 Visualizing Dynamics of Spindle Oscillation During Metaphase

3.2.1 Visualizing Cortical Dynein and the Spindle

As cortical dynein generates the pulling forces on astral microtubules to control spindle position [10, 29], visualizing cortical dynein localization is a key component in the study of spindle positioning. Dynein consists of a large catalytic heavy chain (DHC), and non-catalytic subunits such as the intermediate chain (DIC), the light intermediate chain (DLIC), and the light chain (DLC) [30]. Although there are nine major phylogenetic classes of DHC [31], the cytoplasmic dynein 1 heavy chain (DYNC1H1 in humans) is critical for nearly all mitotic processes [30]. Additionally, there are two genes for each class of the non-catalytic subunit (e.g., DIC-1, DIC-2). These non-catalytic subunits display almost identical subcellular localization patterns to DYNC1H1 (hereafter called DHC), although the expression of DIC-1 is limited in neuronal tissue [32, 33], and the light chain subunit, LC8, can function in other protein complexes in HeLa cells [34].

Due to its significant molecular mass (typically ~500 kDa), it has been difficult to express GFP-tagged DHC in living cells. However, the use of a BAC technique enables the expression of GFP-tagged mouse DHC in cultured human HeLa cells [9] (Fig. 1). As mouse DHC is highly similar to human DHC, mouse DHC-GFP forms a dynein complex with non-catalytic human subunits [21]. This hybrid complex displays an identical localization pattern with other GFP-tagged, human non-catalytic subunits of dynein [34]. In addition, the dynactin complex, a dynein-binding partner, displays an identical localization pattern to dynein during mitosis [33, 34]. Thus, mouse DHC-GFP, GFP-TcTex3 (DLC), or GFP-tagged dynactin subunits such as GFP-Arp1A are widely used to visualize cortical dynein during mitosis [10, 35, 36]. Cortical dynein and dynactin localization can be largely maintained through fixation, but this can compromise the fluorescent intensity and cell morphology. Paraformaldehyde (PFA) fixation, rather than formaldehyde (FA) fixation, is recommended to maintain the localization of cortical dynein and the integrity of cortical cell shape.

3.2.2 Analyzing Spindle Oscillation During Metaphase

As dynein accumulates at both the spindle pole and the cell cortex during metaphase, fluorescent labeling of dynein alone is sufficient to monitor the dynamics of both cortical dynein localization and spindle movement (Fig. 1). Spindle oscillations coupled with cortical dynein asymmetry during metaphase can be analyzed as follows.

1. Plate DHC-GFP or other dynein or dynactin subunit expressing HeLa cells in a glass bottom dish (e.g., MatTek P35G-1.5-20-C) and culture in standard tissue culture medium (e.g., DMEM supplemented with 10 % FBS and 1 % penicillin–streptomycin) for 1 day.
2. Replace medium with CO_2-independent medium (Life Technologies), supplemented with 10 % FBS and 1 % penicillin–streptomycin containing 20 μM MG132 and 50–100 ng/ml Hoechst 33342 (optional) to inhibit anaphase initiation and visualize chromosomes, respectively.
3. Incubate for 30 min in a CO_2 incubator.

4. Move to the pre-warmed microscope stage (37 °C), locate metaphase cells, and mark positions.
5. Start time-lapse experiments with 5-min intervals at multiple positions for ~1 h.
6. Select the time lapse sequence for a given cell and deconvolve the images [37] to increase the clarity of cortical dynein signals. Adjust image contrast using consistent procedures between cells and generate a kymograph with the appropriate software (e.g., Photoshop, Metamorph, ImageJ, *see* **Note 4**.)

3.3 Combining Cell-Cycle Synchronization and RNAi-Mediated Knockdown

Synchronization of the cell cycle is useful in observing and imaging a population of mitotic cells simultaneously in a single field. By combining double thymidine block synchronization with RNAi-mediated depletion of proteins of interest, mitotic phenotypes can be more extensively analyzed.

1. (Day 1) Plate cells in a glass bottom dish (e.g., MatTek P35G-1.5) at about 5 % confluency (*see* **Note 5**), and culture in standard tissue culture medium (e.g., DMEM supplemented with 10 % FBS and 1 % penicillin–streptomycin) for 1 day.
2. (Day 2) Add 2 mM thymidine (stock concentration 200 mM) and incubate in a CO_2 incubator for 17 h.
3. (Day 3) Wash three times with standard tissue culture medium, then add 3 ml siRNA-containing medium. Store the remaining volume in the CO_2 incubator with the culture dish.
   Preparing 6 ml siRNA-containing medium
   - Warm DMEM with 10 % FBS (without penicillin –streptomycin).
   - Mix 500 μl Opti-MEM with 8 μl Lipofectamine RNAiMAX, and pipette gently.
   - Mix 500 μl Opti-MEM with 80 pmol siRNA (stock concentration 20 μM. *see* **Note 6**).
   - Combine the two mixtures, pipette gently, and incubate at room temperature for 20 min.
   - To the Opti-MEM mixture, add 5 ml DMEM with 10 % FBS (total of 6 ml).
4. Add 2 mM thymidine (stock concentration 200 mM) 9 h after release.
5. Culture for 15 h.
6. (Day 4) Wash three times with pre-warmed DMEM containing 10 % FBS and add the remaining 3 ml siRNA-containing medium (from Day 3).
7. Replace the medium with CO_2-independent medium at 6 h after release.

8. Culture for 0.5–1 h in a CO_2 incubator.
9. Move the dish onto the pre-warmed microscope stage (37 °C).
10. Locate and mark appropriate positions (*see* **Note 5**) before starting time-lapse imaging. Generally, multiple positions (e.g., ten different positions) are recorded to increase the number of cells that can be sampled in a single experiment.

3.4 Spindle Orientation Assay Using Micropatterned Coverslips

Spindle orientation on the *x*–*y* plane is mediated by extrinsic signals [13]. Recently, uniquely shaped micropatterned chips have been developed [13] and are available for purchase from CYTOO (https://cytoo.com/, Fig. 2). By combining live cell imaging, RNAi, synchronization, and micropatterned coverslips, genes that are necessary for proper spindle orientation can be determined [10, 13, 38].

1. (Day 1–Day 3) *See* Subheading 3.3 protocol.
2. (Day 4) Prepare a chamber with an L-shaped micropatterned coverslip (Fig. 2a).
3. Wash the cultured cells three times with pre-warmed DMEM containing 10 % FBS.
4. Wash once with PBS and add 200 μl trypsin–EDTA to the dish.
5. Place the dish in a CO_2 incubator at 37 °C for 3–5 min.
6. Pipette cells gently to separate them from each other.
7. Add 2 ml pre-warmed DMEM containing 10 % FBS, and pipette cells gently.

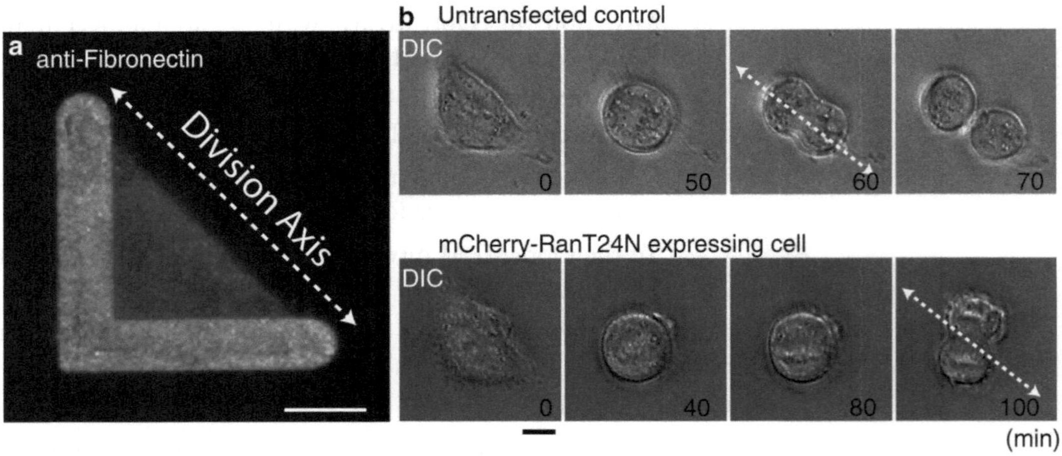

Fig. 2 Spindle orientation assay using L-shaped micropatterned coverslips. (**a**) Fluorescence micrograph showing the analysis of spindle orientation on L-patterned fibronectin-coated coverslips. An interphase HeLa cell cultured on the L-patterned coverslip was fixed and stained with anti-fibronectin antibodies. (**b**) Images from time-lapse movies of control cells (*top*, properly aligned), or cells expressing mCherry-RanT24N mutant (*bottom*, misaligned). Scale bars, 10 μm. Adapted from Ref. [10]

8. Add cells to the chamber with the micropatterned chip (220 μl/well for 4-well chamber, ~1 ml for 1-well chamber). Pipette gently to prevent cells from aggregating in the center of the chamber.
9. Culture in a CO_2 incubator at 37 °C for 1 h. Normally, HeLa cells will adhere to the coverslips within 1 h. However, incubation time may be modified as necessary.
10. Remove the culture medium and wash with pre-warmed DMEM containing 10 % FBS to remove non-adherent cells.
11. Examine the dish on the microscope. If there are still non-adherent cells, repeat **step 10**.
12. Remove the medium and add siRNA containing medium. Culture for 6 h.
13. Replace culture medium with pre-warmed CO_2-independent medium.
14. Incubate for 0.5–1 h in a CO_2 incubator at 37 °C.
15. Move the chamber to the pre-warmed microscope stage.
16. Locate appropriate positions and begin time-lapse imaging.

The methods for analyzing the spindle orientation have been previously described [13, 38].

3.5 Analyzing Membrane Elongation in Mitotic Cells

3.5.1 Analyzing Spindle Positioning and Polar Membrane Elongation Simultaneously During Anaphase

During anaphase, the polar membrane elongates, symmetrically or asymmetrically, to control spindle position by expanding the cellular boundaries as the cleavage furrow ingresses [4]. Thus, analyzing the dynamics of polar membrane elongation is also critical in the study of spindle positioning. To visualize dynein, chromosomes, and membrane elongation simultaneously, mCherry-tagged LifeAct and mCherry-histone H2B as well as DHC-GFP are stably expressed in HeLa cell (Fig. 3). As both chromosomes and membrane are visualized with the same color, this facilitates the ability to analyze the dynamics of chromosomes and membranes during mitosis. Synchronization using a double thymidine block (as described in Subheading 3.3) is also useful for observing a large number of anaphase cells at once in the same field. The dynamic behaviors of polar membrane elongation can be more clearly presented by aligning images of anaphase cells at different time-points (Fig. 3a) and/or generating kymograph (Fig. 3b).

3.5.2 Analyzing Membrane Elongation in Mitotic Cells Without a Mitotic Spindle

Unexpectedly, the membrane elongates in the absence of a mitotic spindle by artificially inducing mitotic exit with CDK inhibitor, flavopiridol, in several cultured cells (e.g., HeLa, RPE1, and BHK) (Fig. 4) [4]. Combining CDK inhibitor with other drug- or RNAi-based functional inhibitions provides the ability to analyze the mechanisms of the membrane expansion.

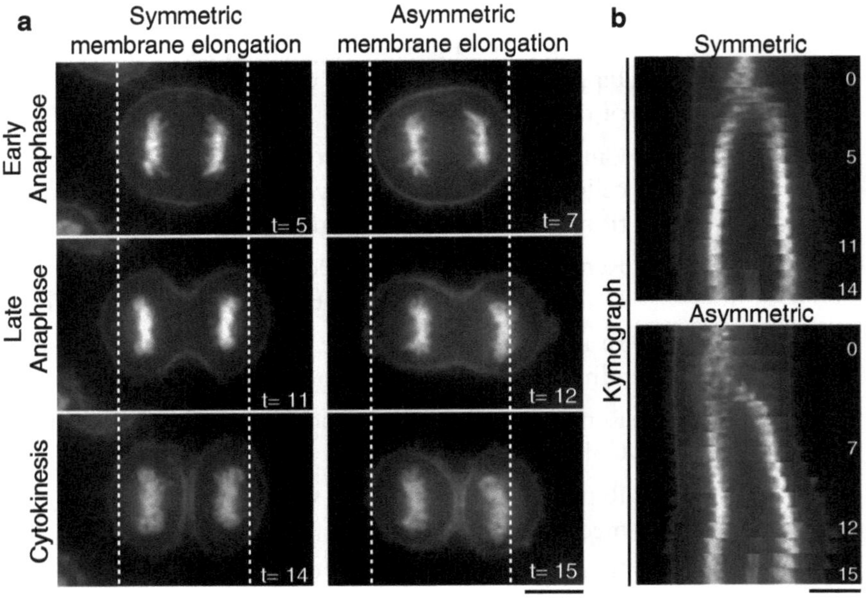

Fig. 3 Polar membrane elongation controls spindle position during anaphase. (**a**) Time-lapse images of symmetrically or asymmetrically elongating cells. Chromosomes and the cell cortex are visualized by mCherry-Histone H2B and Lifeact-mCherry, respectively. *Dashed lines* indicate cellular boundaries at early anaphase. (**b**) Kymographs showing the movement of chromosomes and the cell cortex at 1 min intervals. The polar cell cortex elongates symmetrically (*top*) and asymmetrically (*bottom*) in response to spindle position at the end of early anaphase to center the spindle. Scale bars, 10 μm. Adapted from Ref. [4]

1. (Day 1) Plate cells stably expressing GFP-LGN in a glass bottom dish (e.g., MatTek P35G-1.5-20-C) at ~20–30 % confluency, and culture in standard tissue culture medium (e.g., DMEM supplemented with 10 % FBS and 1 % penicillin–streptomycin) for 1 day. GFP-LGN is used as a marker to analyze chromosome position as GFP-LGN is excluded from the cell cortex near chromosomes [10] (*see* **Note 7**).
2. (Day 2) Add 100 ng/ml nocodazole and incubate in a CO_2 incubator for 5 h (*see* **Note 8**).
3. Replace medium with 1.5 ml CO_2-independent medium containing 100 ng/ml nocodazole and 50–100 ng/ml Hoechst 33342 and incubate for 0.5–1 h.
4. Move the sample to the pre-warmed microscope stage and locate mitotic cells with a chromosomal mass near the cell cortex and asymmetric GFP-LGN cortical signals. Mark these points and image each position.
5. Add 1.5 ml CO_2-independent medium containing 10 μM flavopiridol, 200 ng/ml nocodazole, and 100–200 ng/ml Hoechst 33342 to the sample dish (already containing 1.5 ml media) and mix using a pipette. The final concentration of fla-

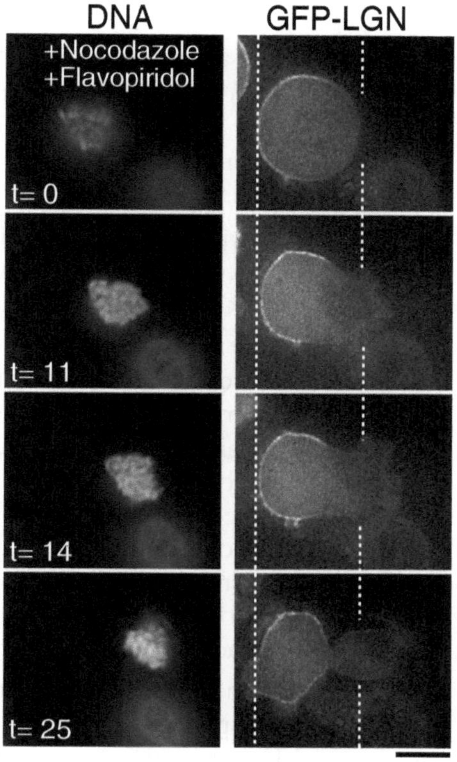

Fig. 4 Membrane elongation assay in mitotic cells without a spindle. Time-lapse images of nocodazole-arrested HeLa cells showing DNA (Hoechst 33342 stain), and GFP-LGN. Nocodazole and flavopiridol were added at *t* = 0. The membrane near the chromosomal mass displayed asymmetric elongation following the addition of flavopiridol. *Dashed lines* indicate cellular boundaries at *t* = 0. Scale bar, 10 μm. Adapted from Ref. [4]

vopiridol, nocodazole, and Hoechst 33342 will be 5 μM, 100 ng/ml, and 50–100 ng/ml, respectively. As medium addition changes the focal plane, adjust the z-position at each of the recorded points.

(Optional) To inhibit the Polo-like kinase 1 (Plk1) or myosin, BI2536 or blebbistatin may be added at the final concentration 10 μM or 100 μM, respectively. In cases where RNAi-mediated knockdown is desired, siRNA can be transfected prior to nocodazole treatment for an appropriate period.

6. Start time-lapse imaging to obtain Hoechst 33342, GFP-LGN, and bright-field images.
7. Deconvolve and align images of cells from different time-points using Photoshop or another appropriate software (Fig. 4).

4 Notes

1. To reduce photo toxicity during UV excitation, an excitation filter with a narrow peak at 380 nm (Chroma Technology or others) can be used for the live cell Hoechst 33342 imaging [16].
2. As typical microscope objectives are designed for use with No. 1.5 coverslips (0.17 mm thickness), No. 1.5 (0.16–0.19 mm thick) is recommended rather than No. 1 (0.13–0.16 mm thick).
3. Hoechst 33342 does not need to be washed out of the medium before observation. However, in cases where Hoechst 33342 remains in the medium, and the incubation time in the CO_2 incubator (**step 4**) is around 30 min, the intensity of chromosome staining may increase due to ongoing incorporation of the dye during time-lapse experiments. In cases where the observed cells display incomplete separation of anaphase chromosomes, reduce the concentration of Hoechst 33342 and replace the medium prior to imaging.
4. Both spindle poles are typically in the same focal plane when the spindle is positioned in the center of metaphase HeLa cells. However, when the spindle comes close to the cell cortex during spindle oscillation, the spindle pole near the cell cortex inclines towards the substrate and tends to be out of focus. Thus, reduced intensity of dynein at the spindle pole near the cell cortex in the kymograph is not due to the dissociation of dynein from the spindle pole, but rather, is due to its position out of the focal plane. In some cases, the spindle orients perpendicularly to the substrate during observation. Collagen or fibronectin coating on glass dish may be useful to keep the spindle parallel to the substrate [39].
5. Optimize the number of cells to avoid 100 % confluency at 4 days. Trypsinize cells rigorously to obtain single cells before seeding. Generally, the depletion efficiency of cells in the center of large colonies is relatively low. In many cases, the knockdown efficiency of an individual cell will be unclear during live cell imaging, so it is recommended to observe cells in small colonies or cells located at the periphery of larger colonies.
6. For single, double, and triple depletions, 40, 40×2, or 30×3 pmol of siRNA is used while maintaining an equivalent amount of RNAiMAX.
7. The GFP-LGN plasmid (pTK14), which is suitable for generating stable cell lines via retroviral transfection [26], is deposited in Addgene: Cheeseman Lab collection (http://www.addgene.org/Iain_Cheeseman/).

8. To increase the number of mitotic-arrested cells, it is possible to extend the period of nocodazole incubation. However, longer nocodazole treatment promotes membrane blebbing in HeLa cells, which makes it difficult to analyze membrane elongation. In addition, prolonged arrest of cells (for more than 18 h) with nocodazole typically results in cell death. In contrast, RPE1 and BHK cells display little membrane blebbing after nocodazole treatment, though the reason for these differences is unknown. However, this characteristic makes it easier to analyze membrane elongation in RPE1 and BHK cells following treatment with CDK inhibitor.

Acknowledgments

I would like to thank Iain Cheeseman and Jagesh Shah for their critical comments and suggestions. The author's research is supported by JST, PRESTO program and JSPS KAKENHI (grant number 26840063). T.K is a recipient of a research grant from The Uehara Memorial Foundation, The Nakajima Foundation, and a Career Development Award of Human Frontier Science Program.

References

1. Cheeseman IM (2014) The kinetochore. Cold Spring Harb Perspect Biol 6:a015826
2. Morin X, Bellaiche Y (2011) Mitotic spindle orientation in asymmetric and symmetric cell divisions during animal development. Dev Cell 21:102–119
3. Kiyomitsu T (2015) Mechanisms of daughter cell-size control during cell division. Trends Cell Biol 25:286–295
4. Kiyomitsu T, Cheeseman IM (2013) Cortical dynein and asymmetric membrane elongation coordinately position the spindle in anaphase. Cell 154:391–402
5. Grill SW, Hyman AA (2005) Spindle positioning by cortical pulling forces. Dev Cell 8:461–465
6. Gonczy P (2008) Mechanisms of asymmetric cell division: flies and worms pave the way. Nat Rev Mol Cell Biol 9:355–366
7. Ou G, Stuurman N, D'Ambrosio M, Vale RD (2010) Polarized myosin produces unequal-size daughters during asymmetric cell division. Science 330:677–680
8. Connell M, Cabernard C, Ricketson D, Doe CQ, Prehoda KE (2011) Asymmetric cortical extension shifts cleavage furrow position in Drosophila neuroblasts. Mol Biol Cell 22:4220–4226
9. Poser I, Sarov M, Hutchins JR, Heriche JK, Toyoda Y, Pozniakovsky A, Weigl D, Nitzsche A, Hegemann B, Bird AW et al (2008) BAC TransgeneOmics: a high-throughput method for exploration of protein function in mammals. Nat Methods 5:409–415
10. Kiyomitsu T, Cheeseman IM (2012) Chromosome- and spindle-pole-derived signals generate an intrinsic code for spindle position and orientation. Nat Cell Biol 14:311–317
11. Neumann B, Walter T, Heriche JK, Bulkescher J, Erfle H, Conrad C, Rogers P, Poser I, Held M, Liebel U et al (2010) Phenotypic profiling of the human genome by time-lapse microscopy reveals cell division genes. Nature 464:721–727
12. Schmitz MH, Gerlich DW (2009) Automated live microscopy to study mitotic gene function in fluorescent reporter cell lines. Methods Mol Biol 545:113–134
13. Thery M, Racine V, Pepin A, Piel M, Chen Y, Sibarita JB, Bornens M (2005) The extracellular matrix guides the orientation of the cell division axis. Nat Cell Biol 7:947–953
14. Minc N, Burgess D, Chang F (2011) Influence of cell geometry on division-plane positioning. Cell 144:414–426

15. Kanda T, Sullivan KF, Wahl GM (1998) Histone-GFP fusion protein enables sensitive analysis of chromosome dynamics in living mammalian cells. Curr Biol 8:377–385
16. Haraguchi T, Kaneda T, Hiraoka Y (1997) Dynamics of chromosomes and microtubules visualized by multiple-wavelength fluorescence imaging in living mammalian cells: effects of mitotic inhibitors on cell cycle progression. Genes Cells 2:369–380
17. Chenn A, McConnell SK (1995) Cleavage orientation and the asymmetric inheritance of Notch1 immunoreactivity in mammalian neurogenesis. Cell 82:631–641
18. Riedl J, Crevenna AH, Kessenbrock K, Yu JH, Neukirchen D, Bista M, Bradke F, Jenne D, Holak TA, Werb Z et al (2008) Lifeact: a versatile marker to visualize F-actin. Nat Methods 5:605–607
19. Skene JH, Virag I (1989) Posttranslational membrane attachment and dynamic fatty acylation of a neuronal growth cone protein, GAP-43. J Cell Biol 108:613–624
20. Kennedy MJ, Hughes RM, Peteya LA, Schwartz JW, Ehlers MD, Tucker CL (2010) Rapid blue-light-mediated induction of protein interactions in living cells. Nat Methods 7:973–975
21. Hutchins JR, Toyoda Y, Hegemann B, Poser I, Heriche JK, Sykora MM, Augsburg M, Hudecz O, Buschhorn BA, Bulkescher J et al (2010) Systematic analysis of human protein complexes identifies chromosome segregation proteins. Science 328:593–599
22. Ai HW, Shaner NC, Cheng Z, Tsien RY, Campbell RE (2007) Exploration of new chromophore structures leads to the identification of improved blue fluorescent proteins. Biochemistry 46:5904–5910
23. Shaner NC, Campbell RE, Steinbach PA, Giepmans BN, Palmer AE, Tsien RY (2004) Improved monomeric red, orange and yellow fluorescent proteins derived from Discosoma sp. red fluorescent protein. Nat Biotechnol 22:1567–1572
24. Filonov GS, Piatkevich KD, Ting LM, Zhang J, Kim K, Verkhusha VV (2011) Bright and stable near-infrared fluorescent protein for in vivo imaging. Nat Biotechnol 29:757–761
25. Yu D, Gustafson WC, Han C, Lafaye C, Noirclerc-Savoye M, Ge WP, Thayer DA, Huang H, Kornberg TB, Royant A et al (2014) An improved monomeric infrared fluorescent protein for neuronal and tumour brain imaging. Nat Commun 5:3626
26. Besschetnova TY, Roy B, Shah JV (2009) Imaging intraflagellar transport in mammalian primary cilia. Methods Cell Biol 93:331–346
27. Chiba K, Hockemeyer D (2015) Genome editing in human pluripotent stem cells using site-specific nucleases. Methods Mol Biol 1239:267–280
28. Ran FA, Hsu PD, Wright J, Agarwala V, Scott DA, Zhang F (2013) Genome engineering using the CRISPR-Cas9 system. Nat Protoc 8:2281–2308
29. Laan L, Pavin N, Husson J, Romet-Lemonne G, van Duijn M, Lopez MP, Vale RD, Julicher F, Reck-Peterson SL, Dogterom M (2012) Cortical dynein controls microtubule dynamics to generate pulling forces that position microtubule asters. Cell 148:502–514
30. Roberts AJ, Kon T, Knight PJ, Sutoh K, Burgess SA (2013) Functions and mechanics of dynein motor proteins. Nat Rev Mol Cell Biol 14:713–726
31. Wickstead B, Gull K (2007) Dyneins across eukaryotes: a comparative genomic analysis. Traffic 8:1708–1721
32. Crackower MA, Sinasac DS, Xia J, Motoyama J, Prochazka M, Rommens JM, Scherer SW, Tsui LC (1999) Cloning and characterization of two cytoplasmic dynein intermediate chain genes in mouse and human. Genomics 55:257–267
33. Raaijmakers JA, Tanenbaum ME, Medema RH (2013) Systematic dissection of dynein regulators in mitosis. J Cell Biol 201:201–215
34. Schmidt JC, Kiyomitsu T, Hori T, Backer CB, Fukagawa T, Cheeseman IM (2010) Aurora B kinase controls the targeting of the Astrin-SKAP complex to bioriented kinetochores. J Cell Biol 191:269–280
35. Collins ES, Balchand SK, Faraci JL, Wadsworth P, Lee WL (2012) Cell cycle-regulated cortical dynein/dynactin promotes symmetric cell division by differential pole motion in anaphase. Mol Biol Cell 23:3380–3390
36. Kotak S, Busso C, Gonczy P (2013) NuMA phosphorylation by CDK1 couples mitotic progression with cortical dynein function. EMBO J 32:2517–2529
37. Swedlow JR (2013) Quantitative fluorescence microscopy and image deconvolution. Methods Cell Biol 114:407–426
38. Saadaoui M, Machicoane M, di Pietro F, Etoc F, Echard A, Morin X (2014) Dlg1 controls planar spindle orientation in the neuroepithelium through direct interaction with LGN. J Cell Biol 206:707–717
39. Toyoshima F, Nishida E (2007) Integrin-mediated adhesion orients the spindle parallel to the substratum in an EB1- and myosin X-dependent manner. EMBO J 26:1487–1498

Chapter 16

Quantification of Mitotic Chromosome Alignment

Cindy Fonseca and Jason Stumpff

Abstract

The alignment of chromosomes during metaphase is a hallmark of mitosis. For this reason, chromosome alignment has served as an informative functional assay for evaluating mitotic fidelity. The common approach of quantifying the number of mitotic cells with unaligned chromosomes within a population has led to the identification of many proteins required for this conserved process. However, more sensitive assays are now required to dissect the complex molecular control of chromosome alignment. In this chapter, we describe a microscopy-based method for objectively quantifying the distribution of fluorescently labeled chromosomes within the mitotic spindle that can be used to evaluate the extent of chromosome alignment within individual mitotic cells.

Key words Chromosome alignment, Mitosis, Mitotic spindle, Congression, Kinetochore

1 Introduction

The question of how chromosomes align at the center of the mitotic spindle has puzzled scientists since the process was first described over a century ago [1]. While many important contributions have been made towards understanding how this conspicuous event occurs, efforts to understand its molecular control are ongoing. Mitotic chromosome alignment depends both on stable attachments between chromosomes and mitotic spindle microtubules, via specialized protein structures called kinetochores, and the regulation of kinetochore microtubule lengths. Thus, chromosome alignment has frequently been used as a phenotypic reporter for abnormal kinetochore microtubule attachments or dynamics [2–9].

Mitotic chromosome alignment is typically quantified via visual inspection of chromosome organization within the mitotic spindle. This approach provides a binary data set reporting the percentage of mitotic cells with aligned and unaligned chromosomes, which limits determination of intermediate phenotypes. Furthermore, because of the subjective determination of what constitutes

Paul Chang and Ryoma Ohi (eds.), *The Mitotic Spindle: Methods and Protocols*, Methods in Molecular Biology, vol. 1413,
DOI 10.1007/978-1-4939-3542-0_16, © Springer Science+Business Media New York 2016

alignment, reproducibility can also be an issue. To address these issues, we have developed a quantitative assay for chromosome alignment that measures the distribution of fluorescently labeled kinetochores within the spindle [10–12]. This objective approach, which is described in detail below, yields reproducible results with the necessary sensitivity to detect partial alignment of mitotic chromosomes. This method can be used to measure the synergistic or antagonistic effects of multiple regulators on alignment, and can be utilized for structure–function analyses of individual mitotic proteins [11, 12].

2 Materials

2.1 Materials for Cell Culture and siRNA Treatment

1. Coverslips: Purchase high quality 12 mm diameter round glass coverslips from a reputable source.
2. Tissue culture plates: Sterile 10 cm and 24-well tissue culture plates are required.
3. Cultured cells: This assay is expected to work with any proliferating adherent cell line. We have specifically tested the method with human retinal pigment epithelial (RPE1) cells and HeLa cells.
4. Cell culture: Culture RPE1 and HeLa cells using standard sterile tissue culture techniques in Minimal Essential Medium Alpha (MEMα) supplemented with 10 % fetal bovine serum and grow in a 37 °C, 5 % CO_2 incubator.
5. Phosphate Buffered Saline (PBS): Prepare or purchase a 1× stock solution of 0.2 M monobasic sodium phosphate, 0.2 M dibasic sodium phosphate and 150 mM NaCl in ddH_2O at pH 7.4. Sterilize by autoclaving.
6. Trypsin: Use tissue culture grade trypsin.
7. Opti-MEM: Use a reduced serum medium such as Opti-MEM (Gibco) for transfection of siRNAs.
8. 1 M hydrochloric acid (HCl): Dilute concentrated HCl drop wise into ddH_2O to a final concentration of 1 M.
9. MG132: Prepare a 10 mM MG132 stock in DMSO and store at −20 °C.

2.2 Materials for Cell Fixation and Immunofluorescence

1. 1 % paraformaldehyde/methanol fix: Dilute high quality paraformaldehyde to a final concentration of 1 % in ice-cold methanol. Total volume should be 80 mL in a 250 mL beaker. The fixative should be made fresh just before use.
2. Tris Buffered Saline (TBS) 1×: Dissolve 8 g NaCl, 0.2 g KCl, 3 g Tris base in 800 mL ddH_2O and pH with HCl to 7.4. Bring volume up to 1 L and sterilize by autoclaving.

3. Antibody dilution buffer (Abdil): Prepare a solution containing 1 % BSA (IgG-free, protease-free), 0.1 % Triton X-100, and 0.02 % sodium azide in 1× TBS.
4. Primary antibody working solutions: Dilute Human anti-centromere antibodies (ACA, Antibodies Inc.) to 2 μg/mL and mouse anti-γ-tubulin antibodies (Sigma) to 1 μg/mL in Abdil. Store working stocks at 4 °C or prepare fresh just before use.
5. Secondary antibody mix: Dilute Alexa Fluor 647 goat anti-mouse IgG (H + L) and Alexa Fluor 594 goat anti-human IgG (Life Technologies) to 4 μg/mL each in Abdil.
6. Cell mounting media: Use an anti-fade cell mounting media containing a DNA stain such as DAPI (e.g., Prolong Gold with DAPI, Life Technologies).

2.3 Materials for Image Acquisition and Analysis

1. Fluorescence microscopy: A standard wide field fluorescence microscope with a CCD camera, high quality 60× or 100× objective lens and filter sets for imaging at least two fluorophores is required.
2. Image analysis software: An image analysis software package that allows quantification of fluorescence intensity in user-defined regions of interest will be needed. We routinely use ImageJ, which is freely available at http://rsb.info.nih.gov/ij. The specific commands given for quantifying fluorescence in Subheading 3.3.2 are for the Macintosh version of ImageJ 1.48d.

3 Methods

3.1 Cell Culture and siRNA Treatment

3.1.1 Prepare Acid Washed Coverslips

Acid washing coverslips increases cell adherence to the glass.

1. In a glass beaker, incubate glass coverslips in 1 M HCl at room temperature for 12–18 h. Cover beaker with Parafilm during the incubation.
2. Carefully remove the HCl by decanting and wash coverslips five times with ddH_2O.
3. In the same glass beaker, wash coverslips five times with 95 % ethanol.
4. Dry coverslips on Kimwipes making sure they are spread out to prevent them from sticking to one another.
5. Place coverslips in a sterile petri dish and keep sterile for future use.

3.1.2 Seed Cells onto Coverslips

When performing cell culture, clean all equipment with 70 % ethanol and work inside a biosafety cabinet using sterile technique to prevent contamination.

1. Grow adherent cells to 70 % confluency in a 10 cm plate containing 10 mL of culture media with antibiotics.
2. Gently suction off media and wash cells with 10 mL of 1× PBS. This is necessary to remove residual serum, which will inactivate trypsin and prevent cell dissociation from the culture dish.
3. Gently remove PBS and add 1 mL of 37 °C trypsin to the 10 cm plate. Gently rock side-to-side making sure the entire culture is bathed in trypsin. Incubate cells in trypsin for 5–10 min until all cells have dissociated from the plate.
4. Once cells have dissociated, add 4 mL of antibiotic-free, 37 °C cell culture medium to the cells. Very slightly tip the culture dish at an angle and mix the cells by pipetting the entire 5 mL 10–15 times with a 10 mL serological pipette. This will pool all the cells in one area of the dish and break up cell clusters.
5. Immediately after pipetting the cells, count them using a hemocytometer or similar method (*see* **Note 1**).
6. Using forceps, place one 12 mm acid washed coverslip into the well of a 24-well culture plate. Double check to make sure the coverslip is sitting flat on the bottom of the dish. Repeat for each condition being tested in the experiment (*see* **Note 2**).
7. Dilute cells to an appropriate concentration to seed at 50 % confluency in a 24-well plate. For HeLa and RPE1 cells, dilute cells to 1.0×10^5 per mL and pipette 0.5 mL (0.5×10^4 cells) directly onto the center of the 12 mm coverslip (*see* **Note 3**).
8. Place the cells in an incubator until they have adhered to the coverslip and started to spread. Incubation time can vary from 1 h to overnight depending on the cell type.

3.1.3 Treat Cells with siRNAs

Cells should be in antibiotic free media when siRNAs are added, and siRNA reagents should be warmed up to room temperature.

1. For each coverslip of cells being treated, dilute and gently mix the following reagents separately in microfuge tubes containing reduced serum media (*see* **Note 4**).
   - 50 μL Opti-MEM + 1.5 μL RNAiMax.
   - 50 μL Opti-MEM + 30 pmol siRNA.
2. Combine siRNA and RNAiMax solutions and allow to complex at room temperature for 15 min.
3. Add each complexed siRNA solution to the appropriate coverslip in the 24-well plate, and gently shake the plate to mix.
4. Incubate the culture dish for 24–72 h in a 37 °C incubator with 5 % CO_2 (*see* **Note 5**).

5. *Optional*: To increase the number of metaphase cells, treat cells with the proteasome inhibitor MG132 for 1–2 h prior to fixation. Add MG132 to a final concentration of 20 μM in each well (*see* **Note 6**).

3.2 Cell Fixation and Immunofluorescence

3.2.1 Fix Cells

1. Using two pairs of forceps, move coverslips with cells into a coverslip staining rack and submerge the rack into a beaker of 1 % paraformaldehyde/methanol fixative (*see* **Note 7**).
2. Fix for 10 min on ice.
3. Wash coverslips three times in 100 mL of 1× TBS for 5 min each (*see* **Note 8**). Perform washes in 250 mL beakers and use caution when lowering the coverslip rack to avoid displacing coverslips.

3.2.2 Immunofluorescently Label Centrosomes and Kinetochores

1. Prepare a staining chamber from a 25 mm Petri dish. Cover the bottom of the dish with Parafilm and place wet Kimwipes around the sides of dish. Replace the lid for all incubation steps to create a humidified chamber (*see* **Note 9**).
2. Place coverslips onto the Parafilm in the chamber with the cell side facing up.
3. To reduce nonspecific binding of antibodies, block cells by adding 30 μL of 20 % goat serum in AbDil to each coverslip. Incubate at room temperature for 1 h on a horizontal shaker.
4. Move coverslips back to staining rack, and wash three times in 1× TBS for 5 min each.
5. Place coverslips cell side up in staining chamber, and add 30 μL of one of the diluted primary antibody solutions to each coverslip. For mouse anti-γ-tubulin antibodies, which label centrosomes, incubate at room temperature for 1–4 h. For human ACA antibodies, which label kinetochores, incubate overnight at 4 °C.
6. Remove excess liquid by dabbing the edge of the coverslip on an absorbant wipe and move coverslips back to the staining rack. Wash three times in 1× TBS for 5 min each.
7. Repeat **steps 5** and **6** for the remaining primary antibody.
8. Move coverslips back to staining rack, and wash three times in 1× TBS for 5 min each.
9. Add 30 μL of diluted secondary antibody mix to each coverslip in the staining chamber and incubate for 1 h at room temperature with shaking. Cover staining chamber with foil to reduce light exposure during incubation.
10. Remove excess liquid by dabbing the edge of the coverslip on an absorbant wipe and wash three times in 1× TBS for 5 min each.

11. Remove excess liquid from coverslips by touching the edge of the coverslip to an absorbant wipe and place onto a glass slide cell side down in a drop of cell mounting media with DAPI.
12. Gently remove the excess mounting media by blotting the top of the coverslip/slide with an absorbant wipe.
13. Allow the mounting media to cure in the dark at room temperature overnight. Slides can be stored at 4 °C or –20 °C.

3.3 Image Acquisition and Analysis

3.3.1 Acquire Images of Mitotic Cells

1. Acquire single focal plane images of fluorescently labeled centrosomes, kinetochores and DNA using a wide field fluorescent microscope (Fig. 1). Image acquisition settings should be optimized and kept consistent when imaging each experimental condition. The criteria for choosing which cells to image should include (a) late prometaphase and metaphase cells that have both poles in the same plane of focus and (b) cells with kinetochores that are attached to microtubules and under tension (i.e., the distance between paired sister kinetochores is higher than is observed at resting length during prophase) (*see* **Note 10**).

3.3.2 Image Analysis to Measure Kinetochore Distribution

The analysis methods described here can be performed within a number of software programs. We describe specific steps for performing the analysis with ImageJ, which is freely available. We have also written a custom macro to partially automate this process in ImageJ, which we will distribute upon request.

1. Export all images as TIFF files from the microscope acquisition software.
2. Open the centrosome and kinetochore images from a cell in ImageJ.

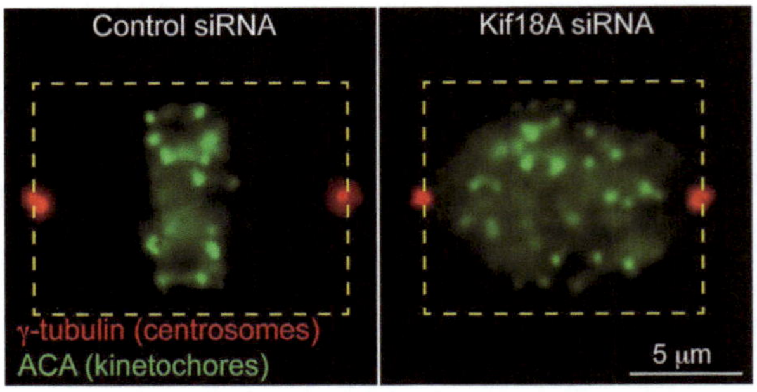

Fig. 1 Example centrosome and kinetochore images for analysis of kinetochore distribution. After a 24 h incubation with control siRNAs or siRNAs targeting the kinesin-like motor Kif18A, RPE1 cells were fixed and immunofluorescently labeled with anti-centromere (ACA, *green*) and anti-γ-tubulin antibodies (*red*). The *dashed yellow boxes* illustrate the region of each cell that was analyzed for ACA fluorescence distribution

3. Use the line tool to draw a single line from the centroid of one γ-tubulin focus to the centroid of the other.
4. Run the measure command under the Analyze menu, which will report the angle and length of the pole-to-pole line.
5. Rotate the kinetochore and the centrosome images based on the angle of the line drawn above so that the pole-to-pole axis is horizontal (Image menu, Transform, Rotate).
6. Using the rectangular selection tool, draw a box that has two sides centered on the centroids of the γ-tubulin foci (*see* yellow boxes in Fig. 1). Adjust the height of the box so it will surround all of the labeled foci in the kinetochore image (*see* **Note 11**). Make note of this box height and use it for all subsequent image analyses.
7. Add the box to the Region of Interest (ROI) Manager (Analyze menu, Tools, ROI Manager), then place the region of interest onto the kinetochore image.
8. Using the Plot Profile command (Analyze menu), measure the average fluorescence intensity in each pixel column within the ROI in the kinetochore image. Save the values from this measurement as a text file and export or copy them into a graphing software program (e.g., Excel).
9. Normalize the average kinetochore fluorescence intensity values (y values) to a zero to one scale, and convert the distances along the pole-to-pole axis (x values) from pixels to microns (Fig.2a, b).
10. Plot the normalized fluorescence values as a function of the distance along the pole-to-pole axis. Fit a Gaussian function to this distribution and use the full width at half maximum (FWHM) as a metric to quantify chromosome alignment in that cell. $\text{FWHM} = 2 \times \text{sqrt}(2 \times \ln 2) \times \sigma$, or $\sim 2.354 \times \sigma$, where σ is the standard deviation of the Gaussian fit function.
11. Alternatively, a ratio of the kinetochore fluorescence near the poles to the kinetochore fluorescence at the center of the spindle, which we call the kinetochore distribution ratio, can be used as a metric. For this analysis, divide the spindle into quarters and sum the average kinetochore fluorescence values in the two quarters near the poles (γ_1 and γ_2) and divide by the sum of the fluorescence in the middle two quarters (ε) ($r = \gamma_1 + \gamma_2 / \varepsilon$). Normalizing the spindle length to one is useful for this analysis (Fig. 2d–e).
12. Repeat **steps 2–11** for each cell imaged and evaluate the distributions of FWHM and kinetochore distribution ratios (Fig. 2c, f).

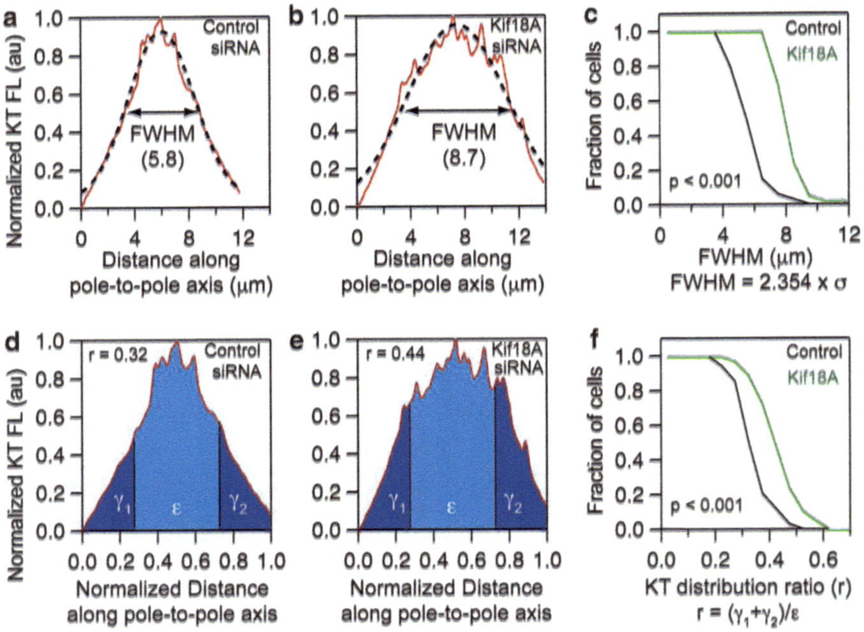

Fig. 2 Two methods for quantifying kinetochore distribution within the spindle. (**a**, **b**) Plots of normalized ACA fluorescence (*red* line and data points) from single RPE1 cells treated with control (**a**) or Kif18A (**b**) siRNAs. Plots were fit with Gaussian functions (*dashed black lines*). The full width at half maximum (FWHM) values calculated from each Gaussian fit are reported in parentheses. (**c**) Plot of the FWHM values measured in a population of RPE1 cells treated with control (*black*) or Kif18A (*green*) siRNAs. (**d**, **e**) The same fluorescence distribution data shown in A and B color-coded to indicate the portion of each plot used to calculate a kinetochore (KT) distribution ratio (*r*). The ratios (*r*) calculated for each cell are reported. (**f**) Plot of KT distribution ratios measured from the same population of cells analyzed in (**c**)

4 Notes

1. The solution of cells should be homogenous before counting and seeding onto coverslips. The cells settle out of solution quickly, so make sure to mix cells well by pipetting before counting and again before plating.
2. The 24-well culture plate can be prepared while cells are trypsinizing to save on time.
3. This suggested density is based on our experience with HeLa and RPE1 cells incubated with siRNAs for 24–48 h. The number of cells plated may need to be adjusted for other cell types or incubation times to prevent overgrowth.
4. This step can be performed with any transfection reagent optimized for delivery of siRNAs. The optimal ratio of RNAiMax and siRNA may need to be determined empirically for each siRNA target.

5. The optimal incubation time for each siRNA must be determined empirically by quantitative evaluation of the target protein via Western blot or immunofluorescence.
6. MG132 is a potent inhibitor of the proteasome that arrests mitotic cells in metaphase [13, 14]. For this assay, it increases the number of cells at the correct stage for analysis and facilitates comparisons between cells at the same point in mitosis. This is especially useful when it is difficult to distinguish prometaphase from metaphase cells by morphology after siRNA depletion of a target. Arresting cells in metaphase also increases the sensitivity of the chromosome alignment assay when analyzing cells that spend more time in prometaphase than metaphase (e.g., RPE1 cells), as asynchronously dividing populations of these cell types will have a low percentage of mitotic cells with aligned chromosomes. However, one should be aware that a prolonged mitotic delay has the potential to reduce the severity of some chromosome alignment phenotypes by providing extra time for the process to occur.
7. Any staining rack that will hold 12 mm round coverslips can be used for this step. We typically use cover glass staining outfits from Thomas Scientific. Carefully keep track of which side of the cover glass the cells are adhered to. This can be facilitated by placing all coverslips in the same orientation with the cell side facing towards a piece of tape on one side of the handled coverslip holder. It is also important to prevent the cells from drying during the entire fixing and staining process.
8. For more stringent washing, include 0.01 % Triton X-100 in the TBS.
9. We use gridded 25 mm dishes and number the squares of the grid with a permanent marker. Placing each coverslip over a number assists with keeping track of individual coverslips through the staining process.
10. Accurate measurement of kinetochore distribution within the spindle depends greatly on the signal to noise ratio of the kinetochore image. A good goal is to optimize imaging conditions to obtain an ACA fluorescence signal at kinetochores that is 2-3 fold higher than the fluorescence signal in the cytoplasm of the cell.
11. If a box drawn from pole-to-pole does not contain all of the kinetochores in the cells being analyzed (e.g., if there are kinetochores on both sides of one pole), the width of the box can be increased to include a region that runs from the cell membrane near one centrosome to the cell membrane near the other centrosome. However, it is critical to define the size of the box consistently for each cell analyzed within one data set (e.g., either by the length of the spindle or the diameter of the cell).

References

1. Mitchison TJ, Salmon ED (2001) Mitosis: a history of division. Nat Cell Biol 3(1):E17–E21
2. Amaro AC, Samora CP, Holtackers R, Wang E, Kingston IJ, Alonso M, Lampson M, McAinsh AD, Meraldi P (2010) Molecular control of kinetochore-microtubule dynamics and chromosome oscillations. Nat Cell Biol 12(4):319–329. doi:10.1038/ncb2033
3. Cai S, O'Connell CB, Khodjakov A, Walczak CE (2009) Chromosome congression in the absence of kinetochore fibres. Nat Cell Biol 11(7):832–838. doi:10.1038/ncb1890
4. Foltz DR, Jansen LE, Black BE, Bailey AO, Yates JR 3rd, Cleveland DW (2006) The human CENP-A centromeric nucleosome-associated complex. Nat Cell Biol 8(5):458–469. doi:10.1038/ncb1397
5. Goshima G, Vale RD (2003) The roles of microtubule-based motor proteins in mitosis: comprehensive RNAi analysis in the Drosophila S2 cell line. J Cell Biol 162(6):1003–1016
6. Lampson MA, Kapoor TM (2005) The human mitotic checkpoint protein BubRl regulates chromosome-spindle attachments. Nat Cell Biol 7(1):93–98. doi:10.1038/ncb1208
7. McCleland ML, Gardner RD, Kallio MJ, Daum JR, Gorbsky GJ, Burke DJ, Stukenberg PT (2003) The highly conserved Ndc80 complex is required for kinetochore assembly, chromosome congression, and spindle checkpoint activity. Genes Dev 17(1):101–114. doi:10.1101/gad.1040903
8. Putkey FR, Cramer T, Morphew MK, Silk AD, Johnson RS, McIntosh JR, Cleveland DW (2002) Unstable kinetochore-microtubule capture and chromosomal instability following deletion of CENP-E. Dev Cell 3(3):351–365
9. Zhu C, Zhao J, Bibikova M, Leverson JD, Bossy-Wetzel E, Fan JB, Abraham RT, Jiang W (2005) Functional analysis of human microtubule-based motor proteins, the kinesins and dyneins, in mitosis/cytokinesis using RNA interference. Mol Biol Cell 16(7):3187–3199
10. Bissonette S, Stumpff J (2014) Quantifying mitotic chromosome dynamics and positioning. J Cell Physiol 229(10):1301–1305. doi:10.1002/jcp.24634
11. Kim H, Fonseca C, Stumpff J (2014) A unique kinesin-8 surface loop provides specificity for chromosome alignment. Mol Biol Cell 25(21):3319–3329. doi:10.1091/mbc.E14-06-1132
12. Stumpff J, Wagenbach M, Franck A, Asbury CL, Wordeman L (2012) Kif18A and chromokinesins confine centromere movements via microtubule growth suppression and spatial control of kinetochore tension. Dev Cell 22(5):1017–1029. doi:10.1016/j.devcel.2012.02.013
13. Genschik P, Criqui MC, Parmentier Y, Derevier A, Fleck J (1998) Cell cycle-dependent proteolysis in plants. Identification Of the destruction box pathway and metaphase arrest produced by the proteasome inhibitor mg132. Plant Cell 10(12):2063–2076
14. Lee DH, Goldberg AL (1996) Selective inhibitors of the proteasome-dependent and vacuolar pathways of protein degradation in Saccharomyces cerevisiae. J Biol Chem 271(44):27280–27284

Chapter 17

Imaging Mitosis in the Moss *Physcomitrella patens*

Moé Yamada, Tomohiro Miki, and Gohta Goshima

Abstract

At first glance, mitosis in plants looks quite different from that in animals. In fact, terrestrial plants have lost the centrosome during evolution, and the mitotic spindle is assembled independently of a strong microtubule organizing center. The phragmoplast is a plant-specific mitotic apparatus formed after anaphase, which expands centrifugally towards the cell cortex. However, the extent to which plant mitosis differs from that of animals at the level of the protein repertoire is uncertain, largely because of the difficulty in the identification and in vivo characterization of mitotic genes of plants. Here, we discuss protocols for mitosis imaging that can be combined with endogenous green fluorescent protein (GFP) tagging or conditional RNA interference (RNAi) in the moss *Physcomitrella patens*, which is an emergent model plant for cell and developmental biology. This system has potential for use in the high-throughput study of mitosis and other intracellular processes, as is being done with various animal cell lines.

Key words *Physcomitrella patens*, Mitosis, Kinesin, Caulonemal apical cell, GFP tagging

1 Introduction

Live-cell microscopy combined with gene perturbation or GFP (or a variant) tagging to a certain protein has become a standard approach in cell biology that has provided insights into gene function. In animal cells, this approach has been applied at the whole-genome level; genome-wide RNAi screening or comprehensive GFP tagging has identified a number of genes required for certain intracellular processes, such as mitosis [1–3]. On the other hand, such a high-throughput approach has not been easily undertaken in seed plants, such as *Arabidopsis* or tobacco, because of technical difficulties.

The moss *Physcomitrella patens* is an emerging model plant for cell and developmental biology [4–6]. The majority of the life cycle is in a haploid state. The full genome sequence is available, and very efficient homologous recombination allows easy genome manipulation, such as knocking out genes or GFP tagging to the endogenous protein. RNAi also works in *P. patens*. When a plasmid

Paul Chang and Ryoma Ohi (eds.), *The Mitotic Spindle: Methods and Protocols*, Methods in Molecular Biology, vol. 1413, DOI 10.1007/978-1-4939-3542-0_17, © Springer Science+Business Media New York 2016

expressing dsRNAs is transiently introduced to the protoplast, the regeneration of the protoplast can be assessed with RNAi knockdown [7]. More recently, we developed a conditional RNAi system that makes it possible to knockdown a single or multiple paralogous genes simultaneously in an inducible manner [8]. Thus, we can make an inducible RNAi transgenic line for any gene of interest, including genes that are essential for the viability of the organism. *P. patens* is also a good system for live microscopy, in particular the "protonemata" tissue of the moss. The protonemal cells grow and divide in a single layer and are amenable to high-resolution microscopy [8–10].

Plants have a large number of kinesin genes (78 in *P. patens* and 61 in *Arabidopsis thaliana*) [11–13]. In order to identify kinesins required for mitosis (mitotic kinesins), we have recently performed a high-throughput study in *P. patens*, where live-cell microscopy was combined with whole kinesin RNAi screening as well as endogenous Citrine tagging of kinesin genes (Citrine is a GFP variant) [12, 14, 15]. Here, we discuss protocols for mitosis imaging in *P. patens*.

2 Materials

2.1 Plasmid Construction for Citrine Tagging

1. Genomic DNA purification kit (MACHEREY-NAGEL Nucleo Spin Plant Midi kit).
2. Plasmid DNA purification kit (e.g., Invitrogen's Midiprep kit).
3. 3 M sodium acetate.
4. 100 % ethanol.
5. 70 % ethanol (at −20 °C).

2.2 Transformation

All the material necessary for transformation is described in the PHYSCOmanual, which was written by the Mitsuyasu Hasebe laboratory at National Institute for Basic Biology, Okazaki, Japan (http://www.nibb.ac.jp/evodevo/PHYSCOmanual/00Eindex.htm). Here, we restate the description.

1. 50 ml conical tube.
2. 15 ml conical tube.
3. 15 ml round-bottom tube.
4. 50 ml round-bottom, glass tube.
5. 0.22 μm syringe-driven filter unit and 10 and 20 ml syringes.
6. Funnel with a sheet of 50 μm nylon mesh.
7. 6-cm petri dish.
8. 2 g PEG6000 in a 50 ml vial and a small stir bar (autoclave).
9. 8 % mannitol.

10. 1 M $Ca(NO_3)_2$.
11. 1 M Tris–HCl (pH 8.0).
12. 1 % MES (pH 5.6).
13. 1 M $MgCl_2$.
14. Following stock solutions.
    (a) Stock A (0.5 M $Ca(NO_3)_2$, 4.5 mM $FeSO_4$).
    (b) Stock B (0.1 M $MgSO_4$).
    (c) Stock C (184 mM KH_2PO_4, pH 6.5).
    (d) Stock D (1 M KNO_3, 4.5 mM $FeSO_4$) (no autoclave).
    (e) Alternative TES (0.22 mM $CuSO_4$, 10 mM H_3BO_3, 0.23 mM $CoCl_2$, 0.1 mM Na_2MoO_4, 0.19 mM $ZnSO_4$, 2 mM $MnCl_2$, 0.17 mM KI).
    (f) 500 mM ammonium tartrate.
    (g) 50 mM $CaCl_2$.
15. Plating medium.
    (a) BCDAT (2000 ml).

Stock B/C/D	20 ml each
500 mM ammonium tartrate	20 ml
50 mM $CaCl_2$	40 ml
Alternative TES	2 ml
Agar	16 g

    (b) PRM plate (2000 ml).

Stock B/C/D	20 ml each
500 mM ammonium tartrate	20 ml
Alternative TES	2 ml
Mannitol	120 g
$CaCl_2$	2.2 g
Agar	16 g

    (c) BCD plate (2000 ml).

Stock B/C/D	20 ml each
50 mM $CaCl_2$	40 ml
Alternative TES	2 ml
Agar	16 g

(d) Protoplast liquid medium (100 ml)

Stock A/B	1 ml each
Stock C	0.1 ml
500 mM ammonium tartrate	54 μl
Mannitol	6.6 g
Glucose	0.5 g

(e) PRM/T solution (200 ml).

Stock B/C/D	2 ml each
500 mM ammonium tartrate	2 ml
Alternative TES	0.2 ml
Mannitol	16 g
$CaCl_2$	0.217 g
Agar (e.g., Sigma A9799; should not be hardened at 45 °C)	1.6 g

2.3 Transgenic Line Selection

The BCDAT medium is supplemented with the following drugs for selection.

G418	Final conc. 20 mg/l
Hygromycin B	Final conc. 30 mg/l
Zeocin	Final conc. 50 mg/l
Nourseothricin (NTC)	Final conc. 75 mg/l
Blasticidin S	Final conc. 75 mg/l

2.4 Transgenic Line Confirmation

1. 10× PCR Buffer (670 mM Tris–HCl, 160 mM $(NH_4)_2SO_4$, 0.1 % Tween 20, pH 8.8).
2. 96-well PCR plate (Sorenson ultraAmp PCR plate 96-well natural), seal (4titude PCR seal).
3. Beads (2 mm diameter, AS ONE).
4. Tissue lyser (TissueLyser II, QIAGEN).
5. PCR kit (KOD FX NEO PCR kit, TOYOBO).
6. Mortar and pestle.
7. Liquid nitrogen.
8. 2× SDS sample buffer.
9. β-mercaptoethanol.
10. Heat Block (100 °C).

11. Centrifuge (LC-200, TOMY).
12. Anti-GFP antibody (e.g., JL-8, CLONTECH).

2.5 Live Cell Imaging

1. Marker lines (*see* Table 1, which lists many published lines).
2. 35-mm glass-bottom dish (MatTek).
3. 6-well, glass-bottom plate (IWAKI).
4. BCD agar medium (*see* Subheading 2.2 or **Note 1**).
5. β-Estradiol (when RNAi is induced in the conditional RNAi system).
6. Microscopy (spinning-disc confocal microscope [100× objective lens, EMCCD camera, 488/561 nm lasers], epifluorescence microscope [10× lens, CCD camera]).

Table 1
***P. patens* marker lines for imaging**

Marker line	Note	Reference
YFP-ARPC4		Perroud and Quatrano [19]
GFP-talin		Finka et al. [20]
GFP-α-tubulin KINID1-Citrine or GFP/mRFP-α-tubulin		Hiwatashi et al. [9]
BRK1-YFP		Perroud and Quatrano [21]
GFP-AtVAM3 GFP-AtVAM3/mRFP-α-tubulin	Vacuolar membrane marker	Oda et al. [22]
Lifeact-GFP		Vidali et al. [10]
For2A-GFP		Vidali et al. [23]
GFP-myoXIa		Vidali et al. [24]
Lifeact-GFP/AIP1-mCherry		Augustine et al. [25]
γATPase-mRFP	Mitochondrion marker	Uchida et al. [26]
GFP-α-tubulin/tdTomato-talin		Yamashita et al. [27]
YFP-Man GFP-COX4 CFP-SKL	Golgi marker Mitochondrion marker Peroxisome marker	Furt et al. [28]
GFP-α-tubulin/histoneH2B-mRFP Aug2-Citrine Aug4-Citrine γ-Tubulin-b-Citrine		Nakaoka et al. [8]

(continued)

Table 1
(continued)

Marker line	Note	Reference
MAP65-Citrine MAP65-Citrine/mCherry-α-tubulin Citrine-MAP65/mCherry-α-tubulin EB1-Citrine/mCherry-α-tubulin	Five MAP65 genes	Kosetsu et al. [16]
EB1-Citrine EB1-Citrine/TagRFP-α-tubulin KINID1-Cerulean/EB1-Citrine		Hiwatashi et al. [29]
mCherry-α-tubulin Kinesin-Citrine/mCherry-α-tubulin Kinesin 7-III-Citrine/Mis12-mCherry	 72 Kinesin genes (43 mitotic kinesin genes)	Miki et al. [12]
Myo8A-GFP/Lifeact-mCherry Myo8A-GFP/mCherry-α-tubulin For2A-GFP/mCherry-α-tubulin Lifeact-GFP/mCherry-α-tubulin		Wu and Bezanilla [30]
Kinesin 7-IIb-Citrine/Mis12-mCherry		Naito and Goshima [14]
PpXMAP215 (PpMOR1)-a-Citrine / mCherry-α-tubulin γ-Tubulin-b-Citrine/mCherry-α-tubulin GFP-α-tubulin/ER-mCherry GFP-α-tubulin/Man1-mRFP GFP-α-tubulin/γATPase-mRFP GFP-α-tubulin/mCherry-SKL	 ER marker Golgi marker Mitochondrion marker Peroxisome marker	Nakaoka et al. [18]

3 Methods

After spore germination, the moss *Physcomitrella patens* develops protonemata tissue, which shows filamentous growth. Protonemata consist of two types of cells, "chloronemata" and "caulonemata." Chloronemal cells are characterized by large chloroplasts. The chloronemal apical cells grow slowly (2–10 μm/h), and mitosis takes place every 22–26 h. Side-branched apical cells in the protonemata are mostly chloronemal. During a few cell cycles, chloronemal apical cells gradually develop the features of caulonemal cells, which are longer and have small chloroplasts, large vacuoles, and oblique cell plates. The caulonemal apical cells grow faster (25–40 μm/h), and mitosis can be observed every ~6 h. Thus, the caulonemal apical cell is the primary target of mitosis imaging [8, 9, 12, 14, 16]. With some frequency, the protonemal cells form a "bud," from which the "gametophore" tissue (a shoot that bears the sexual organ) develops through asymmetric cell division.

The detailed protocols on the basic techniques using moss (culture, maintenance, transformation, etc.) as well as available moss vectors are described in the PHYSCOmanual (http://www.nibb.ac.jp/evodevo/PHYSCOmanual/00Eindex.htm). This web site has comprehensive information on moss handling. In this chapter, we focus on our protocol on Citrine transgenic line selection and mitosis imaging, which are not fully covered in the PHYSCOmanual.

3.1 Plasmid Construction for Citrine Tagging

The moss *Physcomitrella patens* has an unusually high homologous recombination rate. It is therefore easy to tag a gene encoding a fluorescent protein to the endogenous locus of a target gene. This allows us to observe localization and dynamics of the endogenous protein. Importantly, tagging can be scaled up to dozens of genes with reasonable effort and in a reasonable time frame; we have recently established endogenous Citrine (a GFP variant) tagging lines for all 78 moss kinesins and performed comprehensive time-lapse imaging of mitosis [12]. Here, we discuss the tagging strategy with Citrine. For RNAi transgenic line construction, see [8] and [15]; a total of >60 conditional RNAi transgenic lines have been constructed in our laboratory in the past few years.

3.1.1 Plasmid Design

Plasmid design for tagging a fluorescent protein to the N- or C-terminus of the target gene is shown in Fig. 1. Briefly, for N-terminal tagging, the tag (e.g., *Citrine*) is flanked by ~1 kb 5′ UTR sequences and ~1 kb N-terminal gene sequences. Selection markers are not inserted. For C-terminal tagging, ~1 kb C-terminal sequences of the gene of interest and ~1 kb 3′ UTR sequences are attached to the vector that contains *Citrine*, the terminator sequence, and the selection marker (e.g., *nptII* gene for G418 resistance).

3.1.2 Purification of Genomic DNA

Genomic DNA is purified from protonemata using the Plant Midi Kit according to the manufacturer's instructions.

1. Collect sonicated protonemal cells (day 6 or 7) from three BCDAT plates (10 cm diameter, cellophane on top), gently squeeze out water by sandwiching cells between paper towels, and place cells into the mortar.

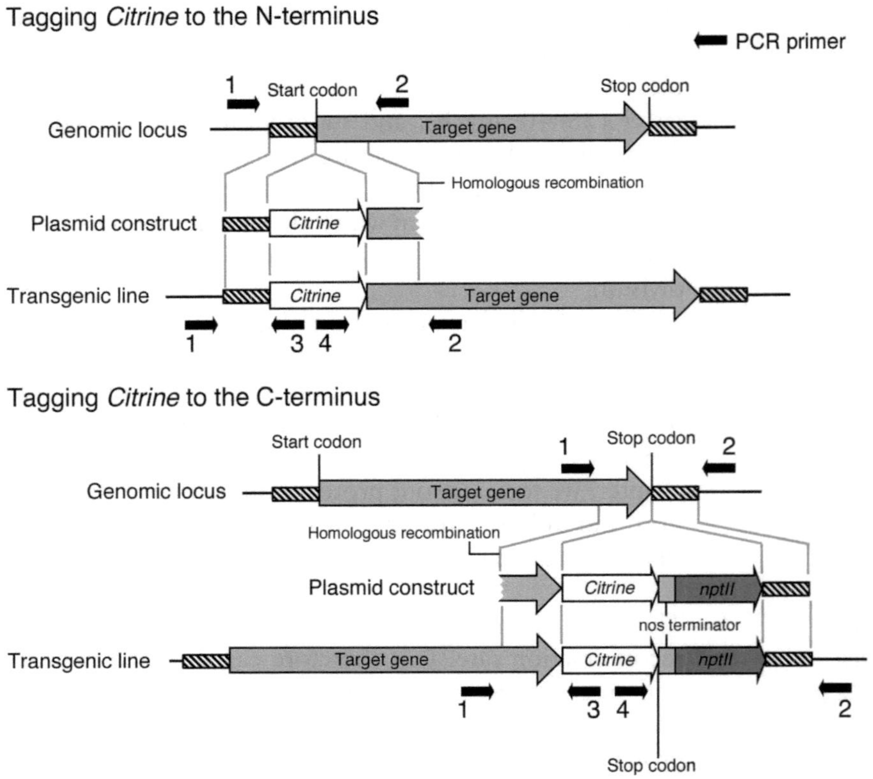

Fig. 1 Strategy for Citrine tagging to the N- or C-terminus of a target gene. A key advantage of using the moss *P. patens* is its unusually high efficiency of homologous recombination. Note that N-terminal tagging is performed without integration of any sequences other than the *Citrine* gene. *Arrows* indicate the locations of the PCR primers used for strain confirmation (primers are numbered #1–#4). The *nptII* gene confers resistance to G418

2. Add liquid nitrogen and grind cells with a pestle. Repeat ~10 times to completely crush the cells.
3. Add 5.1 ml Buffer PL1 and wait 30–60 min for melting.
4. Add 75 μl RNase (10 mg/ml).
5. Transfer the lysate to a 15-ml tube.
6. Heat it in a water bath (65 °C) for 30 min.
7. Centrifuge for 10 min at 4500 × *g* and transfer the supernatant to a new 15-ml tube.
8. Add 6.9 ml Buffer PC and mix immediately by vortexing for 30 s.
9. Load sample on a NucleoSpin Plant II Midi Column.
10. Centrifuge for 2 min at 4500 × *g* and discard the flowthrough (*see* **Note 2**).
11. Wash with 1 ml Buffer PW1.
12. Wash with 3 ml Buffer PW2.

13. Place the NucleoSpin Plant II Midi Column into a collection tube.
14. Add 150 μl Buffer PE and heat it at 65 °C for 30 min.
15. Centrifuge for 2 min at 4500 × *g*.
16. Repeat **steps 14** and **15**.

Final concentration is expected to be ~40 ng/μl.

3.1.3 PCR and Ligation

Standard procedures for cloning can be applied.

3.1.4 Purification

The plasmid is typically purified with the Invitrogen Midiprep kit, starting from a 40-ml culture of *E. coli* in the rich medium. After purification, 30 μg plasmid DNA is digested with appropriate restriction enzymes, which linearize the plasmid (Fig. 1; see "plasmid construct"), followed by ethanol precipitation and dissolving the precipitate with sterile water to a concentration of 1 μg/μl. This DNA solution is then ready for transformation. With the Midiprep kit, we can purify enough of plasmids for several transformations. However, plasmid preparation can be done on a smaller scale (e.g., with a miniprep kit) if only a single transformation is planned.

3.2 Transformation

We follow exactly the protocol available in the PHYSCOmanual. Here, we summarize the procedure with listing several possible problems one might encounter.

1. Preparation of the driselase solution. Add 0.5 g of driselase in a 50 ml conical tube and add 25 ml of 8 % mannitol solution. Mix well for >30 min at room temperature. Centrifuge at 2150 × *g* for 7 min. Transfer the supernatant to a 50 ml syringe with a 0.22-μm filter-unit and filtrate into a 50 ml centrifuge tube. Discard the pelleted driselase.
2. Preparation of the PEG solution. Add 1 ml of 1 M $Ca(NO_3)_2$ and 100 μl of 1 M Tris–HCl (pH 8.0) into 9 ml of 8 % (w/v) mannitol solution and mix. Filter the solution with a 0.22-μm filter. Add 5 ml of the filtered solution to the autoclaved 2 g PEG6000. Stir for >1 h at room temperature and dissolve PEG completely.
3. Preparation of the MMM solution. Mix 910 mg of mannitol, 150 μl of 1 M $MgCl_2$, 1 ml of 1 % MES (pH 5.6), and 8.85 ml of H_2O. Filter the solution with a 0.22 μm filter.
4. Transfer the propagated, healthy protonemata (3–4 g fresh weight, which are collected from 10 culture plates) into the driselase solution with sterile tweezers. Then incubate in the dark at 25 °C for 30 min. Mix gently every 5 min. The healthy protonemata are often obtained 5–6 days after sonication (*see* **Note 3**). With this amount of protonemata, ~10 different plasmids can be transformed.

5. Filter the protonemata through a sheet of 50-μm nylon-mesh that is attached to a funnel.
6. Centrifuge the filtered protoplasts at 180 × *g* for 2 min at room temperature and discard the supernatant with pipette. But do not discard the solution entirely.
7. Rotate very gently the tube with a hand and suspend the centrifuged protoplasts with the remaining 1–2 ml solution.
8. Resuspend gently in 20 ml of 8 % (w/v) mannitol. Repeat this washing procedure twice.
9. Count the number of resuspended protoplasts with hemacytometer and resuspend at 1.6×10^6 ml^{-1} in MMM solution after centrifugation.
10. Add 30 μg of DNA (often linearized by restriction enzyme digestion) into a 15 ml round-bottom tube. Then, add 300 μl of the protoplast suspension and 300 μl of the PEG solution to the tube (*see* **Note 4**). Mix gently by tapping the tube 2–3 times.
11. Incubate the tubes containing the transformation mixture in the 45 °C water bath for 5 min, then in the 20 °C water bath for 10 min.
12. Dilute the transformation mixture by sequentially adding the protoplast liquid medium (300 μl, 500 μl, 700 μl, 1 ml, and 4 ml at 3-min intervals). Upon each addition, mix the solution very gently by tilting the tube.
13. Pour the diluted protoplast solution into a 6 cm petri dish, seal it with Parafilm, and incubate at 25 °C overnight in darkness.
14. On the next day, prepare the PRM/T solution.
15. Overlay a sheet of cellophane on a PRM plate (*see* **Note 5**). Three or six plates are needed for C-terminal (Subheading 3.3) or N-terminal (Subheading 3.4) tagging, respectively.
16. Transfer the protoplast suspension into a 15 ml conical tube and centrifuge at 180 × *g* for 2 min at room temperature.
17. Discard the supernatant and add 9 ml (C-terminal tagging) or 12 ml (N-terminal tagging) of the PRM/T medium stored at 45 °C. Resuspend the protoplasts gently by pipetting.
18. Pour the protoplast suspension on the cellophane layered on a PRM plate (3 ml × 3 plates or 2 ml ×6 plates).
19. Seal the petri dish using a surgical tape. Incubate the plate at 25 °C under continuous white light (*see* **Note 6**).

3.3 Line Selection (C-Terminal Tagging)

We basically follow the PHYSCOmanual for the selection of the C-terminal tagging lines.

1. At 4–6 days after culturing on the PRM plates, the cellophane on which regenerated cells are present is transferred to the drug-containing PRM plate.

2. After 3–5 days, the cellophane is transferred to the drug-free PRM plate. This extra step facilitates selection of the drug-resistant line in the next step.
3. After 4–6 days, the cellophane is transferred again to the drug-containing BCDAT plate.
4. After 3–7 days, 50–100 colonies are selected and inoculated on the drug-free BCDAT plate.
5. Culture for 10–14 days.
6. A piece of moss of each colony is transferred to the drug-containing BCDAT plate.
7. Drug-resistant lines will be identified in ~7 days. Proceed to the genotyping step (Subheading 3.5). In most cases, ≥2 independent transgenic lines are finally obtained.

3.4 Line Selection (N-Terminal Tagging)

In N-terminal tagging, we use a modified selection process because the plasmid construct does not have a selection marker [17].

1. Prepare the circular plasmid in which a gene that confers drug resistance is inserted (*see* **Note 7**).
2. Mix 30 μg linearized plasmid prepared in Subheading 3.1.4 and 5 μg circular plasmid.
3. Carry out transformation following the PHYSCOmanual procedure, which is summarized in Subheading 3.2.
4. In the last step of transformation, protoplasts are plated onto six PRM plates, not three as recommended in the PHYSCOmanual, because individual colonies are hard to pick up after regeneration of densely plated protoplasts.
5. After 4–6 days on the PRM medium, the cellophanes on which cells are growing are transferred to the drug-containing BCDAT medium (e.g., 20 mg/l G418).
6. After culture for ~2 weeks, colonies then become visible.
7. Pick up ~200 colonies and inoculate onto the drug-free BCDAT medium (*see* **Note 8**).
8. Grow the lines for 10–14 days. Some lines will have the linear Citrine construct inserted at the target site. We select those lines by the method described in the next section.

3.5 Transgenic Line Confirmation

If we desire to make a transgenic line via homologous recombination, the candidate line should be confirmed by PCR-based genotyping and immunoblotting.

3.5.1 Genotyping by "Green" PCR

The whole cell extract obtained from the moss protonemata is used for genotyping PCR. This PCR is called "green PCR," since the reaction solution turns green. We typically check ~25 candidate lines for C-terminal tagging and ~200 lines for the N-terminal tagging and obtain ≥2 successfully integrated line.

1. Primers of >25 nucleotides are designed at four locations as described in Fig. 1 (*see* **Notes 9** and **10**).
2. Two 96-well PCR plates are prepared.
3. Put a bead and 30 μl of 10× Green PCR buffer in each well (*see* **Note 11**).
4. Add the moss protonemata (~2 mm in diameter) to each well (*see* **Note 12**).
5. Seal the plate tightly (*see* **Note 13**).
6. Crush the cells and obtain the extract by using the tissue lyser (20 frequency/s for 1 min).
7. Centrifuge the plate at 510×*g* for 15 min.
8. The supernatant can be directly used as a PCR template. The remaining extract can be stored at −20 °C or 4 °C (*see* **Note 14**).
9. PCR is performed with the KOD FX NEO kit under the following conditions (*see* **Note 15**).

Reaction mixture (20 μl)	
2× KOD FX Neo buffer	10 μl
dNTPs (2 mM each)	4 μl
Primer forward	125 nM
Primer reverse	125 nM
Template (prepared in Subheading 3.5.1, **step 7**)	1 μl
Polymerase	0.2–0.25 μl
H_2O	to 20 μl

PCR cycle		
94 °C	2 min	
98 °C	10 s	30 cycles
63 °C	30 s	30 cycles
68 °C	1 min/kb	30 cycles
68 °C	5 min	

10. Agarose gel electrophoresis (1–3 μl loading for each well).
11. We perform three rounds (a total of four sets) of PCRs in the following order.
    (a) Outside–outside (primers #1 and #2 in Fig. 1).
    This will lead to identification of the candidate line.
    (b) Outside–outside (primers #1 and #2).

We re-purify genomic DNA from the same colony to use as the new PCR template. By doing so, a colony composed of mixed wild-type and tagged cells can be excluded.

(c) Inside–outside (primers #1 and #3) and outside–inside (primers #2 and #4).

For the candidate lines, we perform these two PCRs to search for the single-copy Citrine integrant. Either template (original or new) can be used.

12. We pick up multiple lines that give rise to DNA bands with expected length in all four PCRs above, as the candidates of the desired transgenic line (*see* **Note 16**).

3.5.2 Immunoblotting

We recommend performing immunoblotting to make sure that the tagged protein of the expected size is expressed in the transgenic line. Some proteins may not be expressed abundantly, so the band may not be detectable by standard immunoblotting. However, a more important purpose of this check is to eliminate the lines that express the truncated Citrine fusion protein; we suspect that expression of the truncated fusion protein could happen when the Citrine construct is nonhomologously inserted into an extra site and a strong promoter nearby induces the expression. If the truncated protein is more strongly expressed than the full-length one, the observed localization would not represent the endogenous full-length protein.

1. Put 3 μl β-mercaptoethanol to a 1.5 ml tube on ice.
2. Put the moss colony (4 mm or longer in diameter) into a mortar.
3. Add liquid nitrogen and grind colony with a pestle.
4. Add 150 μl of 2× SDS sample buffer and wait for melting (~3 min).
5. Put the extract into the β-mercaptoethanol-containing tube.
6. Heat to 100 °C, 5 min (heat block).
7. Centrifuge at 19,000 ×*g*, 10 min at room temperature.
8. Collect ~70 μl supernatant. Take only the clear part, which contains proteins.
9. Run 10–15 μl for SDS-PAGE.
10. Conduct Western blotting using the anti-GFP antibody, which also recognizes the Citrine protein (*see* **Note 17**).

3.6 Live-Cell Imaging

We have achieved two different types of imaging for mitosis (described in Subheadings 3.6.1 and 3.6.2). Thus far, we have observed only the protonemal tissue, where the actively dividing caulonemal apical cell is the major target (cell cycle duration is ~6 h). A number of moss lines expressing GFP, RFP, or their variants have been published (Table 1). Image sequences are available in the recently published papers (HistoneH2B-mRFP, GFP-tubulin [8, 9], kinesin-GFP [9, 12, 14], MAP65-GFP and GFP-MAP65 [16]). We are also undertaking mitosis imaging in the gametophore (Kosetsu and Goshima, unpublished data). Here, we describe the way we observe kinesin tagged with Citrine together with mCherry-tubulin in caulonemal apical cells. However, the same protocol can be used for imaging other tagged lines. Mitosis observation using an inducible RNAi system is also performed similarly, except that β-estradiol (final concentration 1 μM) is added to the medium for 3–7 days to induce expression of dsRNAs.

3.6.1 High-Resolution Imaging by a Spinning-Disc Confocal Microscope

See videos in [12].

1. Melt the BCD agar medium with a microwave oven, and pour 3 ml to a 35-mm glass-bottom dish (*see* **Note 18**).
2. Wait for a minimum of 30 min, and the medium will become solid.
3. Cut off the central part of the agar medium (~1 cm × ~1 cm).

   Pour ~80 μl of the molten BCD medium onto the cut area so that a thin agar layer can be made on the glass (*see* **Notes 19–21**).
4. Inoculate a piece of protonemata onto the agar pad and culture for 5–6 days.
5. Identify the caulonemal apical cell in prophase under a microscope. There are four criteria in identification of a cell that is ready to break the nuclear envelope and assemble the spindle (Fig. 2a).
   (a) The length of the apical cell is ~50 % longer than the neighboring sub-apical cell.
   (b) A large vacuole is positioned right next to the nucleus (on the basal side) (*see* **Note 22**).
   (c) If microtubules are fluorescently labeled, one will see that the nucleus is surrounded by microtubules, in particular on the apical surface of the nucleus.
   (d) If time-lapse imaging is being performed, change in nuclear shape is another indicator. Caulonemal apical cells in interphase have a round-shaped nucleus. When the mitotic phase is approaching, the nucleus becomes diamond-shaped. However, it still takes more than 15 min until the nuclear envelope breaks down (NEBD). During prophase, the nucleus becomes round again.

   On the other hand, a cell with the following characteristics does not immediately enter mitosis: (1) The nucleus is located close to the cell cortex. (2) Very clear microtubule bundles are detected on the basal side of the nucleus (Fig. 2a).

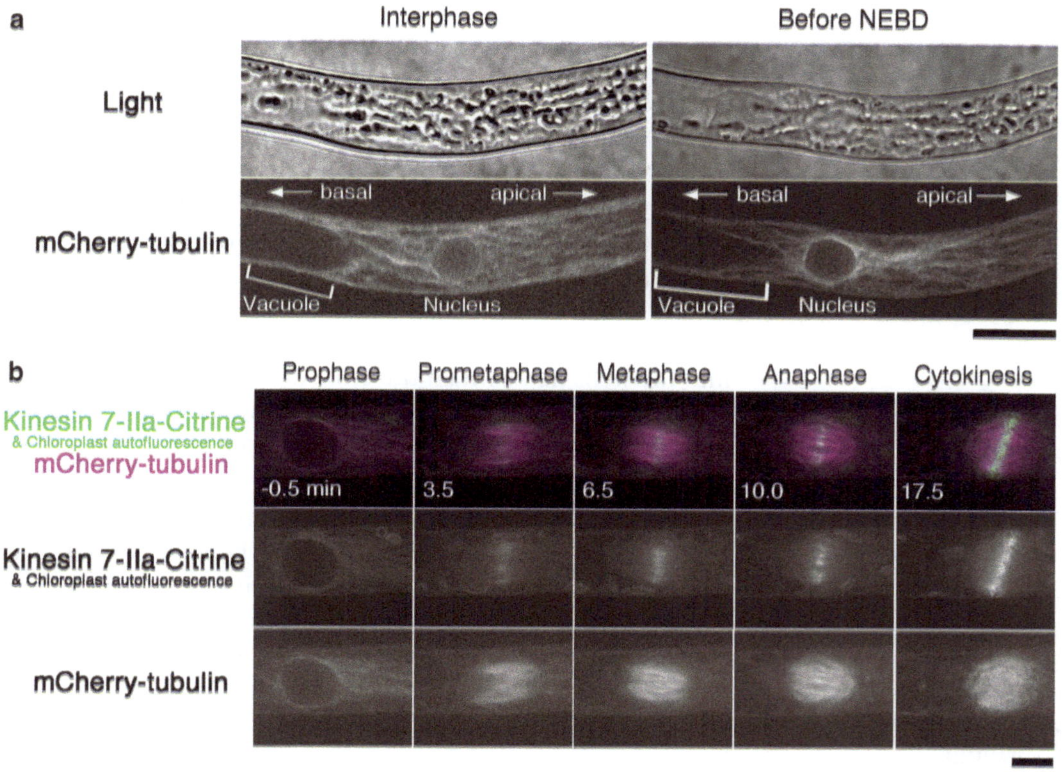

Fig. 2 Imaging mitosis of caulonemal apical cells. (**a**) Caulonemal apical cells in interphase (*left*) and late prophase (*right*). Cells were imaged with transmission light in the *top* panels. Dark particles observed inside the cells are chloroplasts. The nucleus is often embedded by chloroplasts during interphase. *Bottom* panels are mCherry-tubulin images acquired with 561-nm laser by spinning-disc confocal microscopy. The differences in microtubule networks and vacuole positioning between interphase and late prophase cells are described in Subheading 3.6.1, **step 5**. Bar, 20 μm. (**b**) Live imaging of caulonemal cells with spinning-disc confocal microscopy. A cell expressing mCherry-tubulin (*magenta*) and kinesin 7-IIa-Citrine (*green*) is displayed as an example. Kinesin 7-IIa-Citrine is localized to the spindle/phragmoplast midzone. Time 0 corresponds to NEBD. Bar, 10 μm.

6. Kinesin-Citrine localization during mitosis is monitored by time-lapse imaging at 30-s intervals with a spinning-disc confocal microscope. It takes ~1 h from NEBD to completion of cytokinesis (as indicated by the disappearance of the phragmoplast microtubule). Time intervals could be shortened, unless cells die by over-irradiation.

See videos in [8].

3.6.2 Long-Term Imaging by a Wide-Field Microscope (Low Magnification Lens)

The cell-cycle duration of caulonemal apical cells is ~6 h. Long-term imaging, such as image acquisition every 3 min for ~12 h, is often used for phenotype screening. For example, RNAi screening of ~60 kinesin genes for nuclear positioning or cytokinesis was performed with this microscopy [14, 15]. For long-term wide-field imaging, we use 6-well, glass-bottom plates to observe

>2 transgenic lines at the same time (e.g., one control line and multiple RNAi lines).

1. **Steps 1–4** of Subheading 3.6.1 are followed with some modifications. **Step 1**. 4.5 ml of the medium is poured into 1–3 wells; **Step 3**. A larger piece of the agar medium is cut off; and **Step 4**. Two lines are inoculated into each well.
2. To prevent the agar pad from drying, the 6-well plate is sealed with surgical tape. In addition, 2 ml sterile water is added to each of three empty wells. Finally, on the day of observation (5–6 days after inoculation), ~500 μl water is gently poured into the wells that contain the cells. Imaging should be delayed for 1 h. Imaging immediately after adding water should be avoided because the samples move out of focus during imaging.
3. Multi-site imaging is performed with a 10× objective lens. To activate photosynthesis, white light can be used to illuminate the samples in between image acquisition; otherwise, cell growth may be gradually retarded. However, illumination cannot be performed if many sites are imaged at each time point.

4 Notes

1. PHYSCOmanual ver. 2.0 http://www.nibb.ac.jp/evodevo/PHYSCOmanual/00Eindex.htm. PHYTOZOME—plant genome sequence database, including *P. patens* http://www.phytozome.net.
2. This step should be repeated ~3 times, since the lysate exceeds the loading limit of the column.
3. In our laboratory, failure in transformation is usually associated with failure in protoplast preparation (e.g., cells are burst during driselase treatment or the subsequent washing). We think that the failure is largely due to use of "unhealthy" protonemal cells; the cells are sometimes not in good condition, even though they appear to be growing normally in the culture medium. Preparing fresh cells from the stored plates at 4 °C usually overcomes this problem. In fact, for unknown reasons, when cells "rested" at 4 °C for a night are homogenized and plated, those cells are usually in a good condition compared to those continuously passed through homogenization at 25 °C. For protonemal cells stored for long at 4 °C, homogenization and plating are repeated twice, and cells typically become in a good condition. Moreover, protoplasts should be carefully treated, and physical stress on protoplasts should be minimized during the transformation process, such as during treating with driselase, mixing with PEG-DNA solution, or washing.

4. When protoplasts are mixed with DNA and the PEG solution, DNA and protoplasts should be mixed first, followed by the PEG solution addition. For unknown reasons, mixing DNA with PEG at first sometimes results in transformation failure.
5. We used the cellophane that we received as a gift from Futamura Chemical Industry (PL #300, 100 mm × 1000 m). We assume that other commercially available cellophanes would work as well.
6. When we identify bacterial contamination, the corresponding part of the cellophane is cut off by a sterile razor blade. If the host lines are resistant to certain antibiotics (e.g., G418), the remaining part of cellophane is transferred to the drug-containing medium. Typically, this procedure completely gets rid of bacteria.
7. We have used five drugs for the selection process: hygromycin, G418, zeocin, NTC (Nourseothricin), and blasticidin S. Hygromycin, G418, and zeocin work very well; most of the drug-resistant lines identified in the first round of the selection process actually possess the resistant gene. However, zeocin appears to be unstable, and the zeocin-containing plate should not be kept at 4 °C for a long period. NTC and blasticidin are less efficient, and quite frequently, an initially "resistant" line turns out to be non-resistant during the second selection process. Therefore, many candidate lines should be initially selected.
8. The remaining colonies can be kept at 4 °C after the cellophane on the drug-containing BCDAT plate is transferred to the drug-free BCDAT or PRM plate. In case the desired transgenic line cannot be obtained among the 200 candidates in the first round, more colonies can be picked up later.
9. We have been able to amplify up to 13 kb in length by green PCR. However, we typically design primers so that <3-kb fragments are efficiently amplified.
10. Since many *P. patens* genes have paralogues, it should be ascertained that the primers are uniquely designed and not cross-reactive with the paralogous genes.
11. One bead is enough for cell extraction, but addition of two or three beads also works.
12. The extract derived from the gametophore can be also used. However, the yield of extracts is not as good as from the protonemata.
13. Tight sealing is important; otherwise, the moss extract can be cross-contaminated during lysis.
14. Long-term storage should be done at −20 °C. However, repeated freezing/thawing should be avoided.

15. As of December 2014, the green PCR amplification of long DNA sequences works best with this kit in our laboratory. However, other kits would probably work as well.
16. For unknown reasons, we sometimes encounter a case in which an N-terminal tagging line confirmed by PCR shows a strange localization pattern. It is particularly important to select multiple lines for N-terminal tagging and to further check the lines by immunoblotting or other methods.
17. We have used a few different anti-GFP antibodies, including the commercially available JL-8 [Clontech]. However, for all of them, a ~50 kDa band is nonspecifically detected. In case a 50 kDa fusion protein needs to be detected or the target protein is not abundantly expressed, an immunoprecipitation step may be added to concentrate the tagged protein [18].
18. Chloronema-to-caulonema conversion takes place more rapidly on the BCD medium than on the BCDAT medium.
19. It is convenient to keep some BCD agar media in a 1.5 ml tube, since the tube can be heated in the heat block and the medium becomes molten quickly (100 °C, 2 min).
20. A thinner layer of the agar medium is better for clear imaging. However, the thin medium dries rapidly. In such a case, sterile water (~200 μl) should be added at day 3 or 4 to avoid desiccation.
21. The alternative way to make the agar pad is pouring an excess amount of medium into the dish, tilting the dish, and sucking up the medium at the corner, leaving an amount of medium just sufficient for covering the glass surface. This method allows the medium to cover the glass uniformly.
22. This method of assessment does not work in chloronemal cells because a number of large chloroplasts hinder the identification of the vacuole position.

Acknowledgements

We are indebted to Mitsuyasu Hasebe, Yuji Hiwatashi, and other Hasebe laboratory members for all the available and shared reagents for cell biology in moss. They also have provided us valuable information regarding the techniques associated with moss culturing and imaging. We thank Yuki Nakaoka and Ken Kosetsu for developing protocols and critical reading of this chapter. Work on moss in our laboratory is supported by Human Frontier Science Program, the TORAY Science Foundation, and Grants-in-Aid for Scientific Research (15H01227, 15K14540, and 26711012; MEXT, Japan).

References

1. Goshima G, Wollman R, Goodwin N, Zhang JM, Scholey JM, Vale RD, Stuurman N (2007) Genes required for mitotic spindle assembly in Drosophila S2 cells. Science 316:417–421
2. Sonnichsen B, Koski LB, Walsh A, Marschall P, Neumann B, Brehm M, Alleaume AM, Artelt J, Bettencourt P, Cassin E et al (2005) Full-genome RNAi profiling of early embryogenesis in Caenorhabditis elegans. Nature 434:462–469
3. Maliga Z, Junqueira M, Toyoda Y, Ettinger A, Mora-Bermudez F, Klemm RW, Vasilj A, Guhr E, Ibarlucea-Benitez I, Poser I et al (2013) A genomic toolkit to investigate kinesin and myosin motor function in cells. Nat Cell Biol 15:325–334
4. Cove D (2005) The moss Physcomitrella patens. Annu Rev Genet 39:339–358
5. Cove D, Bezanilla M, Harries P, Quatrano R (2006) Mosses as model systems for the study of metabolism and development. Annu Rev Plant Biol 57:497–520
6. Prigge MJ, Bezanilla M (2010) Evolutionary crossroads in developmental biology: Physcomitrella patens. Development 137:3535–3543
7. Vidali L, Augustine RC, Kleinman KP, Bezanilla M (2007) Profilin is essential for tip growth in the moss Physcomitrella patens. Plant Cell 19:3705–3722
8. Nakaoka Y, Miki T, Fujioka R, Uehara R, Tomioka A, Obuse C, Kubo M, Hiwatashi Y, Goshima G (2012) An inducible RNA interference system in Physcomitrella patens reveals a dominant role of augmin in phragmoplast microtubule generation. Plant Cell 24:1478–1493
9. Hiwatashi Y, Obara M, Sato Y, Fujita T, Murata T, Hasebe M (2008) Kinesins are indispensable for interdigitation of phragmoplast microtubules in the moss Physcomitrella patens. Plant Cell 20:3094–3106
10. Vidali L, Rounds CM, Hepler PK, Bezanilla M (2009) Lifeact-mEGFP reveals a dynamic apical F-actin network in tip growing plant cells. PLoS One 4:e5744
11. Shen Z, Collatos AR, Bibeau JP, Furt F, Vidali L (2012) Phylogenetic analysis of the Kinesin superfamily from physcomitrella. Front Plant Sci 3:230
12. Miki T, Naito H, Nishina M, Goshima G (2014) Endogenous localizome identifies 43 mitotic kinesins in a plant cell. Proc Natl Acad Sci U S A 111:E1053–E1061
13. Reddy AS, Day IS (2001) Kinesins in the Arabidopsis genome: a comparative analysis among eukaryotes. BMC Genomics 2:2
14. Naito H, Goshima G (2015) NACK kinesin is required for metaphase chromosome alignment and cytokinesis in the moss Physcomitrella patens. Cell Struct Funct 40:31–41
15. Miki T, Nishina M, Goshima G (2015) RNAi screening identifies the armadillo repeat-containing kinesins responsible for microtubule-dependent nuclear positioning in Physcomitrella patens. Plant Cell Physiol 56:737–749
16. Kosetsu K, de Keijzer J, Janson ME, Goshima G (2013) Microtubule-associated protein65 is essential for maintenance of phragmoplast bipolarity and formation of the cell plate in Physcomitrella patens. Plant Cell 25:4479–4492
17. Jonsson E, Yamada M, Vale RD, Goshima G (2015) Clustering of a kinesin-14 motor enables processive retrograde microtubule-based transport in plants. Nat Plants 1(7):pii, 15087
18. Nakaoka Y, Kimura A, Tani T, Goshima G (2015) Cytoplasmic nucleation and atypical branching nucleation generate endoplasmic microtubules in Physcomitrella patens. Plant Cell 27:228–242
19. Perroud PF, Quatrano RS (2006) The role of ARPC4 in tip growth and alignment of the polar axis in filaments of Physcomitrella patens. Cell Motil Cytoskeleton 63:162–171
20. Finka A, Schaefer DG, Saidi Y, Goloubinoff P, Zryd JP (2007) In vivo visualization of F-actin structures during the development of the moss Physcomitrella patens. New Phytol 174: 63–76
21. Perroud PF, Quatrano RS (2008) BRICK1 is required for apical cell growth in filaments of the moss Physcomitrella patens but not for gametophore morphology. Plant Cell 20:411–422
22. Oda Y, Hirata A, Sano T, Fujita T, Hiwatashi Y, Sato Y, Kadota A, Hasebe M, Hasezawa S (2009) Microtubules regulate dynamic organization of vacuoles in Physcomitrella patens. Plant Cell Physiol 50:855–868
23. Vidali L, van Gisbergen PA, Guerin C, Franco P, Li M, Burkart GM, Augustine RC, Blanchoin L, Bezanilla M (2009) Rapid formin-mediated actin-filament elongation is essential for polarized plant cell growth. Proc Natl Acad Sci U S A 106:13341–13346

24. Vidali L, Burkart GM, Augustine RC, Kerdavid E, Tuzel E, Bezanilla M (2010) Myosin XI is essential for tip growth in Physcomitrella patens. Plant Cell 22:1868–1882
25. Augustine RC, Pattavina KA, Tuzel E, Vidali L, Bezanilla M (2011) Actin interacting protein1 and actin depolymerizing factor drive rapid actin dynamics in Physcomitrella patens. Plant Cell 23:3696–3710
26. Uchida M, Ohtani S, Ichinose M, Sugita C, Sugita M (2011) The PPR-DYW proteins are required for RNA editing of rps14, cox1 and nad5 transcripts in Physcomitrella patens mitochondria. FEBS Lett 585:2367–2371
27. Yamashita H, Sato Y, Kanegae T, Kagawa T, Wada M, Kadota A (2011) Chloroplast actin filaments organize meshwork on the photorelocated chloroplasts in the moss Physcomitrella patens. Planta 233:357–368
28. Furt F, Lemoi K, Tuzel E, Vidali L (2012) Quantitative analysis of organelle distribution and dynamics in Physcomitrella patens protonemal cells. BMC Plant Biol 12:70
29. Hiwatashi Y, Sato Y, Doonan JH (2014) Kinesins have a dual function in organizing microtubules during both tip growth and cytokinesis in Physcomitrella patens. Plant Cell 26:1256–1266
30. Wu SZ, Bezanilla M (2014) Myosin VIII associates with microtubule ends and together with actin plays a role in guiding plant cell division. Elife. doi:10.7554/eLife.03498

Chapter 18

Small Molecule Approach to Study the Function of Mitotic Kinesins

Naowras Al-Obaidi, Johanna Kastl, and Thomas U. Mayer

Abstract

Mitotic motor proteins of the kinesin superfamily are critical for the faithful segregation of chromosomes and the formation of the two daughter cells during meiotic and mitotic M-phase. Of the 45 human kinesins, roughly a dozen are involved in the assembly of the bipolar spindle, alignment of chromosomes at the spindle equator, chromosome segregation, and cytokinesis. The functions of kinesins in these processes are highly diverse and include the transport of cargo molecules, sliding and bundling of microtubules, and regulation of microtubule dynamics. In light of this multitude of diverse functions and the complex functional interplay of different kinesins during M-phase, it is not surprising that one of the greatest challenges in cell biology is the functional dissection of individual motor proteins. Reversible and fast acting small molecules are powerful tools to accomplish this challenge. However, the validity of conclusions drawn from small molecule studies strictly depends on compound specificity. In this chapter, we present methods for the identification of small molecule inhibitors of a motor protein of interest. In particular, we focus on a protein-based large throughput screen to identify inhibitors of the ATPase activity of kinesins. Furthermore, we provide protocols and guidelines for secondary screens to validate hits and select for specific inhibitors.

Key words Small molecule inhibitors, Large throughput screening, Protein-based screen, Mitotic motor proteins, Kinesins, Microtubule, Malachite green assay, Enzyme-coupled assay

1 Introduction

The coordinated migration of chromosomes to the spindle equator, their subsequent alignment midway between the two spindle poles, and their concise separation into the two daughter cells are among the most impressive processes in biology [1]. The appealing beauty of this process—which has been termed "the dance of chromosomes"—stems from the highly dynamic and precisely synchronized movements of chromosomes [2, 3]. These movements require the precisely coordinated activities of multiple motor proteins as well as

*These authors contributed equally.

Paul Chang and Ryoma Ohi (eds.), *The Mitotic Spindle: Methods and Protocols*, Methods in Molecular Biology, vol. 1413,
DOI 10.1007/978-1-4939-3542-0_18, © Springer Science+Business Media New York 2016

the tightly controlled dynamic behavior of microtubules [4, 5]. Yet it is exactly the complex interplay of different motor proteins and the high dynamicity of the different processes during M-phase that makes the functional dissection of the underlying molecular mechanisms difficult. A common approach to elucidate protein function is RNA interference (RNAi) mediated protein depletion. The great advantage of this approach is its built-in specificity, but it typically takes days to efficiently deplete a protein of interest [6]. This lack of temporal resolution makes it very difficult to study for example a potential post-metaphase function of a motor protein if its depletion results in a metaphase arrest due to the activation of the spindle assembly checkpoint. Luckily, these technical issues can be circumvented through the use of small molecules inhibitors, which due to their fast and often reversible mode of action allow the modulation of protein function on a fast time scale [7–11].

One important class of mitotic motor proteins are the kinesins, ATPases that convert the energy released by ATP hydrolysis into mechanical force. In human cells, the kinesin superfamily comprises about 40 different kinesins which fulfill highly diverse functions such as cargo transport, modulation of microtubules dynamics as well as the assembly of the mitotic spindle and segregation of chromosomes. Screening for small molecule inhibitors of kinesins [12] should begin with a clear idea of the benefit of such a compound, i.e., which experiments could be performed that were not amenable to classical approaches. The next steps involve the development of a robust assay, the miniaturization of this assay to a 384-well format, and performing the small molecule screen. It might come as a surprise, but the screen itself is the least time consuming step. In fact, a major focus should lie on the development of a robust assay as well as secondary screens to sort out unspecific compounds [13]. Shortcuts herein and rushing into screening typically come with higher costs than benefits in the form of assay variability or hits that cannot be reproduced and/or act in an unspecific manner. Two general approaches are taken for screening. These involve cell based assays and in vitro assays using pure protein. One major benefit of cell based assays is that they select for membrane permeable compounds and have the potential to identify inhibitors that target a whole pathway instead of an individual protein [14–17]. However, cell based screens require expensive equipment such as automated microscopes, an understanding of image analysis and familiarity with sophisticated image analysis algorithms to quantify the relevant phenotype in an unbiased manner. Thus, cell based assays are costly and more complex to perform than pure enzyme assays, which are important considerations to consider prior to performing the screen. While these obstacles can be overcome, *the* major challenge of any cell-based screen is the identification of the cellular target of the identified compounds [18]. This issue becomes especially apparent when considering that a typical hit from a primary screen has a half maximal effective concentration

(EC_{50}) in the low to median μM concentration range. The best cellular screens assay a phenotype that is unambiguous. A good example of an effective cellular small molecule screen is the one that identified monastrol, a specific inhibitor of the mitotic kinesin Eg5 [19]. Eg5 inhibition results in a very distinct phenotype—mono-astral spindles with chromosomes attached at the tips of radially arranged microtubules. This characteristic phenotype facilitated the identification of Eg5 as the target of monastrol [19]. However, inhibition of most mitotic kinesins will not results in such a pronounced phenotype, and therefore many cell-based assays do not get beyond the identification of compounds with interesting phenotypes. As a result, my laboratory decided to focus on in vitro, protein-based small molecule screens for inhibitors of kinesins.

A protein-based small molecule screen for inhibitors of kinesins involves the following steps:

- Purification of kinesin-of-interest, full-length or motor-domain.
- Establishment of a robust assay to quantify ATPase activity.
- Miniaturization of assay to 384-well format and implementation of positive and negative controls.
- Small molecule screen.
- Hit validation and characterization, e.g., testing of initial screening hit and resynthesized compound, IC_{50} determination, specificity analyses, mode-of-action analyses.

Despite their simplicity, pure protein screens also have caveats. Performing a protein-based screen does not necessarily mean that the protein of interest will be the relevant target of the small molecule hits, once cells are treated. Therefore rigorous specificity analyses must be performed. This is particularly important if one member of a highly homologous protein family, such as the kinesins is targeted. It will be important to confirm that putative inhibitors are selective for your target protein. Another important consideration for pure protein in vitro screens is that unbiased, high throughput screens often require the screening of many compounds—large compound libraries comprising tens to hundreds of thousands of compounds. This is because protein based screens target a specific protein in contrast to cellular screens that can target a whole pathway that contains many potential targets.

A protein based screen requires large amounts of purified protein. The choice of the appropriate protein is an essential determinant for the success of a small molecule screen and, therefore, the following statement should be taken seriously: "you get what you screen for." While this statement sounds trivial at a first glance, it can have a severe impact on the applicability of the identified compounds and, therefore, the decision regarding full-length versus truncated protein, the nature of the affinity tag, and the appropriate expression system should be carefully thought out. Thus, how you generate this protein is an important consideration. For example, the production of full-length kinesins typically requires an insect cell expression system, which is time consuming, laborious, and yields only limited amounts of recombinant protein [20–22]. In many cases, screening for inhibitors of just the enzymatic domain is sufficient, and these smaller domains can be more easily expressed and purified in the high yield *E.coli* system. The motor-domain of most kinesins can be easily purified from bacteria and this fragment is commonly sufficient to reliably measure microtubule-stimulated ATPase activity [23, 24].

ATP hydrolysis can be quantified by an enzyme-coupled assay (ECA, [25]) or a malachite green based assay (MGA, [26]) and therefore these assays are well suited to screen compound libraries for small molecule inhibitors of kinesins. The ECA regenerates hydrolyzed ATP by coupling the ATPase reaction to two other enzymatic reactions that ultimately result in the consumption of NADH whose concentration can be measured spectrophotometrically at 340 nm (Fig. 1a). The assay is performed such that the ATPase reaction is the rate limiting step, allowing the determination of the ATPase rate by measuring the rate of NADH consumption. The benefit of ECA is that ATP consumption can be continuously monitored without disturbing the system. In the MGA, the free orthophosphate product of the ATPase reaction is incorporated into a malachite green-phosphomolybdate complex that absorbs at 620–650 nm (Fig. 1b). In contrast to ECA, MGA provides an endpoint measurement after chemical denaturation of the reaction with perchloric acid. However, given that the MGA is more easily adaptable to large throughput screening and does not bear the risk of identifying compounds that target enzymes of the ECA reaction rather than the kinesin of interest, we focus in this chapter on an MGA-based protocol for compound screening. The following protocol provides a step-to-step instruction for a MGA-based screen for small molecule kinesin inhibitors.

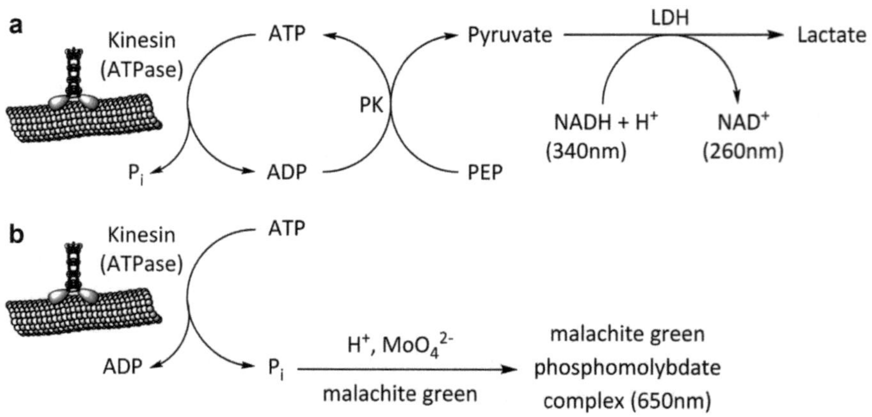

Fig. 1 Schematic representation of (**a**) the enzyme-coupled assay and (**b**) the malachite green assay. *PK* pyruvate kinase, *LDH* lactic dehydrogenase, *PEP* phospho(enol) pyruvate, *NAD* β-nicotinamide adenine dinucleotide

2 Materials

2.1 General Reagents and Buffers

All chemicals are from Sigma-Aldrich, unless otherwise stated.

1. BRB80 (Brinkley Reassembly Buffer): 20 mM K-PIPES pH 6.8, 1 mM $MgCl_2$, 1 mM EGTA (prepare 5× stock, filter, and store at 4 °C).
2. Porcine brain tubulin (10 mg/ml in BRB80; commercial or from own purification): aliquot in 1 ml aliquots and store at −80 °C.
3. Recombinant kinesin motor domain: generally prepared as poly-histidine or glutathione-S-transferase (GST) tagged fusion proteins by one-step affinity chromatography, aliquoted, and stored at −80 °C (*see* **Note 1**).
4. Adenosine 5′-triphosphate magnesium salt (ATP): 100 mM aqueous stock solution (pH ~7), aliquot and store at −80 °C. This 100 mM stock solution is used to prepare the ATP solutions for the MGA and ECA.
5. Guanosine 5′-triphosphate sodium salt hydrate (GTP): 100 mM aqueous stock solution (pH ~7), aliquot, and store at −80 °C.
6. Glycerol: 100 %.
7. Guanidine hydrochloride: 6 M aqueous solution, store at 4 °C.
8. Paclitaxel (Taxol, from *Taxus brevifolia*): 10 mM stock solution in DMSO, aliquot, and store at −80 °C.
9. Dimethyl sulfoxide (DMSO): molecular biology grade.

2.2 Reagents for Malachite Green Assay (MGA)

1. Bovine serum albumin (BSA): 10 mg/ml aqueous stock solution, sterile-filtered, store at 4 °C.
2. $BRB80^{++}$: BRB80 (1×) + 0.1 μM taxol + 0.1 mg/ml BSA (prepare fresh).
3. Malachite green solution: 1.3 M HCl, 8.5 mM ammonium molybdate, 362.5 μM malachite green.
4. ATP stock solution: prepare 300 μM stock in BRB80.
5. Perchloric acid stock solution: prepare 1.2 and 1.0 M aqueous solutions.

2.3 Reagents for Enzyme-Coupled Assay (ECA)

1. Enzyme-coupled assay buffer (ECAB): 10 mM imidazole, 5 mM potassium acetate, 2 mM EGTA, 4 mM magnesium acetate, pH to 7.2 with 5 % acetic acid. Store at 4 °C.
2. $ECAB^t$: 20 μM taxol in ECAB. Prepare fresh.
3. ATP stock solution: prepare 10 mM stock in $ECAB^t$.
4. Phospho(enol)pyruvic acid monopotassium salt (PEP): 50 mM stock solution in ECAB, aliquot and store at −80 °C. Final concentration in assay: 5 mM.

5. Pyruvate kinase from rabbit muscle (PK, Type II, ammonium sulfate suspension).
6. l-lactic dehydrogenase from rabbit muscle (LDH, Type II, ammonium sulfate suspension).
7. β-nicotinamide adenine dinucleotide (NADH): 10 mM stock solution in ECAB, aliquot, and store at −80 °C.
8. Enzyme-substrate 10x mix (ES10x): 40 U/ml PK, 80 U/ml LDH, and 2 mM NADH in ECAB^t. Prepare fresh.
9. Bovine serum albumin (BSA): 5 mg/ml stock solution in ECAB, store at 4 °C.
10. Kinesin dilution buffer: 1 mg/ml BSA in ECAB^t. Prepare fresh.

2.4 Small Molecules

1. Small molecule library in 384-well format. Compounds dissolved in DMSO at 10 mM.

2.5 Plastic Ware

1. Reagent reservoirs (e.g., from *VWR*).
2. 384-Well microplate with flat and UV light transparent well bottom (e.g., Greiner Bio-One μCLEAR).
3. 1.5 ml centrifuge tubes, e.g., Beckman Microfuge Tube Polyallomer (for TLA-55).

2.6 Instruments

1. Tabletop ultracentrifuge (e.g., Beckman Coulter Optima MAX-E equipped with TLA-55 rotor).
2. Water bath or thermomixer.
3. Photometer with cuvettes suitable for absorbance measurements at 280 nm.
4. Centrifuge suitable for microplates (e.g., Thermo Scientific Multifuge 3S with microplate carrier).
5. Cooled Eppendorf centrifuge.
6. Microplate reader suitable for absorbance measurements (e.g., Tecan infinite F500 equipped with absorbance filters 340 and 650 nm).
7. Multichannel pipette (e.g., Brand).
8. Microplate dispenser (e.g., Multidrop, Thermo Scientific).

3 Methods

3.1 Preparation of Taxol-Stabilized Microtubules (MTs)

1. Thaw 1 ml porcine brain tubulin (10 mg/ml) on ice, and transfer to 1.5 ml TLA-55 centrifuge tube.
2. Clarify by ultracentrifugation using a TLA-55 rotor (10 min, 136,000 × *g*, 4 °C).

3. After centrifugation, incubate the TLA-55 rotor at 35 °C for the next centrifugation step. Take half of the tubulin supernatant (SN) (500 μl) and transfer to a TLA-55 centrifuge tube containing 285 mg glycerol and 334 μl 1× BRB80. Repeat this step with the second half of the SN.
4. Add 10 μl of 100 mM GTP (final concentration: 1 mM) to each tubulin mixture, mix well and polymerize by incubation in a water bath for 30 min at 37 °C.
5. Add 4 μl of 10 mM taxol (final concentration: 40 μM) to each MT mix, mix well, but gently (*see* **Note 2**) and incubate for another 30 min at 37 °C.
6. Pellet the MTs by ultracentrifugation using a TLA-55 rotor (30 min, 136,000 ×*g*, 35 °C)
7. Meanwhile, prepare resuspension buffer by mixing 2 μl of 10 mM taxol with 800 μl BRB80 (1×).
8. Discard the supernatant, i.e., unpolymerized tubulin, and resuspend the MT pellet in a total of 200 μl of resuspension buffer for both tubes using a cutoff pipette tip (*see* **Note 2**). Use the 200 μl to first resuspend the microtubules from one tube and resuspend the MTs from the second tube herein.
9. Measure the concentration (c) of the pooled MT suspension by preparing 1:100 and 1:50 dilutions in 6 M guanidine hydrochloride solution and reading the absorbance at 280 nm (A_{280}) against blank buffer. The concentration can be calculated according to

$$c = \frac{A_{280}}{115000\ \mathrm{M}^{-1}\mathrm{cm}^{-1}} \times \text{dilution factor.}$$

   Calculate the mean of both dilutions to estimate the concentration of polymerized tubulin.
10. Add the required volume of resuspension buffer, to obtain a taxol-stabilized MT mix corresponding to a tubulin concentration of 100 μM.

3.2 Malachite Green Assay

The rate of ATP hydrolysis depends on the concentrations of the ATPase itself, i.e., the kinesin, the substrate ATP and pseudo-substrate microtubules (pseudo-substrate because they are not turned over). For the initial test of kinesin activity, ATPase activity is measured using the MGA at varying kinesin concentrations, while the concentrations of ATP and MTs are kept constant at 100 and 1 μM, respectively. In our hands, these concentrations of ATP and MTs are a good starting point to get a robust and reliable readout. To determine the screening concentration of your kinesin of interest, titrate its concentration in 11 steps ranging from 1 nM to 1 μM, and measure its activity at five time points (0, 6, 12, 18, and 24 min; *see* **Note 3**).

1. Calculate the number of required conditions to determine the total volume of buffer that has to be prepared. The final assay volume of 60 µl per reaction is obtained by addition of 20 µl ATP stock solution (300 µM) to 40 µl of master mix containing kinesin and MTs in assay buffer (BRB80^{++}). In total, we measure 12 reactions: 11 kinesin concentrations and one reaction without kinesin as control to ensure that ATP hydrolysis is mediated by the kinesin of interest (background reaction).
2. Prepare the reaction mixes: first dilute 9 µl of 100 µM taxol-stabilized MT stock solution in 591 µl BRB80^{++} and distribute each 50 µl into 11 reaction tubes (1.5 ml).
3. Centrifuge the thawed kinesin solution using a cooled Eppendorf centrifuge (20,000 × *g*, 10 min, 4 °C) to pellet potentially denatured protein and obtain a homogeneous protein solution.
4. For the kinesin dilutions, prepare 100 µl buffer containing 1.5 µM MTs and 1.5 µM kinesin in BRB80^{++}. This mix is for your reaction with the highest kinesin concentration (the final concentration of 1 µM is obtained after addition of ATP). Now, perform 10 serial 1:2 dilutions using the reaction tubes prepared in **step 2**. To this end, transfer 50 µl of the higher concentrated kinesin mix to a tube containing 50 µl of 1.5 µM MTs but no kinesin in BRB80^{++}, mix well, and repeat nine times. The eleventh tube prepared in **step 2** is the control mix containing 50 µl of BRB80^{++} with only MTs for the kinesin-free reaction. Now, you should have one tube containing the highest kinesin concentration, ten tubes containing the serial dilutions, plus one tube for the kinesin-free control (12 tubes in total).
5. For each condition, five wells (for 5 time points) of a 384-well plate are filled with 40 µl of 1 M perchloric acid (12 reactions × 5 time points = 60 wells).
6. Fill 40 µl from each of the 12 tubes (**step 4**) into separate wells of a 384-well plate.
7. Start the ATPase reaction by addition of 20 µl ATP stock solution (final concentration in 60 µl assay volume is 100 µM). Start synchronously by adding ATP with a multichannel pipette simultaneously—this allows best temporal control—to all wells and mix the reaction volume thoroughly.
8. To stop the reactions, transfer 10 µl of each of the 12 reaction mixes into the first 12 wells containing perchloric acid. Do this simultaneously for all 12 reactions using a multichannel pipette and mix vigorously. For the 0 min time point, do it immediately after starting the reaction by the addition of ATP. Repeat this step after 6, 12, 18, and 24 min.

7. When the last time point is taken, 40 µl of malachite green solution is added to all stopped reaction mixes using a multichannel pipette and mixed vigorously.

8. After 20 min incubation at RT, the absorbance at 650 nm is measured with a plate reader.
9. Subtract the absorbance of the background reactions (kinesin-free reaction) from the corresponding kinesin-dependent values to obtain background-corrected absorbances for each time point. According to the Beer–Lambert Law, this value is proportional to the concentration of orthophosphate produced by the ATPase, i.e., the kinesin activity. If you wish to determine the ATPase activity, you need to additionally measure phosphate standards to translate absorbance values into actual concentration values.

Once you can reproducibly measure the ATPase activity of your kinesin (*see* **Note 4**), choose a kinesin concentration within the linear range of your assay, i.e., using half or double concentration would lead to half or double activity, respectively. Using this kinesin concentration, titrate the concentration of ATP while keeping MTs at saturating concentrations and determine the K_m for ATP [27, 28]. Repeat this assay with constant saturating ATP concentrations and determine the $K_{1/2}$ for MTs. Following the statement "you get what you screen for," consider that the concentration of ATP and MTs in your screening assay impinges on how likely ATP and/or MT competitive inhibitors will be identified. In this protocol, 100 μM ATP in combination with a compound concentration of 10 μM was chosen to favor identification of ATP uncompetitive or noncompetitive inhibitors (*see* **Note 5**). Once the optimal concentrations for the kinesin, ATP, and MTs have been determined, the assay has to be optimized for high throughput screening (HTS).

3.3 Optimization for HTS

The identification of an appropriate small molecule inhibitor typically involves the screening of tens to hundreds of thousands of compounds. This order of magnitude requires that the assay provides a sufficiently robust readout to compensate for undesired assay variability and can be (partially) automated. For the optimization of an established assay for screening conditions it is therefore key to find the perfect compromise between availability and cost of reagents, maximal automation, and robustness of the readout. For example, you can reduce the amount of kinesin required for an MGA-based screen by prolonging the reaction time in order to obtain similar levels of readout signal.

To maximize reproducibility, calculate the amounts of reagents needed for a complete screen and prepare the corresponding amounts. Be generous in the calculation of the required volumes so that one screen can be accomplished with one batch of reagents even when several compound plates have to be rescreened (*see* **Note 6**). Typically, compounds are dissolved in 100 % DMSO and stored in 384-well plates with the last two columns (column 23 and 24) being empty. This layout allows the inclusion of negative controls, i.e., reaction mix without kinesin (column 23) and positive controls, i.e., complete reaction mix containing the compound vehicle control DMSO (column 24).

3.3.1 Assay Automation

In the following protocol, we describe the protocol for automated liquid handling and compound transfer using a Tecan Freedom EVO® robot (Fig. 2). This protocol uses final concentrations of 100 and 1 μM for ATP and MTs, respectively. Keep in mind that you have to replace these values by the concentrations gained from your own assay optimization. Likewise, you have to use the optimized values for kinesin concentration and ATPase reaction time.

1. Prepare per 384-well plate:
   - 7.36 ml reaction mix with kinesin: kinesin (at 1.5 × final concentration) + 1.5 μM MTs in BRB80^{++} in a 50 ml Falcon tube.
   - 320 μl reaction mix without kinesin (1.5 μM MTs in BRB80^{++}).
   - 3.84 ml ATP stock solution (300 μM) in a reservoir.
   - 15.36 ml 1.2 M perchloric acid in a 50 ml Falcon tube.
   - 23.04 ml malachite green solution in a 50 ml Falcon tube.
   - Thaw one 384-well plate of the screening library (10 mM compound in 100 % DMSO), (*see* **Note 7**).
2. Distribute 20 μl/well of the reaction mix without kinesin in negative control wells (column 23) of an empty 384-well plate.

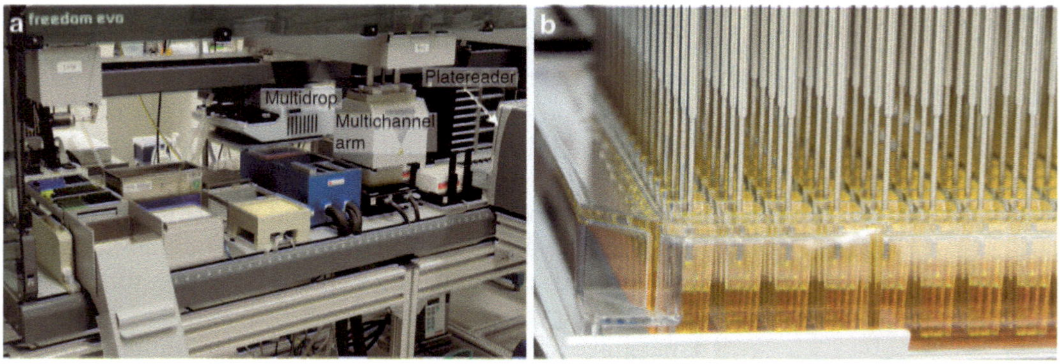

Fig. 2 (**a**) Freedom Evo® robotic platform (Tecan) including a multichannel arm for liquid handling using a 384-fixed tips- or disposable-tips adapter and integrated multidrop and plate reader devices. The fully software operated workstation can be used to fully automate plate filling, compound transfer, incubation, and measurement. (**b**) Close-up on the Multichannel arm aspirating liquid from a 384-well plate using the 384-fixed tips adapter

3. Place the plate on the robot platform and use the multidrop device to fill wells in columns 1–22 and 24 (except for the negative control wells in column 23) with 20 μl reaction mix containing the kinesin.
4. Transfer 30 nl of compounds in DMSO from the stock plate to the reaction plate using the liquid handling device of your robot.
5. Start the reaction simultaneously in all wells by adding 10 μl ATP stock solution from a reservoir into the reaction plate using the liquid handling device of your robot.
6. Move the plate on a shaker device to mix briefly, then move it back to a position on the robot platform and incubate for the previously optimized reaction time at RT (*see* **Note 8**).
7. Stop the reaction simultaneously in all wells by adding 60 μl perchloric acid (1.2 M, final concentration in 90 μl assay volume is 0.8 M) from a reservoir into the reaction plate using the liquid handling device of your robot.
8. Transfer 30 μl of these reactions to an empty 384-well-plate and supplement each of these wells with 40 μl of malachite green solution. Mix and incubate for 20 min at RT.
9. Move the plate into the plate reader and measure absorbance at 650 nm.
10. Calculate ATPase rates as done before.

3.3.2 Measuring Assay Quality with the Z' Value

Automation and miniaturization naturally come at the cost of increased variability and diminished reproducibility. Once the general screening procedure has been established it is therefore critical to ensure that the readout of kinesin activity under these conditions is robust enough for HTS. A measure for HTS suitability is the *Z'* value [29] which can be determined by preparing and measuring several wells of a positive control (PC: reaction mix with kinesin + DMSO) and a negative control (NC: reaction mix without kinesin + DMSO) using the following equation:

$$Z' = 1 - (3 \times \sigma_{NC} + 3 \times \sigma_{PC}) / (\mu_{NC} - \mu_{NC})$$

μ is the average and σ is the standard deviation of ATPase rates measured for the wells containing the positive control (PC) and negative control (NC). Assays with *Z'* values ≥0.5 are generally considered suitable for HTS.

3.4 Secondary Screen Using an Enzyme-Coupled Assay (ECA)

Once an initial hit has been identified, the next tasks include hit validation, the determination of the half maximal inhibitory concentration (IC_{50}) and the mode of inhibition, i.e., competitive, uncompetitive, or noncompetitive. Since the ECA allows continuous measurements without disturbing the system, it is a powerful approach for initial hit characterization. Furthermore, a readout unrelated to the one used for initial hit identification helps to exclude compounds that interfere with the screening assay rather than targeting the kinesin of interest.

3.4.1 Setting Up the ECA

Like for the MGA, the following protocol describes the optimization of the ECA, i.e., titration of concentrations of kinesin, ATP, and MTs to obtain a robust and reliable readout.

1. Calculate the number of required conditions to determine the total volume of buffer that has to be prepared. The assay volume of 100 μl per reaction is obtained by mixing 1 μl DMSO, 2.25 μl BSA, 10 μl PEP, 10 μl ES10x, 1 μl ATP, 1 μl MTs, 5 μl kinesin, 69.75 μl $ECAB^t$.
2. Thaw stocks and frozen reagents on ice and keep everything but MTs on ice.
3. Centrifuge the thawed kinesin solution in a cooled Eppendorf centrifuge (10 min, 20,000 × *g*, 4 °C) to pellet potentially denatured protein and to obtain a homogeneous protein solution.
4. As outlined for the MGA, prepare a serial dilution of your kinesin starting with 20 μM as the highest concentration (i.e., 1 μM final concentration when 5 μl of this are used in 100 μl assay volume). Use $ECAB^t$ supplemented with 1 mg/ml BSA for the dilutions. Now, you should have one tube containing the highest kinesin concentration, ten tubes containing the serial dilutions, plus one tube for the kinesin-free control (12 tubes in total).
5. Transfer 5 μl of each of the 12 tubes to 12 wells of a 384-well plate. Keep the plate on ice.
6. Prepare on ice the following master mix containing for each reaction 69.75 μl $ECAB^t$, 10 μl PEP, 2.25 μl BSA (5 mg/ml), 10 μl ES10x, 1 μl DMSO, and 1 μl ATP (10 mM). After warming up the master mix to RT, add 1 μl MTs for each reaction. Incubate the 384-well plate containing the kinesin dilutions and negative control at RT as well.
7. For each reaction, transfer 95 μl volume of the master mix to the wells containing the kinesin dilutions and negative control to yield the complete reaction volume of 100 μl. Mix thoroughly, yet carefully by pipetting up and down (avoid bubbles). Work quickly as the ATPase reaction is fairly quick and accordingly, NADH is consumed within a short time period.
8. Centrifuge the microtiter plate (800 × *g*, 30 s, RT) to get rid of bubbles and settle fluid levels.
9. Immediately, start reading the absorbance every 30 s at 340 nm for 10 min using a plate reader.

10. Calculate the steady-state rates of the ATPase reactions by reading out the slope of the absorbance measurements using a linear fit. Subtract the slope of the background reaction from the kinesin-dependent slopes to obtain background-corrected rates. Use the Beer–Lambert Law to transform the change in absorbance to a change in ATP concentration (extinction coefficient for NADH is $\varepsilon_{340} = 6220\ M^{-1}\ cm^{-1}$).

As for the MGA, choose an appropriate kinesin concentration and then titrate ATP and MTs to optimize the assay. To this end, prepare appropriate stocks of ATP or MTs, transfer defined volumes of these to empty wells of a 384-well-plate, and complete the reaction mixes by addition of a master mix containing the other assay components (at constant concentrations).

3.4.2 ECA Inhibitory Studies

Once the optimal concentrations for the kinesin, ATP, and MTs have been determined the assay can be used to characterize a small number of hits at different concentrations. To this end, the identified hits should be prepared as 100× stock solutions in DMSO.

1. Calculate the number of required conditions to determine the total volume of buffer that has to be prepared. The assay volume of 100 μl per reaction is obtained by mixing 1 μl DMSO, 2.25 μl BSA, 10 μl PEP, 10 μl ES10x, 1 μl ATP, 1 μl MTs, 1 μl kinesin, 73.75 μl $ECAB^t$.
2. Thaw stocks and frozen reagents on ice and keep everything but MTs on ice.
3. Centrifuge the thawed kinesin solution in a cooled Eppendorf centrifuge (10 min, 20,000 × *g*, 4 °C) to pellet potentially denatured protein and to obtain a homogeneous protein solution.
4. Using the optimized concentrations of kinesin, ATP, and MTs determined in Subheading 3.4.1, prepare 100× stock solutions in $ECAB^t$. For the kinesin, use $ECAB^t$ supplemented with 1 mg/ml BSA.
5. Transfer 1 μl of each compound stock (100×) to each a well of a 384-well plate. Additionally, two wells should contain 1 μl DMSO each, one for the background reaction (negative control, i.e., kinesin-free reaction), and one for the uninhibited enzymatic reaction (positive solvent vehicle control).
6. For the negative control, transfer 1 μl of $ECAB^t$ supplemented with 1 mg/ml BSA to one well containing 1 μl DMSO.
7. Prepare on ice the following master mix containing for each reaction 73.75 μl $ECAB^t$, 10 μl PEP, 2.25 μl BSA (5 mg/ml), 10 μl ES10x, and 1 μl ATP.
8. After warming up the master mix to RT, add 1 μl MTs for each reaction.

9. Transfer 98 μl of the master mix to the well containing the negative control (kinesin-free reaction) to yield the complete reaction volume of 100 μl. Mix thoroughly, yet carefully by pipetting up and down (avoid bubbles).
10. For each kinesin-containing reaction, add 1 μl of the 100× kinesin stock solution to the remaining master mix to obtain the desired final concentration. Mix thoroughly. From now on, work quickly as the ATPase reaction is fast.
11. For each reaction, transfer 99 μl of the master mix to each a well of the microtiter plate containing compounds or DMSO (positive control) and mix thoroughly.
12. Centrifuge the microtiter plate (800 × *g*, 30 s, RT) to remove bubbles and settle fluid levels.
13. Immediately, start reading the absorbance every 30 s at 340 nm for 10 min using a plate reader.
14. Calculate the steady-state rates of the ATPase reactions as mentioned before.

The ECA provides a robust and easy approach to validate hits, determine their IC_{50} values, and gain first insights into their mode of action, e.g., ATP competitive mode of action. Furthermore, compound selectivity can be preliminarily analyzed by testing different kinesins. Obviously, this type of approach provides specificity information only with respect to the tested proteins. Therefore, to explicitly address compound specificity, more complex systems, i.e., cellular systems, must be applied.

4 Notes

1. Kinesins can be obtained by recombinant expression as polyhistidine or GST-tagged fusion proteins and subsequent 1-step purification by affinity chromatography according to standard protocols [30]. For inhibitory studies of the ATPase activity, it is sufficient to purify the globular kinesin motor domains, which, in most cases can be obtained by bacterial expression. If you wish to study inhibition of full-length kinesins, we recommend expressing the proteins in insect cells. Either way it is critical to avoid freeze–thaw cycles and prolonged storage on ice as this can significantly decrease enzymatic activity. We recommend preparing small aliquots of the purified kinesin, which can be individually thawed on ice shortly before each experiment. Before starting inhibitory studies with your kinesin, we recommend testing its activity. A well behaving kinesin features an enormous boost of its ATPase activity upon binding to MTs. This should be controlled by comparing the enzymatic

rates of the reaction with and without MTs. Inhibition of the MT-stimulated ATPase activity of a kinesin provides an additional level of specificity, ruling out the possibility of false positives by inhibiting a co-purified contaminant ATPase.

2. Resuspend the polymerized MTs thoroughly with a lot of patience, as MT clumps will later perturb assay performance. When working with polymerized MTs, use cut-off pipette tips and remember to avoid vortexing as shear forces will cause damage to the protofilaments. Do not put MTs on ice as they will depolymerize when exposed to cold. We generally store and use them at room temperature (up to a few days). Since kinesins, like many other enzymes, significantly suffer from being kept at temperatures above 4 °C, we recommend that you keep them on ice until needed, and to elevate the temperature of the assay mix just prior to addition of MTs, adding the MTs as late as possible. Once MTs are added, work quickly.
3. Measuring the reaction at different time points can yield valuable information about potentially limiting factors of the enzymatic reaction. For the screen, you should choose a time point at which constant enzymatic activity of the kinesin is guaranteed.
4. You might need to go through several rounds of adjustment and assay optimization to obtain the ideal conditions for your kinesin. Multiple factors such as kinesin purity and activity impact optimal assay conditions. However, as aforementioned, assay optimization is worthwhile to ensure reproducible hits.
5. Kinesins are characterized by a highly conserved motor domain which comprises the binding sites for ATP and microtubules. Therefore, compounds that for example bind to the ATP binding site are likely to inhibit multiple kinesins. This caveat can be reduced by choosing assay conditions, e.g., high ATP concentration, favoring allosteric inhibitors.
6. Consider the dead volume, which depends on your liquid handling device and your exact settings and has to be added to the actual volume of required reagents to ensure that all reactions contain the exactly calculated volumes. Be generous! It is bothersome to repeat a whole assay just because a few microliters of a buffer are missing.
7. Screening compounds are typically dissolved in 100 % DMSO. DMSO is hygroscopic, i.e., it adsorbs moisture from the environment. It is therefore advisable to thaw compound plates in a desiccator chamber.
8. Depending on how stable the room temperature can be controlled, it might be advisable to incubate the assay plate in a temperature-controlled chamber on your robot platform.

Acknowledgements

The authors would like to thank Tim Mitchison and Stuart Schreiber who founded the Institute of Chemistry and Cell Biology (ICCB) at Harvard-Medical School Boston where TUM started his research on kinesin inhibitors. Many thanks to Lars Henschke and Silke Müller, head of the screening facility at the University of Konstanz, for their support during the preparation of this chapter. This work was supported by funding from the Konstanz-Research School Chemical Biology (KoRS-CB) and the Collaborative Research Center 969 "Chemical and Biological Principles of Cellular Proteostasis" of the German Research Foundation (DFG).

References

1. Walczak CE, Heald R (2008) Mechanisms of mitotic spindle assembly and function. Int Rev Cytol 265:111–158. doi:10.1016/S0074-7696(07)65003-7
2. Walczak CE, Cai S, Khodjakov A (2010) Mechanisms of chromosome behaviour during mitosis. Nat Rev Mol Cell Biol 11(2):91–102. doi:10.1038/nrm2832
3. Civelekoglu-Scholey G, Cimini D (2014) Modelling chromosome dynamics in mitosis: a historical perspective on models of metaphase and anaphase in eukaryotic cells. Interface Focus 4(3):20130073. doi:10.1098/rsfs.2013.0073
4. Glotzer M (2009) The 3Ms of central spindle assembly: microtubules, motors and MAPs. Nat Rev Mol Cell Biol 10(1):9–20, doi:nrm2609 [pii] 10.1038/nrm2609
5. Drummond DR (2011) Regulation of microtubule dynamics by kinesins. Semin Cell Dev Biol 22(9):927–934. doi:10.1016/j.semcdb.2011.09.021
6. Martin SE, Caplen NJ (2007) Applications of RNA interference in mammalian systems. Annu Rev Genomics Hum Genet 8:81–108. doi:10.1146/annurev.genom.8.080706.092424
7. Mayer TU (2003) Chemical genetics: tailoring tools for cell biology. Trends Cell Biol 13(5):270–277, doi:S0962892403000771 [pii]
8. Eggert US, Field CM, Mitchison TJ (2006) Small molecules in an RNAi world. Mol Biosyst 2(2):93–96. doi:10.1039/b515335b
9. Crews CM, Shotwell JB (2003) Small-molecule inhibitors of the cell cycle: an overview. Prog Cell Cycle Res 5:125–133
10. Lampson MA, Kapoor TM (2006) Unraveling cell division mechanisms with small-molecule inhibitors. Nat Chem Biol 2(1):19–27. doi:10.1038/nchembio757
11. Mitchison TJ (2002) Probing cell division with "chemical genetics". Harvey Lect 98:19–40
12. Good JA, Skoufias DA, Kozielski F (2011) Elucidating the functionality of kinesins: an overview of small molecule inhibitors. Semin Cell Dev Biol 22(9):935–945. doi:10.1016/j.semcdb.2011.09.023
13. Baell J, Walters MA (2014) Chemistry: chemical con artists foil drug discovery. Nature 513(7519):481–483. doi:10.1038/513481a
14. Eggert US (2013) The why and how of phenotypic small-molecule screens. Nat Chem Biol 9(4):206–209, doi: 10.1038/nchembio.1206
15. Carpenter AE (2007) Image-based chemical screening. Nat Chem Biol 3(8):461–465. doi:10.1038/nchembio.2007.15
16. Korn K, Krausz E (2007) Cell-based high-content screening of small-molecule libraries. Curr Opin Chem Biol 11(5):503–510. doi:10.1016/j.cbpa.2007.08.030
17. Lokey RS (2003) Forward chemical genetics: progress and obstacles on the path to a new pharmacopoeia. Curr Opin Chem Biol 7(1):91–96
18. Schenone M, Dancik V, Wagner BK, Clemons PA (2013) Target identification and mechanism of action in chemical biology and drug discovery. Nat Chem Biol 9(4):232–240. doi:10.1038/nchembio.1199
19. Mayer TU, Kapoor TM, Haggarty SJ, King RW, Schreiber SL, Mitchison TJ (1999) Small

molecule inhibitor of mitotic spindle bipolarity identified in a phenotype-based screen. Science 286(5441):971–974, doi:7948 [pii]

20. Contreras-Gomez A, Sanchez-Miron A, Garcia-Camacho F, Molina-Grima E, Chisti Y (2014) Protein production using the baculovirus-insect cell expression system. Biotechnol Prog 30(1):1–18. doi:10.1002/btpr.1842

21. Tao L, Scholey JM (2010) Purification and assay of mitotic motors. Methods 51(2):233–241. doi:10.1016/j.ymeth.2010.01.019

22. Hirokawa N, Noda Y (2001) Preparation of recombinant kinesin superfamily proteins using the baculovirus system. Methods Mol Biol 164:57–63

23. Stock MF, Hackney DD (2001) Expression of kinesin in Escherichia coli. Methods Mol Biol 164:43–48

24. Thormahlen M, Muller J, Mandelkow E (2001) Crystallization of kinesin. Methods Mol Biol 164:223–233

25. Lindsley JE (2001) Use of a real-time, coupled assay to measure the ATPase activity of DNA topoisomerase II. Methods Mol Biol 95:57–64

26. Baykov AA, Evtushenko OA, Avaeva SM (1988) A malachite green procedure for orthophosphate determination and its use in alkaline phosphatase-based enzyme immunoassay. Anal Biochem 171(2):266–270

27. Friel CT, Bagshaw CR, Howard J (2011) Analysing the ATP turnover cycle of microtubule motors. Methods Mol Biol 777:177–192. doi:10.1007/978-1-61779-252-6_13

28. Gilbert SP, Mackey AT (2000) Kinetics: a tool to study molecular motors. Methods 22(4): 337–354. doi:10.1006/meth.2000.1086

29. Zhang JH, Chung TD, Oldenburg KR (1999) A simple statistical parameter for use in evaluation and validation of high throughput screening assays. J Biomol Screen 4(2):67–73

30. Catarinella M, Gruner T, Strittmatter T, Marx A, Mayer TU (2009) BTB-1: a small molecule inhibitor of the mitotic motor protein Kif18A. Angew Chem Int Ed Engl 48(48):9072–9076. doi:10.1002/anie.200904510

Part VI

Novel Approaches to Study Spindle Function and Regulation

Chapter 19

Identification and Characterization of Mitotic Spindle-Localized Transcripts

Amy B. Emerman, Ashwini Jambhekar, and Michael D. Blower

Abstract

RNAs associate with the mitotic spindle in a variety of organisms, where they can spatially regulate protein production, ensure their proper segregation during cell division, or perform translation-independent roles in spindle formation. The identification of spindle-associated RNAs is an important first step in understanding the biological consequences of this phenomenon. In this chapter, we describe a method to use *Xenopus laevis* egg extracts to assemble and isolate mitotic spindles and to identify the spindle-associated RNAs. The method described here can be used in combination with immunodepletions, the addition of inhibitors, or other perturbations to investigate factors that affect RNA localization to the spindle. Finally, we describe a method to assess the consequences of ablating RNA in the extract on spindle formation.

Key words RNA localization, Mitotic Spindle, *Xenopus laevis*, RNaseA

1 Introduction

The mitotic spindle is a large macromolecular complex that segregates chromosomes during each eukaryotic cell division. This highly dynamic structure is regulated by a plethora of finely tuned interactions between molecules. While most studies in this area have focused on interactions between proteins, it is becoming appreciated that RNA also plays an essential role in regulating spindle architecture. RNA has been reported on spindles in numerous systems, including human cells [1], frog oocytes and eggs [2–6], snail embryos [7, 8], fly embryos [9, 10], and surf clams [11, 12].

A variety of studies have uncovered three major functions of RNA localization to the spindle. Some transcripts were found to localize to spindles in early embryos in order to ensure their equal or unequal partitioning into daughter cells [7, 11], where they would be utilized later during development. Other transcripts, such as Cyclin B [2, 9], Bub3 [2], and Xkid [13], were shown to be locally

*These authors contributed equally.

Paul Chang and Ryoma Ohi (eds.), *The Mitotic Spindle: Methods and Protocols*, Methods in Molecular Biology, vol. 1413,
DOI 10.1007/978-1-4939-3542-0_19, © Springer Science+Business Media New York 2016

translated on spindles at the sites where their protein products would be needed. Subsequently, RNA was found to play a more direct, translation-independent role in spindle assembly. In these studies, complete ablation of RNA, but not inhibition of translation, severely compromised spindle assembly [3, 4]. Additional studies revealed noncoding roles of spindle-associated RNA, including promoting the assembly of centromeres [14] and directly regulating at least one essential mitotic kinase [5, 15, 16]. These studies highlight the concept that RNA can play multiple important roles in regulating spindle architecture and cell-cycle progression.

Xenopus laevis egg extracts have historically been used to study spindle formation and regulation. The eggs are arrested in metaphase II of meiosis due to high Cyclin B protein levels, and methods for preparing undiluted egg cytoplasm are well established [17]. The cytoplasmic extract preserves many intermolecular interactions that are present in whole cells, and readily assembles spindles around exogenously added sperm nuclei or chromatin. The open nature of this system facilitates perturbation or direct visualization of components of interest. For instance, the essential roles of molecular motors in spindle assembly were elucidated by adding function-blocking antibodies to egg extracts [18], and the addition of fluorescently labeled proteins or transcripts has been used to assay molecular localization to the spindle [4, 6, 13].

This system offers a unique opportunity to isolate spindles and assess transcripts that localize there. Spindles can be generated in large quantities and isolated without their disruption, preserving endogenous spindle-RNA interactions and yielding high concentrations of spindle-associated RNA, both of which are difficult to accomplish in cell culture-based systems. Furthermore, the mechanisms underlying RNA localization to the spindle can be investigated by monitoring spindle-localized RNAs following the addition of relevant drugs to the extract, or by immunodepleting proteins of interest [6].

In this chapter, we describe methods for isolating spindles from *Xenopus laevis* egg extracts, purifying and sequencing the pool of bound transcripts, and analyzing and validating the sequencing results. We have compared two different mRNA purification approaches for library preparation—selection for the mRNA 5′ m^7G cap through a high affinity mutant of eIF4E (5′ Cap-capture) [19, 20], and selection for the polyA tail through LNA oligo-dT isolation—and find that they isolate largely overlapping, but some distinct, sets of transcripts [21]. In this protocol, we provide strategies for normalizing both the high-throughput sequencing data and the qPCR validation to identify the subset of transcripts that are underrepresented or overrepresented on the spindle. The method described in this chapter improves upon previously described taxol-based methods to analyze microtubule-associated RNAs from *Xenopus* eggs [4, 22] by isolating RNA from fully formed mitotic spindles, which include both chromatin and centrosomes in addition to microtubules. Finally, we present a protocol for enzymatically depleting RNA from extract and evaluating the effects on spindle formation.

2 Materials

All reagents, equipment, and benchtops should be RNase-free. Filtered pipette tips should be used throughout the protocol, standard 1.5 ml Eppendorf tubes are used for all procedures.

2.1 Extract Cycling and Mitotic Spindle Assembly

1. CSF-arrested *Xenopus laevis* egg extract, prepared as described [17, 23].
2. 5× BrB80: 400 mM PIPES, 5 mM $MgCl_2$, 5 mM EGTA, (pH to 6.8 with KOH).
3. 20× MMR: 100 mM HEPES, 2 mM EDTA, 2 M NaCl, 40 mM KCl, 20 mM $MgCl_2$, 40 mM $CaCl_2$, (pH to 7.8 with NaOH).
4. Sperm dilution buffer: 75 mM KCl, 0.5 mM spermidine trihydrochloride, 0.2 mM spermine tetrahydrochloride, 250 mM sucrose (pH to 7.3–7.5 with NaOH).
5. Sperm nuclei at 100,000/μl, prepared as described [17].
6. Rhodamine-labeled tubulin at 2 mg/ml in 1× BrB80, 1 mM GTP.
7. 10 mM $CaCl_2$ in Sperm dilution buffer.
8. Spindle fix: 1× MMR, 48 % glycerol, 11 % formaldehyde, 1 μg/ml DAPI.
9. Glass slides and coverslips.
10. Fluorescence microscope.

2.2 Spindle Pelleting

1. Spindle dilution solution: 1× BrB80, 30 % glycerol, 0.5 % Triton X-100.
2. 60 % spindle cushion: 1× BrB80, 60 % glycerol.
3. 14 ml round-bottom glass tubes (Fisher Scientific 45500-15).
4. Sorvall HB-6 rotor with adapters.
5. Ultrapure water.
6. Plastic Pasteur transfer pipettes (no need for these to be RNase-free).
7. RNeasy RNA extraction kit (Qiagen, cat. # 74104).
8. Buffer RLT of the RNeasy RNA extraction kit, supplemented with 143 mM (1 % v/v) β-mercaptoethanol.
9. 200-proof ethanol.
10. 0.5× TBE.
11. 1 % agarose gel in 0.5× TBE.

2.3 Library Preparation

1. Recombinant His-S-GFP-eIF4E$_{K119A}$ (pMB 628) expressed in bacteria and purified as described (stored in 50 μg aliquots in PBS+10 % glycerol at −80 °C) [21].
2. GFP-nAB™ magnetic agarose (Allele Biotechnologies, cat # ABP-nAb-GFPA100).
3. Magnetic stand.
4. Buffer A: 10 mM potassium phosphate buffer (pH 8.0), 100 mM KCl, 2 mM EDTA, 5 % glycerol, 0.005 % Triton X-100, 1.3 % poly(vinyl) alcohol (98–99 % hydrolyzed).
5. 1× PBS pH 7.4.
6. Dynabeads® MyOne™ Streptavidin C1 (Life Technologies, cat. # 65001).
7. LNA enhanced oligo-T20 capture probe, 5′-biotin labeled (Exiqon, cat. # 300100-03).
8. 2× Binding buffer: 40 mM Tris (pH to 7.5 with HCl), 0.2 M NaCl, 2 mM EDTA, 0.2 % *N*-lauryl sarcosine.
9. TE: 10 mM Tris (pH to 8 with HCl), 1 mM EDTA.
10. TruSeq Stranded RNA Sample Prep Kit (Illumina, cat. # RS-122-2101).

2.4 Data Analysis

1. fastQC (http://www.bioinformatics.babraham.ac.uk/projects/fastqc/).
2. fastX toolkit (http://hannonlab.cshl.edu/fastx_toolkit/).
3. A reference sequence for alignment, *see* **step 1** of Subheading 3.3.2 for suggestions.
4. Bowtie [24].
5. Tophat [25].
6. Cufflinks [26].
7. JGI genome annotations (Xenbase.org).
8. Computer with at least 8 GB of RAM. The best options are computers that run a Unix-variant operating system (Mac OS X or Linux) as most sequencing software is designed to run on Unix machines. A typical sequencing run can generate 50 gigabytes of data per lane, so it is a good idea to have a computer with a 1 TB hard drive.

2.5 Data Validation

1. Primers designed for qPCR against gene of interest. For a positive control we use AKAP9 (Fw: GGCACCAGCCTGTTACAGAT, Rv: GAGGCATTTCCCCCTTGGAA), for a negative control we use Sec61β (Fw: TACCGTGGCCAGTACAACAC, Rv: TTGAGAAGAACGCGCAGGAT), and for ribosomal RNA we use 28 s rRNA (Fw: GACGCGATGTGATTTCTGCC, Rv: GCCTCACCGGGTAAGTGAAA).
2. SuperScript® III Reverse Transcriptase kit (Life Technologies, cat. # 18080044).

3. Random Primers.
4. 10× dNTPs (10 mM each of dGTP, dATP, dCTP, dTTP in ultrapure water).
5. RNAse inhibitor.
6. RNaseH.
7. iQ™ SYBR® Green Supermix (Bio-Rad, cat. # 170-8880).

2.6 RNase Treatment

1. All reagents listed in Subheadings 2.1 and 2.2.
2. RNaseA, Type XII-A (Sigma-Aldrich, cat. # R5500) at 10 mg/ml in water.
3. 40 % spindle cushion: 1× BrB80, 40 % glycerol.
4. Methanol at −20 °C.
5. 14 ml round-bottom glass tubes (Fisher Scientific 45500-15) filled with a small amount of epoxy at the bottom to create a flat surface inside of the tube (*see* **Note 1**).
6. 12 mm round glass coverslips.
7. 1 % agarose gel in 0.5× TBE + 0.5 μg/ml ethidium bromide.
8. Vectashield® (Vector Laboratories, cat. # H-1000) containing 1 μg/ml DAPI.

2.7 RNase Treatment on Beads

1. CNBr-activated Sepharose 4B.
2. RNaseA, Type XII-A (Sigma-Aldrich, cat. # R5500) at 10 mg/ml in water.
3. Recombinant GFP (pMB156) or other negative control protein of choice.
4. 1 M Tris–HCl (pH to 8 with HCl).
5. 5 M NaCl.
6. 1 M $NaHCO_3$ (pH to 8.3 with NaOH/HCl).
7. Coupling buffer: 0.1 M $NaHCO_3$, 0.5 M NaCl (pH to 8.3 with NaOH/HCl).
8. Acid wash buffer: 0.1 M sodium acetate, 0.5 M NaCl (pH to 4 with acetic acid).
9. Base wash buffer: 0.1 M Tris–HCl, 0.5 M NaCl (pH to 8 with HCl).
10. 1 mM HCl at 4 °C.
11. 0.2 M glycine (pH to 2 with HCl).
12. 1× PBS.
13. 1× XB: 100 mM KCl, 1 mM $MgCl_2$, 50 mM sucrose, 10 mM HEPES (pH to 7.7 with KOH).
14. Poly-prep chromatography column, 15 ml size (Bio-Rad, cat. # 731-1550).

15. Micro Bio-Spin Chromatography Column, 1 ml size (Bio-Rad, cat. # 732-6204).
16. Parafilm M.
17. RNase inhibitor.

3 Methods

3.1 Purification of Mitotic Spindle-Associated RNA

In this protocol, mitotic spindles are formed in *Xenopus* egg extract and pelleted through a 60 % glycerol cushion with detergent, which isolates spindles from other large organelles, membranes, and free cytoplasmic components. RNA is then extracted from the pelleted spindles, and can be analyzed by gel electrophoresis, sequencing, or PCR-based assays. Prior to embarking on a detailed analysis of the spindle-associated RNAs, pilot experiments should be performed to ensure that the spindle pelleting conditions are sufficiently selective (*see* **Note 2**).

3.1.1 Prepare CSF with Rhodamine-Labeled Tubulin for Cycled Spindles

1. Prepare cytostatic factor arrested egg extract (CSF extract) as described [23]. CSF extract should be freshly prepared for this experiment. Begin the experiment immediately after preparing CSF extract (*see* **Note 3**).
2. Determine the amount of cycled extract that will be needed for the experiment. We find that at least 150 μl of cycled extract is necessary to obtain sufficient RNA for analysis. Include an additional 10–20 % of extract to account for loss in pipetting.
3. Aliquot the amount of extract determined in **step 2** into a new 1.5 ml Eppendorf tube. Add rhodamine-labeled tubulin to a final concentration of 5–10 μg/ml. Use this labeled extract for all remaining steps of the protocol.

3.1.2 Prepare Cycled Spindles

This protocol takes advantage of the fact that adding calcium results in Cyclin B degradation, thus releasing metaphase-arrested extracts into interphase. Following replication of the DNA during interphase, extracts can be rearrested in metaphase by adding an equal volume of fresh CSF extract (without calcium treatment).

1. Aliquot half of the volume determined in Subheading 3.1.1, **step 2**, plus an additional 10 % of labeled extract to a new Eppendorf tube (*see* **Note 4**).
2. To this aliquot, add sperm nuclei to a final concentration of 500 nuclei/μl. Incubate at 20 °C for 5 min (*see* **Note 5**).
3. Add $CaCl_2$ to a final concentration of 0.6 mM to induce cell cycle progression of the extract into interphase. Incubate at 20 °C for 90 min. Store remaining, labeled CSF extract at 12 °C until needed.

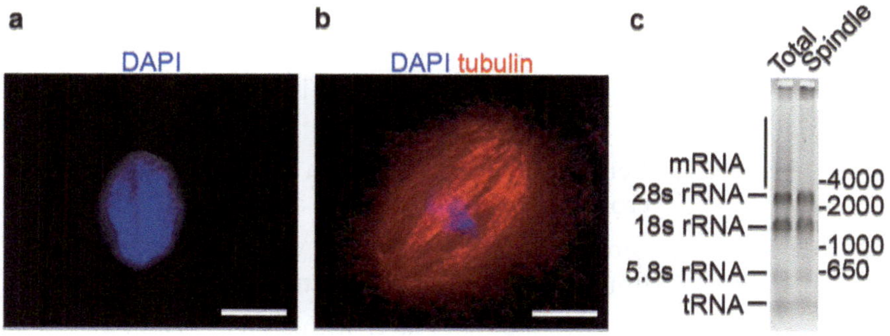

Fig. 1 Representative images throughout the progression of extract cycling and spindle-RNA isolation. (**a**) Sperm nuclei in interphase, 60 min after the addition of calcium. (**b**) Mitotic spindles 60 min after rearresting the extract in metaphase. Chromatin is visualized with DAPI (*blue*) and the microtubules are visualized with rhodamine-labeled tubulin (*red*). Scale bar = 10 μm. (**c**) Total RNA and spindle-associated RNA run on a SYBR® Gold-stained 1 % agarose gel. A DNA ladder in base pairs is shown for reference

4. Monitor progression into interphase by spotting 1 μl of the extract onto a glass slide, then overlay with 3 μl of spindle fix and a coverslip. Observe under a fluorescence microscope. Once the extract is in interphase, fully assembled nuclei should be visible via DAPI stain (Fig. 1a) (*see* **Note 6**).
5. After 90 min, mix equal volumes of interphase extract with labeled CSF extract. This will induce progression of the cell cycle into mitosis. Incubate at 20 °C for 60 min, flicking the tubes every 10–15 min to avoid aggregation of spindles.
6. Monitor progression into mitosis by spotting 1 μl of the extract onto a glass slide, then overlay with 3 μl of spindle fix and a coverslip. Ensure that spindles have assembled by visualizing the rhodamine-labeled tubulin and DAPI-stained DNA under a fluorescence microscope. At this stage, bipolar spindles should have formed around >80 % of the sperm nuclei (Fig. 1b). Extract quality can be assessed by examining the percentage of bipolar spindles vs. defective, non-bipolar spindles. A good quality extract should contain >70 % bipolar spindles after 60 min in mitosis. Extracts with a lower efficiency of spindle assembly or a high percentage of monopolar spindles are unlikely to produce high-quality spindle-associated RNAs.

3.1.3 Pellet Spindles

1. Set aside 2 μl of cycled extract for use as a total RNA sample. Keep this sample on ice until you are ready to proceed with RNA isolation.
2. Add spindle dilution solution to the remaining cycled extract at a ratio of 1 ml of solution per 50 μl of cycled extract. Mix gently, but thoroughly.
3. Prepare round-bottom glass tubes with 10 ml of 60 % spindle cushion. Carefully layer 1 ml of the diluted extract from **step 2**

on top of each cushion until all of the diluted extract is used (three tubes are required for 150 μl of cycled extract).

4. Spin the tubes at 17,000 × *g* in a Sorvall HB-6 rotor for 10 min at 20 °C.
5. Remove the spindle dilution solution from the top of the glycerol cushion with a vacuum aspirator.
6. Rinse the sides of the tube with approximately 2 ml of water using a plastic Pasteur pipette.
7. Remove the water layer with a vacuum aspirator, and continue to remove approximately 6 ml of the glycerol cushion in a circular motion.
8. Rinse again with 2 ml of water, being careful not to let the water mix to the bottom of the glass tube.
9. Remove the second water rinse with the vacuum aspirator, and continue to remove the remaining glycerol cushion without disturbing the bottom of the tube. Tip the tube to draw off any remaining glycerol (*see* **Note 7**).

3.1.4 Isolate Total RNA and Spindle RNA

1. Add 350 μl of Buffer RLT containing β-ME to one spindle pellet.
2. Resuspend the pellet by pipetting the Buffer RLT up and down.
3. Transfer the Buffer RLT to the next pellet to pool the spindle RNA, and continue to vortex and transfer the Buffer RLT until all of the pellets are collected.
4. In parallel, add 350 μl of Buffer RLT containing β-ME to the reserved total RNA sample and mix well.
5. Continue with the RNeasy kit following the manufacturer's protocol, starting at **step 4** of the "Purification of Total RNA from Animal Cells" of the RNeasy kit (Addition of an equal volume of 70 % Ethanol). At the final step, elute the RNA in 30 μl of water (*see* **Note 8**).
6. Determine the concentration of the eluted RNA (*see* **Note 9**).
7. Check an aliquot of each RNA sample on a 1 % agarose gel using UV imaging of ethidium bromide or SYBR Gold to confirm that the RNA is not degraded. The ribosomal RNA bands should appear crisp in all samples (Fig. 1c) (*see* **Note 10**).

3.2 Library Preparation

At this step, libraries of the total RNA and spindle RNA will be prepared for Illumina Sequencing. Typically 200 ng of total or spindle-associated RNA is sufficient to prepare a high quality sequencing library. We typically barcode our libraries and sequence 4–8 libraries per lane on a HiSeq 2000. This results in ~10–20 million unique reads per library. Prior to library preparation, the mRNAs must be isolated to remove ribosomal RNA. We have used two methods to select for mRNAs and remove ribosomal RNA: 5′ Cap-capture, as previously described with modifications [19, 20], and LNA oligo-dT selection. The 5′ Cap-capture method may be preferable if there is concern that some mRNAs may have short polyA tails. We find that the 5′ Cap-capture method isolates some mRNAs more efficiently than LNA oligo-dT, but that the datasets prepared by both methods largely overlap [21].

3.2.1 Option #1: Isolate mRNAs by 5′ Cap-Capture

1. Transfer 200 ng of total RNA and 200 ng of spindle RNA into separate Eppendorf tubes.
2. Heat the RNA at 70 °C for 10 min to denature RNA secondary structure, then place on ice.
3. Prepare a GFP-eIF4E$_{K119A}$ protein master mix: 40 μl Buffer A, 0.8 μl RNase Inhibitor, and 1.6 μg GFP-eIF4E$_{K119A}$ per RNA sample.
4. Add one aliquot of GFP-eIF4E$_{K119A}$ protein master mix to each RNA sample and incubate on ice for 45 min.
5. While the GFP-eIF4E$_{K119A}$ is incubating with RNA, aliquot 20 μl of GFP-nAB™ magnetic agarose to a new Eppendorf tube for each RNA sample and wash these beads twice with 1 ml of Buffer A.
6. At the end of the RNA/GFP-eIF4E$_{K119A}$ incubation, remove the final Buffer A wash from the beads and add the RNA/GFP-eIF4E$_{K119A}$ mix to the beads.
7. Incubate on ice for 1 h, agitating the beads every 5 min by flicking the tube.
8. Wash the beads three times with 100 μl of Buffer A.
9. After the final Buffer A wash, wash once with PBS and move the samples to PCR tubes.
10. Collect the beads on a magnetic stand and remove the final wash.
11. Resuspend the beads in 19.5 μl of the Fragment, Prime, Finish mix from the TruSeq Stranded mRNA Sample Prep Kit.
12. Incubate at 94 °C for 8 min, then hold at 4 °C.
13. Collect the beads on a magnetic stand and transfer the supernatant to new PCR tubes.
14. Proceed immediately with the "Synthesize First Strand cDNA" step of the TruSeq kit and continue to follow the manufacturer's instructions.

3.2.2 Option #2: Isolate mRNAs by LNA Oligo-dT Selection

These instructions are given for one sample; scale beads and the LNA oligo-dT accordingly for multiple samples. The wash volumes are appropriate for 5–50 μl of beads.

1. Collect 5 μl of MyOne Streptavidin Dynabeads on a magnetic stand and remove the supernatant.
2. Add 20 pmol of the LNA oligo-dT plus TE to a final volume of 10 μl.
3. Incubate at 37 °C for 10 min, shaking at 400 rpm.
4. Wash the beads once with 100 μl TE.
5. Wash the beads once with 100 μl 1× binding buffer.
6. Resuspend the beads one more time in 100 μl 1× binding buffer, then aliquot the beads into separate tubes for each sample.
7. Dilute 200 ng of each RNA sample into 50 μl of water.
8. Add 50 μl of 2× binding buffer.
9. Incubate at 65 °C for 10 min, then keep on ice.
10. Place the beads on a magnetic stand and remove the buffer.
11. Add each RNA sample to one aliquot of beads.
12. Incubate at 37 °C for 10 min, shaking at 400 rpm.
13. Wash each sample three times with 1 ml 1× binding buffer. In the last wash, move the beads to PCR tubes.
14. Remove the final wash, and continue at **step 11** of Subheading 3.2.1.

3.3 Data Analysis

The output from the high-throughput sequencing run will be a file of sequences (hundreds to hundreds of millions of reads) in the fastq format, which contains information about the location of each read on the flowcell, the bases read, and the quality of each base read. The ultimate goal is to convert the sequencing file into a table containing a count of the reads per gene in each sample using several bioinformatics steps. For most sequencing experiments normalized read counts are expressed at Reads/Fragments Per Kilobase per Million mapped (RPKM, FPKM). This normalization takes into account both library size and transcript length and is currently the standard normalization for all RNA-seq experiments. The final table can be used to identify transcripts that are differentially represented on spindles compared to total RNA.

3.3.1 Sequence Quality Control and Processing

1. Examine the overall quality of the library using the fastQC tool. Ensure the absence of systematic problems with base qualities or library complexity prior to embarking on a detailed analysis.
2. Remove all reads containing bases with a low quality score using the fastx toolkit. A Q-score of 20 represents a 99 % chance of an accurate base call, and we typically remove any reads with scores lower than 20.

3. Remove duplicate reads using the fastx toolkit. Duplicate reads are assumed to arise during PCR amplification of library products, rather than from independent mRNA fragments; therefore, we take the conservative approach of removing all duplicates.

3.3.2 Sequence Alignment, Counting, and Data Analysis

1. Select a reference sequence for alignment. The options for *X. laevis* include RefSeq (a conservative collection of ~11,000 well-characterized and annotated full-length mRNAs) or NCBI Unigene (a larger collection of ~35,000 less well-annotated sequences encompassing the RefSeq dataset). These options only allow for analysis of previously defined transcripts. Alternatively, reads can be aligned to the *X. laevis* genome (available at Xenbase.org), a more time- and computation-intensive process, but one which allows discovery of novel RNAs. Some transcripts represented in NCBI Unigene may not be included in the current version of the *X. laevis* genome.
2. Align reads to the reference sequence using Bowtie (if aligning to RefSeq or Unigene) or Tophat (if aligning to the genome) which generate ungapped and gapped alignments, respectively. Select appropriate values for the number of tolerated mismatches (0–3) (*see* **Note 11**) and the number of allowed alignments (*see* **Note 12**). The output of the alignments will be a . sam or .bam file containing the name, sequence, alignment position, and quality of each aligned read.
3. If the reads were aligned to the *X. laevis* genome, you will need a set of genome coordinates (in .gtf or .gff3 format) that describes the locations of genes. Existing genome annotations identified by the JGI can be downloaded from Xenbase.org. These annotations have an excellent representation of protein-coding genes; however, they have poorly annotated untranslated regions and are lacking some lncRNAs and splice variants. For a more complete gene model, we use a function in Cufflinks that predicts transcripts from several of our own sequencing files and merges these transcripts with the JGI annotations.
4. Count the reads using the output of **Step 2** and Cufflinks. If the reads were aligned to a transcriptome (such as RefSeq or NCBI Unigene), then counting the reads is straightforward. If the reads were aligned to the *X.laevis* genome, then you will also need to use the genome coordinates from **step 3**. The output of Cufflinks consists of several files that summarize the raw and normalized counts of reads per gene (the latter is defined as RPKM, or reads per kilobase per million mapped). Cufflinks outputs these statistics on a per-transcript or per gene basis, so isoform-specific behavior and weighted read counting can be evaluated if the reads were aligned to the genome.

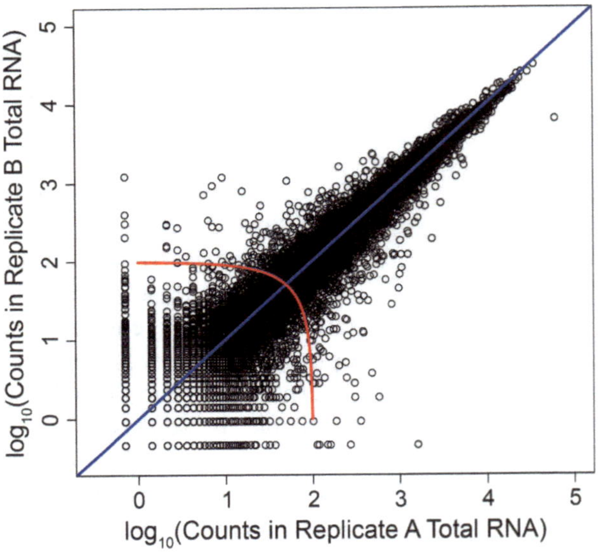

Fig. 2 Relationship of read count and reproducibility. Illumina libraries of total RNA were prepared and sequenced from two *Xenopus* egg extracts. The scatterplot depicts the read counts from each extract, the *blue line* indicates equal reads in each extract, and the *red curve* illustrates a 100 read threshold. The majority of the variability in read counts between the two extracts is in genes with less than 100 total reads

5. Filter the data to exclude transcripts with low-confidence measurements. We typically remove transcripts with fewer than 100 reads, a filter shown to yield high-confidence data [21, 27] (Fig. 2).
6. Perform a general analysis of the results. Compare the selection methods by plotting normalized reads in the oligo-dT vs the 5′ Cap-capture selected libraries (Fig. 3a). Examine spindle enrichment by plotting the normalized reads in spindle RNA vs total RNA libraries (Fig. 3b), and by generating a histogram of spindle enrichment across the dataset (Fig. 3c). Finally, examine the correlation of spindle enrichment between biological replicates (Fig. 3d) (*see* **Note 13**). mRNAs that are significantly enriched on spindle microtubules can be identified using standard differential gene expression analysis tools (such as EdgeR), but require biological triplicates for analysis.

3.4 Data Validation

In this step spindle RNA candidates will be validated by quantitating RNA levels in purified spindle versus total cytoplasm using quantitative PCR (qPCR). We determine an enrichment value for each transcript as the abundance in spindle RNA divided by its abundance in total RNA. We normalize this value to 28 s ribosomal RNA, which we set to 1. We begin by isolating total and spindle-associated RNAs as described in Subheading 3.1.

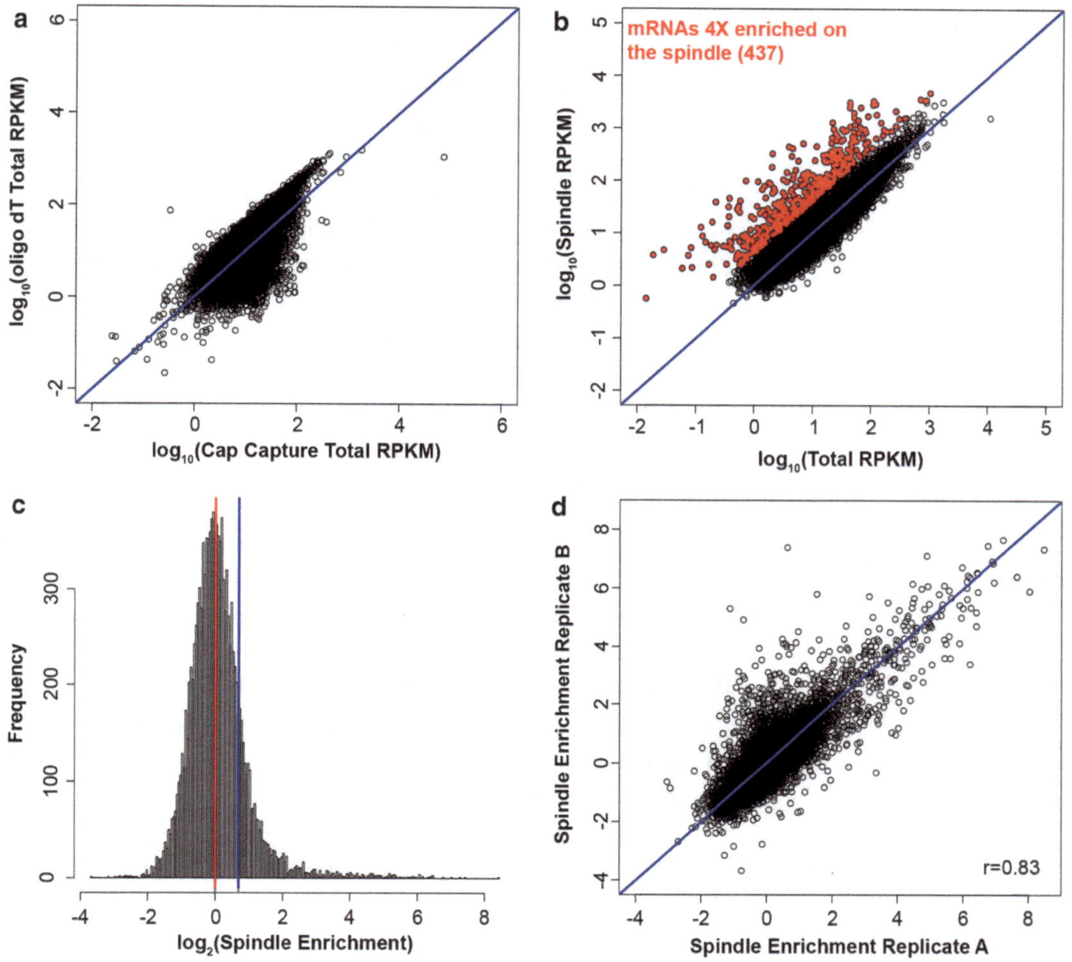

Fig. 3 Analysis of the sequencing results. (**a**) Comparison of normalized read counts of mRNAs purified from the total RNA of a single extract using either oligo-dT or Cap-capture selections. Genes to the right of the diagonal blue line have a higher representation in the Cap-capture libraries. (**b**) Scatterplot showing enrichment of a subset of transcripts (*red*) on mitotic spindles. All genes with a combined read count of >100 in the spindle and total RNA libraries are shown. (**c**) Histogram of the same genes as depicted in panel **b**, demonstrating the long shoulder of spindle-enriched genes. Spindle enrichment was calculated as RPKM in the spindle RNA library/RPKM in total. The *red line* depicts the median spindle enrichment and the *blue line* depicts that mean spindle enrichment. (**d**) Reproducibility of spindle enrichment scores between biological duplicates (two independent extracts). The Pearson correlation coefficient is 0.83

3.4.1 Reverse Transcription

1. Aliquot 200 ng of total RNA and 200 ng of spindle RNA into PCR tubes.
2. Perform reverse transcription (RT) with SuperScript® III using random primers at a final concentration of 2.5 ng/μl and according to the manufacturer's instructions.
3. After completion of the reaction, add 1 μl of RNaseH to a 20 μl RT reaction. Incubate at 37 °C for 20 min.

3.4.2 Quantitative PCR

1. Perform qPCR with the cDNA from the previous step using iQ™ Sybr® Green Master Mix. Include qPCRs for positive and negative control RNAs and rRNA. We use AKAP9 as our positive control, Sec61β as our negative control, and 28 s rRNA for normalization.
2. Include a standard curve for each transcript being queried by pooling together aliquots of each cDNA sample. Serially dilute the pooled cDNA tenfold three times to obtain 4 dilutions (*see* **Note 14**).
3. cDNA samples are typically diluted between 1/20 and 1/50 for qPCR analysis (depending on the abundance of the transcript being assayed), and 5 μl of the dilution is used for each 25 μl qPCR reaction. All unknown and standard-curve samples are assayed in triplicate.

3.4.3 Analyze qPCR Results

1. Confirm that all of the samples have a low standard deviation across the triplicate reactions for each transcript.
2. Determine the starting quantity of each transcript in the total and spindle samples based on the Ct value of each sample and the standard curve of the corresponding transcript.
3. For each transcript, determine an enrichment score by dividing the starting quantity in the spindle RNA by the starting quantity in the corresponding total RNA.
4. Divide the enrichment scores of each transcript by the corresponding enrichment score of ribosomal RNA to normalize the samples (Fig. 4).

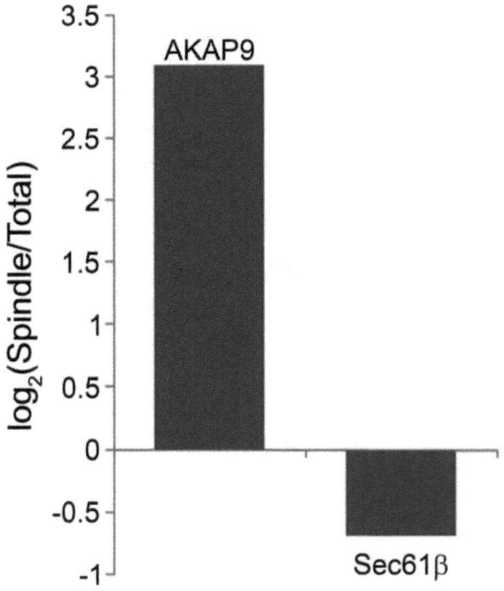

Fig. 4 Typical qPCR results of spindle enrichment for positive (AKAP9) and negative (Sec61β) control RNAs. The spindle enrichment values were normalized to 28 s rRNA as described in Subheading 3.4.3

3.5 RNaseA Treatment of Spindles

Xenopus egg extracts afford a unique opportunity to study both coding and noncoding functions of transcripts by ablating all RNAs in the system. Unlike cell-based systems, in which RNA is typically depleted by chemically inhibiting one or more of the RNA polymerases (and relying on natural degradation of preexisting transcripts), RNA can be directly degraded in *Xenopus* egg extracts by the addition of RNaseA. Because the extracts maintain a high protein concentration and relatively low levels of translation, RNA-depleted extracts retain several of their typical functionalities. In this protocol, RNA is depleted from extracts and the effects on spindle formation and morphology are examined. To distinguish between translation-dependent and—independent effects, control reactions containing the translation inhibitor puromycin are analyzed (*see* **Note 15**).

3.5.1 Prepare RNaseA-Treated CSF Spindles

CSF spindles, also referred to as half-spindles, form when unreplicated sperm nuclear DNA is added to mitotically arrested CSF. At early time points, monopolar spindles assemble from the single set of kinetochores and centrosome found at each sperm nucleus. Subsequently, these spindles convert to a bipolar form through activation of the chromatin-dependent spindle assembly pathway [28].

1. Prepare the desired amount of extract plus rhodamine-labeled tubulin as described in Subheading 3.1.1. CSF spindles require 25 μl per reaction. Start with 10 % additional volume for diluting reagents and to compensate for loss during pipetting.
2. Aliquot 25 μl of labeled extract into a new Eppendorf tube for each reaction (control and RNaseA).
3. Add RNaseA to a final concentration of 0.1 mg/ml. If the volume of the added RNaseA is too small to pipette accurately, first dilute the RNaseA in a small amount of reserved extract, and use this stock to achieve the desired final concentration. In parallel, prepare a control sample containing an equivalent volume of water. Incubate both reactions at 20 °C for 1 h.
4. At the end of the hour, reserve 1 μl of each sample for RNA analysis. Extract can be snap-frozen in liquid nitrogen and stored at −80 °C, or added directly to Buffer RLT containing β-ME for RNA extraction (see below).
5. To the remainder of each sample, add sperm nuclei to a final concentration of 500 nuclei/μl. If the volume of sperm nuclei is too small to pipette accurately, first dilute it in reserved extract. Use the diluted sperm nuclei immediately; otherwise they may begin to initiate spindle assembly prior to addition to experimental extracts. Incubate at 20 °C for 1 h, flicking tubes every 10–15 min.
6. Monitor progress of spindle assembly at 10, 30, and 50 min by spotting 1 μl of the reaction on a glass slide, then overlay with

3 μl of spindle fix and a coverslip. Observe under a fluorescence microscope. Monopolar and bipolar spindles should be visible at 30 min, and most spindles will be bipolar at 50 min.

7. Dilute each reaction in spindle dilution solution. Layer each reaction on top of 5 ml of 40 % spindle cushion in a glass tube containing an adaptor overlayed with a round coverslip (*see* **Note 1**).
8. Follow Subheading 3.1.3, **steps 4–8**, taking care that washes with water do not touch the coverslip.
9. Following the final wash, remove most of the liquid with a vacuum aspirator, until the liquid just covers the coverslip. Extract each coverslip with a metal hook, taking care to monitor the orientation of each coverslip (*see* **Note 1**).
10. Fix in cold methanol at −20 °C for 10 min.
11. Dry the coverslips at room temperature in a coverslip rack.
12. Mount the coverslips in Vectashield® containing 1 μg/ml DAPI. Seal the edges of the coverslip with nail polish.

3.5.2 Prepare RNaseA-Treated Cycled Spindles

1. Prepare the desired amount of extract plus rhodamine-labeled tubulin as described in Subheading 3.1.1. Cycled spindles require 60 μl of extract per reaction, plus an additional 10–20 % for preparing dilutions and to compensate for loss in pipetting.
2. Cycle 20 μl of extract per reaction plus 10 % extra into interphase as described in Subheading 3.1.2, **steps 2–4**.
3. When the reactions have been incubating for 30 min, prepare extracts for rearresting the reactions in metaphase. Add RNaseA or water to 40 μl of labeled extract for RNaseA-treated or control samples, respectively, as described in Subheading 3.5.1, **step 3**. Incubate both at 20 °C for 1 h.
4. When the interphase extracts have incubated for 90 min, split the reaction into 20 μl aliquots. Add 40 μl of control or RNaseA-treated extract, adding additional RNaseA to maintain the final concentration at 0.1 mg/ml. Add an equivalent amount of water to the control reactions. Incubate at 20 °C for 1 h. Monitor spindle assembly as described in Subheading 3.1.2, **step 6**. Bipolar spindles should be visible after 45 min.
5. At the end of the hour, reserve 1 μl of each sample for RNA analysis as described in Subheading 3.5.1, **step 4**.
6. Follow Subheading 3.5.1, **steps 7–12**, to isolate spindles on coverslips and prepare for microscopy.

3.5.3 Observe Spindles by Microscopy

1. Observe rhodamine-labeled tubulin and DAPI-stained chromosomes on a fluorescence microscope. Control CSF spindle reactions will display spindles around >80 % of the chromatin, with the majority being bipolar and the rest monopolar.

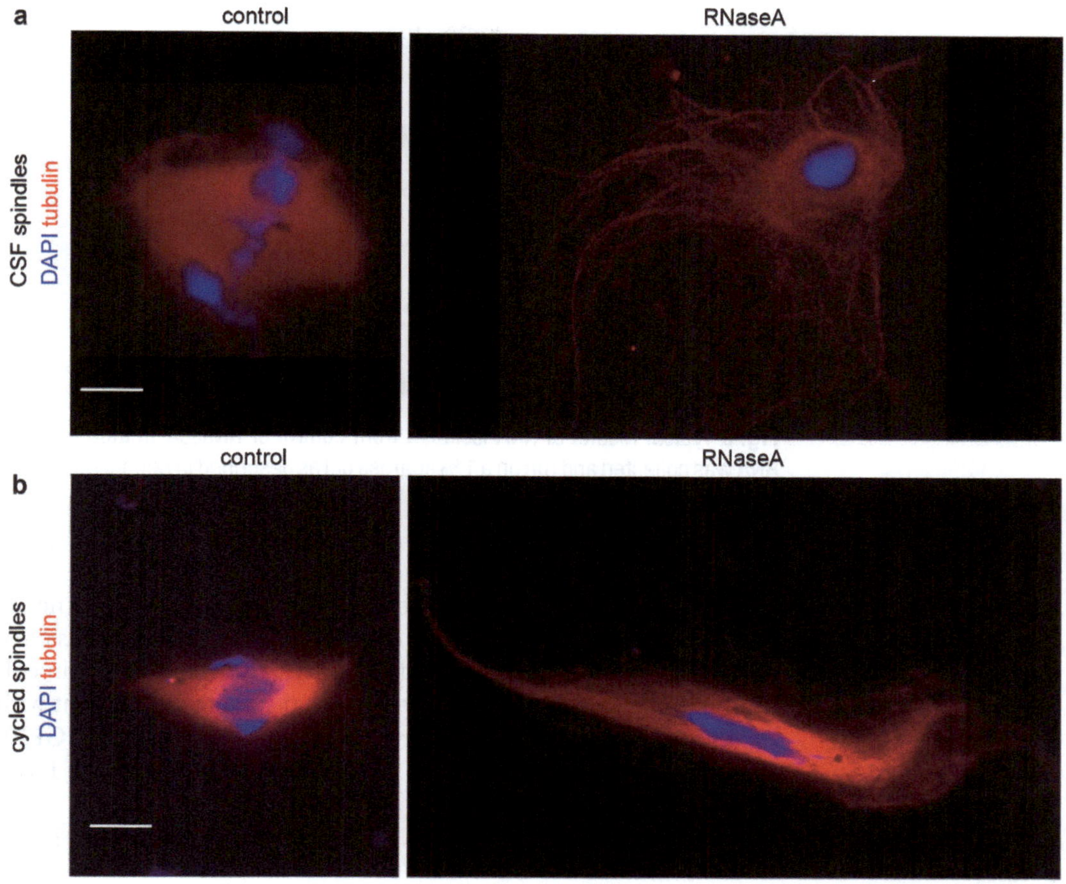

Fig. 5 Typical spindles formed in control or RNaseA-treated extracts. (**a**) CSF spindles and (**b**) cycled spindles. Chromatin is visualized with DAPI (*blue*) and the microtubules are visualized with rhodamine-labeled tubulin (*red*). Scale bar = 10 μm

Control cycled spindles will display >80 % frequency of bipolar spindles. In both cases, RNaseA-treated spindles will form at a lower frequency, display disorganized microtubules, and have reduced tubulin density. In RNaseA-treated cycled spindles, chromosome misalignment will also be evident (Fig. 5).

3.5.4 RNA Analysis

1. Isolate RNA from the reserved 1 μl aliquots of extract as described in Subheading 3.1.4, **Steps 4–5**. Typical yields are 5–10 μg of RNA per 1 μl of extract.
2. Run 100 ng of each RNA on a 1 % agarose gel with a DNA ladder for reference. RNA from untreated samples should reveal two sharp bands of rRNA migrating at the position of 1.2 and 1.8 kb DNA markers. RNA from RNaseA-treated extracts should reveal a heavy smear centered around a 300 bp DNA marker (Fig. 6).

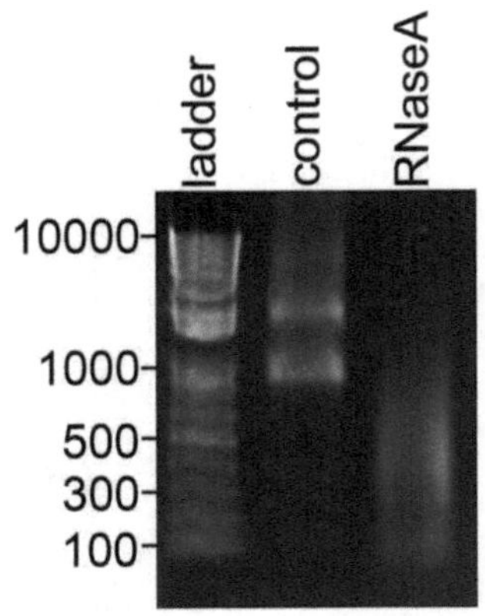

Fig. 6 Typical results of RNA isolation from control or RNaseA-treated extracts. RNA was collected and run on a 1 % agarose gel as described in Subheading 3.5.4. A DNA ladder is shown for reference

3.6 Treatment of Extracts With RNaseA on Beads

Unlike the methods described above, this protocol allows for nearly complete removal of RNaseA by coupling it to sepharose beads prior to mixing with extract. The beads are then removed with a spin column. In parallel, it is important to perform a control treatment with a column bound to an inert protein (we use recombinant GFP), because the beads treatment can be toxic to extract (*see* **Note 16**).

3.6.1 Couple RNaseA to Beads

1. Weigh out 0.2–0.3 g of CNBr Sepharose and place in a 15 ml column.
2. To activate the beads, add 2.5 ml of cold 1 mM HCl, rotate at 4 °C for 3–5 min with caps on the column top and bottom, then allow the HCl to drain. Repeat this wash three more times (four washes total).
3. Wash once with 5 ml of cold coupling buffer as in **step 2**.
4. Cap the column bottom. Add 1 ml of RNaseA or an equivalent amount of GFP protein plus 150 μl of 5 M NaCl and 150 μl of 1 M $NaHCO_3$ to achieve a final concentration of 1× of coupling buffer. Cap both ends of the column and rotate at room temperature for 4 h.
   All subsequent steps are performed at room temperature.
5. Wash with 5 ml of coupling buffer by adding the buffer to the column and allowing the buffer to drip out.
6. Cap the column bottom and add 1 ml of 0.1 M Tris pH 8, cap the column top, and rotate at room temperature for 2 h. Drain the solution.
7. Wash with 3 ml of base wash buffer (as in **step 5**), then with 3 ml of acid wash buffer. Repeat these two alternating washes three more times, for a total of eight washes.

8. Wash with 10 ml of 0.2 M glycine pH 2.
9. Wash with 2.5 ml of 1× PBS. Store beads at 4 °C in PBS as a 50 % slurry (*see* **Note 16**).

3.6.2 Treat Extract with Bead-Coupled RNaseA

1. Transfer 30–50 μl of the bead slurry into a 1 ml column (the exact amount will have to be optimized for your purposes). Allow the buffer to drain, then wash with 1 ml of 1× XB.
2. Cap the bottom of the column and add 50 μl of extract. Cap the top and seal both ends with Parafilm to prevent leakage. Rotate at 20 °C for 30 min.
3. Place the column in a 1.5 ml Eppendorf tube and centrifuge at 1000 ×*g* to collect the extract. Reserve 1–2 μl to check for RNA degradation, as described in Subheading 3.5.4. These RNA samples should look similar when run on an agarose gel to those described in Subheading 3.5.4 and in Fig. 6.
4. Depending on the application, RNase inhibitor can be added at a ratio of up to 1/50 to inhibit any residual RNase activity.

4 Notes

1. See Hannak and Heald [23] for detailed diagrams of preparing tubes with coverslips and extracting the coverslips following centrifugation.
2. We recommend performing two independent control experiments to confirm the selectivity of the pelleting conditions. First, resuspend the spindle pellets from **step 9** of Subheading 3.1.3 in protein loading buffer instead of Buffer RLT + β-ME in order to perform a western blot analysis to ensure that spindle associated proteins, but not protein markers for other subcellular structures, pellet under these conditions. Second, include 10 μM nocodazole in the reactions at **step 5** of Subheading 3.1.2 to inhibit microtubule polymerization, and ensure that no RNA or spindle-associated proteins pellet following the 60 % spindle cushion.
3. Immunodepletions may be desired to assay the effect of a factor on RNA localization to the spindle. If performing such an experiment, immunodeplete the extract at this step as previously described [6, 23].
4. Throughout the protocol, if the total volume of extract exceeds 100 μl, split the reaction into multiple tubes containing at most 100 μl each. This step is necessary to ensure that sufficient oxygen is available in the tube.
5. When mixing reagents into CSF extract, it is important to mix gently, either by flicking gently or by pipetting. It is especially important to avoid generating air bubbles while mixing.

6. We recommend monitoring progression through interphase at 60 and 80 min after the addition of calcium. In meiosis and mitosis, the nuclei appear condensed, thin, and stringy. At 60 min into interphase, the nuclei appear large and round. By 80 min, the nuclei will appear round, but spongy as the chromosomes begin to condense again.
7. Following centrifugation, the spindle pellet will be at the bottom of the tube. However, the spindle pellet will not be visible by eye.
8. We have had the most success with the RNeasy kit. However, other RNA isolation methods (e.g., TRIzol) may be used.
9. From 150 μl of cycled extract, we typically recover about 200–500 ng of spindle-associated RNA. If the amount recovered is drastically more than this amount, it is likely that the spindle sample was contaminated with total RNA. If the amount is drastically less, it is likely that the spindles failed to form properly, or that the pelleted spindles were lost during the washes.
10. To conserve the spindle RNA, which may be limiting, we recommend running as little as 20 ng of RNA on the gel. Low amounts of RNA can be visualized by post-staining the agarose gel with SYBR® Gold and imaging the gel on a Typhoon scanner at 502 nm.
11. The number of tolerated mismatches affects the efficiency and accuracy of alignment: a lower mismatch allowance yields a smaller number of alignments, but those alignments will have a higher confidence. For a non-isogenic organism such as *X. laevis*, we typically allow up to two mismatches.
12. Sequences that align to multiple locations in the reference may either be omitted (the most conservative approach, and the one we most commonly use), or used to perform a weighted count of the reads.
13. Microsoft Excel may be used for these analyses; however, we prefer to use R, which has more robust statistical analysis and graphing tools available.
14. It is important to pool cDNA from all of the samples together for the standard curve because some transcripts may only be measurable in one sample and not others.
15. To test translation-dependence, a parallel sample containing the translation inhibitor puromycin at 0.1 mg/ml (diluted from a 10 mg/ml stock in 10 mM HEPES pH 7.2) can be prepared [5]. Puromycin treatment of CSF spindles should proceed for 30 min before the addition of sperm nuclei. For cycled spindles, puromycin treatment of the aliquot used to rearrest the reaction in metaphase should proceed for 30 min prior to mixing with the interphase extract. The final concen-

tration in this case should always be maintained at 0.1 mg/ml. Puromycin-treated spindles will appear largely indistinguishable from control spindles, with the only defect being a minor reduction in tubulin density.

16. RNaseA-coupled beads may be stored at 4 °C for up to 2 weeks. Over time the bound proteins become unfolded and are toxic to the extract. Therefore, it is imperative to prepare a control column of beads simultaneously with the RNaseA beads to monitor toxicity to the extract.

References

1. Hussain S, Benavente SB, Nascimento E et al (2009) The nucleolar RNA methyltransferase Misu (NSun2) is required for mitotic spindle stability. J Cell Biol 186:27–40. doi:10.1083/jcb.200810180
2. Groisman I, Huang YS, Mendez R et al (2000) CPEB, maskin, and cyclin B1 mRNA at the mitotic apparatus: implications for local translational control of cell division. Cell 103:435–47. doi:10.1038/35043029
3. Blower MD, Nachury M, Heald R, Weis K (2005) A Rae1-containing ribonucleoprotein complex is required for mitotic spindle assembly. Cell 121:223–34. doi:10.1016/j.cell.2005.02.016
4. Blower MD, Feric E, Weis K, Heald R (2007) Genome-wide analysis demonstrates conserved localization of messenger RNAs to mitotic microtubules. J Cell Biol 179:1365–73. doi:10.1083/jcb.200705163
5. Jambhekar A, Emerman AB, Schweidenback CT, Blower MD (2014) RNA stimulates Aurora B kinase activity during mitosis. PLoS ONE 9, e100748. doi:10.1371/journal.pone.0100748
6. Schweidenback CT, Emerman AB, Jambhekar A, Blower MD (2014) Evidence for multiple, distinct ADAR-containing complexes in Xenopus laevis. RNA. doi:10.1261/rna.047787.114
7. Lambert JD, Nagy LM (2002) Asymmetric inheritance of centrosomally localized mRNAs during embryonic cleavages. Nature 420:682–6. doi:10.1038/nature01241
8. Kingsley EP, Chan XY, Duan Y, Lambert JD (2007) Widespread RNA segregation in a spiralian embryo. Evol Dev 9:527–39. doi:10.1111/j.1525-142X.2007.00194.x
9. Raff JW, Whitfield WG, Glover DM (1990) Two distinct mechanisms localise cyclin B transcripts in syncytial Drosophila embryos. Development 110:1249–61
10. Lécuyer E, Yoshida H, Parthasarathy N et al (2007) Global analysis of mRNA localization reveals a prominent role in organizing cellular architecture and function. Cell 131:174–87. doi:10.1016/j.cell.2007.08.003
11. Alliegro MC, Alliegro MA, Palazzo RE (2006) Centrosome-associated RNA in surf clam oocytes. Proc Natl Acad Sci U S A 103:9034–8. doi:10.1073/pnas.0602859103
12. Alliegro MC, Alliegro MA (2008) Centrosomal RNA correlates with intron-poor nuclear genes in Spisula oocytes. Proc Natl Acad Sci U S A 105:6993–7. doi:10.1073/pnas.0802293105
13. Eliscovich C, Peset I, Vernos I, Méndez R (2008) Spindle-localized CPE-mediated translation controls meiotic chromosome segregation. Nat Cell Biol 10:858–65. doi:10.1038/ncb1746
14. Rošić S, Köhler F, Erhardt S (2014) Repetitive centromeric satellite RNA is essential for kinetochore formation and cell division. J Cell Biol 207:673. doi:10.1083/jcb.20140409711122014c
15. Ferri F, Bouzinba-Segard H, Velasco G et al (2009) Non-coding murine centromeric transcripts associate with and potentiate Aurora B kinase. Nucleic Acids Res 37:5071–80. doi:10.1093/nar/gkp529
16. Ideue T, Cho Y, Nishimura K, Tani T (2014) Involvement of satellite I noncoding RNA in regulation of chromosome segregation. Genes Cells 19:528–38. doi:10.1111/gtc.12149
17. Murray AW (1991) Cell cycle extracts. Methods Cell Biol 36:581–605
18. Walczak CE, Vernos I, Mitchison TJ et al (1998) A model for the proposed roles of different microtubule-based motor proteins in establishing spindle bipolarity. Curr Biol 8:903–13
19. Choi YH, Hagedorn CH (2003) Purifying mRNAs with a high-affinity eIF4E mutant

identifies the short 3′ poly(A) end phenotype. Proc Natl Acad Sci U S A 100:7033–8. doi:10.1073/pnas.1232347100

20. Bajak EZ, Hagedorn CH (2008) Efficient 5′ cap-dependent RNA purification : use in identifying and studying subsets of RNA. Methods Mol Biol 419:147–60. doi:10.1007/978-1-59745-033-1_10
21. Blower MD, Jambhekar A, Schwarz DS, Toombs JA (2013) Combining different mRNA capture methods to analyze the transcriptome: analysis of the Xenopus laevis transcriptome. PLoS ONE 8, e77700. doi:10.1371/journal.pone.0077700
22. Sharp JA, Plant JJ, Ohsumi TK et al (2011) Functional analysis of the microtubule-interacting transcriptome. Mol Biol Cell 22:4312–23. doi:10.1091/mbc.E11-07-0629
23. Hannak E, Heald R (2006) Investigating mitotic spindle assembly and function in vitro using Xenopus laevis egg extracts. Nat Protoc 1:2305–14. doi:10.1038/nprot.2006.396
24. Langmead B, Trapnell C, Pop M, Salzberg SL (2009) Ultrafast and memory-efficient alignment of short DNA sequences to the human genome. Genome Biol 10:R25. doi:10.1186/gb-2009-10-3-r25
25. Trapnell C, Pachter L, Salzberg SL (2009) TopHat: discovering splice junctions with RNA-Seq. Bioinformatics 25:1105–11. doi:10.1093/bioinformatics/btp120
26. Trapnell C, Williams BA, Pertea G et al (2010) Transcript assembly and quantification by RNA-Seq reveals unannotated transcripts and isoform switching during cell differentiation. Nat Biotechnol 28:511–5. doi:10.1038/nbt.1621
27. Ingolia NT, Ghaemmaghami S, Newman JR, Weissman JS (2009) Genome-wide analysis in vivo of translation with nucleotide resolution using ribosome profiling. Science 324:218–23. doi:10.1126/science.1168978
28. Sawin KE, Mitchison TJ (1991) Mitotic spindle assembly by two different pathways in vitro. J Cell Biol 112:925–40

Chapter 20

Probing Mitosis by Manipulating the Interactions of Mitotic Regulator Proteins Using Rapamycin-Inducible Dimerization

Edward R. Ballister and Michael A. Lampson

Abstract

Inducible dimerization is a general approach to experimentally manipulate protein–protein interactions with temporal control. This chapter describes the use of rapamycin-inducible dimerization to manipulate mitotic regulatory proteins, for example to control kinetochore localization. A significant feature of this method relative to previously described protocols is the depletion of endogenous FKBP12 protein, which markedly improves dimerization efficiency.

Key words Rapamycin, Kinetochore, Dimerization

1 Introduction

Protein–protein interactions underlie the regulation of essentially every process in biology, and mitosis is no exception. These interactions can be experimentally controlled through inducible dimerization, a family of techniques that has been applied to diverse biological processes. One of the most widely used tools for inducible dimerization is the small molecule rapamycin, which simultaneously binds two different proteins, FKBP12 (hereafter FKBP) and mTOR, with very high affinity (~2 nM) [1]. Rapamycin is used as a general tool for heterodimerization by genetically fusing proteins of interest (POIs) to FKBP12 and a minimal rapamycin-binding fragment of mTOR (FRB) [2, 3]. In this way, rapamycin can be applied to any pair of POIs (POI1 and POI2) amenable to genetic manipulation. An advantage of rapamycin over other chemical dimerizers is that FRB binds only to the rapamycin-FKBP complex, which prevents formation of unproductive rapamycin-FRB complexes.

Paul Chang and Ryoma Ohi (eds.), *The Mitotic Spindle: Methods and Protocols*, Methods in Molecular Biology, vol. 1413,
DOI 10.1007/978-1-4939-3542-0_20, © Springer Science+Business Media New York 2016

Inducible dimerization has been used to probe mitosis in different ways. In the example described in this method, the Spindle Assembly Checkpoint (SAC) protein Mad1 is recruited to kinetochores during metaphase, a time when SAC proteins are normally removed from kinetochores [4, 5]. Returning Mad1 to kinetochores at metaphase reactivates the SAC. Inducible dimerization has also been used to control cohesin ring closure [6], manipulate SUMO isopeptidase activity at kinetochores [7], and rapidly remove TACC3-ch-TOG-clathrin complexes from kinetochore-microtubule fibers by sequestering the complexes to mitochondria, where they cannot perform their normal functions [8].

One factor which limits the effectiveness of rapamycin-induced dimerization is the presence of endogenous FKBP, a highly abundant cytosolic protein in yeast and animal cells which competes against exogenous POI1-FKBP for ternary complex formation with POI2-FRB upon addition of rapamycin [4, 9]. FKBP is not essential, at least in tissue culture cells [10–13], so this problem can be overcome by depleting endogenous FKBP using RNAi, using either siRNA transfection or expression of an engineered shRNA/miRNA as described here [4]. FKBP null mice develop fatal cardiac defects [14], but tissue-specific FKBP knockout later in development can be well tolerated [11].

The overall method involves construction and expression of a pair of fusion proteins (POI1-FKBP and POI2-FRB), RNAi depletion of endogenous FKBP, and addition of rapamycin as the experimental perturbation (Fig. 1). We use Recombinase Mediated Cassette Exchange (RMCE) [15] to generate cell lines stably expressing POI1-FKBP and POI2-FRB fusion proteins plus shRNA against endogenous FKBP. The specific protocol used here (HiLo RMCE) allows rapid generation of stable cell lines constitutively and/or inducibly expressing a number of cDNAs and miRNA-based shRNAs [16]. The reagents are generally available for use in other academic or nonprofit environments at no cost.

2 Materials

2.1 Generation and Culture of Stable Cell Lines

1. HiLo acceptor cell line (*see* **Note 1**).
2. RMCE donor plasmid capable of expressing FKBP- and FRB-fusion proteins as well as shRNA against endogenous FKBP (in this case, Mis12-GFP-FKBPx3 and FRB-mCherry-Mad1) (*see* **Notes 2** and **3**).
3. Cre expression plasmid, preferably with nuclear localization sequence (*see* **Note 4**).

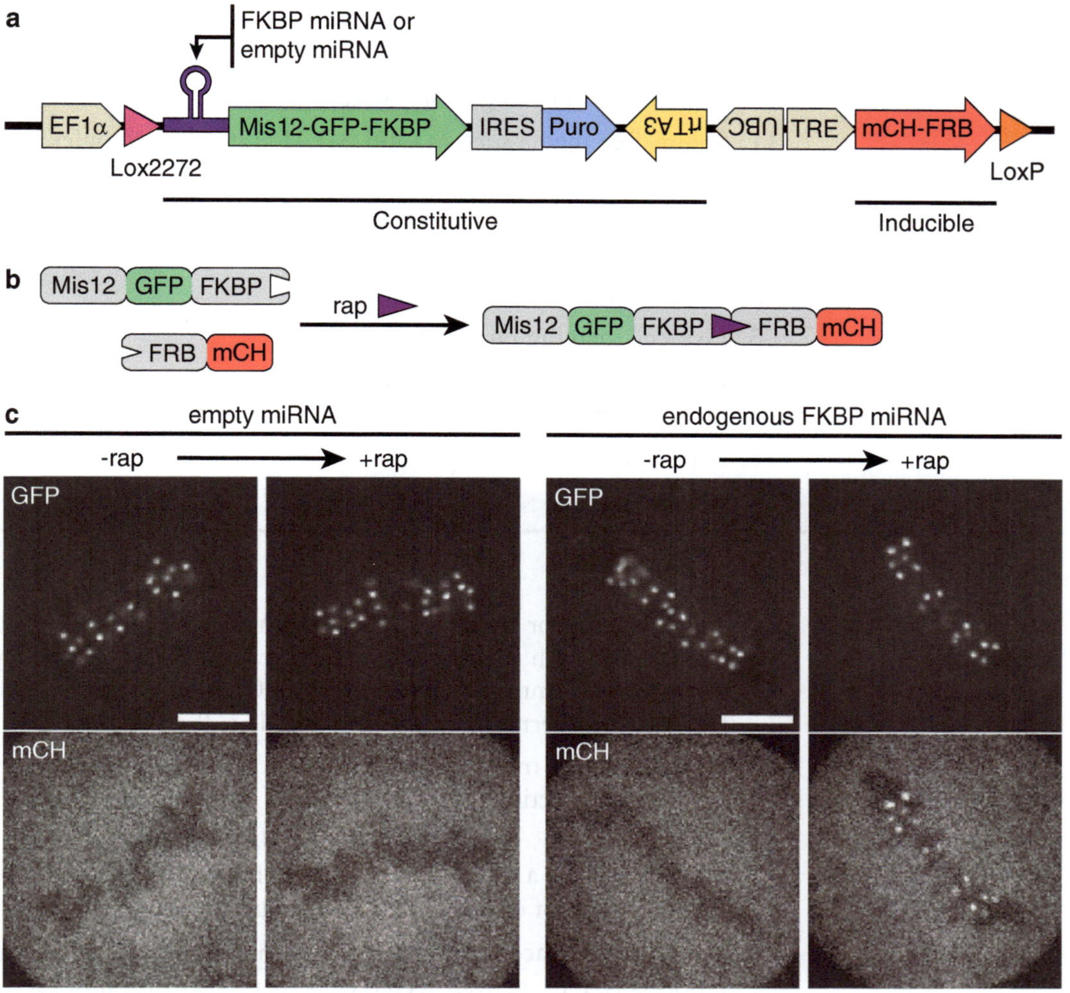

Fig. 1 Rapamycin-mediated recruitment to kinetochores. (**a**) Diagram of a DNA cassette used to constitutively express Mis12-GFP-FKBP and miRNA, and inducibly express mCherry-FRB. The cassette is integrated between Lox acceptor sites downstream of the EF1a promoter [16]. (**b**) Schematic representation of rapamycin-mediated recruitment of mCherry-FRB to kinetochore-localized Mis12-GFP-FKBP. (**c**) HeLa cells expressing Mis12-GFP-FKBP, mCherry-FRB and either an empty miRNA backbone or miRNA against the 3′ UTR of endogenous FKBP were imaged before and ~1 min after the addition of 500 nM rapamycin (rap). This figure was originally published in The Journal of Cell Biology (Ballister et al. [4] Fig. 1)

4. Puromycin ready-made solution 10 mg/mL.
5. Growth medium: Dulbecco's modified Eagle's medium (DMEM) supplemented with 1 % PenStrep and 10 % fetal bovine serum (FBS).
6. Transfection medium: DMEM supplemented with 10 % FBS.
7. DMSO.
8. Cryopreservation medium: 50 % FBS, 40 % DMEM, 10 % DMSO.

9. Selection medium: growth medium supplemented with 1 μg/mL puromycin (*see* **Note 4**).
10. FuGENE 6 (Promega) (*see* **Note 6**).
11. Serum-free media such as Opti-MEM (Life Technologies).
12. Doxycycline hydrochloride.

2.2 Live Cell Imaging

1. Microscope suitable for live-cell fluorescent imaging (*see* **Note 7**).
2. Poly-lysine coated #1.5 coverslips.
3. Live-cell imaging chamber.
4. Imaging medium Leibovitz-15 (L15) CO_2-independent media, supplemented with 1 % PenStrep and 10 % FBS.
5. Rapamycin.

3 Methods

3.1 Production and Maintenance of Stable Cell Line

1. Culture acceptor cells in a 35 mm dish (or 1 well of a 6-well plate) in growth medium until they reach ~60 % confluency (seeding a 35 mm plate with ~125,000 HeLa acceptor cells, followed by overnight culture is typically sufficient).
2. Replace growth medium with transfection medium for >30 min prior to transfection.
3. Cotransfect cells with RMCE donor plasmid and Cre expression plasmid at a 100:1 mass ratio of donor plasmid and Cre plasmid (1 μg of donor plasmid + 10 ng of Cre plasmid).
4. After 24 h, replace transfection medium with growth medium. Optionally, passage transfected cells to a larger dish if desired.
5. 48 h after transfection, replace growth medium with selection medium. Cells may be quite dense at this stage.
6. After 24 h of puromycin selection, many cells should die. Remove these by rinsing the plate vigorously 2× with selection medium, continue to culture in selection medium.
7. After an additional 24 h, the vast majority of cells should have died, but ~100–1000 microcolonies of resistant cells should be apparent. Change media again to remove dead cells.
8. Monitor culture until colonies merge and become confluent, or until the cell density in the interior of the colonies becomes too high (usually ~7–10 days after transfection). At this point, passage 80 % of cells to a 10 cm dish, and seed the remaining 20 % on a coverslip to check for appropriate expression and localization of fluorescently tagged fusion proteins.
9. Culture cells in selection medium until they reach 80 % confluence.
10. Harvest cells and freeze several aliquots in cryopreservation medium. At this point, generating the stable cell line is complete.

3.2 Inducible Dimerization Live Cell Imaging Experiment

1. Seed a sufficient number of cells on coverslips so that they will reach the desired density after 48 h of culture.
2. If using dox-inducible expression: 48–30 h prior to experiment, add doxycycline to a final concentration of 200 ng/mL.
3. Set up microscope and environmental control apparatus, if the experiment requires their use.
4. Warm imaging medium to 37 °C.
5. Prepare a stock of imaging medium + 750 μM rapamycin, hold at 37 °C.
6. Replace growth medium with pre-warmed imaging medium.
7. Transfer coverslip to imaging chamber and quickly add 500 μL pre-warmed imaging medium.
8. Mount chamber on microscope and locate mitotic cells.
9. Image cells to document "pre-dimerization" state. In this example, kinetochores should be clearly identifiable at the metaphase plate in the GFP channel, and FRB-mCherry-Mad1 should be diffuse in the cytosol (and concentrated at any unattached or misaligned kinetochores which may be present).
10. Add 1 mL of imaging medium + rapamycin to the chamber, mixing quickly but gently to achieve a final concentration of 500 μM rapamycin.
11. Resume imaging to document effects of rapamycin on FRB-POI fusion localization and cellular phenotype, relative to controls. In this example, FRB-mCherry-Mad1 signal accumulates at kinetochores within 1 min, and most cells treated in this manner remain arrested in metaphase [4].

4 Notes

1. RMCE involves recombination between a donor plasmid and a specially designed acceptor locus in the genome of the cells being genetically modified. Typically this acceptor locus is present at single copy at a unique site in the genome of the acceptor cells. In the HiLo system, the donor plasmid contains two nonidentical Lox sites (LoxP and Lox2272) which flank the "cassette" of DNA which is to be exchanged. The acceptor locus contains the same pair of Lox sites, with an EF1α promoter immediately upstream of the Lox2272 site. This promoter is used to drive transcription of a selection marker immediately downstream of the Lox2272 element (within the acceptor cassette). 11 monoclonal acceptor cell lines are available, including human (HeLa, HeLa-S3, A549, HT1080, HEK293T, and U2OS) and mouse (NIH3T3, CAD, L929, N2a, and P19) cell lines.

2. Depending on the particular proteins being dimerized, it may be preferable for both to be expressed constitutively, or one or both to be expressed inducibly. In this example, Mis12-GFP-FKBPx3 is expressed constitutively and FRB-mCherry-Mad1 is expressed inducibly.
3. In the plasmids described in this Method, an shRNA against endogenous FKBP is transcribed and processed as part of an engineered miRNA nested within an intron in the same RNA PolII transcript as Mis12-GFP-FKBPx3. This obviates the need for a distinct RNA PolIII promoter for shRNA transcription. The shRNA sequence used here is 5′-AUAUGGAU UCAUGUGCACAUGGUUUUGGCCACUGACUGA CCAUGUGCAUGAAUCCAUAU-3′ which targets a region (5′-CAUGUGCACAUGAAUCCAUAU-3′) within the human FKBP 3′ untranslated region (3′ UTR). Because the endogenous FKBP 3′ UTR is not included in the exogenous FKBP transgene transcript, it is not necessary to add silent mutations to the exogenous FKBP transgene to protect it from targeting by the shRNA. siRNA transfection has also proved effective in our hands for FKBP knockdown, targeting the endogenous 3′ UTR with the following pair of annealed siRNA oligos: 5′-GCACAAGUGGUAGGUUAACdTdT-3′ + 5′-GUUAACCUACCACUUGUGCdTdT-3′ [10].
4. The HiLo system uses Cre recombinase to catalyze recombination between donor and acceptor Lox sites. Plasmids for transient Cre expression (including a Cre construct containing a nuclear import sequence) are available from the Makeyev lab and repositories such as Addgene.
5. An appropriate concentration for the selection agent must be determined empirically from a kill-curve experiment. 1 μg/mL puromycin works well for HeLa cells.
6. This protocol uses Fugene 6 for transfection, but other DNA transfection reagents or electroporation may be used instead.
7. In this example, a microscope (DM4000; Leica) with a 100× 1.4 NA objective, an XY Piezo-Z stage (Applied Scientific Instrumentation), a spinning disk (Yokogawa), an electron multiplier charge-coupled device camera (ImageEM; Hamamatsu Photonics), and a laser merge module equipped with 488- and 593-nm lasers (LMM5; Spectral Applied Research) controlled by MetaMorph software (Molecular Devices) equipped with an environmental chamber for temperature control (Incubator BL; PeCon GmbH).

References

1. Banaszynski LA, Liu CW, Wandless TJ (2005) Characterization of the FKBP.rapamycin.FRB ternary complex. J Am Chem Soc 127:4715–4721. doi:10.1021/ja043277y
2. Rivera VM, Clackson T, Natesan S, Pollock R, Amara JF, Keenan T et al (1996) A humanized system for pharmacologic control of gene expression. Nat Med 2:1028–1032
3. Putyrski M, Schultz C (2012) Protein translocation as a tool: the current rapamycin story. FEBS Lett 586:2097–2105. doi:10.1016/j.febslet.2012.04.061
4. Ballister ER, Riegman M, Lampson MA (2014) Recruitment of Mad1 to metaphase kinetochores is sufficient to reactivate the mitotic checkpoint. J Cell Biol 204:901–908. doi:10.1083/jcb.201311113
5. Kuijt TEF, Omerzu M, Saurin AT, Kops GJPL (2014) Conditional targeting of MAD1 to kinetochores is sufficient to reactivate the spindle assembly checkpoint in metaphase. Chromosoma. doi:10.1007/s00412-014-0458-9
6. Gruber S, Arumugam P, Katou Y, Kuglitsch D, Helmhart W, Shirahige K et al (2006) Evidence that loading of cohesin onto chromosomes involves opening of its SMC hinge. Cell 127:523–537. doi:10.1016/j.cell.2006.08.048
7. Cubeñas-potts C, Goeres JD, Matunis MJ, Bloom KS (2013) SENP1 and SENP2 affect spatial and temporal control of sumoylation in mitosis. Mol Biol Cell 24:3483–3495. doi:10.1091/mbc.E13-05-0230
8. Cheeseman LP, Harry EF, McAinsh AD, Prior IA, Royle SJ (2013) Specific removal of TACC3-ch-TOG-clathrin at metaphase deregulates kinetochore fiber tension. J Cell Sci 126:2102–2113. doi:10.1242/jcs.124834
9. Coutinho-Budd JC, Snider SB, Fitzpatrick BJ, Rittiner JE, Zylka MJ (2013) Biological constraints limit the use of rapamycin-inducible FKBP12-Inp54p for depleting PIP2 in dorsal root ganglia neurons. J Negat Results Biomed 12:13. doi:10.1186/1477-5751-12-13
10. Weiwad M, Edlich F, Kilka S, Erdmann F, Jarczowski F, Dorn M et al (2006) Comparative analysis of calcineurin inhibition by complexes of immunosuppressive drugs with human FK506 binding proteins. Biochemistry 45:15776–15784. doi:10.1021/bi061616p
11. Hoeffer CA, Tang W, Wong H, Santillan A, Patterson RJ, Martinez LA et al (2008) Removal of FKBP12 enhances mTOR-Raptor interactions, LTP, memory, and perseverative/repetitive behavior. Neuron 60:832–845. doi:10.1016/j.neuron.2008.09.037
12. De Angelis B, Dotti G, Quintarelli C, Huye LE, Zhang L, Zhang M et al (2009) Generation of Epstein-Barr virus-specific cytotoxic T lymphocytes resistant to the immunosuppressive drug tacrolimus (FK506). Blood 114:4784–4791. doi:10.1182/blood-2009-07-230482
13. Gerard M, Deleersnijder A, Daniëls V, Schreurs S, Munck S, Reumers V et al (2010) Inhibition of FK506 binding proteins reduces alpha-synuclein aggregation and Parkinson's disease-like pathology. J Neurosci 30:2454–2463. doi:10.1523/JNEUROSCI.5983-09.2010
14. Shou W, Aghdasi B, Armstrong DL, Guo Q, Bao S, Charng MJ et al (1998) Cardiac defects and altered ryanodine receptor function in mice lacking FKBP12. Nature 391:489–492. doi:10.1038/35146
15. Turan S, Galla M, Ernst E, Qiao J, Voelkel C, Schiedlmeier B et al (2011) Recombinase-mediated cassette exchange (RMCE): traditional concepts and current challenges. J Mol Biol 407:193–221. doi:10.1016/j.jmb.2011.01.004
16. Khandelia P, Yap K, Makeyev EV (2011) Streamlined platform for short hairpin RNA interference and transgenesis in cultured mammalian cells. PNAS 108:12799–12804. doi:10.1073/pnas.1103532108

Chapter 21

Studying Kinetochore Kinases

Adrian T. Saurin and Geert J.P.L. Kops

Abstract

Mitotic kinetochores are signaling network hubs that regulate chromosome movements, attachment error-correction, and the spindle assembly checkpoint. Key switches in these networks are kinases and phosphatases that enable rapid responses to changing conditions. Describing the mechanisms and dynamics of their localized activation and deactivation is therefore instrumental for understanding the spatiotemporal control of chromosome segregation.

Key words Kinetochore, Kinase, Small molecule inhibitor, Antibody, Signaling, Phosphorylation, Spindle checkpoint

1 Introduction

Named by Lester Sharp in the early 1930s for their apparent ability to power chromosome movements, kinetochores are the business ends of chromosomes during mitosis. They are large protein assemblies that connect chromatin to the mitotic spindle which is tasked with dragging identical, duplicated chromatids in opposite directions. Kinetochores are therefore the focal points of signaling networks that regulate chromosome–spindle attachments and cell cycle progression [1–3].

Prometaphase is the time of mitosis during which the chromosomes attempt to become bioriented, that is, to have each sister from a pair of duplicate chromosomes attach to microtubules from opposite spindle poles. In mammalian cells, this process is quite error-prone. In fact, many initial attachments are of the so-called syntelic kind and are in need of correcting [4–6]. Consequently, error-correction mechanisms have evolved that enable iterative rounds of detachment–attachment until a chromosome is bioriented. All the while, anaphase initiation cannot be permitted. Surveillance of attachment/biorientation and transmission of that information to the cell cycle machinery is the primary job of the spindle assembly checkpoint (SAC). The combined actions of the

Paul Chang and Ryoma Ohi (eds.), *The Mitotic Spindle: Methods and Protocols*, Methods in Molecular Biology, vol. 1413, DOI 10.1007/978-1-4939-3542-0_21, © Springer Science+Business Media New York 2016

error-correction and SAC machineries therefore prevent erroneous chromosome segregations [7, 8].

As with most signaling networks wired to achieve speed and responsiveness, kinases and phosphatases are at the heart of attachment error-correction and the SAC. The kinases Aurora B, MPS1, BUB1, PLK1, and CDK1, and the phosphatases PP1 and PP2A-B56 all have major contributions. Several excellent reviews have summarized our current understanding of their molecular workings, and we therefore refer the reader to those for detailed insights [1–8]. For the purpose of this chapter, however, the following is relevant; localization of BUB1 to human kinetochores is crucial for SAC activity as it ensures localized production of the wait-anaphase signal. BUB1 recruitment is induced by MPS1, which phosphorylates multiple repeated motifs in the kinetochore scaffold KNL1 to create phospho-docking sites for the BUB1 adaptor BUB3 (Fig. 1). The axis Aurora B–MPS1 –BUB1 is not only at the heart of the SAC, it is also the core of attachment error-correction: Aurora B phosphorylates microtubule-binding proteins such as NDC80/HEC1 to allow destabilization of erroneous attachments (Fig. 1). MPS1 ensures localized Aurora B activity and likely has other targets required for efficient error-correction. Finally, BUB1 promotes Aurora B localization by phosphorylating histone H2A on T120. Therefore, the error-correction and SAC pathways are extensively entwined, and this is further exemplified by the fact that the phosphatases PP1 and PP2A-B56 counteract both error-correction and the SAC to stabilize chromosome spindle attachments and allow anaphase onset [9–14].

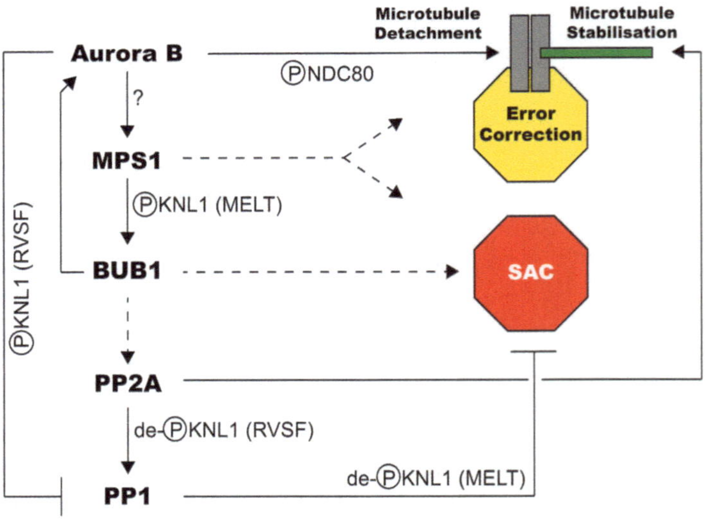

Fig. 1 Network model of kinase/phosphatase signaling at kinetochores. A selected set of kinases and phosphatases that control error-correction and the SAC are indicated along with some of their relevant substrates. *Solid lines* indicate direct connections, while *dashed lines* indicate indirect ones

Recent years have seen much advance in our mechanistic understanding of localized activation and inactivation of kinetochore kinase signaling pathways. This is largely due to reagents that allow immediate inhibition of kinases, and reagents to monitor kinase output at high spatial and temporal resolution. This chapter describes how reagents such as small molecule inhibitors and phospho-specific antibodies can be used to draw meaningful conclusions about kinetochore kinase function.

2 Material

2.1 Visualizing and Quantifying Kinase Localization and Activity: Fixed Analysis

1. High precision 1.5H 12 mm coverslips.
2. Poly-l-lysine solution (Sigma-Aldrich). Dilute 1:10 in PBS (0.01 % w/v) just prior to coating wells.
3. Nocodazole (Millipore or Sigma-Aldrich). Make 1 mg/ml stock in DMSO and store at −20 °C.
4. Paclitaxel (Taxol; LC Labs). Make 1 mM stock in DMSO and store at −20 °C.
5. Noscapine (Tocris Biosciences). Make 50 mM stock in ddH_2O and store at −20 °C.
6. S-Trityl-l-cysteine (STLC; Sigma-Aldrich). Make 10 mM stock in DMSO and store at −20 °C.
7. Monastrol (Tocris Biosciences). Make 200 mM stock in DMSO and store at −20 °C.
8. MG-132 (Sigma-Aldrich). Make 5 mM stock in DMSO and store at −20 °C.
9. Pro-TAME (Boston Biochem). Provided as a 20 mM stock in DMSO.
10. Thymidine (Sigma-Aldrich). Make 200 mM in H_2O, store at room temperature and use within 2 weeks (*see* **Note 1**).
11. Pre-extraction media (PEM). 100 mM PIPES (pH 6.8), 1 mM $MgCl_2$, 5 mM EGTA, 0.2 % Triton X-100 (all from Sigma-Aldrich). Mix fresh immediately prior to use.
12. Paraformaldehyde (Sigma-Aldrich). Use the prilled version to prevent noxious powder from being released when weighing (*see* **Note 2**).
13. 245 mm square plates (Corning). For use as an antibody incubation chamber.
14. Parafilm. 100 mm wide for antibody incubations.
15. Various primary antibodies, listed in Table 1.
16. Alexa Fluor® range of fluorescent secondary antibodies (Life Technologies). Use highly cross-adsorbed versions wherever possible to limit cross-reaction.

Table 1
A list of kinetochore antibodies that have been verified to work by immunofluorescence

Type	Antibody	Species	Dilution	Company	Cat number/reference
Kinase	Aurora B (AIM-1, Clone 6)	Mouse	1:1000	BD Transduction	611083
Kinase	Bub1	Rabbit	1:1000	Bethyl	A300-373A-1
Kinase	Cyclin B1 (Clone GNS1)	Mouse	1:500	Santa Cruz	sc-245
Kinase	Cyclin B1	Rabbit	1:500	Santa Cruz	sc-752
Kinase	MPS1, NT (Clone 3-472-1)	Mouse	1:1000	Millipore	05-682
Kinase	Plk1	Rabbit	1:1000	Custom	[15]
Aurora B autophosphorylation site	p-Aurora B (Thr232)	Rabbit	1:4000	Rockland	600-401-677
Aurora B chromatin substrate	p-Histone H3 (Ser10)	Rabbit	1:10000	Millipore	06-570
Aurora B inner kinetochore substrate	p-CenpA (Ser7)	Rabbit	1:1000	Millipore	07-232
Aurora B outer kinetochore substrate	p-NDC80 (Ser55)	Rabbit	1:1000	Genetex	GTX70017
BUB1 substrate	p-Histone H2A (Thr120)	Rabbit	1:4000	Active motif	39391
CDK1 kinetochore substrate	p-BUBR1 (Ser670)	Rabbit	1:2000	Custom	[12]
MPS1 autophosphorylation site	p-MPS1 (Thr676)	Rabbit	1:1000	Custom	[16]
MPS1 kinetochore substrate	pMELT-KNL1 (pThr943/Thr1155)	Rabbit	1:2000	Custom	[12]
MPS1 kinetochore substrate	pMELT-KNL1 (pThr875)	Rabbit		Custom	[14, 17]
PLK1 substrate	p-BUBR1 (Thr680)	Rabbit	1:1000	Custom	[11]
Unattached kinetochore marker	Mad1 (Clone BB3-8)	Mouse	1:1000	Millipore	MABE867
Kinetochore marker	KNL1	Rabbit	1:1000	Abcam	ab-70537
Kinetochore marker	NDC80 (Clone 9G3)	Mouse	1:5000	Abcam	ab-3613
Kinetochore marker	ACA	Human	1:2000	Fitzgerald	90c-CS1058
Kinetochore marker	CenpC	Guinea pig	1:5000	Caltag Medsystems	PD030

17. 4′,6-diamidino-2-phenylindole dihydrochloride (DAPI; Sigma-Aldrich). Make 1 mg/ml stock in H_2O and store at −20 °C. Store working stock at 4 °C for up to 3 months.
18. Microscope slides (72 × 25 mm).
19. ProLong® Gold antifade mounting media (Life Technologies).
20. Deltavision Elite deconvolution microscope (Applied Precision) equipped with a 100 × 1.4 numerical aperture (NA) objective and a Coolsnap HQ2 (Photometrics) CCD camera. Other similar automated microscope setups capable of multi-position imaging and high precision Z-sectioning could perform equally well.
21. Fiji ImageJ. Open source image analysis software (download from http://fiji.sc/Fiji).

2.2 Visualizing and Quantifying Kinase Localization: Live Imaging

1. 8-well chamber μ-slide (Ibidi).
2. Syringe and 23 G needle.
3. Leibovitz-15 (L-15) medium supplemented with 10 % FBS and 1 % PenStrep (all from Life Technologies).

2.3 Modulating Kinetochore Kinase Activity

A number of different kinase inhibitors have been developed to target kinetochore kinases. In addition, various gatekeeper kinase mutants have also been developed using the Shokat approach to achieve specific kinase inhibition (*See* Jallepalli et al., Chapter 22) [18]. These are listed in Table 2.

3 Methods: Fixed Cells Analysis

3.1 Determining the Correct Experimental Conditions

Kinetochore kinase activity is always affected (directly or indirectly) by kinetochore–microtubule attachment or tension. Therefore, as mitosis progresses the kinase activity at individual kinetochores will change as kinetochore–microtubule attachments form. For this reason, it is usually advisable to reduce this variability by performing experiments in the presence of nocodazole to inhibit microtubule polymerization (*see* **Note 3**).

To measure the effects of microtubule attachment on kinetochore kinase activity directly, various inhibitors can be used to enrich cells at different stages of kinetochore–microtubule attachment (as outlined in Table 3).

If required, the total number of mitotic cells can be further enriched by thymidine synchronization prior to inhibitor treatment (*see* **Note 5**).

1. Arrest cells for 24 h in full growth media and thymidine (2 mM).
2. Wash cells 1× in PBS and release into full growth media for 6 h.

Table 2
A list of inhibitors against different kinetochore kinases

Kinase	Small molecule inhibitor	IC50	[work] in HeLa	Reference
Aurora B	ZM447439	130 nM	2 μM	[19]
	Hesperadin	~250 nM	100 nM	[20]
	various others, in (pre)clinical trials			[21]
MPS1	Mps1-IN1	367 nM	10 μM	[22]
	Reversine	98.5 nM	1 μM	[23]
	AZ3146	35 nM	2 μM	[24]
	NMS-P715	65 nM	1 μM	[25]
	Various others, in (pre)clinical trials			
PLK1	BI2536	0.83 nM	100 nM	[26]
	TAL	19 nM	1 μM	[27]
	Various others, in (pre)clinical trials			[21]
BUB1	None			
CDK1	RO-3306	35 nM	10 μM	[28]
	Note that various other CDK inhibitors are available, including roscovitine, flavopiridol, and olomoucine, but they generally inhibit a broader range of CDK enzymes.			
'shokat' alleles				
AurB-as (L154A;H250Y)	NA-PP1	ND	2 μM	[29]
Mps1-as (M602G and M602A)	23-dMB-PP1/3 MB-PP1	ND	1/10 μM	[30, 31]
PLK1-as	3 MB-PP1	ND	10 μM	[32]
BUB1: is natural gatekeeper mutant (G866)	2OH-BNPP1	250 nM	Only in vitro	[33]
CDK1-as (chicken, none reported in humans)	1NM-PP1	ND	10 μM	[34]

3. Add fresh media containing required drugs (except for MG-132 which prevents mitotic entry and so must be added 8–12 h after the thymidine release).
4. Cells will be enriched in mitosis after a total of 8–12 h following thymidine release (depending on cell type).

Table 3
A list of Inhibitors that can be used to modulate kinetochore–microtubule attachment status

Drug	Working concentration (in HeLa cells)	Effect	Uses
Nocodazole	3.3 μM	Unattached kinetochores	To visualize unattached kinetochores and determine maximal kinase activity [35, 36]
Taxol	1 μM	Few unattached kinetochores	Generates a few Aurora-B dependent unattached kinetochores due to lack of tension [37]
Nocodazole	20–80 nM	Partial chromosome alignment	To visualize attached kinetochores under tension and unattached or unaligned kinetochores within the same cell.
Noscapine	2–25 μM	Partial chromosome alignment	To visualize attached kinetochore under tension and unattached or unaligned kinetochores within the same cell.
STLC	10 μM	Monotelic attachments	To visualize attached and unattached kinetochores within the same cell (using Mad1 to mark unattached kinetochores)
Monastrol	200 μM	Monotelic attachments or merotelic attachments (after washout)	To visualize attached and unattached kinetochores within the same cell (using Mad1 to mark unattached kinetochores). Monastrol washout leads to increased number of merotelic attachments (i.e., one kinetochore attached to both spindle poles).
MG-132	5 μM	Complete chromosomal alignment	To visualize attached kinetochores under tension (*see* **Note 4**).
Pro-TAME	12 μM	Complete chromosomal alignment	To visualize attached kinetochores under tension without inhibiting proteasome activity (*see* **Note 4**).

3.2 Setting Up a Fixed-Cell Imaging Experiment

1. Transfer coverslips to 24-well plates using a vacuum aspirator to transfer sterile coverslips inside a tissue culture hood.
2. Plate cells onto coverslips at required density in 0.5 ml media and then immediately press coverslip down with a pipette tip to ensure no air is trapped beneath (*see* **Note 6**).
3. Allow at least 24 h for cells to attach, but often longer depending on experimental conditions (siRNA knockdown, synchronization, etc.; *see* **Note 7**).
4. When the particular mitotic state is achieved, cells should be fixed, with or without prior pre-extraction with PEM to decrease the cytosolic signal (*see* **Note 8**).

5. If pre-extraction is required, remove media and add 0.5 ml PEM for 60 s before adding 0.5 ml 4 % PFA for an additional 2 min (*see* **Note 9**).
6. Remove media or PEM/PFA and add 4 % PFA for 10 min (*see* **Note 10**).
7. Wash 3× in PBS.

3.3 Staining with Total and Phospho-specific Antibodies

1. Block coverslips in PBS + 3 % BSA + 0.2 % Triton (0.5 ml/well) for at least 30 min.
2. Replace blocking solution with PBS.
3. Dilute primary antibodies in PBS/3 % BSA (*see* **Note 11**).
4. Add 40 μl of primary antibody mix per coverslip onto Parafilm (*see* **Note 12**). Place coverslip on top (cells facing down) and leave at least 2 h at room temperature, but generally overnight at 4 °C.
5. Transfer coverslips back into the 24-well plate (cells facing upwards) and wash 3× in PBS for 10 min each on a shaking platform.
6. Dilute secondary antibodies (1:1000 of the Alexa Fluor® from Life Technologies) and DAPI (1:1000 of 1 mg/ml stock) in PBS + 3 % BSA. Place coverslip on 40 μl, exactly as for the primary incubation, but just leave at least 2 h at room temperature in the dark (*see* **Note 13**). The kinetochore markers are usually stained with Alexa Fluor® 647, which allows the test proteins to be stained with the visible wavelengths.
7. Transfer back into 24-well plate (cells facing upwards) and wash 3× in PBS for 10 min each.
8. Dip coverslips in absolute ethanol and place cells face up on tissue paper to dry (for about 5 min).
9. Mount coverslips on slide in a small drop of ProLong® Gold antifade (Life Technologies) and leave overnight to set.

3.4 Imaging

1. The experiments were performed on a Deltavision Elite deconvolution microscope described in materials (Applied Precision). Other microscopes equipped with a motorized XY-stage and an accurate Z-stage motor may perform equally well.
2. For each particular stain, allow enough time to image the full set of treatment groups with the same settings on the same day. This will ensure intensity variations are not artificially induced during sampling.
3. Start with the coverslip that is predicted to have the maximal signal intensity (this is often the control sample).
4. Store the XY positions of 10–20 mitotic cells by visualizing the DAPI signal through the eyepiece (*see* **Note 14**).

5. Switch the light path to the camera and change exposure settings for each channel to give a maximal kinetochore signal intensity of 1000–1500 (*see* **Note 15**).
6. Switch to the far-red channel to visualize the Alexa Fluor® 647 kinetochore signal (either ACA or Cenp-C). Move to position 1 and use the software to adjust the Z position to just below the lowest kinetochore and update the position.
7. Repeat for each stored XY position without changing the manual focus on the microscope body.
8. To simplify quantification choose the channels to be scanned in the same order every time. Our macros (*see* Subheading 3.4) are predesigned for the order: DAPI, test antibodies (488 and/or 568), kinetochore marker (647).
9. Take 43 × 0.2 μM sections upwards (from the stored *z*-position) for each *XY* position. Choose the option in Softworx to automatically deconvolve images and make maximum intensity projections of the deconvolved stacks.

3.5 Quantifying Kinetochore Intensities

Kinetochore intensities can be quantified in Fiji imageJ by using a macro to automate the steps highlighted below. The aim is to quantify the kinetochore intensities with the test antibody as a ratio of the kinetochore marker intensity (after correcting each channel for background signal). Contact either the Saurin or Kops labs for macros predesigned for this purpose (as used in [16]).

1. Open the deconvolved maximum intensity projection using Fiji imageJ (*see* **Note 16**).
2. Use crop to remove any non-mitotic cells in view.
3. Create a selection of the DAPI signal using the "auto-threshold" command followed by "create selection". Use "add to manager" to save this selection.
4. Use "duplicate" to open the image again in a new image window.
5. Move to the last channel in the new window, use the "convolve" filter to clearly define the kinetochores and create a selection of these kinetochores using the auto-threshold command, as previously. Enlarge the selection by 2 pixels (*see* **Note 17**) and "add to manager". This is the kinetochore region [KT].
6. Use the "make inverse" command to choose a selection outside of all kinetochores and "combine" with the DAPI selection to define a region outside of the kinetochores but inside the DAPI. This chromatin area can be used as the background region [BG] for most kinetochore kinases and substrates (*see* **Note 18**).
7. Close the duplicated window.

8. In the original image window, measure the mean [KT] and [BG] intensity for all channels.
9. Calculate KT intensities for the test antibody using the formula: $[\text{KT-BG}]_{\text{test antibody}}/[\text{KT-BG}]_{\text{kinetochore marker}}$ (*see* **Note 19**).
10. The change in KT intensity upon treatment can be calculated by: Mean KT intensity (control)/Mean KT intensity (treatment group) (*see* **Note 20**).
11. The mean and standard deviation of this ratio can then be calculated between individual experiments for statistical purposes.

4 Methods: Live Cell Imaging

4.1 Setting Up a Live-Cell Imaging Experiment

1. Plate cells into 8-well chamber slides (Ibidi) using 250 μl of media per well. Plate the same number of cells as for a standard well of a 24-well plate.
2. On the day of the experiment, exchange media to L-15 (Leibovitz) media with serum for at least 3 h prior to imaging (mitotic entry is inhibited for the first few hours after switching to L-15; *see* **Note 21**).
3. Turn on the microscope heating unit to equilibrate the microscope to 37 °C for at least 2 h prior to filming (*see* **Note 22**).
4. Secure the chamber slide into position in the stage holder and allow to equilibrate to temperature (*see* **Note 23**).
5. Determine the required exposure time and, if possible, try to minimize toxicity by sacrificing resolution using binning (*see* **Note 24**).
6. Run a test on a sample field of view to determine whether the chosen settings are suitable and that the level of bleaching is acceptable (*see* **Note 25**).
7. Start experiment and check again after a few time points to ensure there has been no focal drift (*see* **Note 26**).

4.2 Adding Small Molecules During Time-Lapse Imaging

Adding and washing out drugs is possible during a time-lapse imaging experiment, but care must be taken not to move the chamber slide or dislodge the loosely attached mitotic cells. Also, if drugs are to be added to different wells from the start of a time-lapse movie using the ibidi chamber slides, then this should be performed when the slide is positioned securely on the microscope and not in the tissue culture hoods (*see* **Note 27**).

1. Make sure the chamber slide is securely placed in position by ensuring the spring loaded slide holder is under sufficient tension (*see* **Note 28**).

2. Remove the tight fitting lid from the chamber slide and place upside down to allow easy removal with minimum disturbance during filming.
3. Start movie for the desired number of timepoints prior to drug addition and then pause.
4. To remove media use a 5 ml syringe coupled to a 23 G needle to withdraw media from the far corner of the well (as far away from the filming position as possible; *see* **Note 29**).
5. Add 250 μl media containing drugs gently down the same corner of the well using a 1 ml pipette.
6. Replace lid and continue filming (*see* **Note 30**).

5 Notes

1. Check before use and if crystals have formed then make fresh. Store in a warm place to prevent crystallization.
2. Dissolve at 4 % in PBS by incubating for approximately 2 h at 60 °C with occasional stirring. Do not to heat above 60 °C since paraformaldehyde will degrade at high temperatures. Store at 4 °C and use within 2 days. Always warm to room temperature prior to use.
3. This is particularly important if the chosen treatments are known to affect kinetochore/microtubule attachments, since these would otherwise indirectly affect kinase activity.
4. Keep time of drug addition short because extended metaphase durations can lead to premature sister chromatid splitting due to cohesion fatigue [38, 39].
5. This type of enrichment will induce variability in the length of time individual cells have spent in mitosis, which can indirectly impact on phosphorylation levels of some substrates [12]. To reduce variability, treat with inhibitors for a fixed duration in non-synchronized cells.
6. If cells adhere poorly to glass then coat coverslips in poly-l-lysine solution for at least 1 h prior to addition of cells (remove and wash 3× in PBS just before cells are added).
7. If experimental conditions are likely to alter the duration of mitosis then this could also indirectly affect substrate phosphorylation (*see* **Note 5**). If required, treat with the Cdk1 inhibitor RO-3306 (10 μM) for 2 h (to induce mitotic exit and prevent new mitotic entry). Wash cells 3× in PBS and replace with fresh media (−/+ inhibitors, if required) for a fixed duration to standardize the length of time in mitosis (allow at least 15 min for mitotic entry following RO-3306 washout).

8. For kinases or substrates that give a cytoplasmic signal, visualization of the kinetochore signal can be improved by pre-extraction for 1 min in a Triton-containing buffer (PEM) to permeabilize the membrane and remove cytosolic proteins.
9. Take care not to dislodge mitotic cells, which are very loosely attached in the Triton-containing PEM buffer. The addition of PFA before the PEM is removed is important to prevent inadvertent loss of cells.
10. For most antigens we find standard PFA fixation is sufficient, but a small number may benefit from alternative fixation methods. If the signal intensity is low, then try fixation in either methanol or methanol–acetone (50/50) at −20 °C for 5 min. This fixes cells by protein denaturation which may reveal antibody epitopes that are otherwise obscured.
11. A typical experiment will contain a kinetochore marker (human-ACA or Guinea Pig Cenp-C; both give strong signals and allow co-staining with mouse and/or rabbit antibodies) and a total kinase and/or phospho-substrate antibody (*see* Table 1). For phospho-specific antibodies incubate with dephospho-peptide whenever possible, since we find this increases the phospho-specific signal even for antibodies that have been purified by phospho-peptide enrichment.
12. Use a corning 245 mm square tissue culture plate for antibody incubations. Place a strip of Parafilm (100 mm thickness) across the plate and draw a grid to represent a 24-well plate. Place antibody drops directly onto the Parafilm and perform coverslip incubations with the lid on to prevent evaporation.
13. We use the highly cross-adsorbed options (goat anti-guinea pig for Cenp-C or goat anti-human for ACA) because these show no cross-reaction with mouse/rabbit. If using sheep primary antibodies then use secondary antibodies raised in donkey, because we see significant cross-reaction with secondaries raised in goat.
14. Choose cells that are separated from the surrounding cells to simplify the quantification.
15. This allows for a threefold increase in signal before the 12-bit Coolsnap-HQ2 camera is overexposed. If this is set on the treatment group with the highest predicted kinetochore intensity this is usually sufficient to account for inter-cell variability and prevent overexposure (unless variability is particularly high, for example following transient protein overexpression).
16. We routinely use maximum intensity projections, but it is also possible to use sum intensity projections to include all data from every *Z*-projection. This may be especially important if kinetochore volume changes occur between samples/treat-

ments. Note that cytoplasmic background could then begin to obscure KT signals, if this is the case, then this can be reduced by removing sections above/below the kinetochores prior to projection.

17. This selection should now encompass the kinetochore signal in the test antibody channel, but if the test sample is larger (as with Aurora B for example, which gives a diffuse signal around the kinetochore) then increase this number, as required.
18. If the kinase or substrate localizes to chromatin (e.g., Aurora B, pH2A etc.) then choose a ring around the DAPI selection (i.e., cytoplasm) as the background region instead.
19. Expressing the intensities as a ratio of an internal kinetochore control allows normalization against changes in kinetochore size and/or antibody signal intensity, both of which can vary between cells and between experiments.
20. Expressing data as a fold-change following treatment normalizes against variation in the relative signal intensity of target vs kinetochore marker, which can otherwise vary considerably between experiments.
21. L-15 (Leibovitz) is a CO_2-independent media, but standard media (minus phenol red to prevent autofluorescence) can be used if a 95 % air/5 % CO_2 line is available at the microscope.
22. If a large number of live microscopy experiments are performed then keep the microscope unit at 37 °C permanently. This will prevent downtime without affecting the quality of the fixed analysis.
23. Temperature fluctuations will cause the focal point to drift, and therefore, allow enough time for the temperature to stabilize and check the focus is steady for at least 5–10 min before filming.
24. By using 2×2 binning the resolution is decreased fourfold with a corresponding fourfold increase in fluorescence signal. 4×4 binding leads to 16-fold changes, which can be very important to increase low signal intensities if detailed resolution is not required.
25. For example, if running a time series of 100 time points with $3 \times Z$ sections every 5 min, then quickly image one area of cells with those settings (but without the time series delay). Use the data inspector in Softworx to measure fluorescence decay throughout the course of the 100 images and try to ensure this is no more than 10 %.
26. If focal drift has occurred, use the microscope focus dial to update the correct focus in one position. This should correct all the other positions since they are all relative.

27. The ibidi chamber slides have a very low depth to the wells and we have frequently observed mixing between the wells during transport (i.e., from tissue culture room to the microscope). This can be minimized by using very low volumes of media (150 ml per 8-well chamber); however, we have found the safest option is to only add drugs when the chamber slide is securely positioned on the microscope stage.
28. If using different stage inserts, then we find that Blu-Tack (or a similar flexible adhesive) is excellent for preventing movement during filming.
29. Cut off the end of the needle to remove the beveled edge and bend slightly. This will allow full removal of media when placing the needle into the corner of the well so that it just touches the base of the well.
30. If washout is required, replace lid upside down and repeat the same procedure at the time of drug washout.

References

1. Sacristan C, Kops GJPL (2015) Joined at the hip: kinetochores, microtubules, and spindle assembly checkpoint signaling. Trends Cell Biol 25:21. doi:10.1016/j.tcb.2014.08.006
2. London N, Biggins S (2014) Signalling dynamics in the spindle checkpoint response. Nat Rev Mol Cell Biol 15:736–748
3. Foley EA, Kapoor TM (2013) Microtubule attachment and spindle assembly checkpoint signalling at the kinetochore. Nat Rev Mol Cell Biol 14:25–37
4. Carmena M, Wheelock M, Funabiki H, Earnshaw WC (2012) The chromosomal passenger complex (CPC): from easy rider to the godfather of mitosis. Nat Rev Mol Cell Biol 13:789–803
5. Magidson V et al (2011) The spatial arrangement of chromosomes during prometaphase facilitates spindle assembly. Cell 146:555–567
6. Kitajima TS, Ohsugi M, Ellenberg J (2011) Complete kinetochore tracking reveals error-prone homologous chromosome biorientation in mammalian oocytes. Cell 146:568–581
7. Vleugel M, Hoogendoorn E, Snel B, Kops GJPL (2012) Evolution and function of the mitotic checkpoint. Dev Cell 23:239–250
8. Lara-Gonzalez P, Westhorpe FG, Taylor SS (2012) The spindle assembly checkpoint review. Curr Biol 22:R966–R980
9. Liu D et al (2010) Regulated targeting of protein phosphatase 1 to the outer kinetochore by KNL1 opposes Aurora B kinase. J Cell Biol 188:809–820
10. Foley EA, Maldonado M, Kapoor TM (2011) Formation of stable attachments between kinetochores and microtubules depends on the B56-PP2A phosphatase. Nat Cell Biol 13:1265–1271
11. Suijkerbuijk SJE, Vleugel M, Teixeira A, Kops GJPL (2012) Integration of kinase and phosphatase activities by BUBR1 ensures formation of stable kinetochore-microtubule attachments. Dev Cell 23:745–755
12. Nijenhuis W, Vallardi G, Teixeira A, Kops GJPL, Saurin AT (2014) Negative feedback at kinetochores underlies a responsive spindle checkpoint signal. Nat Cell Biol 16:1257–1264
13. Pinsky BA, Nelson CR, Biggins S (2009) Protein phosphatase 1 regulates exit from the spindle checkpoint in budding yeast. Curr Biol 19:1182–1187
14. Espert A et al (2014) PP2A-B56 opposes Mps1 phosphorylation of Knl1 and thereby promotes spindle assembly checkpoint silencing. J Cell Biol 206:833–842
15. Macůrek L et al (2008) Polo-like kinase-1 is activated by aurora A to promote checkpoint recovery. Nature 455:119–123
16. Saurin AT, van der Waal MS, Medema REH, Lens SMA, Kops GJPL (2011) Aurora B potentiates mps1 activation to ensure rapid checkpoint establishment at the onset of mitosis. Nat Commun 2:316–319

17. Yamagishi Y, Yang C-H, Tanno Y, Watanabe Y (2012) MPS1/Mph1 phosphorylates the kinetochore protein KNL1/Spc7 to recruit SAC components. Nat Cell Biol 14: 746–752
18. Bishop AC, Buzko O, Shokat KM (2001) Magic bullets for protein kinases. Trends Cell Biol 11:167–172
19. Ditchfield C et al (2003) Aurora B couples chromosome alignment with anaphase by targeting BubR1, Mad2, and Cenp-E to kinetochores. J Cell Biol 161:267–280
20. Hauf S et al (2003) The small molecule Hesperadin reveals a role for Aurora B in correcting kinetochore-microtubule attachment and in maintaining the spindle assembly checkpoint. J Cell Biol 161:281–294
21. Lens SMA, Voest EE, Medema RH (2010) Shared and separate functions of polo-like kinases and aurora kinasesin cancer. Nat Rev Cancer 10:1–17
22. Kwiatkowski N et al (2010) Small-molecule kinase inhibitors provide insight into Mps1 cell cycle function. Nat Chem Biol 6:359
23. Santaguida S, Tighe A, D'Alise AM, Taylor SS, Musacchio A (2010) Dissecting the role of MPS1 in chromosome biorientation and the spindle checkpoint through the small molecule inhibitor reversine. J Cell Biol 190:73–87
24. Hewitt L et al (2010) Sustained Mps1 activity is required in mitosis to recruit O-Mad2 to the Mad1-C-Mad2 core complex. J Cell Biol 190:25–34
25. Colombo R et al (2010) Targeting the mitotic checkpoint for cancer therapy with NMS-P715, an inhibitor of MPS1 kinase. Cancer Res 70:10255–10264
26. Lénárt P et al (2007) The small-molecule inhibitor BI 2536 reveals novel insights into mitotic roles of polo-like kinase 1. Curr Biol 17:304–315
27. Santamaria A et al (2007) Use of the novel Plk1 inhibitor ZK-thiazolidinone to elucidate functions of Plk1 in early and late stages of mitosis. Mol Biol Cell 18:4024–4036
28. Vassilev LT et al (2006) Selective small-molecule inhibitor reveals critical mitotic functions of human CDK1. Proc Natl Acad Sci U S A 103:10660–10665
29. Hengeveld RCC et al (2012) Development of a chemical genetic approach for human aurora B kinase identifies novel substrates of the chromosomal passenger complex. Mol Cell Proteomics 11:47–59
30. Sliedrecht T, Zhang C, Shokat KM, Kops GJPL, Cimini D (2010) Chemical genetic inhibition of Mps1 in stable human cell lines reveals novel aspects of Mps1 function in mitosis. PLoS One 5:e10251
31. Maciejowski J et al (2010) Mps1 directs the assembly of Cdc20 inhibitory complexes during interphase and mitosis to control M phase timing and spindle checkpoint signaling. J Cell Biol 190:89–100
32. Burkard ME et al (2007) Chemical genetics reveals the requirement for Polo-like kinase 1 activity in positioning RhoA and triggering cytokinesis in human cells. Proc Natl Acad Sci U S A 104:4383–4388
33. Kang J et al (2008) Structure and substrate recruitment of the human spindle checkpoint kinase bub1. Mol Cell 32:394–405
34. Hochegger H et al (2007) An essential role for Cdk1 in S phase control is revealed via chemical genetics in vertebrate cells. J Cell Biol 178:257–268
35. Yang Z, Kenny AE, Brito DA, Rieder CL (2009) Cells satisfy the mitotic checkpoint in Taxol, and do so faster in concentrations that stabilize syntelic attachments. J Cell Biol 186:675–684
36. Santaguida S, Vernieri C, Villa F, Ciliberto A, Musacchio A (2011) Evidence that Aurora B is implicated in spindle checkpoint signalling independently of error correction. EMBO J 30:1508–1519
37. Collin P, Nashchekina O, Walker R, Pines J (2013) The spindle assembly checkpoint works like a rheostat rather than a toggle switch. Nat Cell Biol 15:1378–1385
38. Daum JR et al (2011) Cohesion fatigue induces chromatid separation in cells delayed at metaphase. Curr Biol 21:1018–1024
39. Lara-Gonzalez P, Taylor SS (2012) Cohesion fatigue explains why pharmacological inhibition of the APC/C induces a spindle checkpoint-dependent mitotic arrest. PLoS One 7:e49041

Chapter 22

Engineering and Functional Analysis of Mitotic Kinases Through Chemical Genetics

Mathew J.K. Jones and Prasad V. Jallepalli

Abstract

During mitosis, multiple protein kinases transform the cytoskeleton and chromosomes into new and highly dynamic structures that mediate the faithful transmission of genetic information and cell division. However, the large number and strong conservation of mammalian kinases in general pose significant obstacles to interrogating them with small molecules, due to the difficulty in identifying and validating those which are truly selective. To overcome this problem, a steric complementation strategy has been developed, in which a bulky "gatekeeper" residue within the active site of the kinase of interest is replaced with a smaller amino acid, such as glycine or alanine. The enlarged catalytic pocket can then be targeted in an allele-specific manner with bulky purine analogs. This strategy provides a general framework for dissecting kinase function with high selectivity, rapid kinetics, and reversibility. In this chapter we discuss the principles and techniques needed to implement this chemical genetic approach in mammalian cells.

Key words Bump-hole, Chemical genetics, Analog-sensitive, Kinase engineering, Genome editing

1 Introduction

1.1 General Principles of Chemical Genetics

Analog-sensitive (as) kinase engineering was develop by the Shokat laboratory based on an insightful observation made while studying the interaction between the Src family kinases and the ATP-competitive inhibitor, PP1 [3–5]. The structure of PP1 bound to the Src family kinase, Hck, revealed the inhibitor mimics the adenine ring of ATP in its interactions within the catalytic pocket. The orientation of the inhibitor within the pocket provided the foundation for building PP1 analogs with enlarged substituents capable of specifically targeting kinases containing small glycine or alanine gatekeeper residues [4, 6]. The steric complementation provided by the gatekeeper mutation can be visualized in the structure of the tyrosine kinase Hck bound to PP1 or the bulky PP1 analogs 1-NAPP1, 1-NM-PP1, or 3-MB-PP1 (Fig. 1b, d). Although PP1

Paul Chang and Ryoma Ohi (eds.), *The Mitotic Spindle: Methods and Protocols*, Methods in Molecular Biology, vol. 1413, DOI 10.1007/978-1-4939-3542-0_22, © Springer Science+Business Media New York 2016

can fit within the wild-type Hck kinase, the bulky substituents in the PP1 analogs are oriented toward the gatekeeper and only fit within the pocket created by the gatekeeper mutation [6].

The high conservation of the bulky gatekeeper residue is key to ensuring the specificity and versatility of this approach. The chemical genetic strategy provides significant kinetic and specificity advantages over alternate methods used to study the function of protein kinases (*see* Table 1). Since native kinases are resistant to bulky purine analogs, cells expressing a wild-type kinase of interest (KOI) can be used to control for any off-target effects of the bulky purine analogs, allowing the phenotypic effects of the inhibitors to be linked to the genotype of the kinase, providing a significant advantage over traditional pharmacological inhibitors [7, 8].

The gatekeeper residue is a key determinant of kinase inhibitor sensitivity, and substituting the gatekeeper residue can provide resistance to pharmacological inhibitors directed towards the

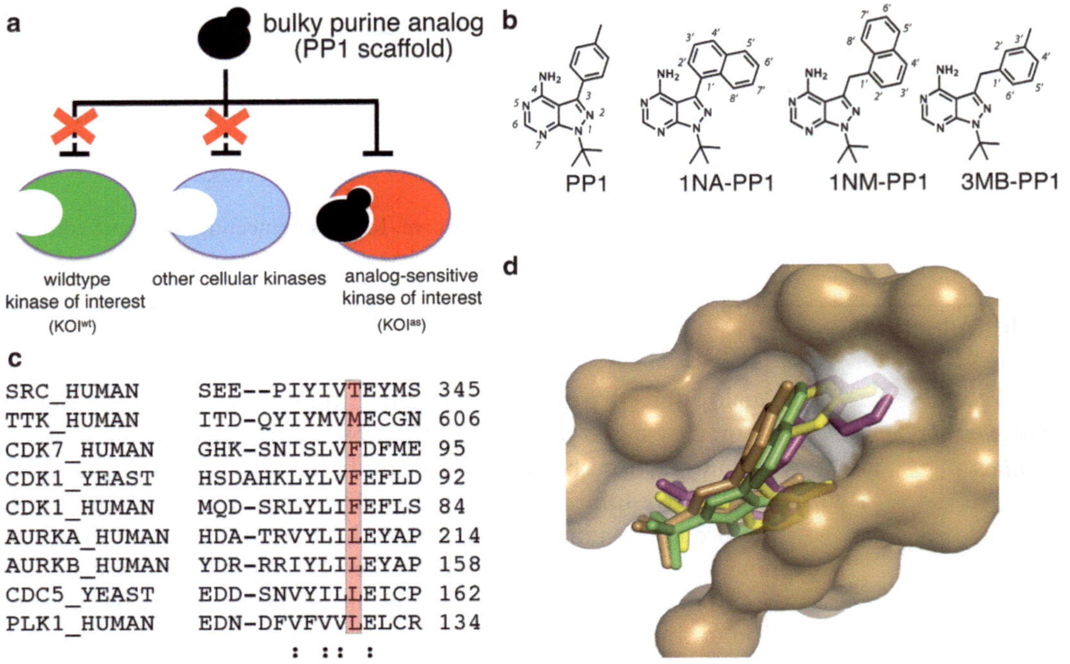

Fig. 1 Selective kinase inhibition through engineered steric complementarity. (**a**) The gatekeeper residue within your favorite kinase is mutated to a smaller residue to enlarge the catalytic pocket. (**b**) Selective inhibition is achieved amongst all other cellular kinases using the PP1 analogs 1-NA-PP1, 1-NM-PP1, or 3-MB-PP1. (**c**) Clustal alignment showing the hinge region and gatekeeper residue for Src and a selection of human and yeast mitotic kinases. (**d**) Steric complementarity visualized by overlaying the Hck kinase structure and the PP1 (*brown*), 1-NA-PP1 (*green*), 1-NM-PP1 (*pink*), and 3-MB-PP1 (*yellow*) inhibitors. The gatekeeper region is displayed in *white*. Adapted from Ref. [6]

Table 1
Comparison of methods for probing kinase function

Method of kinase inactivation	Kinetics	Specificity	Penetrance
RNA interference (RNAi)	Delayed (2–5 days).	High potential for off-target effects.	Varies depending on gene and target sequence.
Dominant negative (DN)	Delayed (12–48 h).	Potential for off-target effects. Requires high levels of exogenous protein expression.	Varies with each kinase. Often no phenotype reported.
Stable knockout (sKO)	Stable phenotype. Not compatible with essential genes.	Highly selective.	Highly penetrant. Potential for adaptation.
Conditional knockout (cKO)	Delayed (2–5 days for Cre recombinase dependent excision).	Highly selective.	Highly penetrant.
Pharmacological inhibitors	Rapid onset (minutes–hours).	Variable Very difficult to exclude off-target effects.	Varies with each chemical. Many highly and weakly penetrant examples.
Chemical genetics	Rapid onset (minutes–hours).	Highly selective (built-in control for specificity).	Many highly penetrant examples.

wild-type kinases. For example, $Plk1^{as}$ cells are not only sensitive to certain bulky purine analogs but also cross-resistant to two other Plk1 inhibitors (BI-536 and TAL) currently being developed as cancer chemotherapeutics [9]. Similarly, the BCR-ABL gatekeeper residue is a hotspot for resistance mutations in patients with chronic myelogenous leukemia who relapse during imatinib (Gleevec) therapy [10]. In this case, acquisition of a larger gatekeeper results in the steric blockade of imatinib from the active site. Inhibitor resistance also provides useful differentiation of the on-target and off-target effects of small molecules, and especially important when assessing kinase inhibitors as therapeutic agents (Fig. 1a) [11, 12].

The chemical genetic strategy has proven effective for protein kinases in diverse model organisms [2, 6]. In mammals a variety of cell types can be used, including bone marrow derived hematopoietic cells, neurons, and tumor cell lines [11, 13–15], allowing for flexibility and alignment with the underlying biological question or downstream applications. This chapter outlines the key steps for generating a chemical genetic system in human somatic cells (Fig. 2).

Design and build analog-sensitive kinase allele

↓

Substitute for endogenous wild-type kinase

↓

Screen allele/analog pairs for penetrant and selective inhibition

↓

Assay kinase-inhibited cells for mitotic defects

Fig. 2 Key steps for generating a chemical genetic system in human somatic cells

2 Materials

2.1 Cell Culture and Production of Retroviral Particles

1. Phoenix cells (ATCC).
2. Medium for Phoenix cells:
   D-MEM supplemented with 10 % fetal bovine serum (FBS) and 0.1 mg/ml penicillin–streptomycin.
3. Hank's balanced salt solution (HBSS).
4. Opti-MEM medium.
5. 0.05 % trypsin–EDTA.
6. Tissue culture-treated T-25 and T-75 flasks.
7. Tissue culture-treated 96-, 12-, and 6-well plates.
8. G418.
9. Puromycin.
10. Blasticidin.
11. 0.45 μm syringe filters.
12. 10 ml syringes.
13. Fugene 6 transfection reagent (Promega).
14. Polybrene.
15. VSV-G plasmid (Addgene #8454).
16. Crystal violet staining solution (Crystal Violet 0.05 % w/v) formaldehyde (1 %) methanol (1 %) in PBS.

2.2 Molecular Biology Reagents

1. Phusion polymerase kit (NEB).
2. Gibson assembly mix (NEB).
3. Primers.
4. 10 mM dNTPs.
5. Mineral oil.
6. Ultrapure water (RNase/DNase-free).

7. Tris–EDTA (TE) buffer.
8. PCR purification kit.
9. Plasmid DNA purification kit.
10. Carbenicillin and kanamycin.
11. Luria broth (LB) media.
12. Standard DNA gel electrophoresis reagents and apparatus.

2.3 Chemical Genetic, Cell Staining, and Protein Reagents

1. 1NM-PP1 and 3 MB-PP1 (EMD Millipore).
2. Staralog inhibitors [2, 16].
3. Phospho-specific substrate antibody (if available).
4. Anti-phospho-Ser/Thr-Pro MPM-2 antibody (EMD Millipore).
5. Alexa fluor 488-labeled goat anti-mouse IgG (H+L) antibody (Life Technologies).
6. Propidium iodide.
7. Bovine serum albumin (BSA).
8. Standard SDS PAGE reagents and apparatus.

3 Method

3.1 Kinase Design and Engineering

The first step in engineering an analog sensitive (AS) kinase allele is to identify the gatekeeper residue. The gatekeeper residue has been annotated for over 7000 protein kinases in different model organisms in the Kinase Sequence Database (http://sequoia.ucsf.edu/ksd/). It can also be identified by aligning the sequence of your KOI with c-Src using Clustal Omega (http://www.clustal.org/omega/) [17]. The gatekeeper corresponds to T338 in the c-Src kinase and is commonly a leucine, methionine, phenylalanine or threonine in other kinases. The residues surrounding the gatekeeper may also help identify the correct position. Two hydrophobic amino acids are located at the -2 and -1 (I/L/M/F and V/F/L/I) positions and an acidic residue (E/D) can be found at the +1 position (Fig. 1c). Atypical kinases may require additional structural modeling to accurately locate the gatekeeper. Detailed guidelines for identifying the gatekeeper within atypical kinases have been provided elsewhere [2].

The next step is to replace the gatekeeper with a smaller amino acid. In most cases, glycine is the preferred substitute, as this offers maximal enlargement of the ATP-binding pocket and opportunity for allele-specific inhibition. However, for 15–30 % of kinases this alteration compromises activity and function unduly, in which case alanine can be used [2]. For example, an alanine substitution (M602A) was better tolerated than glycine for human Mps1 [18]. Some kinases may require a second-site suppressor to restore or improve the function of a gatekeeper-mutated kinase [2, 19]. For example, human Plk1 requires an additional mutation (C67V) to tolerate glycine in place of its normal gatekeeper (L130G) [20]. In cases where both glycine and alanine are poorly tolerated, another option is to engineer an electrophile-sensitive (ES) kinase, which involves replacing the gatekeeper with cysteine, a larger amino acid that is often well tolerated [21]. One limitation of the ES approach is that kinase inhibition is irreversible, precluding experiments that depend on inhibitor washout to reactivate the kinase. Finally, some gatekeeper mutants with reduced activity have also been rescued chemically using the bulky purine analog N6-benzylaminopurine (6-BAP). Addition of 6-BAP to cells expressing the Cdk2 (F80G) AS kinase was shown to restore both cyclin A binding and preserve kinase activity in culture [22].

The gatekeeper mutation(s) can be introduced into the KOI using conventional site-directed mutagenesis [23]. This approach can be more challenging to perform in large mammalian expression vectors, especially plasmids containing long terminal repeats (LTRs) for viral delivery. Here we describe an approach to assemble the desired mutations into a final expression vector by combining conventional overlapping PCR with Gibson assembly. PCR fragments containing the desired gatekeeper mutations are assembled into the final linearized vector using overlapping homology. The example below describes how to generate the gatekeeper mutation and introduce the mutated KOI into a suitable expression vector in one step.

3.2 Gatekeeper Mutagenesis and Cloning via Gibson Assembly

Gibson assembly allows 5 or more fragments to be joined in one step. The approach is flexible and seamless, avoiding the need to introduce new restriction enzyme sites or site-specific recombination sequences. Briefly, the KOI is first amplified in two overlapping N and C terminal fragments using primers that contain ~20 bp of overlap with their neighbors, as well as the desired gatekeeper mutation (Fig. 3a). In the Gibson assembly reaction, three different enzymes (a 5′ exonuclease, a DNA polymerase, and a DNA ligase) serve to cross-anneal, fill in, and ligate the fragments and vector, thus creating the final expression construct (Fig. 3b). A sample protocol is provided below.

1. Using primers that meet the specifications above, amplify the two KOI fragments using a high fidelity PCR enzyme (e.g., Phusion or PFU polymerase with an error rate $<2.0 \times 10^{-6}$).

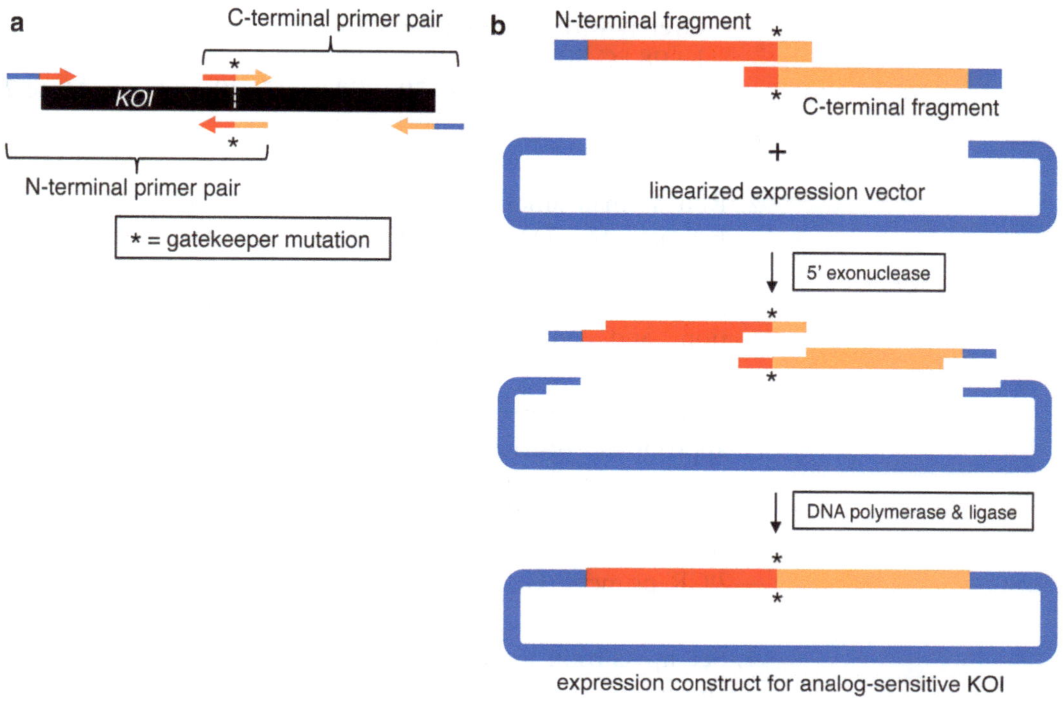

Fig. 3 Construction of analog-sensitive kinase alleles via Gibson assembly. (**a**) Overlapping N and C terminal kinase fragments are generated using PCR. The tails of the forward N-terminal primer and the reverse C-terminal primer need ~20 bp of overlap homology with the expression vector. The N-terminal reverse primer and the C terminal forward primer also overlap and incorporate the desired gatekeeper mutation (*asterisks*). (**b**) The kinase fragments are combined with the linearized expression plasmid in the Gibson assembly reaction mix, which contains three enzymatic activities: (1) an exonuclease that generates single-stranded 3′ overhangs, allowing for strand annealing between the fragments; (2) a DNA polymerase that fills in any remaining gaps; and (3) a DNA ligase that seals nicks to create covalently closed circular DNA, generating the final expression construct

The wild-type kinase can be amplified in a single fragment. The following conditions are suitable for amplification with Phusion polymerase.

	1× reaction (μl)
5× Phusion buffer	4
10 mM dNTPs	0.4
PCR water	12.4
Phusion polymerase	0.2
10 μM F primer	1.0
10 μM R primer	1.0
Template cDNA (10 ng)	1.0
total	**20 μl**

Overlay with mineral oil and cycle:
1× 98 °C 30 s
30×(98 °C 10 s/55 °C 30 s/72 °C 15–30 s/kb) (*see* **Note 2**)
1×(72 °C 20 s)
4 °C hold

2. Run a small amount (~2 μl) on a 1.0 % TAE-agarose gel. Estimate the DNA concentration from band intensity (*see* **Note 3**). If one major product is obtained, use the crude PCR product directly for cloning; otherwise, purify by gel extraction and quantify by UV spectrophotometry or gel electrophoresis.
3. Linearize the expression plasmid at the insertion site using an appropriate restriction enzyme (linearized plasmid should appear as a single distinct band on an agarose gel). The sequences on each side of the unique cleavage site are the regions of overlap (~20 bp) to included into the appropriate PCR primers. Gel purify the linear plasmid and quantify as above.
4. Set up Gibson assembly reactions, together with a negative (no-insert) control (*see* **Note 4**):

	1× reaction (μl)
Linearized vector (0.02 pmol)	1.0
N-terminal kinase fragment (0.06 pmol)	1.0
C-terminal kinase fragment (0.06 pmol)	1.0
2× Gibson assembly mix	3.0
total	**6 μl**

5. Overlay with mineral oil and incubate in thermocycler at 50 °C for 15–60 min.
6. Transform 1–2 μl of the Gibson assembly reaction into DH10B or an alternate *E. coli* cloning strain.

3.3 Replacing the Endogenous Kinase with the Analog-Sensitive Allele

As analog-sensitive kinases are generally recessive, their usage requires the endogenous kinase pool to be completely and permanently removed from the cell. This objective can be accomplished using Adeno-associated virus (AAV)-mediated gene targeting [18, 20, 24], stable RNAi [25], or RNA-guided nucleases [26, 27]. An alternative approach is to "knock in" the gatekeeper substitution at the KOI gene locus, thus maintaining endogenous expression control [15, 22]. A limitation of this approach is that for many kinases it is difficult to predict a priori which gatekeeper substitution will work best (i.e., maintain activity and support cellular function in the absence of analogs, yet be fully and effectively inhibited in their presence). Furthermore, this approach becomes infeasible if second-site suppressor mutations in a distant exon are also necessary. While in vitro kinase assays can provide some guidance on these points, they may not replicate the physiologic situation accurately, due to the use of artificial substrates and reaction conditions, as well as omission of kinase scaffolds, feedback controls, and subcellular localization. Indeed, analog-sensitive versions of human Plk1 [20] and yeast Cdc28/Cdk1 [3] have reduced biochemical activity as measured in vitro but fully reconstitute kinase function (including the quantitative phosphorylation of substrates) in vivo. For these reasons, we prefer to introduce analog-sensitive kinases as transgenes, allowing for rapid in vivo testing of different gatekeeper (and if needed, second-site suppressor) mutations. Depending on the nature of the expression system, constructs can be delivered through stable transfection or virus-mediated transduction. Below we provide a protocol for generating pantropic retroviral particles that can transduce in all cell types.

1. For small-scale retroviral production, transfect a T-25 flask of Phoenix cells. These HEK293 derivatives express the gag and pol genes from Moloney murine leukemia virus (MoMuLV) necessary for virion assembly and replication. Grow cells (in DMEM + 10 % FBS + pen/strep) to approximately 50–60 % confluence.
2. Dilute 3 μg of plasmid DNA (2 μg of retroviral construct + 1 μg of the pVSV-G pantropic envelope vector) into 400 μl of Opti-MEM.
3. Add 9 μl of FuGene 6 to diluted DNA and mix gently.
4. Incubate DNA/FuGene 6 mixture at room temperature for 20 min.
5. Replace medium on cells with 4–5 ml fresh media.
6. Add transfection mix onto cells dropwise. Gently tilt flask to distribute complexes onto the entire cell monolayer.
7. Incubate cells overnight in tissue culture incubator. The Phoenix cells must be handled with BL2 safety precautions.
8. Discard old medium 24 h after transfection and replace with fresh media.
9. Seed target cells into multiple wells in a 12-well plate at a confluency of ~10–15 %. The following day, when the cells reach ~ 20–30 % confluence, they will be ready for retroviral infection.
10. Collect retrovirus-containing supernatant 48 h after transfection into a 15 ml conical tube. Spin out cell debris at 1000×*g* for 5 min. and/or filter through 0.45 μm syringe filter.
11. Replace media on target cells with 1 ml of fresh antibiotic free media and add 1 ml of filtered retrovirus-containing supernatant.

12. Add polybrene to final concentration of 4 μg/ml during transduction (*see* **Note 9**).
13. Incubate for 8 h to overnight.
14. Replace media with fresh complete medium and allow cells to recover after infection and expand into T-25 flask when confluent (~1–3 days).
15. Split cells into a T-75 flask and begin selecting for transduced cells once cells have attached (at least 24 h later). Adding selection sooner may delay their recovery from the viral infection.
16. When using expression constructs encoding GFP, you should be able to detect expression by fluorescence microscopy 1–2 days after infection.
17. Evaluating transgene expression by Western blotting, immunofluorescence microscopy in the bulk population. It is also a good idea to sort GFP-positive cells by flow cytometry or isolate a clonal population with appropriate transgene expression.

3.4 Screening Analog–Allele Pairs for Optimal Specificity and Penetrance

Once the endogenous kinase has been replaced with an analog-sensitive version, the next step is to identify the optimal analog (that is, the compound that best discriminates the analog-sensitive and wild-type kinase alleles). A number of suitable analogs based on the PP1 scaffold (e.g., 1-NA-PP1, 1-NM-PP1, and 3-MB-PP1) have been used successfully. In addition, a new series of Staralog inhibitors (based on staurosporine as a general kinase-inhibiting scaffold) have been developed [16]. Thus there are multiple options for researchers seeking to identify the most penetrant inhibitor among these small molecules. It is possible to in vitro screen a panel of inhibitors using purified wild-type and analog-sensitive kinases with peptide or protein substrates and gamma-^{32}P-ATP. However, the gold standard for screening a panel of inhibitors is to measure kinase inhibition in cells by immunoblotting with a phospho-specific substrate antibody and/or assessing a biologically relevant phenotype. Ideally, the bulky purine analogs will possess greater than tenfold selectivity for the AS kinase. However, in many instances a phospho-specific substrate antibody is not available. In such cases, phenotypic screening provides a facile and physiologically relevant means of identifying the most effective allele–inhibitor pair.

3.5 Assessing Susceptibility to Growth Inhibition Using Crystal Violet Staining

To assess whether the AS cell line is susceptible to growth inhibition by bulky purine analogs, we suggest that researchers begin with a simple growth assay measured by crystal violet stain to assess whether bulky purine analogs can fully abrogate the KOIs function in vivo. This simple assay is extremely easy and informative and can be applied to screen multiple allele and drug combinations for

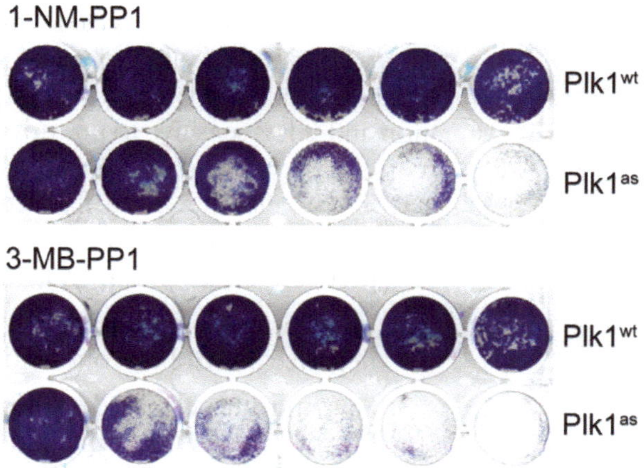

Fig. 4 Visualizing analog-dependent and allele-specific growth inhibition using crystal violet staining. Plk1^{wt} or Plk1^{as} cells were cultured in the presence of 1-NM-PP1 or 3-MB-PP1 (*left* to *right*: 0 μM, 0.078 μM, 0.313 μM, 1.25 μM, 5.0 μM, and 20.0 μM) for 8 days. Adapted from Ref. [20]

dose-dependent growth inhibition. We have used this assay previously to demonstrate the specificity of 3-MB-PP1 and 1-NM-PP1 towards Plk1^{as} and Mps1^{as} cells [18, 20] (Fig. 4).

1. Plate ~2000 cells/well in a 24-well plate. A good starting point is to test serial fourfold dilutions of the bulky purine analogs 1NM-PP1 and 3-MB-PP1 across each row (0 μM, 0.078 μM, 0.3125 μM, 1.25 μM, 5.0 μM, 20.0 μM). At the highest concentration (~20 μM), many cell lines such as HCT116 and RPEs will display mild growth inhibition (*see* **Note 10**).
2. After 6–10 days remove growth media (do not wash cells) and stain cells with 1 ml of crystal violet staining solution for 20 min.
3. Remove staining solution and save.
4. Wash each plate individually by dipping into 5 L beaker of tap water.
5. Air-dry each plate upside down over paper towel.

3.6 Rapid Assessment of M Phase Arrest or SAC Override by Flow Cytometry

Mitotic kinases are generally required to either satisfy or activate the spindle assembly checkpoint (SAC), which blocks anaphase and mitotic exit if kinetochore–microtubule attachments are absent or incorrect. Consequently, when starting to work with a new chemical-genetic system is useful to know whether kinase inhibition results in prolonged mitotic arrest, or alternatively overrides mitotic arrest induced by microtubule poisons or kinesin-specific inhibitors. This can be determined by flow cytometry on fixed cells stained with a mitosis-specific marker (see below) as well as time-lapse microscopy. The former approach is technically easier and scales readily to a large number of samples, while the latter approach provides superior temporal resolution and is less sensitive to perturbations in the rate of mitotic entry.

1. Grow cells in a T-75 flask to approximately 60–70 % confluency. Treat cells with the bulky purine analog of choice in the presence or absence of nocodazole (16 μM) for 8–12 h.
2. In addition to your samples, keep an extra flask to serve as a negative control (second antibody only staining).
3. Save mitotic cells by decanting the tissue culture medium into 50 ml tubes. Wash cells in 10 ml of PBS and collect into the same 50 ml tubes.
4. Remove adherent cells with 2 ml of trypsin and collect the cells by centrifugation at 900 × *g* for 5 min. Combine cells from **steps 2** and **3**.
5. Resuspend the cell pellets with 900 μl PBS. Add 2.0 ml cold (−20 °C) ethanol to cell suspension in a dropwise manner while gently vortexing. Fix cells overnight at −20 °C. Fixed cells can be stored at −20 °C for up to 2 months.
6. Pellet cells at 900 × *g* for 5 min. Remove the ethanol supernatant carefully and resuspend the pellet in 500 μl in PBS + 0.2 % Triton X-100. Transfer cells into a 1.5 ml microfuge tube and collect by centrifugation at 900 × *g* for 5 min.
7. Resuspend the cell pellet with 500 μl blocking buffer (PBS +0.2 % Triton X-100 + 3 % BSA) for 10 min. Centrifuge at 900 × *g* for 5 min. and remove the supernatant.
8. Resuspend cells with 500 μl blocking buffer + 1.0 μg anti-phospho-MPM-2 antibody. Incubate for 1 h at room temperature. Mix gently every ~10 min.
9. Centrifuge at 900 × *g* for 5 min. Remove the supernatant. Wash 3× in 500 μl blocking buffer.
10. Resuspend pellet in 500 μl blocking buffer + 1.0 μg Alexa fluor 488 labeled goat anti-mouse IgG (H+L) antibody. Incubate for 1 h at room temperature in the dark. Mix gently every ~ 10 min.
11. Centrifuge at 900 × *g* for 5 min. Remove the supernatant. Wash 3× in 500 μl blocking buffer.
12. Resuspend in 500 μl PBS + 2 mg RNase A + 200 ng propidium iodide (PI). Incubate at 37 °C for 1 h.
13. Analyze cells by FACS. Be sure to use appropriate doublet discrimination settings. Plot Alexa 488 (FL1-H) on a log scale across the *y*-axis versus PI on a linear scale on the *x*-axis. Mitotic cells will have a 4 N DNA and should stain an order of magnitude higher for Alexa 488.

4 Notes

1. Fragments can also be ordered as synthetic gene blocks, avoiding the need to locate a full-length and mutation-free cDNA clone as a PCR template.
2. For primers having an annealing temperature ≥ 72 °C (as determined for their template-specific 3′ ends), a two-step thermocycling protocol is recommended.
3. NanoDrop readings are unreliable for crude PCR products because of contaminating primers and dNTPs.
4. Calculations for Gibson assembly:

   Vector 0.02 pmol = 13.2 ng × size kb

   N-terminal insert 0.06 pmol = 39.6 ng × size kb

   C-terminal insert 0.06 pmol = 39.6 ng × size kb

5. Choosing an appropriate selection marker is relatively straightforward as it depends on the sensitivity of your chosen cell line. For human telomerase reverse transcriptase (hTERT)–immortalized retinal pigment epithelial (RPE) cells we recommend using G418 (500 μg/ml) selection. Puromycin (10–15 μg/ml) and blasticidin (5–10 μg/ml) can be also be used but are less effective. HCT116 cells sensitive to G418 (500 μg/ml), blasticidin (5–10 μg/ml), and puromycin (2 μg/ml).
6. Optimizing transgene expression can be achieved through promoter selection. Promoters such as CMV and SV40 are capable of high levels of expression and should be chosen for highly abundant kinases. Low-abundance kinases can be expressed using the EF1α promoter. We also recommend screening clonal populations by western blotting for appropriate transgene expression.
7. Although affinity and large localization tags such as GFP and mCherry can in some case cause problems with protein function, we recommend incorporating tags into your transgene. We generally include both an affinity and localization tag, as they can be extremely useful when studying the different properties of the active and inactive kinase.
8. In many cell lines (including HCT116), a large fraction of retroviral integrations that are initially transcriptional active become silenced over time, unless selective pressure is applied. To counteract the silencing of retroviral transgenes we recommend using plasmid where the transgenes are linked to the selection marker (e.g., an ires-neo or ires-puro plasmid such as pQCXIN and pQCXIP).

9. Polybrene is a positively charged polymer that reduces charge repulsion between the virus and the cellular membrane. Polybrene to a final concentration of 2–10 μg/ml is suitable for most human cell lines; however, some cell types are sensitive to Polybrene. The optimum concentration may need to be determined empirically. Exposure to Polybrene (>24 h) will increase its toxicity to cells.
10. The concentration range is a starting point. We advise that follow-up experiments be performed at sample concentrations closer to the effective dose identified in this range. The assay should be stopped once the untreated wells have reached confluence.

References

1. Manning G, Whyte DB, Martinez R, Hunter T, Sudarsanam S (2002) The protein kinase complement of the human genome. Science 298:1912
2. Lopez MS, Kliegman JI, Shokat KM (2014) The logic and design of analog-sensitive kinases and their small molecule inhibitors. Methods Enzymol 548:189
3. Bishop A et al (2000) Unnatural ligands for engineered proteins: new tools for chemical genetics. Annu Rev Biophys Biomol Struct 29:577
4. Liu Y et al (1999) Structural basis for selective inhibition of Src family kinases by PP1. Chem Biol 6:671
5. Bishop AC et al (1998) Design of allele-specific inhibitors to probe protein kinase signaling. Curr Biol 8:257
6. Zhang C et al (2013) Structure-guided inhibitor design expands the scope of analog-sensitive kinase technology. ACS Chem Biol 8:1931
7. Knight ZA, Shokat KM (2007) Chemical genetics: where genetics and pharmacology meet. Cell 128:425
8. Lera RF, Burkard ME (2012) The final link: tapping the power of chemical genetics to connect the molecular and biologic functions of mitotic protein kinases. Molecules 17:12172
9. Burkard ME, Santamaria A, Jallepalli PV (2012) Enabling and disabling polo-like kinase 1 inhibition through chemical genetics. ACS Chem Biol 7:978
10. Gorre ME et al (2001) Clinical resistance to STI-571 cancer therapy caused by BCR-ABL gene mutation or amplification. Science 293:876
11. Wong S et al (2004) Sole BCR-ABL inhibition is insufficient to eliminate all myeloproliferative disorder cell populations. Proc Natl Acad Sci U S A 101:17456
12. Burkard ME, Jallepalli PV (2010) Validating cancer drug targets through chemical genetics. Biochim Biophys Acta 1806:251
13. Soskis MJ et al (2012) A chemical genetic approach reveals distinct EphB signaling mechanisms during brain development. Nat Neurosci 15:1645
14. Miller AL, Zhang C, Shokat KM, Lowell CA (2009) Generation of a novel system for studying spleen tyrosine kinase function in macrophages and B cells. J Immunol 182:988
15. Larochelle S et al (2007) Requirements for Cdk7 in the assembly of Cdk1/cyclin B and activation of Cdk2 revealed by chemical genetics in human cells. Mol Cell 25:839
16. Lopez MS et al (2013) Staurosporine-derived inhibitors broaden the scope of analog-sensitive kinase technology. J Am Chem Soc 135:18153
17. Sievers F et al (2011) Fast, scalable generation of high-quality protein multiple sequence alignments using Clustal Omega. Mol Syst Biol 7:539
18. Maciejowski J et al (2010) Mps1 directs the assembly of Cdc20 inhibitory complexes during interphase and mitosis to control M phase timing and spindle checkpoint signaling. J Cell Biol 190:89
19. Zhang C et al (2005) A second-site suppressor strategy for chemical genetic analysis of diverse protein kinases. Nat Methods 2:435
20. Burkard ME et al (2007) Chemical genetics reveals the requirement for Polo-like kinase 1

activity in positioning RhoA and triggering cytokinesis in human cells. Proc Natl Acad Sci U S A 104:4383

21. Garske AL, Peters U, Cortesi AT, Perez JL, Shokat KM (2011) Chemical genetic strategy for targeting protein kinases based on covalent complementarity. Proc Natl Acad Sci U S A 108:15046
22. Merrick KA et al (2011) Switching Cdk2 on or off with small molecules to reveal requirements in human cell proliferation. Mol Cell 42:624
23. Ho SN, Hunt HD, Horton RM, Pullen JK, Pease LR (1989) Site-directed mutagenesis by overlap extension using the polymerase chain reaction. Gene 77:51
24. Berdougo E, Terret ME, Jallepalli PV (2009) Functional dissection of mitotic regulators through gene targeting in human somatic cells. Methods Mol Biol 545:21
25. Sliedrecht T, Zhang C, Shokat KM, Kops GJ (2010) Chemical genetic inhibition of Mps1 in stable human cell lines reveals novel aspects of Mps1 function in mitosis. PLoS One 5, e10251
26. Gaj T, Gersbach CA, Barbas CF 3rd (2013) ZFN, TALEN, and CRISPR/Cas-based methods for genome engineering. Trends Biotechnol 31:397
27. Cong L, Zhang F (2015) Genome engineering using CRISPR-Cas9 system. Methods Mol Biol 1239:197

Part VII

The Mitotic Spindle and Cancer

Chapter 23

Using Cell Culture Models of Centrosome Amplification to Study Centrosome Clustering in Cancer

Mijung Kwon

Abstract

The link between centrosome amplification and cancer has been recognized for more than a century, raising many key questions about the biology of both normal and cancer cells. In particular, the presence of extra centrosomes imposes a great challenge to a dividing cell by increasing the likelihood of catastrophic multipolar divisions. Only recently have we begun to understand how cancer cells successfully divide by clustering their extra centrosomes for bipolar division. Several hurdles to dissecting centrosome clustering include limitations in the methodologies used to quantify centrosome amplification, and the lack of appropriate cell culture models. Here, we describe how to accurately assess centrosome number and create isogenic cell lines with or without centrosome amplification. We then describe how imaging of cell division in these cell culture models leads to identification of the molecular machinery uniquely required for cells with extra centrosomes. These approaches have led to the identification of molecular targets for selective cancer therapeutics that can kill cancer cells with extra centrosomes without affecting normal cells with two centrosomes. We further anticipate that the approaches described here will be broadly applicable for studying the causes and consequences of centrosome amplification in a variety of contexts across cancer pathophysiology, such as cell migration and metastasis.

Key words Centrosome, Mitosis, Cancer, Multipolar spindle, Cell division, Extra centrosomes, Aneuploidy, Selective cancer therapeutics, Centrosome amplification

1 Introduction

The centrosome is an organelle that nucleates microtubules (MTs) and functions as a microtubule-organizing center (MTOC) within the cell. Centrosome number is tightly controlled, such that centriole duplication is limited to once per cell cycle and occurs in a semiconservative manner similar to DNA replication [1, 2]. Thus, normal cells enter mitosis with two centrosomes, ensuring the formation of bipolar spindle with one centrosome at each spindle pole (Fig. 1a). However, centrosome amplification is a hallmark of cancer, both in cultured cancer cells and in clinical specimens [3, 4]. In the late nineteenth century, Hansemann, Galeotti, and Boveri proposed that multipolar spindles are a signature

Paul Chang and Ryoma Ohi (eds.), *The Mitotic Spindle: Methods and Protocols*, Methods in Molecular Biology, vol. 1413,
DOI 10.1007/978-1-4939-3542-0_23, © Springer Science+Business Media New York 2016

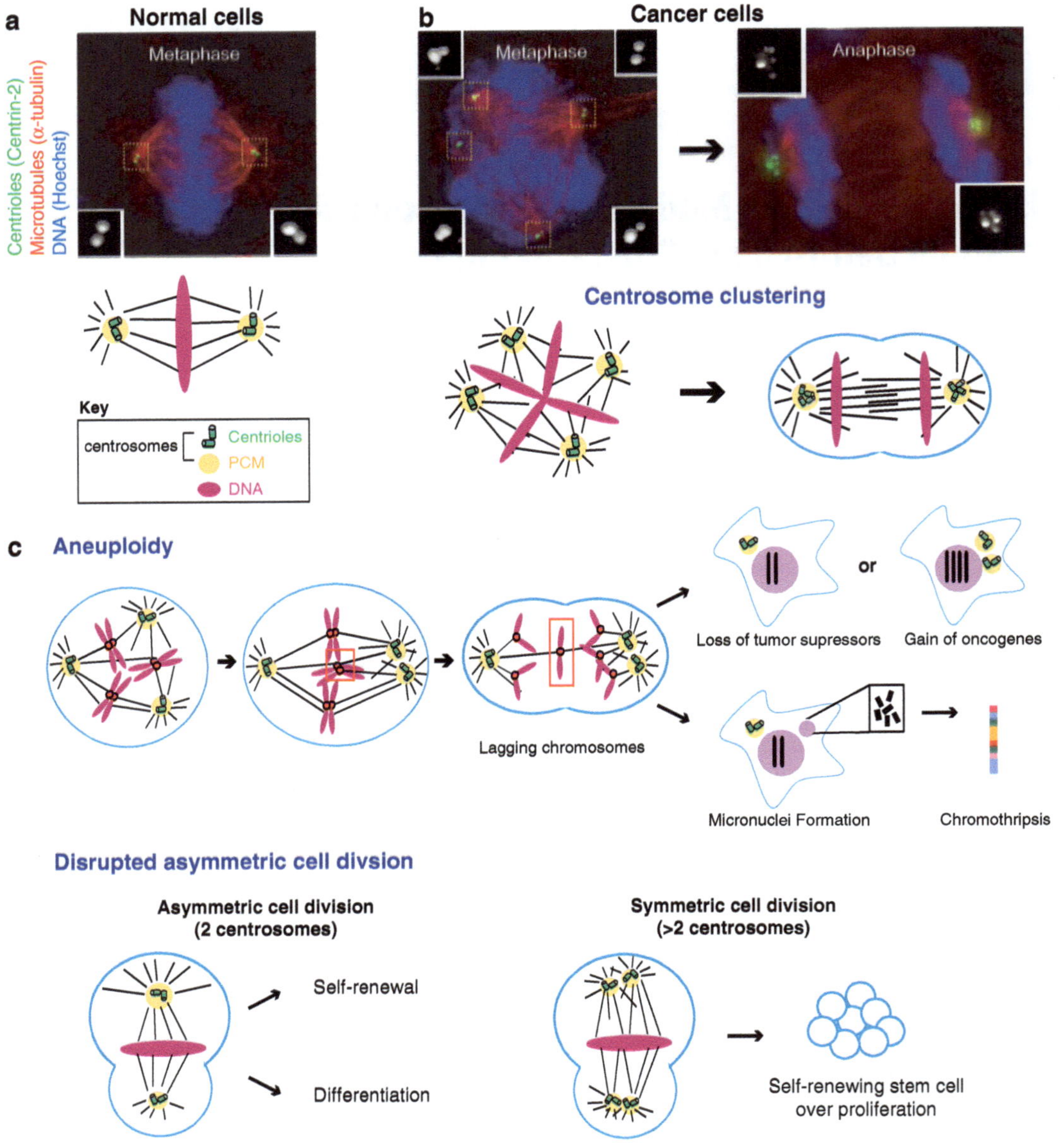

Fig. 1 Cancer cells harboring extra centrosomes successfully divide in a bipolar fashion by clustering extra pairs of centrioles. **(a)** The centrosome consists of an orthogonally oriented pair of centrioles (*green*) that are surrounded by pericentriolar material (PCM, *yellow*). In normal somatic cells, centrosomes duplicate once per cell cycle, and the two centrosomes nucleate MTs to form a bipolar spindle that equally segregates genetic material into two daughter cells during mitosis. Insets show high magnification images of centrioles labeled with anti-centrin 2 antibody in boxed regions. **(b)** Cancer cells with extra centrosomes form pseudo-bipolar spindles by clustering their extra centrosomes (one pole contains clustered multiple pairs of centrioles). Centrosome clustering leads to the conversion of multipolar metaphase spindles into bipolar anaphase spindles. **(c)** Centrosome clustering can drive CIN and defective asymmetric cell division. *Top*: Centrosome clustering promotes bipolar division through a transient multipolar spindle intermediate, generating a high incidence of abnormal “merotelic” chromosome attachments (*box*) that are not detected by the spindle assembly checkpoint [13, 14]. Subsequent lagging chromosomes leading to whole-chromosome mis-segregation can in turn contribute to mutator phenotype by loss of tumor suppressors or gain of oncogenes. Lagging

of pathological mitosis in cancers that could contribute to tumorigenesis [5–7]. However, this proposal is paradoxical. How do cancer cells evade the devastating potential of multipolar divisions that can lead to inviable progenies with gross chromosome instability, and continue to proliferate? Several lines of evidence accumulated over recent decades from a variety of cancer cells revealed that cancer cells employ mechanisms to cope with supernumerary centrosomes. While it is plausible for cancer cells to remove or inactivate extra centrosomes to maintain two functional MTOCs for spindle bipolarity, the most prevalent mechanism in cancer cells is the clustering of extra centrosomes [4, 8, 9]. Thus, despite the presence of extra centrosomes, cancer cells are able to divide successfully by coalescing their centrosomes into two groups to form a functional bipolar spindle during mitosis [8, 10–12] (Fig. 1b).

Mitotic centrosome clustering is not without consequences, and it can contribute to tumorigenesis in at least two ways (Fig. 1c). First, although clustering of extra centrosomes in cancer cells prevents multipolar anaphases that would result in massive chromosome mis-segregation and compromised cell viability, it promotes a low level of chromosomal instability (CIN) through abnormal chromosome attachments [13, 14]. Thus, centrosome clustering allows these cancer cells to proliferate with ongoing genetic instability. Second, centrosome clustering can impair the proper control of asymmetric cell division, a process that determines cells to self-renew or differentiate. Clustering of extra centrosomes disrupts asymmetric behaviors of centrosomes and asymmetric partitioning of cell fate determinants into each daughter cell [15]. Subsequent symmetric division leads to continued expansion of stem cell pools that culminate in tumorous growth. The extent to which CIN contributes to the initiation and progression of tumorigenesis remains unclear, as is the role of extra centrosomes as a driver of human cancers [15, 16]. Nevertheless, it has become evident that centrosome amplification is not merely a by-product of tumorigenesis, but rather may directly contribute to tumorigenesis [15, 17].

The development of a number of methods that accurately assess centrosome amplification and visualize cell division in cancer

Fig. 1 (continued) chromosomes can also be encapsulated into micronuclei within which defective DNA replication and subsequent DNA damage (boxed region) lead to chromothripsis, clustered rearrangements of chromosomes observed in multiple human cancers [39]. *Bottom*: In asymmetrically dividing tissues, such as in fly larval brain, neuroblasts undergo asymmetric division that segregates different cell fate determinants into two daughter cells. This process relies on asymmetric maturation of centrosomes and interaction of astral MTs with the cell cortex, thus determining the pools of dividing stem cells and differentiated neurons that are critical for the maintenance of tissue homeostasis (*left*). Clustering of extra centrosomes impairs asymmetric cell division, and the resulting symmetric division leads to the expansion of dividing stem cell population and tumorous outgrowth (*right*) [15] cells with extra centrosomes has been essential to dissecting centrosome clustering and its impact on tumorigenesis. In numerous past studies, the extent of centrosome amplification and the frequency of centrosome clustering in cancers may have been greatly underestimated. This is due to the limitations of a widely used method that utilizes pericentriolar material (PCM) markers to detect centrosomes in tissue samples and cell lines derived from human cancer patients. Immunostaining of specimens with PCM markers such as γ-tubulin and pericentrin can underestimate supernumerary centrosomes if they are tightly clustered within a PCM mass. Moreover, this method cannot distinguish whether multipolar spindles arise from de-clustering of amplified centrosomes, or from general defects in the maintenance of centrosome integrity, for instance due to fragmentation of PCM or centriole splitting (Fig. 2). Electron microscopy and modern super-resolution microscopy can provide a detailed view of centrosomal defects associated with cancers at the ultra-structural level [18, 19]. However, full characterization of the extent of centrosome amplification and mitotic consequences in cancer specimens would require three-dimensional image acquisition with high temporal and spatial resolution, which is technically difficult and impractical.

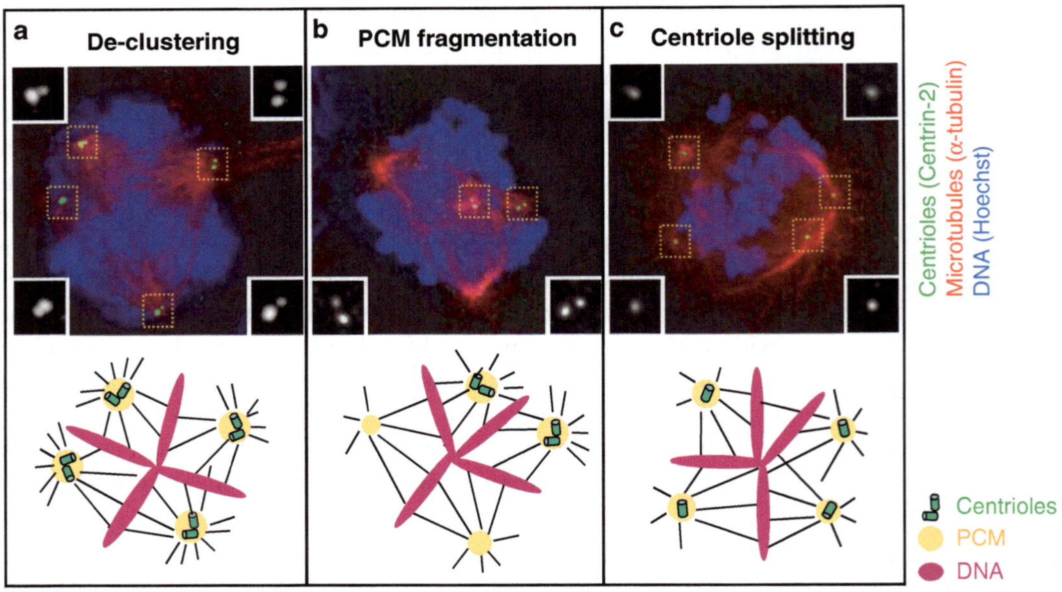

Fig. 2 Multipolar spindles generated by different mechanisms in cells with or without extra centrosomes. Multipolar spindles can be generated by **(a)** De-clustered supernumerary centrosomes with each pole containing one or more pairs of centrioles while maintaining centrosome integrity or, **(b)** PCM fragmentation where some poles contain PCM but lack centrioles, or **(c)** Premature centriole disengagements characterized by split centrioles that each form a separate pole. Notably, PCM fragmentation and centriole splitting can produce spindle multipolarity even in normal cells with two centrosomes. Inducing multipolarity by disrupting centrosome clustering (without affecting centrosome integrity) can form a basis for the selective strategy targeting cancer cells with extra centrosomes

To overcome this, live and fixed imaging of dividing cancer cells labeled with centriolar makers and chromosomes has proven useful. Over the past decade, an approach combining functional perturbation with microscopy-based phenotypic analysis led to the identification of several proteins and forces that play key roles in supernumerary centrosome clustering. Starting with the seminal discovery of the MT motor dynein, and its partner MT-associated protein (MAP) NuMA [10], recent genome-wide RNAi screens in *Drosophila* S2 cells and human cancer cells uncovered key molecular mechanisms that enable bipolar divisions in cancer cells with extra centrosomes [12, 20, 21] (Fig. 3). These pathways depend on a variety of mitotic and cytoskeletal machinery, including the spindle assembly checkpoint (SAC) that controls mitotic timing [12, 22], MAPs and MT motors that coalesce MT minus-ends together [10, 12], actin-dependent forces that anchor the spindle to the cell cortex [12], and proteins that promote tension between kinetochores and spindle MTs [20]. Thus, it is now evident that cancer cells exploit the mechanisms utilized in normal cells in order to achieve bipolar division in the presence of extra centrosomes.

Selective targeting of cancer cells with extra centrosomes without detrimental effects on normal cells with two centrosomes is of great interest, as it is expected to avoid the general cytotoxicity that has been the main hurdle to antimitotic drug therapy. Experimental approaches for targeted inhibition of centrosome clustering can identify mechanisms that are uniquely required for the division of cancer cells with supernumerary centrosomes, but are dispensable for the division of normal cells. Indeed, inhibition of centrosome clustering by siRNA-mediated knockdown of the MT motor, HSET/KIFC1 (Kinesin-14 family), and the integrin-linked kinase (ILK) selectively induces multipolar division and cell death in cancer cells harboring extra centrosomes, while sparing normal cells with two centrosomes [12, 23]. HSET is an especially appealing drug target because kinesins are druggable [24], and HSET is not essential for normal somatic cell division in vitro as well as in vivo [25, 26]. Recently, two small molecule HSET inhibitors, AZ82 and CW069, were developed [27, 28]. They have been shown to selectively induce cell death in cancer cells with extra centrosomes [27, 28] without any bone marrow toxicity, an indicator of side effects for highly proliferative cells [27], thus showing promise as new anticancer therapeutics in clinics. Further therapeutic utility of anti-centrosome clustering agents would greatly benefit from validation of their efficacy in isogenic cell culture models with or without centrosome amplification, along with identification of tumor types which exhibit high sensitivity to such compounds.

In this chapter, we describe methodologies to accurately assess centrosome amplification and mitotic centrosome clustering in both cancerous and non-transformed mammalian cells. We provide an appropriate choice of available cell lines and methods to generate

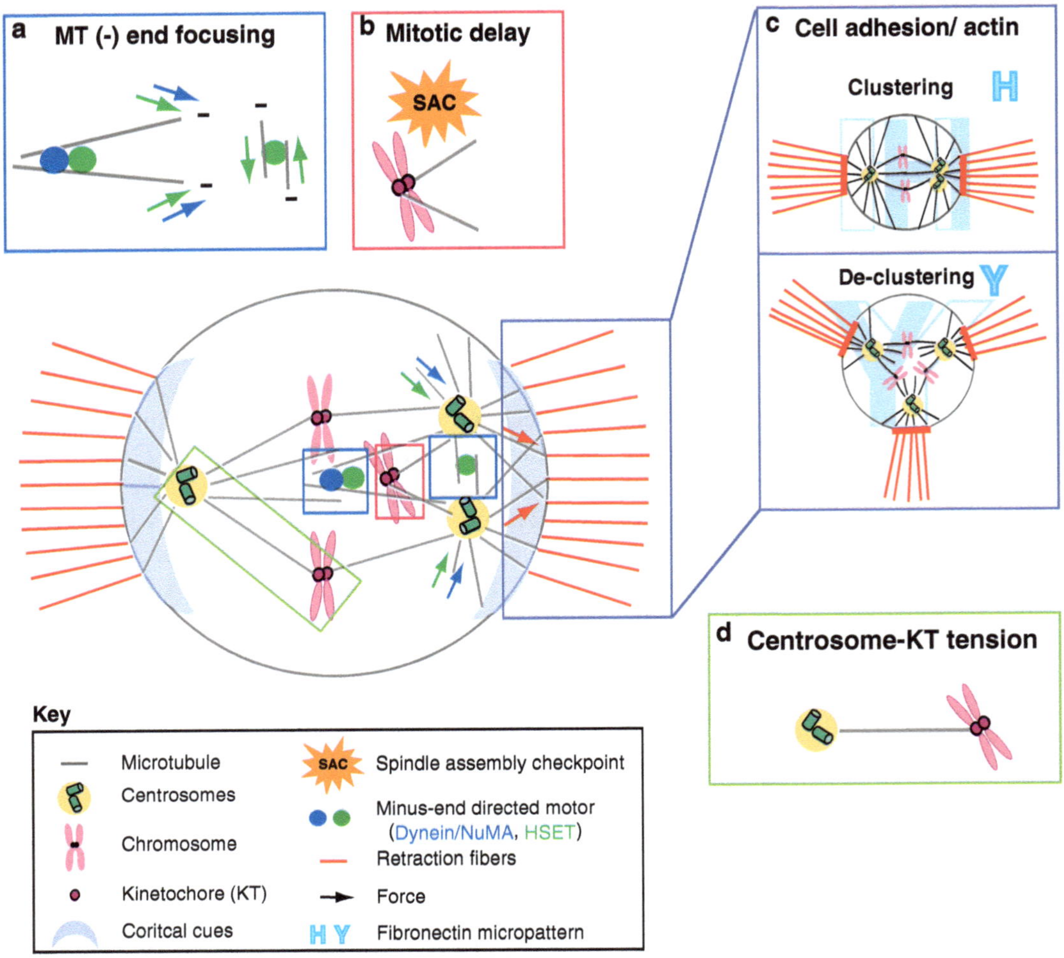

Fig. 3 Multiple pathways cooperate in promoting centrosome clustering. **(a)** MT motors, such as Dynein (with its partner NuMA) and HSET, can bundle MT minus ends together to focus poles (*blue* and *green arrows*) [10, 12]. **(b)** Spindle assembly checkpoint (SAC) senses abnormal kinetochore–MT attachments in cells with extra centrosomes and delays the onset of anaphase until the poles are clustered, thus preventing detrimental multipolar anaphases [12, 22]. **(c)** The actin cortex contributes to pulling forces on astral MTs (*red arrows*), and thus either promotes (H) or inhibits (Y) centrosome clustering depending on the distribution of cell matrix adhesion contacts [12]. Different shapes of fibronectin micropatterns can be used to manipulate interphase cell adhesion patterns. During mitosis, strong adhesion sites from interphase remain attached to form actin-rich structures called retraction fibers [32, 40]. The retraction fibers link rounded mitotic cell bodies to adhesive contacts and provide cortical cues for astral MTs to anchor. Subsequently, localized actin-dependent pulling forces which depends on the MT-binding myosin Myo10 position centrosomes, leading to either bipolar (on H pattern) or multipolar (on Y pattern) division [38]. **(d)** Proteins that function in generating or maintaining tension between kinetochores and centrosomes also promote the clustering of extra centrosomes [20]. Multiple mechanisms thus cooperate to ensure bipolar spindle formation prior to the beginning of anaphase

isogenic cell lines with or without centrosome amplification using two methods: by transient block of cytokinesis, or by overduplication of centrosomes via inducible expression of Plk4, a master regulator of centrosome duplication [1, 29]. Combining functional perturbation (via RNAi or chemical compounds) with fixed and live-cell fluorescence microscopy in these models provides the opportunity to uncover mechanisms that are uniquely required for bipolar division in cells containing extra centrosomes.

In addition to the direct applications in cancer biology outlined above, these approaches can shed light on the general principles of spindle mechanics. For example, female meiotic division in some species relies on the formation of bipolar spindle by focusing acentriolar poles [25]. Interestingly, the forces and molecules holding centrosomes together in cancer cells may have much in common with those that bundle the minus ends of MTs in acentriolar spindle poles during meiotic division [12, 30, 31] (Fig. 3a). Likewise, both normal and cancer cells depend on actin-dependent cortical forces that are regulated by cell adhesion in order to determine the position of centrosomes and cell division plane during mitosis [12, 32, 33, 38] (Fig. 3c). Thus, we also describe methods to study cell division in a controlled cellular microenvironment with the use of fibronectin-coated micropatterns. In particular, we highlight how cortical forces, regulated by cell adhesion patterns, position mitotic centrosomes during spindle orientation and centrosome clustering. The methodologies described in this chapter can therefore identify novel molecular players in both meiotic and mitotic divisions.

2 Materials

2.1 Available Cell Lines with or Without Centrosome Amplification

1. Cell lines: commercially available from American Type Culture Collection (ATCC).
2. Medium: Dulbecco's modified Eagle's medium (DMEM) (NHDF, NIH-3T3, DLD1, HeLa, MCF-7, U2OS, CF-PAC1, N1E-115), McCoy's 5A (HT-29, HCT-116, MDA-231), DMEM:F12 (hTERT-RPE-1), MEM (UPCI:SCC114) or RPMI (T47D, BT-549, NHO2A) containing 10 % fetal bovine serum (FBS), 100 IU/ml penicillin, and 100 μg/ml streptomycin.
3. Tissue culture dishes.
4. Incubator: Maintain cell lines at 37 °C with 5 % CO_2 atmosphere.

2.2 Generation of Isogenic Cell Lines with or Without Centrosome Amplification

1. Cell lines of choice: Cell lines with normal centrosome content [e.g., U2OS, DLD1, hTERT-RPE-1 (human retinal pigment epithelial cell line immortalized with telomerase, RPE-1 hereafter)].

2.2.1 Tetraploid Cells with Extra Centrosomes by Blocking Cytokinesis

1. Dihydrocytochalasin B (DCB, Sigma): 10 mM stock in DMSO, store at −20 °C. An inhibitor of actin polymerization to block cytokinesis.
2. SB203580 (Promega): 10 mM stock in DMSO, store at −20 °C. A selective inhibitor of p38 mitogen-activated protein kinase. In some cell lines, such as non-transformed or primary cells (e.g., RPE-1 cells), tetraploidization induced by cytokinesis failure leads to cell cycle arrest in G1 via activation of p53 and p38-dependent stress responses [34, 35]. Inhibitors of stress response kinase p38 can be used to bypass G1 arrest and allow tetraploid cells with extra centrosomes to enter mitosis.

2.2.2 Centrosome Overduplication by Inducible Plk4 Overexpression

1. Human embryonic kidney 293FT cell line (HEK-293 FT, Invitrogen) to produce viruses.
2. 6-Well tissue culture plates.
3. Tetracycline (Tet)-free FBS: Used to avoid leaky expression of genes under the Tet-inducible promoter that may be induced by Tetracycline present in normal serum.
4. DNA: pLenti-CMV-Tet repressor (R)-Blast (17492, Addgene) and pLenti- CMV/TO-Neo-Dest containing Tet-inducible Plk4 [17]. Lentiviral packaging plasmids [Gag-pol (psPAX2/packaging plasmid, Addgene) and VSV-G (pMD2.G/envelope plasmid, Addgene)].
5. Lipofectamine 2000 (Invitrogen), store at 4 °C.
6. Opti-MEM (Invitrogen), store at 4 °C.
7. Sterile 0.45-μm syringe filter (Corning).
8. 10 ml Luer-Lok™ tip syringe.
9. Polybrene infection/transfection reagent (Millipore): 10 mg/ml stock, store at −20 °C.
10. Blasticidin (Invitrogen): 10 mg/ml stock in water, store at −20 °C.
11. Geneticin/G418 solution (Invitrogen): 50 mg/ml stock, store at 4 °C.

2.3 Assessment of Centrosome Number and Centrosome Clustering by Immunofluorescence Microscopy

2.3.1 Preparations of Specimens for Immunofluorescence Microscopy

1. Glass coverslips: soaked and stored in 70 % ethanol (EtOH).
2. 6-Well tissue culture plates
3. Doxycycline (Sigma): 10 mg/ml stock, store at −20 °C. 2 μg/ml used to induce expression of Plk4 under the Tet-inducible promoter.
4. Phosphate-buffered saline (PBS).
5. ACS spectrophotometric grade methanol (MeOH, Sigma): Fixative used for optimal fixation for centrin immunostaining, store at −20 °C.
6. Extraction buffer: 0.25 % Triton X-100 in PBS.

7. Blocking solution: 5 % bovine serum albumin (BSA), 0.1 % Triton X-100 in PBS.
8. Washing buffer: 0.1 % Triton X-100 in PBS.
9. Antibodies: rabbit anti-centrin2 (sc-27793-R, Santa Cruz) for centriole immunostaining, mouse anti-α-tubulin (DM1a, T9026, Sigma) for MT immunostaining, rabbit anti-pericentrin (ab4448, Abcam) or anti-mouse γ-tubulin (GTU-88, T6557, Sigma) for PCM staining, and species-specific fluorescent secondary antibodies (1:1000, Molecular Probes).
10. Hoechst 33342 (Invitrogen): 10 mg/ml stock, store at 4 °C. 2 μg/ml (1:5000) to stain DNA.
11. ProLong Antifade mounting medium (Molecular Probes).

2.3.2 Imaging of Fixed Specimens to Evaluate Centrosome Number and Centrosome Clustering

1. Yokogawa CSU-X1 spinning disk confocal mounted on a Nikon Ti-E inverted microscope (Nikon Instruments) equipped with a cooled CCD camera (Orca ER, Hamamatsu).
2. MetaMorph Software (Molecular Devices) for image acquisition and analysis.
3. 405 nm, 488 nm, 561 nm and 642 nm lasers.

2.4 Live Cell Imaging of Centrosome Clustering

1. DNA: Lentiviral GFP-Centrin1 (pLentiLox3.7) to image centrioles or Lentiviral GFP-H2B (pLenti6/V5) to image chromosomes.
2. 6-Well tissue culture plates and glass-bottom 12-well tissue culture plates (Mattek).
3. Thymidine (Sigma): 100 mM stock in PBS, store at −20 °C. For synchronization of cells in G1/S phase of the cell cycle.
4. siRNAs: siRNAs for target genes (Dharmacon), non-targeting siRNAs as negative controls (Dharmacon), or/and siRNAs targeting HSET/KIFC1 (human, LQ-004958-00-0002, Dharmacon; mouse, 287750, 287751, and 287752, Ambion) as positive controls.
5. Lipofectamine RNAi Max transfection reagent (Invitrogen).
6. Opti-MEM (Invitrogen).
7. Medium without phenol red supplemented with 10 % FBS, 100 IU/ml penicillin and 100 μg/ml streptomycin.
8. TE2000-E2 inverted Nikon microscope (Nikon Instruments) equipped with a cooled CCD camera (Orca ER, Hamamatsu) and the Nikon Perfect Focus system, enclosed in a temperature- and CO_2-controlled chamber to maintain an atmosphere of 37 °C and 5 % CO_2.
9. NIS-Elements software (Nikon Instruments) for image acquisition and analysis.

2.5 Long-Term Live Cell Imaging on Fibronectin Micropatterns

1. Micropatterned coverslips: CYTOO chips coated with fibronectin (FN) (CYTOO), store at 4 °C. The shape of micropatterns can be chosen according to experimental needs. For example, L-shaped patterns are used to study spindle orientation in cells with two centrosomes, whereas Y-shaped patterns used for cells with extra centrosomes. The appropriate size of micropatterns can be chosen depending on cell size (small, medium and large sizes are available, see manufacturer's guideline). For example, use medium size for RPE-1 and U2OS cells. Alternatively, equivalent micropatterns from other manufacturers or in-house chips can be used.
2. 6-Well tissue culture plates or 35 mm tissue culture dishes.
3. Forceps.
4. CYTOO Chamber.
5. Medium without phenol red supplemented with 10 % FBS, 100 IU/ml penicillin and 100 μg/ml streptomycin.
6. Thymidine (Sigma): 100 mM stock in PBS, store at −20 °C.
7. Microscope components and software, as described in Subheading 2.4 (**items 8** and **9**).
8. Origin Software (OriginLab or equivalent software from other sources): For graphical presentation of angular distributions of spindle orientation.

3 Methods

3.1 Available Cell Lines with or Without Centrosome Amplification

Over the last three decades, increasing numbers of mammalian cancer cell lines from different tissue origins have been characterized for their centrosome content and centrosome clustering efficiency. Figure 4 has a short list of useful mammalian cell culture models to study centrosome clustering, as well as causes and consequences of centrosome amplification. These cell lines are readily available and some are routinely used in laboratories. The percentage of cells with extra centrosomes varies between cell lines. Most cell types listed here tend to cluster extra centrosomes efficiently (with the exception of HeLa cells), and are suitable to study centrosome clustering. For example, nearly 100 % of mouse neuroblastoma N1E-115 cells contain extra centrosomes that can be efficiently clustered to form bipolar spindles [8, 12]. In some cancer cell lines, such as the human breast cancer cell lines MDA-231 and BT-549, about half of all cells contain extra centrosomes [12, 31], providing a facile system to study the consequences of spindle multipolarity induced by defects in centrosome clustering and/or integrity. In contrast, the vast majority (~>90 %) of non-transformed cells, such as RPE-1 and NHDF, contain two centrosomes and can be used as counterparts with normal centrosome

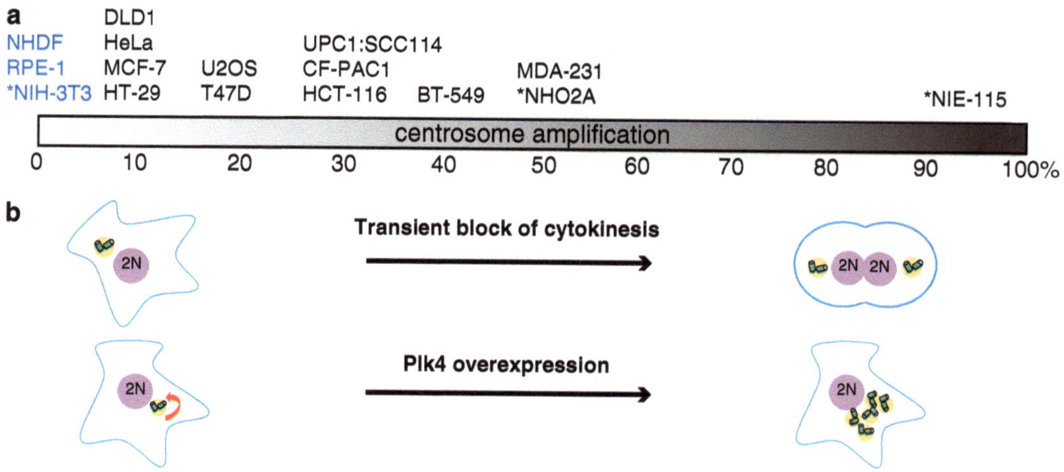

Fig. 4 Useful mammalian cell culture models with or without centrosome amplification. **(a)** Panels of non-transformed (*blue text*) and cancer cell lines (*black text*) are listed in order of increasing centrosome amplification. The percentage of cells with centrosome amplification in each line was previously quantified by immunostaining for centriolar markers. While some cancer cell lines have generally normal centrosome number (e.g., HeLa, U2OS, DLD1), many lines from diverse tissue origins possess supernumerary centrosomes. In contrast, non-transformed cells primarily possess normal centrosome content, and thus can be used as a basis for comparison against centrosome-amplified lines. All lines are derived from human tissues, except for a few mouse lines (*asterisks*). NHDF, human dermal fibroblast [27]; RPE-1, human retinal pigment epithelial cells [12]; NIH-3T3, mouse embryonic fibroblast [12]; DLD1, human colon adenocarcinoma [21]; HeLa, human cervical carcinoma [12]; MCF-7, human mammary adenocarcinoma [12]; HT-29, human colon carcinoma [12]; U2OS, human osteosarcoma [12]; T47D, human mammary ductal carcinoma [12]; UPC1:SCC114, human oral squamous cell carcinoma [10]; CF-PAC1, human pancreatic ductal adenocarcinoma [12]; HCT-116, human colorectal carcinoma [12]; BT-549, human breast ductal carcinoma [12]; MDA-231, human mammary adenocarcinoma [12]; NHO-2A, mouse neuroblastoma [12]; NIE-115, mouse neuroblastoma [8, 12]. (**b**). Using cell lines harboring normal centrosome content, pairs of isogenic cell lines with or without centrosome amplification can be generated in two different ways: by transient block of cytokinesis, or by overduplication of centrioles via inducible overexpression of Plk4

content [12, 27]. Interestingly, experimentally induced extra centrosomes are often efficiently clustered in these cell lines [10, 12, 13, 21], highlighting that centrosome clustering mechanisms exist in normal cells.

3.2 Generation of Isogenic Cell Lines with or Without Centrosome Amplification

In order to directly assess the consequences of centrosomes amplification, it is important to start with cell lines that primarily contain two centrosomes (Fig. 4, for example, RPE-1, U2OS, and DLD1). These cells can be used to create isogenic cells (i.e., they share a genetically identical background) that differ only in centrosome number; the control bears two centrosomes, while the other is induced to carry extra centrosomes. Here, we describe two methods to induce extra centrosomes: (1) transient inhibition of cytokinesis to produce tetraploid cells, which are paired with control diploid cells, and (2) inducible overexpression of Plk4, a master regulator of centrosome duplication, to generate isogenic diploid cells with or without centrosome amplification (Fig. 4b). This approach can also be used as a tool to increase the incidence of whole chromosome mis-segregation events, in order to study the consequences of aneuploidy.

3.2.1 Tetraploid Cells with Extra Centrosomes by Blocking Cytokinesis

1. Treat RPE-1 cells (or your cells of interest with normal centrosome number) with 4 μM DCB for ~16–18 h (*see* **Note 1**).
2. Release cells from DCB by washing five times with pre-warmed, drug-free medium (*see* **Note 2**).
3. Add fresh medium containing 10 μM SB203580 (p38 inhibitor) for ~6–24 h to allow cells to recover from G1 arrest and reenter the cell cycle (*see* **Note 3**).

3.2.2 Centrosome Overduplication by Inducible Plk4 Overexpression

Constitutive overduplication of centrosomes causes lethality or loss of cell proliferation in most cell lines [36]. This toxicity can be avoided by using a tetracycline-inducible Plk4 expression system to drive transient centrosome overduplication [29]. The extent of Plk4 induction can be adjusted to maximize the fraction of cells with extra centrosomes without compromising cell viability. Various studies successfully used this approach to study centrosome duplication. The method described here employs a lentiviral expression system that can be used in a wide variety of cell lines, including those that are difficult to transfect (e.g., primary cell lines).

Production of Viruses

1. Day −1: One day before transfection (day 0), plate HEK-293 FT cells so that cell density will be ~50 % confluent at the time of transduction in 6-well plates with 2 ml of antibiotic-free medium containing 10 % FBS.
2. Day 0: Prepare DNA mix for transfection reaction by mixing 1 μg of the lentiviral vectors pLenti-CMV-TetR-Blast, 1 μg of Gag-pol (psPAX2/packaging plasmid) and 0.5 μg of VSV-G (pMD2.G/envelope plasmid) in 250 μl Opti-MEM.
3. Prepare 10 μl of Lipofectamine 2000 in 250 μl Opti-MEM.
4. Combine the DNA and Lipofectamine solutions, invert to mix, and incubate for 15 min at room temperature (RT).
5. Add 500 μl of transfection mix dropwise to HEK-293 FT cells, and incubate for 6 h.
6. Aspirate the medium, add fresh complete growth medium containing 10 % Tet-free FBS and wait for 2 days for virus production.
7. Day 2: approximately 48 h after transfection, harvest lentivirus by collecting the culture medium (2 ml) and passing it through a 0.45-μm filter using a syringe.

Infection of Target Cells with Viruses

1. Day 1: 1 day before lentivirus collection, seed target cells (RPE-1 or other cell line with normal centrosome number) in antibiotic-free growth medium in 6-well plates so that cell density will be ~25–50 % confluent at the time of infection.
2. Day 2: Infect cells by adding 2 ml of filtered virus (**step** 7 above) with 8 μg/ml polybrene, and return to an incubator for 6 h.
3. Remove lentivirus and add fresh complete growth medium containing 10 % Tet-free FBS.
4. Days 4–18: 2 days post-infection, remove medium and add medium containing Blasticidin (5–10 μg/ml) for 10–14 days to select for Blasticidin-resistant cells. Subculture cells during the selection period. Stable Blasticidin-resistant cells can be pooled.
5. Day 18: Generate lentivirus carrying the Plk4 expression construct by repeating **steps 1**–7 of the previous section "Production of Viruses" 3.2.2.1, substituting pLenti-CMV/TO-Neo-Dest containing Tet-inducible Plk4 for pLenti-CMV-TetR-Blast.
6. Day 20: Infect Blasticidin-resistant stable cells as in **step 2**, using lentivirus containing the inducible Plk4 construct.
7. Remove lentivirus and add fresh complete growth medium containing 10 % Tet-free FBS.
8. Day 22: 2 days post-infection, remove medium and add medium containing Geneticin (200–500 μg/ml) for 10–15 days to select for Geneticin-resistant cells.
9. Days 32–37: Pool stable cells that are resistant to both Blasticidin and Geneticin, and maintain the cultures under selection for both.

3.3 Assessment of Centrosome Number and Centrosome Clustering by Immunofluorescence Microscopy

The following protocol describes the use of isogenic cell lines with inducible Plk4 expression (described in Subheading 3.2.2). This method for labeling cells with centrin, a centriolar marker, can be applied to other cell lines of choice (Fig. 4), as well as for tetraploid cells (described in Subheading 3.2.1). The majority of centrin protein, at both endogenous and overexpressed levels, exists in a soluble cytoplasmic pool. This feature obscures visualization of the insoluble, centriole-associated pool in specimens fixed by formaldehyde that are widely used to preserve many cytoskeletal structures. Here, we use extraction and fixation of cells with cold MeOH to successfully visualize centriole-associated centrin.

3.3.1 Preparations of Specimens for Immunofluorescence Microscopy

1. Maintain stable cell lines carrying inducible Plk4 in complete growth medium containing Tet-free FBS.
2. Trypsinize cells, and plate them in duplicate on glass coverslips placed in 6-well plates for treatment with doxycycline or vehicle (*see* **Note 4**).

3. Next day, add vehicle or 2 2 μg/ml doxycycline to induce Plk4 expression and return to an incubator for 16–18 h (*see* **Note 5**).
4. Wash cells four times with PBS.
5. Add fresh medium and incubate for additional 24 h (*see* **Note 6**).
6. Rinse cells with PBS and fix cells by adding cold (−20 °C) MeOH; keep the plates at −20 °C for 10 min (*see* **Note** 7).
7. Gently aspirate MeOH, and rehydrate cells by addition of PBS for 5–10 min
8. Incubate for 5 min with 0.25 % Triton X-100 in PBS.
9. Rinse with PBS, and proceed to immunostaining or store at 4 °C to continue later.
10. Add blocking solution (5 % BSA, 0.1 % Triton X-100 in PBS) and incubate for 30 min at RT.
11. Incubate with primary antibodies in blocking solution for 1 h at RT: Rabbit anti-centrin-2 (1:200), mouse α-tubulin (DM1a, 1:500) (*see* **Note 8**).
12. Wash three times with 0.1 % Triton X-100 in PBS, 10 min each.
13. Incubate with fluorescent conjugated secondary antibodies, diluted 1:1000 in blocking solution, for 1 h at RT.
14. Wash three times with 0.1 % Triton X-100 in PBS, 10 min each.
15. Add Hoechst 33542, diluted 1:5000 in 0.1 % Triton X-100 in PBS, for 10 min.
16. Wash three times with 0.1 % Triton X-100 in PBS, 10 min each.
17. Mount coverslips with ProLong Antifade mounting medium (Molecular Probes) and leave to dry for 2–3 h. Store the slides at 4 °C in the dark.

3.3.2 Imaging of Fixed Specimens to Evaluate Centrosome Number and Centrosome Clustering

Centrioles are small cylindrical structures, 100–250 nm in diameter and 100–500 nm in length that are embedded in PCM. As these dimensions are near the resolution limit of light microscopes, it is critical to perform high-resolution three-dimensional imaging to accurately count the number of centriole pairs per cell and per mitotic spindle pole. To this end, we use a Yokogawa CSU-X1 spinning disk confocal on a Nikon Ti inverted microscope (Nikon Instruments) equipped with a 100× Plan Apo objective with high numerical aperture (e.g., 1.4 NA). Rounded mitotic cells are imaged in a series of 0.2–0.5 μm optical sections recorded with an Orca ER CCD camera (Hamamatsu Photonics). Acquisition parameters, shutters, filter positions and focus are controlled by MetaMorph 7 software (Molecular Devices). Combining siRNA-mediated depletion or small molecule inhibitor approaches with the methods described here can provide powerful insights into the roles played by specific proteins or processes in maintaining centrosome integrity or clustering for bipolar spindle assembly (Fig. 2).

3.3.3 Analysis of Centrosome Number and Mitotic Phenotypes

1. Quantify the percentage of mitotic cells with two centrosomes (two pairs of centrin-positive dots) and with extra centrosomes (three or more pairs of centrin-positive dots) for each condition.
2. Within each population of cells defined above, quantify the fraction of bipolar and multipolar spindles using the α-tubulin signal and chromosome configuration. This analysis yields several classes of cells, listed as follows: (1) A bipolar spindle with two centrosomes has one pair of centrin-positive dots per α-tubulin pole; (2) A multipolar spindle with two centrosomes exhibits fragmented PCM (additional centrin-negative acentriolar poles, *see* Fig. 2b) and/or split centrioles (multiple poles with one centrin dot each, *see* Fig. 2c); (3) A bipolar spindle with clustered supernumerary centrosomes is characterized by one or more pairs of centrin-positive dots clustered into two poles (tightly clustered into two poles or partially clustered into two groups consisting of multiple poles with bi-oriented chromosomes); (4) A multipolar spindle with de-clustered supernumerary centrosomes is characterized by multiple pairs of centrin-positive dots distributed among multiple poles (*see* Fig. 2a).
3. Score approximately ~100 metaphase-like and anaphase figures for each experiment (*see* **Note 9**).

3.4 Live Cell Imaging of Centrosome Clustering

Here, we describe procedures to quantify bipolar and multipolar anaphases in isogenic cells with or without centrosome amplification by long-term live cell imaging (Fig. 5a, b). While differential interference contrast (DIC) imaging alone is often sufficient to distinguish bipolar and multipolar anaphases, inclusion of GFP-H2B or GFP-Centrin has additional advantages. A GFP-Centrin marker can be used to directly visualize the process of multiple centriole pairs being coalesced into each mitotic pole during centrosome clustering, but it requires high-resolution imaging (60× or 100× objective) to resolve GFP-positive centrin dots in a single cell. Imaging of GFP-H2B does not directly visualize centrosomes, but can be very informative when combined with analysis on fixed specimens using centriole markers (described in Subheading 3.3). Bipolar and multipolar anaphases can be inferred from chromosome configurations indicated by GFP-H2B even at low magnification (10× or 20× objective), allowing simultaneous imaging of many mitotic cells in a single field of view. This approach enables higher throughput for rapid screening of many experimental conditions, and is even suitable for large-scale screening applications. Moreover, live imaging allows detection of additional phenotypes

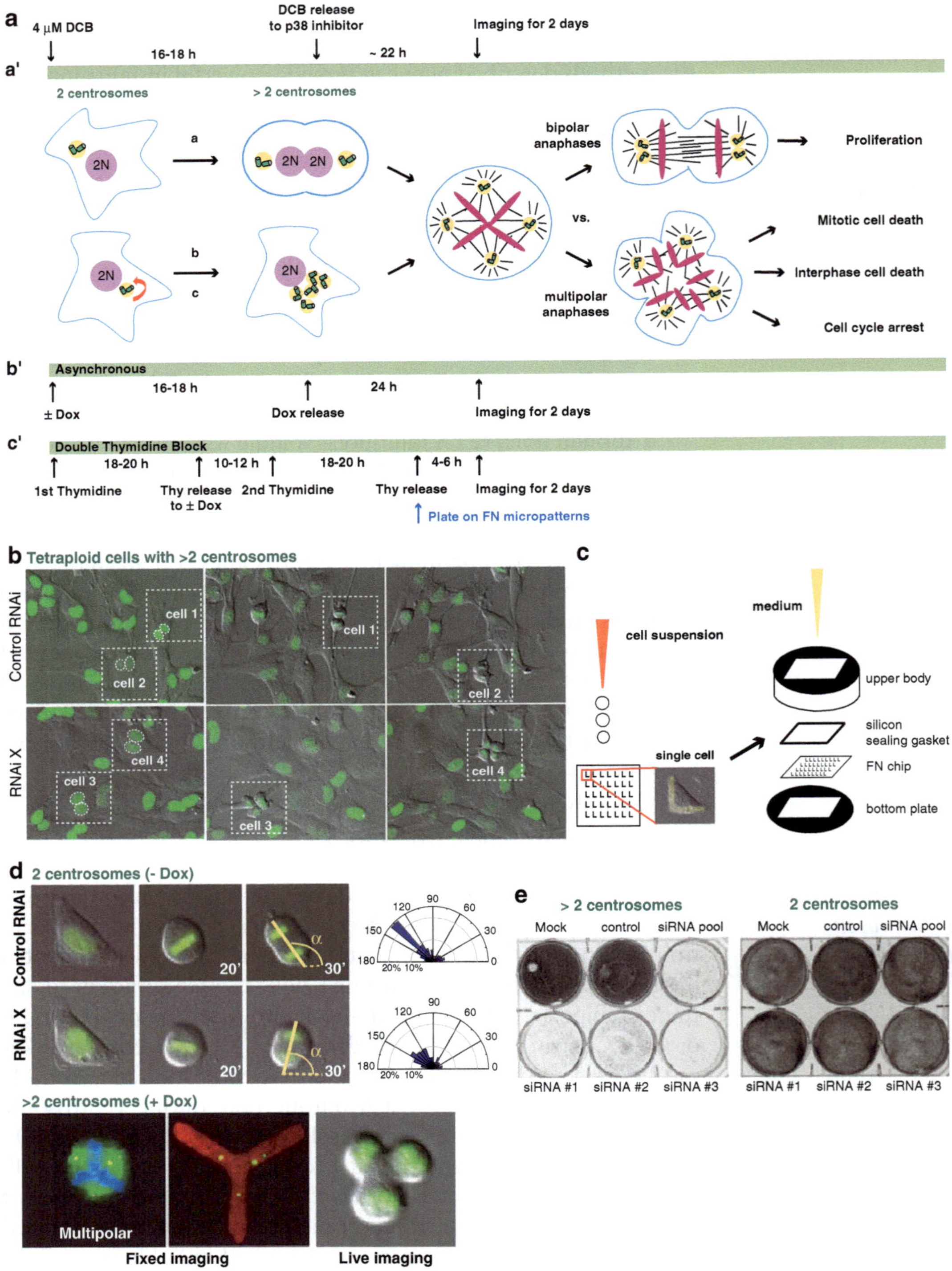

Fig. 5 Live-cell imaging for the analysis of cell division in isogenic cell culture models with or without centrosome amplification. (**a**). Schematic outline of the experimental design for long-term imaging of cells with 2 or >2 centrosomes: tetraploid cells generated by cytokinesis block (a), or cells with (+Dox) or without (−Dox) extra centrosomes induced by Plk4 expression without (b) or with (c) synchronization by double thymidine block. In live-cell imaging experiments, the fraction of bipolar and multipolar anaphases can be quantified from ~2 days of imaging. Further analysis of the cells that undergo multipolar anaphase often requires 3–5 days of imaging. Cell fates can be classified for continued proliferation, cell cycle arrest, mitotic cell death, or interphase cell death. (**b**). Representative time-lapse DIC and GFP-H2B images showing tetraploid cells that undergo

associated with multipolar anaphases that can be overlooked in the analysis of fixed cells. For instance, quantification of parameters such as cell cycle timing, cytokinesis defects, and cell fate after multipolar anaphase (cell cycle arrest or death) can be valuable in understanding the causes and consequences of centrosome amplification in cell division and other contexts [12, 13, 22, 37] (Fig. 5a).

3.4.1 Tetraploid Cells with Extra Centrosomes

1. Grow diploid RPE-1 cells expressing GFP-H2B in 6-well plates in DMEM/F12 phenol red-free medium with 10 % FBS. Alternatively, cells can be plated directly in glass-bottom 12-well tissue culture plates (*see* **Note 10**).
2. Perform RNAi experiments by transfecting cells with siRNAs using Lipofectamine RNAi MAX transfection reagent (Invitrogen) according to the manufacturer's instructions, aiming for maximal depletion of the protein of interest during the course of long-term imaging (Fig. 5A-a) (*see* **Note 11**).
3. Six or more hours after siRNA transfection, follow the instructions in Subheading 3.2.1 to generate tetraploid cells with extra centrosomes (Fig. 5A-a). Briefly, treat cells with DCB for 16–18 h and release into warm medium containing 10 % FBS and p38 inhibitor for an additional ~6–24 h.
4. Set the microscope environmental chamber to maintain an atmosphere of 37 °C and 5 % humidified CO_2. The experiment described here uses a Nikon Ti-E inverted microscope equipped with a 20× objective, a precision motorized stage (Bioprecision, Ludl), cooled CCD camera (Coolsnap HQ2, Photometrics), and the Nikon Perfect Focus system, controlled by the NIS-Elements software (Nikon Instruments, Inc.).

Fig. 5 (continued) bipolar or multipolar anaphases following depletion of target protein (X) by RNAi. Nuclear morphology was used to distinguish tetraploid (binucleated) from diploid (mononucleated) cells in mixed populations. (**c**). Cell preparation procedure for long-term imaging of division on fibronectin (FN) micropatterns. An L-shape micropattern is depicted here, but cells can be plated on other micropattern shapes as desired. FN chips are assembled into a chamber. The mounted chamber is then filled with culture medium and transferred to a microscope for imaging. (**d**). Example time-lapse DIC and GFP-H2B images used for the analysis of centrosome positioning during spindle orientation (cells plated on "L" patterns, *top*), and centrosome clustering ("Y" patterns, *bottom*). *Top*: Spindle orientation is determined by the angle (α) of chromosome segregation axis (*yellow line*) relative to the reference *X* axis (*yellow dotted line*) at anaphase onset, and is presented as an angular distribution. *Bottom*: A cell with extra centrosomes undergoing multipolar division on a "Y" pattern. Parallel immunofluorescence microscopy experiment shows centrosomes, marked by pericentrin, in *yellow*. (**e**). Colony formation assay shows selective cell death resulting from siRNA-mediated inhibition of centrosome clustering. In order to rule out potential off-target effects, multiple independent siRNA duplexes were tested individually and pooled together to determine whether the extent of target protein depletion (not shown) correlates with antiproliferative outcomes

5. Acquire images for DIC and GFP-H2B in multiple fields of view at 10-min intervals for 2 days. We find that tetraploid RPE-1 cells with extra centrosomes typically have peak number of cells in mitosis ~30 h after DCB release.
6. Analyze movies using NIS-Elements software: follow both binucleated tetraploid cells and mononucleated diploid cells that enter mitosis based on H2B morphology (*see* Fig. 5a-a' and b).
7. Quantify the fraction of bipolar and multipolar anaphases by chromosome (GFP-H2B) and cell (DIC) morphology.

3.4.2 Centrosome Overduplication by Inducible Plk4 Overexpression

This protocol utilizes stable cell lines carrying a tetracycline-inducible Plk4 expression system, and has been used successfully with both asynchronously growing cells (Fig. 5a-b') and cells synchronized by double thymidine block (Fig. 5a-c').

Asynchronous Cells

1. Plate cells in glass-bottom 12-well tissue culture dishes (Mattek) in duplicate for treatment with doxycycline or vehicle (experimental and control, respectively) (Fig. 5a-b') (*see* **Note 12**).
2. On the following day, perform RNAi experiments, aiming for maximal depletion of the protein of interest during long-term imaging (*see* **Note 11**).
3. Add fresh medium containing 2 μg/ml doxycycline for 16–18 h to induce Plk4 expression (*see* **Note 5**).
4. Remove doxycycline by washing four times with warm PBS.
5. Add fresh medium and incubate for additional 24 h (*see* **Note 6**).
6. Set up microscope to acquire images for DIC and GFP-H2B in multiple fields of view at 10-min intervals for 2 days, as described in the Subheading 3.4.1, **steps 4** and **5**.
7. Quantify bipolar and multipolar anaphases by chromosome (GFP-H2B) and cell (DIC) morphology.

Synchronization with Double Thymidine Block

1. Plate cells in glass-bottom 12-well tissue culture dishes (Mattek) in duplicate for treatment with doxycycline or vehicle (experimental and control, respectively) (Fig. 5a-c') (*see* **Note 12**).
2. On the following day, perform RNAi experiments, aiming for maximal depletion of the protein of interest during long-term imaging (*see* **Note 11**).
3. Add medium containing 2.5 mM thymidine to cells for 18–20 h (first thymidine block).
4. Release by gently washing three times with warm PBS (*see* **Note 13**).
5. Add fresh medium containing 2 μg/ml doxycycline for 10–12 h to induce Plk4 expression.

6. Gently remove doxycycline by washing three times with warm PBS.
7. Add medium containing 2.5 mM thymidine and incubate for 18–20 h (second thymidine block).
8. Release by gently washing three times with warm PBS.
9. 4–6 h after release from the second thymidine block, set up microscope to acquire images for DIC and GFP-H2B in multiple fields of view at 10-min intervals for 2 days, as described in the Subheading 3.4.1, **steps 4** and **5**.
10. Quantify bipolar and multipolar anaphases by chromosome (GFP-H2B) and cell (DIC) morphology.

3.5 Long-Term Live Cell Imaging on Fibronectin Micropatterns

Imaging of cells dividing on fibronectin (FN) micropatterns is useful to study how the distribution of cellular adhesive contacts and the associated cortical forces determine the positioning of centrosomes and the cell division plane [12, 32, 33, 38]. FN, a major component of the ECM that interacts with integrins, can be printed on a glass coverslips in a variety of shapes to define the adhesion geometry of cells (Fig. 5c, d). This technique was employed to show that the distribution of cell adhesion contacts can either promote or interfere with centrosome clustering in cells containing extra centrosomes, thus influencing whether cancer cells undergo bipolar or multipolar divisions [12, 38] (Fig. 3c). With the appropriate choice of micropattern shapes, this assay can further serve as a platform to identify molecular mechanisms that control mitotic centrosome positioning during spindle orientation and clustering of supernumerary centrosomes. We describe protocols for live imaging to study spindle orientation in cells with two centrosomes using an L-shaped pattern, and the clustering of extra centrosomes using a Y-shaped pattern (Fig. 5d).

1. Plate cells expressing GFP-H2B in 6-well plates.
2. Synchronize cells using double thymidine block, as described in Subheading 3.4.2 (Fig. 5a-c').
3. Upon release from the second thymidine block, trypsinize and count cells, and prepare a suspension of 2×10^4 cells/ml in warm complete growth medium (Fig. 5a-c', blue arrow, and C).
4. Prepare a 35-mm dish or 6-well plates containing 1 ml of medium.
5. Transfer one FN micropatterned coverslip, patterned side up, to the 35-mm dish or each well using forceps. Gently swirl the plate to immerse the coverslip in medium.
6. Add 3 ml of cell suspension (6×10^4 cells) to 35-mm dish or each well. Gently mix to distribute cells evenly across the coverslip surface (avoid circular swirling movements to prevent cells from concentrating in the center).

7. Let cells settle at RT in the tissue culture hood for 10 min.
8. Move the plate into an incubator at 37 °C.
9. 30–60 min post attachment, check cells under a light microscope. Ideally, each patterned area should be occupied by a single cell (Fig. 5c). For most cell lines, 1 h is sufficient for attachment; some cells adopt the shape of the FN micropattern, while others settle but maintain a round morphology.
10. Gently remove the medium and add fresh medium (be careful not to allow the micropatterned surface to dry out).
11. Some micropatterns may be occupied by multiple cells. Remove these extra cells by gently pipetting the medium up and down, while monitoring the cells under a microscope. Repeat until the majority of micropatterns are occupied by a single cell.
12. Add fresh medium and incubate at 37 °C to allow cells to adhere for an additional 3–5 h. Most cells should adopt the shape of the micropattern by this time.
13. Pick up a micropatterned coverslip with forceps and assemble into CYTOO chamber (Fig. 5c, arrow) (*see* **Notes 14** and **15**).
14. Immediately add 2–2.5 ml of medium to the chamber. Place the chamber in the incubator and monitor to make sure it is not leaking (*see* **Note 16**).
15. Set up microscope to acquire images for DIC and GFP-H2B in multiple fields of view at 10-min intervals for 2 days, as described in Subheading 3.4.1, **steps 4** and **5**. For cell division axis measurements, it is convenient to orient the chamber on the microscope stage such that the *X*-axis of the "L" pattern is parallel to the reference line (Fig. 5d, dotted yellow line).
16. Analyze the first round of mitotic division, following only individual interphase cells that are properly attached to a micropattern.
17. Measure spindle orientation (α) inferred by the axis of chromosome separation (GFP-H2B) and cell elongation (DIC) at anaphase onset using NIS-Elements or other software of choice (Fig. 5d).
18. Log the data into a spreadsheet (~100 cells or more is preferred).
19. Calculate the fraction of cells that are distributed throughout 0–180° and generate angular distribution of spindle orientation using Origin (Origin Lab) or other equivalent software (Fig. 5d). The normal spindle orientation angle (α) on an "L" pattern is 135°.

3.6 Assessment of Cell Fates and Cell Viability

In order to test if inhibition of centrosome clustering can selectively kill cancer cells with extra centrosomes or impede their proliferation, several methodologies can be used in a variety of cell culture models (Fig. 4). Disrupted cell proliferation or induced cell death can be quantified in cell populations using a luminescence-based CellTiter-Glo assay (Promega) to measure cellular ATP levels [21, 37], cell counting assay [12, 20], and colony formation assay -an in vitro cell survival assay based on the ability of single cells to grow into colonies [12, 23] (Fig. 5e). In addition, long-term imaging and subsequent lineage tracing can be performed for each cell to determine if multipolar anaphase leads to cell cycle arrest or cell death [13, 23] (Fig. 5a).

4 Notes

1. If used for live-cell imaging, start with a stable cell line that expresses GFP-H2B (or mCherry-H2B to visualize chromosomes) to generate tetraploid cells. This marker reports on cell cycle progression and mitotic stages, bipolar versus multipolar anaphase, as well as the fate of daughter cells after multipolar division (for example, arrest or death). Alternatively, a stable cell line that expresses GFP-centrin and/or mCherry-H2B can be used to visualize clustering of extra centriole pairs.
2. When washing out DCB, use pre-warmed complete growth medium instead of warm PBS. This helps to prevent cell cycle arrest of daughter cells that otherwise appears to result from cellular stress responses. It also promotes both the recovery of tetraploid cells from G1 arrest and the proliferation of their progeny.
3. Transient treatment with p38 inhibitor is one way to bypass tetraploidy-induced G1 arrest. Alternatively, siRNA-mediated knock down of p53 can be used to bypass the G1 arrest. Transfect cells with p53 siRNA (SMARTpool, Dharmacon) using Lipofectamine RNAi Max (Invitrogen) prior to DCB treatment.
4. A control with no doxycycline treatment should always be performed. Comparison against this control condition provides essential information including: (1) background level of centrosome amplification in a given cell line under experimental conditions, (2) degree of centrosome amplification specifically resulting from doxycycline-induced Plk4 expression, and (3) in drug or siRNA studies, any effect of the target gene or pathway on centrosome biogenesis or numbers.
5. For any given cell line, the duration of doxycycline treatment should be titrated to obtain highest possible fraction of cells with extra centrosomes without observed toxicity. Several factors must be considered, including doubling time, and the tolerance of a given cell line to the presence of overduplicated centrosomes. In stable U2OS cells with inducible Plk4 [29],

we observe ~15–20 % background level of cells with extra centrosomes prior to doxycycline treatment, which increases up to ~90 % following a 16–18 h doxycycline treatment.

6. During doxycycline treatment, Plk4 overexpression induces centriole duplication, which manifests as multiple procentrioles joined to a mother centriole in a rosette pattern [29]. During the first mitotic division, these cells divide in a bipolar fashion with two rosettes, each comprising multiple centrioles that still remain engaged as a unit. Incubation for an additional 24-h period allows separation of amplified centriole pairs and the second round of division, during which cells divide by clustering of amplified centriole pairs.
7. Soaking glass coverslips in 70 % EtOH is sufficient to support cell attachment in most cell lines. However, some cell lines that attach loosely tend to float away or are stripped off as a single sheet from the glass surface upon addition of MeOH. To avoid cell loss, keep cells at optimal (not over confluent) density, and add MeOH to plates gently. As necessary, coverslips coated with FN (Sigma, 10 ng/ml) or poly-L-lysine (BD Biosciences) can be used to further enhance cell attachment. For FN coating, coat glass coverslips that were stored in 70 % EtOH with FN solution (10 ng/ml in PBS) at 37 °C for 1–2 h. Wash the coverslips with PBS once and block with 1 % BSA in PBS (which was previously heat-inactivated at 55 °C for 30 min) at 37 °C for 1 h. Aspirate the blocking solution and air-dry coated coverslips for 30 min under UV light without washing again.
8. α-Tubulin staining could serve as an important indicator of centrosome functionality in MT nucleation. In addition, staining for PCM markers (e.g., γ-tubulin or pericentrin) can allow direct visualization of centrosome integrity defects (Fig. 2).
9. In most cases, centriole number can be scored visually through the eyepiece using a 100×, 1.4 NA objective by checking each spindle pole (stained for α-tubulin, γ-tubulin, or pericentrin). When there is uncertainty in resolving centrin dots, collect Z-series of images and confirm centriole number both in the Z-stacks and in maximum intensity projections. Because confocal microscopy is collecting approximately a 0.5-μm Z-plane volume in each image with a 100×, 1.4 NA objective, the use of a 0.5-μm optical section thickness produces a nonoverlapping Z-series. Taking 0.2-μm steps could produce overlapping information as the same centrin spot could appear in two adjacent Z-planes (often it is in focus in one plane, and out of focus in the next). However, if z projection images are carefully evaluated to make sure not to recount the same centrin spot, the overlapping information is often very helpful to resolve multiple centrin spots that are in close proximity.

10. Start with stable cell lines expressing GFP-H2B or GFP-Centrin before DCB treatment. Unlike some cancer cells that maintain a high percentage of population with extra centrosomes (Fig. 4), non-transformed cells or those that contain normal centrosome content tend to lose experimentally induced extra centrosomes over time in culture (>6–10 days) [13]. We successfully use stable cell lines expressing GFP-H2B or GFP-Centrin generated by lentiviral infection. Imaging experiments are performed with freshly prepared tetraploid cells, within <6–10 days after inducing extra centrosomes. However, loss of extra centrosomes by longer-term culturing can be used effectively as a means to generate control tetraploid cells with normal centrosome content [13, 17, 21]. These cells can be isolated as a pure population to be used as a control for freshly generated tetraploid cells with extra centrosomes. Briefly, perform sequential fluorescence-activated cell sorting (FACS) of DCB-treated cells to obtain 8c tetraploid cells by DNA content. Culture sorted 8c cells for >10 days and repeat the sequential FACS to select for pure 8c tetraploid cells. The sorted 8c cells should be subsequently screened to verify loss of extra centrosomes.
11. The time required to deplete a target protein depends on its turnover rate, as well as cell doubling time. Protocol modifications should be employed as necessary to achieve more complete depletion (for example, by performing two consecutive rounds of siRNA transfection during longer period) or partial knockdown (by shortening the duration of RNAi-mediated protein depletion to obtain primary phenotype prior to terminal phenotype, e.g., cell death). In particular, long-term imaging experiments often require lasting effects of protein depletion over several days (Fig. 5). Be sure to maintain optimal confluency during multiple rounds of RNAi by subculturing as needed. Furthermore, cells can be split in duplicate to facilitate parallel analysis of both fixed and live cells.
12. It is important to set up parallel experiments for immunofluorescence microscopy of centrin staining, as described in Subheading 3.3, to quantify the degree of centrosome amplification in each experiment. Small day-to-day variations or changes in the experimental conditions can be normalized based on comparing the incidence of multipolar anaphases (by live cell imaging) relative to the fraction of cells that contain extra centrosomes (by fixed imaging).
13. Cells attach loosely to glass-bottom dishes compared to tissue culture dishes or multi-well plates. To minimize cell loss,

aspirate gently from the edge of each plate or well using low vacuum speed. For cells that do not attach to glass, use dishes pre-coated with FN or poly-L-Lysine.

14. When using multiple experimental conditions (e.g., control vs. RNAi) in a single experiment, consider using multi-well chambers with four separate sealed sections. Using the same FN-coated micropatterns, assemble a multi-well chamber first and then add cell suspension into each section. For a 4-well chamber, add 100 μl of 6×10^4 cells/ml for 10 min at RT, and then fill the chamber with medium. Follow instructions as described in Subheading 3.5, **steps 7–12**.
15. Alternatively, FN chips can be prepared for immunofluorescence microscopy. To preserve actin and retraction fiber morphology, as well as to label actin with Alexa Fluor phalloidin, we use a paraformaldehyde (PFA) fixation method instead of using MeOH, as follows. Wash cells with pre-warmed PBS and fix with 4 % PFA in PBS for 15 min at RT. Permeabilize cells with 0.25 % Triton X-100 in PBS for 5 min, block with 5 % BSA, 0.1 % Triton X-100 in PBS for 30 min, and stain for primary antibodies [e.g., anti-pericentrin (1:1500, Abcam), anti-α-tubulin DM1A (1:500, Sigma)] for 1 h. During the secondary antibody incubation, include Alexa Fluor phalloidin (1:250, Molecular Probes), and subsequently stain for DNA with Hoechst 33342 (1:5000, Invitrogen) in PBS.
16. Place the assembled chamber on top of Kimwipes and check for leaking repeatedly over a period of ~15 min. If the chamber leaks, it must be reassembled. Disassemble the chamber, pick up the FN chip with forceps, and transfer it back to a 6-well plate or a 35 mm dish containing medium. Then, reassemble the chamber, making sure the FN chip is perfectly centered inside of the chamber and the rubber gasket is not loose.

Acknowledgements

This protocol is currently used in David Pellman laboratory and has been refined over the years by several people including Drs. S. Godinho and N. Ganem. We thank N. Umbreit, A. Spektor, Y. Kaplan, H. Zhang for discussions and/or comments on the manuscript; all members of the Pellman laboratory for advice and suggestions. M.K was supported by a special fellow award from the Leukemia and Lymphoma Society and a Susan G. Komen grant. Our work was supported by the Howard Hughes Medical Institute and the NIH (GM061345).

References

1. Bettencourt-Dias M, Glover DM (2007) Centrosome biogenesis and function: centrosomics brings new understanding. Nat Rev Mol Cell Biol 8(6):451–463.
2. Nigg EA (2007) Centrosome duplication: of rules and licenses. Trends Cell Biol 17(5):215–221.
3. Chan JY (2011) A clinical overview of centrosome amplification in human cancers. Int J Biol Sci 7(8):1122–1144.
4. Nigg EA (2002) Centrosome aberrations: cause or consequence of cancer progression? Nat Rev Cancer 2(11):815–825.
5. Boveri T (2008) Concerning the origin of malignant tumours by Theodor Boveri. Translated and annotated by Henry Harris. J Cell Sci 121(Suppl 1):1–84.
6. Galeotti G (1893) Beitrag zum Studium des Chromatins in den Epithelzellen der Carcinome. Beitr Pathol Anat Allg Pathol 14:249–271.
7. Hardy PA, Zacharias H (2005) Reappraisal of the Hansemann-Boveri hypothesis on the origin of tumors. Cell Biol Int 29(12): 983–992.
8. Ring D, Hubble R, Kirschner M (1982) Mitosis in a cell with multiple centrioles. J Cell Biol 94(3):549–556.
9. Brinkley BR (2001) Managing the centrosome numbers game: from chaos to stability in cancer cell division. Trends Cell Biol 11(1):18–21.
10. Quintyne NJ, Reing JE, Hoffelder DR, Gollin SM, Saunders WS (2005) Spindle multipolarity is prevented by centrosomal clustering. Science 307(5706):127–129.
11. Acilan C, Saunders WS (2008) A tale of too many centrosomes. Cell 134(4):572–575.
12. Kwon M, Godinho SA, Chandhok NS, Ganem NJ, Azioune A, Thery M, Pellman D (2008) Mechanisms to suppress multipolar divisions in cancer cells with extra centrosomes. Genes Dev 22(16):2189–2203.
13. Ganem NJ, Godinho SA, Pellman D (2009) A mechanism linking extra centrosomes to chromosomal instability. Nature 460(7252): 278–282.
14. Silkworth WT, Nardi IK, Scholl LM, Cimini D (2009) Multipolar spindle pole coalescence is a major source of kinetochore mis-attachment and chromosome mis-segregation in cancer cells. PLoS One 4(8), e6564.
15. Basto R, Brunk K, Vinadogrova T, Peel N, Franz A, Khodjakov A, Raff JW (2008) Centrosome amplification can initiate tumorigenesis in flies. Cell 133(6):1032–1042.
16. Schvartzman JM, Sotillo R, Benezra R (2010) Mitotic chromosomal instability and cancer: mouse modelling of the human disease. Nat Rev Cancer 10(2):102–115.
17. Godinho SA, Picone R, Burute M, Dagher R, Su Y, Leung CT, Polyak K, Brugge JS, Thery M, Pellman D (2014) Oncogene-like induction of cellular invasion from centrosome amplification. Nature 510(7503):167–171.
18. Lingle WL, Lutz WH, Ingle JN, Maihle NJ, Salisbury JL (1998) Centrosome hypertrophy in human breast tumors: implications for genomic stability and cell polarity. Proc Natl Acad Sci U S A 95(6):2950–2955.
19. Mennella V, Agard DA, Huang B, Pelletier L (2014) Amorphous no more: subdiffraction view of the pericentriolar material architecture. Trends Cell Biol 24(3):188–197.
20. Leber B, Maier B, Fuchs F, Chi J, Riffel P, Anderhub S, Wagner L, Ho AD, Salisbury JL, Boutros M, Kramer A (2010) Proteins required for centrosome clustering in cancer cells. Sci Transl Med 2(33):33ra38.
21. Drosopoulos K, Tang C, Chao WC, Linardopoulos S (2014) APC/C is an essential regulator of centrosome clustering. Nat Commun 5:3686.
22. Yang Z, Loncarek J, Khodjakov A, Rieder CL (2008) Extra centrosomes and/or chromosomes prolong mitosis in human cells. Nat Cell Biol 10(6):748–751.
23. Fielding AB, Lim S, Montgomery K, Dobreva I, Dedhar S (2011) A critical role of integrin-linked kinase, ch-TOG and TACC3 in centrosome clustering in cancer cells. Oncogene 30(5):521–534.
24. Jackson JR, Patrick DR, Dar MM, Huang PS (2007) Targeted anti-mitotic therapies: can we improve on tubulin agents? Nat Rev Cancer 7(2):107–117.
25. Endow SA, Komma DJ (1998) Assembly and dynamics of an anastral:astral spindle: the meiosis II spindle of Drosophila oocytes. J Cell Sci 111(Pt 17):2487–2495.
26. Mountain V, Simerly C, Howard L, Ando A, Schatten G, Compton DA (1999) The kinesin-related protein, HSET, opposes the activity of Eg5 and cross-links microtubules in the mammalian mitotic spindle. J Cell Biol 147(2):351–366.
27. Watts CA, Richards FM, Bender A, Bond PJ, Korb O, Kern O, Riddick M, Owen P, Myers RM, Raff J, Gergely F, Jodrell DI, Ley SV (2013) Design, synthesis, and biological evaluation of an allosteric inhibitor of HSET that targets cancer cells with supernumerary centrosomes. Chem Biol 20(11):1399–1410.
28. Wu J, Mikule K, Wang W, Su N, Petteruti P, Gharahdaghi F, Code E, Zhu X, Jacques K, Lai

Z, Yang B, Lamb ML, Chuaqui C, Keen N, Chen H (2013) Discovery and mechanistic study of a small molecule inhibitor for motor protein KIFC1. ACS Chem Biol 8(10):2201–2208.

29. Kleylein-Sohn J, Westendorf J, Le Clech M, Habedanck R, Stierhof YD, Nigg EA (2007) Plk4-induced centriole biogenesis in human cells. Dev Cell 13(2):190–202.
30. Breuer M, Kolano A, Kwon M, Li CC, Tsai TF, Pellman D, Brunet S, Verlhac MH (2010) HURP permits MTOC sorting for robust meiotic spindle bipolarity, similar to extra centrosome clustering in cancer cells. J Cell Biol 191(7):1251–1260.
31. Kleylein-Sohn J, Pollinger B, Ohmer M, Hofmann F, Nigg EA, Hemmings BA, Wartmann M (2012) Acentrosomal spindle organization renders cancer cells dependent on the kinesin HSET. J Cell Sci 125(Pt 22):5391–5402.
32. Thery M, Bornens M (2006) Cell shape and cell division. Curr Opin Cell Biol 18(6):648–657.
33. Fink J, Carpi N, Betz T, Betard A, Chebah M, Azioune A, Bornens M, Sykes C, Fetler L, Cuvelier D, Piel M (2011) External forces control mitotic spindle positioning. Nat Cell Biol 13(7):771–778.
34. Ganem NJ, Pellman D (2007) Limiting the proliferation of polyploid cells. Cell 131(3): 437–440.
35. Andreassen PR, Lohez OD, Lacroix FB, Margolis RL (2001) Tetraploid state induces p53-dependent arrest of nontransformed mammalian cells in G1. Mol Biol Cell 12(5):1315–1328.
36. Holland AJ, Fachinetti D, Zhu Q, Bauer M, Verma IM, Nigg EA, Cleveland DW (2012) The autoregulated instability of Polo-like kinase 4 limits centrosome duplication to once per cell cycle. Genes Dev 26(24): 2684–2689.
37. Kwiatkowski N, Jelluma N, Filippakopoulos P, Soundararajan M, Manak MS, Kwon M, Choi HG, Sim T, Deveraux QL, Rottmann S, Pellman D, Shah JV, Kops GJ, Knapp S, Gray NS (2010) Small-molecule kinase inhibitors provide insight into Mps1 cell cycle function. Nat Chem Biol 6(5):359–368.
38. 2013.0467Kwon M, Bagonis M, Danuser G, Pellman D (2015) Direct microtubule-binding by Myosin-10 orients centrosomes towards retraction fibers and subcortical actin clouds. Dev Cell 34(3):323–37.
39. Crasta K, Ganem NJ, Dagher R, Lantermann AB, Ivanova EV, Pan Y, Nezi L, Protopopov A, Chowdhury D, Pellman D (2012) DNA breaks and chromosome pulverization from errors in mitosis. Nature 482(7383):53–58.
40. Mitchison TJ (1992) Actin based motility on retraction fibers in mitotic PtK2 cells. Cell Motil Cytoskeleton 22(2):135–151.

Chapter 24

Generation and Purification of Tetraploid Cells

Elizabeth M. Shenk and Neil J. Ganem

Abstract

Tetraploid cells are genetically unstable and have the capacity to promote the development and/or progression of human malignancies. It is now estimated that ~40 % of all solid tumors have passed through a tetraploid intermediate stage at some point during their development. Understanding the biological characteristics of tetraploid cells that impart oncogenic properties is therefore a highly relevant and fundamentally important aspect of cancer biology. Here, we describe strategies to efficiently generate and purify tetraploid cells for use in cell biological studies.

Key words Cytokinesis, Mitotic slippage, Polyploid, Hippo, FUCCI, Centrosome

1 Introduction

The vast majority of non-transformed human cells contain two copies of each chromosome and are termed diploid. In preparation for cell division, all chromosomes are replicated so that they may be evenly distributed to two daughter cells during mitosis. However, catastrophic failures in mitosis or cytokinesis can give rise to tetraploid cells, which have a doubled DNA content (four copies of each chromosome). The generation of tetraploid cells through non-programmed mechanisms can have significant consequences, as spontaneously arising tetraploid cells are chromosomally unstable and have the capacity to promote tumorigenesis [1–9]. It is now recognized that tetraploidization events are common in solid tumors, and correlate with poor prognosis [10, 11].

Studying the biology of tetraploid cells can allow us to more fully understand how they promote tumor progression. However, given their rarity in mixed populations of cells, rapid and efficient methods are required to generate and purify them. Several methods currently exist to generate tetraploid cells. These include fusing diploid cells using polyethylene glycol [12]; inhibiting cytokinesis using cell permeable small molecules that prevent furrow ingression and cleavage [4, 12–15]; and promoting mitotic

Paul Chang and Ryoma Ohi (eds.), *The Mitotic Spindle: Methods and Protocols*, Methods in Molecular Biology, vol. 1413, DOI 10.1007/978-1-4939-3542-0_24, © Springer Science+Business Media New York 2016

slippage by using small molecules that prevent satisfaction of the spindle assembly checkpoint [16]. Of these methods, inhibiting cytokinesis is the simplest and most effective way to generate tetraploid cells. In contrast to cell fusion experiments, which are highly inefficient and time consuming, or induction of mitotic slippage, which first requires a dramatically prolonged mitosis that produces extensive DNA damage and even cell death [17, 18], inhibiting cytokinesis with small molecules can be achieved on large populations of cells without inducing mitotic abnormalities or DNA damage [15, 19, 20]. The most commonly used small molecules used to inhibit cytokinesis include blebbistatin, a myosin II inhibitor, and the cytochalasins, which disrupt actin polymerization [14, 21]. These compounds prevent cytokinetic furrow ingression and cell cleavage following anaphase, and thus lead to the formation of binucleated tetraploid cells.

While several simple methods exist to generate tetraploid cells, purifying tetraploids from a predominantly diploid population of cells poses a significant technical challenge. One approach is size separation. However, although tetraploid cells are larger than diploid cells, purification strategies based on cell size are inefficient (our unpublished data). Moreover, DNA content alone cannot distinguish diploid from tetraploid cells because diploid cells in G_2/M phase of the cell cycle (which have replicated their chromosomes) have the same amount of DNA as tetraploid cells in G_1 phase of the cell cycle (both contain 4C DNA content). While tetraploid cells that progress to G_2/M and possess 8C DNA content can be readily distinguished from diploids, these cells are relatively rare because non-transformed tetraploids exhibit markedly reduced proliferation [13, 15].

Here, we describe an approach to generate and purify tetraploid cells with high efficiency using the fluorescent, ubiquitin-based, cell cycle indicator (FUCCI) system [22]. FUCCI consists of two fluorescently labeled proteins, truncated forms of hCdtl (consisting of amino acids 30–120) and hGeminin (consisting of amino acids 1–110), whose expression alternates based on cell cycle progression. hCdtl (fused to a red-fluorescent protein) is a DNA replication licensing factor that is present during G_1 phase but is ubiquitinated by SCF^{Skp2} and degraded during S/G_2/M phases. hGeminin (fused to a green-fluorescent protein) is a negative regulator of DNA licensing that is present during S/G_2/M phases but is ubiquitinated by APC^{Cdh1} and degraded at the end of mitosis and throughout G_1 (*see* **Note 1**). Thus, FUCCI provides a simple, fluorescence readout of cell cycle position. This system provides a critical tool in overcoming the technical barrier of isolating tetraploids from diploids: FUCCI can be used to discriminate G_1 tetraploids from G_2/M diploids, both of which possess 4C DNA content, because G_1 tetraploids emit red fluorescence while G_2/M diploids emit green fluorescence.

2 Materials

The following materials and equipment are used in this protocol:

2.1 Cell Culture

1. 10 cm^2 polystyrene tissue culture plates.
2. 15 cm^2 polystyrene tissue culture plates.
3. Phenol red-free DMEM:F12 media supplemented with 10 % fetal bovine serum (FBS), 100 IU/ml penicillin, and 100 μg/ml streptomycin (*see* **Note 2**).
4. 0.25 % trypsin–EDTA.
5. Sterile phosphate-buffered saline (PBS).
6. Cell line of interest expressing the FUCCI reporter system (here we use the telomerase-immortalized human retinal pigment epithelial cell line RPE-1, from ATCC). It is also important to have unlabeled cells, cells expressing hCdt1-RFP alone, cells expressing hGem-GFP alone, and cells expressing the complete FUCCI system (both hCdt1-RFP and hGem-GFP, which we refer to as RPE-1 FUCCI) (*see* **Note 3**).
7. Dihydrocytochalasin B (10 mM stock in DMSO).

2.2 Fluorescence Activated Cell Sorting

1. Hoechst 33342 (10 mg/ml in water).
2. Polystyrene round bottom tubes with cell strainer caps (35 μm nylon mesh).
3. Sterile 15 ml conical tubes.
4. Sterile 50 ml conical tubes.
5. 0.05 % trypsin–EDTA.
6. Aluminum foil.

2.3 Equipment

1. Tissue culture incubator set at 37 °C with 5 % humidified CO_2.
2. Tissue culture hood.
3. Hemocytometer.
4. High-speed centrifuge.
5. Phase-contrast microscope equipped with a 10× objective for cell counting.
6. Standard epifluorescence microscope equipped with excitation and emission filters necessary to visualize fluorescence in three colors (e.g., DAPI/FITC/TRITC).
7. FACS machine equipped with a UV laser (355 nm), a 488 nm laser, and a 561 nm laser, and a 100 μm nozzle (*see* **Note 4**).

3 Methods

The following protocol describes the method used to generate and purify tetraploid RPE-1 cells using the FUCCI reporter system. In addition to growing RPE-1 FUCCI cells (hCdt-mCherry and hGem-AzamiGreen), control RPE-1 cells expressing hCdt-mCherry alone, hGem-AzamiGreen alone, and unlabeled RPE-1 cells must also be carried (*see* **Note 5**). These are used to calibrate the FACS machine.

3.1 Cell Culture

1. Use freshly thawed and early passage cells for all experiments. The cells should be maintained on 10 cm^2 polystyrene tissue culture plates in phenol red-free DMEM:F12 media supplemented with 10 % fetal bovine serum (FBS), 100 IU/ml penicillin, and 100 μg/ml streptomycin. Maintain the cells at 40–80 % confluence. If the cells grow beyond 80 % confluence, discard them and thaw fresh cells. Maintain cells for only 1–2 months before thawing fresh cells.
2. To set up a tetraploid purification experiment, expand RPE-1 FUCCI cells from 10 cm^2 tissue culture maintenance dishes into 15 cm^2 dishes. Aspirate medium from each 10 cm^2 tissue culture dish, wash with sterile PBS, and add 2.5 ml of 0.25 % trypsin–EDTA. Incubate the cells in a tissue culture incubator at 37 °C for ~5 min or until most of the cells have detached.
3. Collect the cells in 10 ml of complete medium to inactivate the trypsin, and then pellet the cells for 5 min at 280 × *g* in a high-speed centrifuge.
4. Thoroughly resuspend the cells in 10 ml of complete medium. Count the cells with a hemocytometer.
5. Plate 6×10^6 RPE-1 FUCCI cells per 15 cm^2 dish. In general, one seeded 15 cm^2 dish will ultimately yield ~$0.5–1 \times 10^6$ purified tetraploid cells following FACS. Scale up the number of dishes as needed. In addition, plate 2×10^6 unlabeled RPE-1 cells, RPE-1 cells expressing hCdt-mCherry, and RPE-1 cells expressing hGem-AzamiGreen into separate 10 cm^2 dishes. These are used to calibrate the FACS machine.

3.2 Generating Tetraploid Cells

Multiple approaches have been developed to generate tetraploid cells in vitro. These include inhibiting cytokinesis, promoting mitotic slippage, or fusing diploid cells. This protocol will focus on use of the cell permeable mycotoxin dihydrocytochalasin B (DCB), which disrupts actin polymerization and thus causes cytokinetic cleavage furrow regression, cytokinesis failure, and tetraploidy. This approach is beneficial for a number of reasons. First, this highly potent compound can be added to an entire population of proliferating cells in order to generate a significant number of tetraploid cells. Second, this compound does not disrupt mitotic spindle assembly or the efficiency of chromosome segregation, and does not produce DNA damage. This is in contrast to drugs (e.g., Taxol, nocodazole), which induce tetraploidy by promoting prolonged mitotic arrest and mitotic slippage by preventing inactivation of the spindle assembly checkpoint. Finally, DCB is a reversible drug that can be washed from cells, thus enabling the inhibition of cytokinesis during only a single cell cycle. This is a major benefit over knocking out or knocking down the expression of genes essential for cytokinesis, which will induce repeated cytokinesis failures. To generate tetraploid cells:

1. Dilute DCB to 4 μM (from a 10 mM stock) in 20 ml complete growth medium and mix thoroughly. Add the growth medium containing DCB to each of the RPE-1 FUCCI cells plated on 15 cm^2 dishes the previous day (using a 10× objective on a phase-contrast microscope, confirm that cells are 50–60 % confluent at the time of DCB addition) (*see* **Note 6**).
2. Incubate the cells in DCB for 16 h in a 37 °C tissue culture incubator (*see* **Note 7**). It should be noted that since DCB disrupts the actin cytoskeleton, the morphology of the cells will be dramatically and visibly altered: the cells will have ruffled edges and will be less spread on the tissue culture dish.
3. Aspirate the medium and gently rinse the cells with sterile PBS.
4. Wash the cells five times (5 min each time) in pre-warmed growth medium. These washes are essential to completely remove residual DCB from the cells. The DCB-treated cells should flatten out and assume a normal looking morphology immediately following the completion of these washes. At this point, ~40–60 % of the cells should be visible as binucleated tetraploids under a phase-contrast microscope.

3.3 Purifying Tetraploid Cells by FACS

1. Following the last wash, add complete growth medium containing 4 μg/ml Hoechst dye (to label the DNA) to the RPE-1 FUCCI cells. In addition, add growth medium containing 4 μg/ml Hoechst dye to the 10 cm^2 dishes seeded with the unlabeled control RPE-1 cells, RPE-1 cells expressing hCdt-mCherry, and RPE-1 cells expressing hGem-Azami-Green. Incubate all dishes at 37 °C for 30 min.
2. Aspirate the medium from each tissue culture dish, wash with sterile PBS, and add 6 ml of 0.05 % trypsin–EDTA to each 15 cm^2 dish (2.5 ml of trypsin to the 10 cm^2 control dishes). Incubate the cells in the tissue culture incubator at 37 °C for ~5 min or until the cells have detached (*see* **Note 8**).
3. Collect the cells in 10 ml of complete medium to inactivate the trypsin, then pellet the cells for 5 min at 280×*g* in a high-speed centrifuge.

4. Resuspend the pelleted cells in complete medium containing 4 μg/ml Hoechst (for cells from one 10 or 15 cm^2 dish, resuspend in 200 μl of medium; for each additional 15 cm^2 dish, add 50 μl to the resuspension volume). Resuspend the pellets thoroughly by slowly pipetting up and down with a p1000 pipette tip (~50 times).

5. To remove clumps, strain the resuspended cells through 35 μm nylon strainer caps into round bottom FACS tubes wrapped in aluminum foil. The cells are now ready to be FACS-sorted.

6. Calibrate the FACS machine with the control cell lines. First, run the unlabeled diploid RPE-1 cells stained for Hoechst in order to calibrate the UV laser (355 nm) and produce sharp 2C and 4C peaks (Fig. 1a). Second, run RPE-1 cells expressing hGem-AzamiGreen to define and gate AzamiGreen$^+$ cells (use the 488 nm laser). Third, run RPE-1 cells expressing hCdt1-mCherry alone to define and gate mCherry$^+$ cells (use the 561 nm laser).

7. Once the FACS machine is properly calibrated, sort mCherry$^+$/AzamiGreen$^-$ 2C cells (these represent G_1 diploids) into one 15 ml conical tube containing 5 ml of complete growth medium (Fig. 1b). Sort mCherry$^+$/AzamiGreen$^-$ 4C cells (these represent G_1 tetraploids) into a separate 15 ml conical tube containing 5 ml of complete growth medium (Fig. 1b). Both collection tubes should be wrapped in aluminum foil. To ensure maximum viability, the FACS sorting should take no longer than 1 h (*see* **Note 9**). To limit clumping and/or settling of cells during the FACS sort, the cellular resuspension should be briefly vortexed every 5 min.

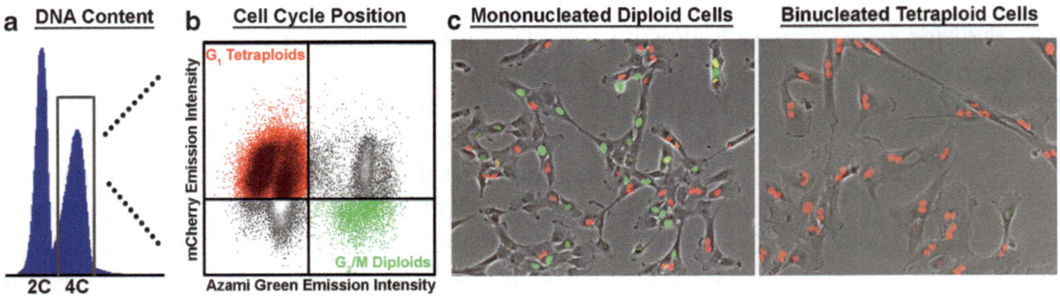

Fig. 1 FACS isolation of tetraploid cells. (a) A representative FACS profile of DNA content from RPE-1 FUCCI cells following 16 h DCB treatment. The 2C peak contains G_1 diploids while the 4C peak contains both G_2/M diploids and G_1 tetraploids. (b) Diploid and tetraploid cells within the 4C peak are distinguished by assessing mCherry (*y*-axis) and AzamiGreen (*x*-axis) fluorescence intensity. Tetraploid cells in G_1 are mCherry$^+$ and AzamiGreen$^-$ (*top left quadrant*), while diploid cells in G_2/M are mCherry$^-$ and AzamiGreen$^+$ (*bottom right quadrant*). (c) Sorted tetraploid cells generated by DCB treatment are binucleated (*right panel*) and can be easily distinguished from mononucleated diploid cells (*left panel*) by phase-contrast and fluorescence microscopy

8. Pellet the sorted cells for 5 min at 280 × *g* in a high-speed centrifuge, resuspend them in complete growth medium, and then count/plate into tissue culture dishes as needed. To assess purity, the sorted cells can be visualized by fluorescence microscopy once they have attached to the tissue culture dish (this takes ~1–3 h). Tetraploid cells, which are mCherry$^+$ and AzamiGreen$^-$, will appear as binucleated cells exhibiting red fluorescence (Fig. 1c, right panel). Because non-transformed tetraploid cells activate the Hippo tumor suppressor pathway and do not proliferate well, there will be few binucleated cells exhibiting green fluorescence (indicative of S-phase entry). By contrast, diploid cells will be mononucleated and exhibit both red and green fluorescence, as these cells resume proliferation immediately following the sorting procedure (Fig. 1c, left panel). In general, the tetraploid population should be 85–95 % pure.

4 Notes

1. The FUCCI system was developed by Dr. Atushi Miyawaki's group at the laboratory for Cell Function and Dynamics at the Riken Brain Science Institute and is commercially available through many sources. Information about the constructs can be found at: http://cfds.brain.riken.jp/Fucci.html.
2. The use of phenol-red free medium is not required, however, the absence of phenol red decreases background fluorescence and improves the efficiency of FACS sorting.
3. Both lentiviral and retroviral vectors expressing components of the FUCCI system can be used to generate stable cell lines. Live-cell imaging should be used to confirm that the reporter constructs are cycling properly. Both hCdt1-mCherry and hGem-AzamiGreen should localize exclusively to the nucleus (except during mitosis, when the nuclear membrane breaks down and hGem-AzamiGreen becomes diffuse throughout the cytoplasm).
4. A variety of nozzle sizes are available for sorting cellular resuspensions on FACS machines. Because of the increased size of tetraploid cells, we have found that sorting with the larger 100 μm nozzle leads to far less nozzle clogging during the sorting procedure.
5. The original FUCCI reporters consisted of hCdt1(30/120) fused to the fluorescent protein Kusabira orange-2, and hGem(1/110) fused to the fluorescent protein AzamiGreen.

Kusabira orange-2 maximally absorbs light at 551 nm and emits light at 565 nm, while AzamiGreen maximally absorbs light at 492 nm and emits light at 505 nm. A second generation version of FUCCI consists of hCdtl(30/120) fused to mCherry (absorbs light at 587 nm and emits at 610 nm), and hGem(1/110) fused to mVenus (absorbs light at 515 nm and emits at 528 nm). We found that pairing hCdtl(30/120)-mCherry with hGem(1/110)-AzamiGreen is the most ideal combination, as the mCherry and AzamiGreen emission spectra are the most spatially separated and thus exhibit the least amount of bleed-through fluorescence.

6. The confluence of cells is critically important at the time of DCB addition. Because DCB disrupts the actin cytoskeleton, sparsely plated cells are more likely to become detached from the tissue culture dish. Conversely, if cells are too dense at the time of DCB addition, cells may reach overconfluence by 16 h.
7. Disruption of the actin cytoskeleton by DCB will prevent cells in early G_1 phase from entering S-phase. Thus, only cells in late G_1, S, G_2, or M phases of the cell cycle will proceed through mitosis and fail cytokinesis following DCB treatment. Incubation of cells with DCB for 16 h is recommended because that is the approximate duration of the cell cycle in RPE-1 cells. Longer treatments in DCB will not produce additional tetraploid cells.
8. We have observed that using 0.05 % trypsin instead of 0.25 % trypsin reduces the amount of cell clumping during FACS.
9. 1 h is generally sufficient time to FACS sort cells from as many as 16 separate 15 cm^2 dishes. However, if more cells are required, it is recommended to use two FACS machines or stagger the experiments in order to limit the amount of time cells are kept in suspension.

Acknowledgments

N.J.G is a Karin Grunebaum Cancer Research Foundation Fellow in the Shamim and Ashraf Dahod Breast Cancer Research Laboratories and is supported by grants from the Richard and Susan Smith Family Foundation, the Searle Scholars Program, the Melanoma Research Alliance, the Skin Cancer Foundation, the Sarcoma Foundation of America, and the NIH/NCI (K99/R00 CA154531-01).

References

1. Coward J, Harding A (2014) Size Does Matter: Why Polyploid Tumor Cells are Critical Drug Targets in the War on Cancer. Front Oncol 4:123. doi:10.3389/fonc.2014.00123
2. Davoli T, de Lange T (2012) Telomere-driven tetraploidization occurs in human cells undergoing crisis and promotes transformation of mouse cells. Cancer Cell 21:765–776. doi:10.1016/j.ccr.2012.03.044
3. Duelli DM, Padilla-Nash HM, Berman D, Murphy KM, Ried T, Lazebnik Y (2007) A virus causes cancer by inducing massive chromosomal instability through cell fusion. Curr Biol 17:431–437. doi:10.1016/j.cub.2007.01.049
4. Fujiwara T, Bandi M, Nitta M, Ivanova EV, Bronson RT, Pellman D (2005) Cytokinesis failure generating tetraploids promotes tumorigenesis in p53-null cells. Nature 437:1043–1047. doi:10.1038/nature04217
5. Ganem NJ, Godinho SA, Pellman D (2009) A mechanism linking extra centrosomes to chromosomal instability. Nature 460:278–282. doi:10.1038/nature08136
6. Ganem NJ, Storchova Z, Pellman D (2007) Tetraploidy, aneuploidy and cancer. Curr Opin Genet Dev 17:157–162. doi:10.1016/j.gde.2007.02.011
7. Lundberg G, Jin Y, Sehic D, Ora I, Versteeg R, Gisselsson D (2013) Intratumour diversity of chromosome copy numbers in neuroblastoma mediated by on-going chromosome loss from a polyploid state. PLoS One 8, e59268. doi:10.1371/journal.pone.0059268
8. Sotillo R, Hernando E, Diaz-Rodriguez E, Teruya-Feldstein J, Cordon-Cardo C, Lowe SW, Benezra R (2007) Mad2 overexpression promotes aneuploidy and tumorigenesis in mice. Cancer Cell 11:9–23. doi:10.1016/j.ccr.2006.10.019
9. Storchova Z, Kuffer C (2008) The consequences of tetraploidy and aneuploidy. J Cell Sci 121:3859–3866. doi:10.1242/jcs.039537
10. Dewhurst SM, McGranahan N, Burrell RA, Rowan AJ, Gronroos E, Endesfelder D, Joshi T, Mouradov D, Gibbs P, Ward RL, Hawkins NJ, Szallasi Z, Sieber OM, Swanton C (2014) Tolerance of whole-genome doubling propagates chromosomal instability and accelerates cancer genome evolution. Cancer Discov 4:175–185. doi:10.1158/2159-8290.CD-13-0285
11. Zack TI, Schumacher SE, Carter SL, Cherniack AD, Saksena G, Tabak B, Lawrence MS, Zhang C-Z, Wala J, Mermel CH, Sougnez C, Gabriel SB, Hernandez B, Shen H, Laird PW, Getz G, Meyerson M, Beroukhim R (2013) Pan-cancer patterns of somatic copy number alteration. Nat Genet 45:1134–1140. doi:10.1038/ng.2760
12. Wong C, Stearns T (2005) Mammalian cells lack checkpoints for tetraploidy, aberrant centrosome number, and cytokinesis failure. BMC Cell Biol 6:6. doi:10.1186/1471-2121-6-6
13. Andreassen PR, Lohez OD, Lacroix FB, Margolis RL (2001) Tetraploid state induces p53-dependent arrest of nontransformed mammalian cells in G1. Mol Biol Cell 12:1315–1328
14. Carter SB (1967) Effects of cytochalasins on mammalian cells. Nature 213:261–264
15. Ganem NJ, Cornils H, Chiu SY, O'Rourke KP, Arnaud J, Yimlamai D, Thery M, Camargo FD, Pellman D (2014) Cytokinesis failure triggers hippo tumor suppressor pathway activation. Cell 158:833–848. doi:10.1016/j.cell.2014.06.029
16. Rieder CL, Maiato H (2004) Stuck in division or passing through: what happens when cells cannot satisfy the spindle assembly checkpoint. Dev Cell 7:637–651. doi:10.1016/j.devcel.2004.09.002
17. Ganem NJ, Pellman D (2012) Linking abnormal mitosis to the acquisition of DNA damage. J Cell Biol 199:871–881. doi:10.1083/jcb.201210040
18. Orth JD, Loewer A, Lahav G, Mitchison TJ (2012) Prolonged mitotic arrest triggers partial activation of apoptosis, resulting in DNA damage and p53 induction. Mol Biol Cell 23:567–576. doi:10.1091/mbc.E11-09-0781
19. Krzywicka-Racka A, Sluder G (2011) Repeated cleavage failure does not establish centrosome amplification in untransformed human cells. J Cell Biol 194:199–207. doi:10.1083/jcb.201101073
20. Panopoulos A, Pacios-Bras C, Choi J, Yenjerla M, Sussman MA, Fotedar R, Margolis RL (2014) Failure of cell cleavage induces senescence in tetraploid primary cells. Mol Biol Cell 25:3105–3118. doi:10.1091/mbc.E14-03-0844
21. Straight AF, Cheung A, Limouze J, Chen I, Westwood NJ, Sellers JR, Mitchison TJ (2003) Dissecting temporal and spatial control of cytokinesis with a myosin II Inhibitor. Science 299:1743–1747. doi:10.1126/science.1081412
22. Sakaue-Sawano A, Kurokawa H, Morimura T, Hanyu A, Hama H, Osawa H, Kashiwagi S, Fukami K, Miyata T, Miyoshi H, Imamura T, Ogawa M, Masai H, Miyawaki A (2008) Visualizing spatiotemporal dynamics of multicellular cell-cycle progression. Cell 132:487–498. doi:10.1016/j.cell.2007.12.033

Chapter 25

Anti-Microtubule Drugs

Stefan Florian and Timothy J. Mitchison

Abstract

Small molecule drugs that target microtubules (MTs), many of them natural products, have long been important tools in the MT field. Indeed, tubulin (Tb) was discovered, in part, as the protein binding partner of colchicine. Several anti-MT drug classes also have important medical uses, notably colchicine, which is used to treat gout, familial Mediterranean fever (FMF), and pericarditis, and the vinca alkaloids and taxanes, which are used to treat cancer. Anti-MT drugs have in common that they bind specifically to Tb in the dimer, MT or some other form. However, their effects on polymerization dynamics and on the human body differ markedly. Here we briefly review the most-studied molecules, and comment on their uses in basic research and medicine. Our focus is on practical applications of different anti-MT drugs in the laboratory, and key points that users should be aware of when designing experiments. We also touch on interesting unsolved problems, particularly in the area of medical applications. In our opinion, the mechanism by which any MT drug cures or treats any disease is still unsolved, despite decades of research. Solving this problem for particular drug–disease combinations might open new uses for old drugs, or provide insights into novel routes for treatment.

Key words Microtubules, Vinblastine, Vincristine, Nocodazole, Paclitaxel, Colchicine, Microtubule drugs, Combretastatin A4

1 Introduction

Natural substances that bind to Tb or MTs have been isolated as extracts from many plants. They have been used as drugs for millennia and in the last five decades, they have been of utmost importance both as cancer therapeutics and as experimental tools to understand microtubule biology. Today, despite the advent of targeted therapies, their importance in cancer treatment remains undiminished. Drugs like paclitaxel, vinblastine, and vincristine are a fundamental pillar of chemotherapy protocols for most solid and hematopoietic neoplasms. Moreover, new, clinically effective, MT binding compounds are continuously discovered and new therapeutic indications are approved. The drugs we discuss in this article are listed in Table 1. They were chosen for their importance in basic research and/or medicine (Table 2). They represent most of

Paul Chang and Ryoma Ohi (eds.), *The Mitotic Spindle: Methods and Protocols*, Methods in Molecular Biology, vol. 1413,
DOI 10.1007/978-1-4939-3542-0_25, © Springer Science+Business Media New York 2016

Table 1
Summary of drugs discussed in this chapter

Drug[a]	Binding site[a]	In vitro IC_{50}[b]	In vivo C_{max}[c]
Colchicine	Colchicine	250 nM (to inhibit migration) [1] 13–250 nM (cell proliferation) [2]	3–10 nM [3]
Colcemid	Colchicine	~100 nM (cell proliferation) [4]	Currently not used in man, despite early successes in cancer and low toxicity
Combretastatin A4	Colchicine	7 nM [5]	5 μM [6]
Nocodazole	Colchicine (?) [7]	50 nM	Not used in man
Benomyl		5 μM [8]	Not used in man
T138067	Binds covalently to Cys-239 of β-tubulin	11–165 nM [9]	Not used in man
Vinblastine	Vinca	0.45 nM [10]	100–500 nM [3, 11]
Vincristine	Vinca	1–5 nM [12]	100–500 nM [3, 11]
Paclitaxel	Taxane	2.5–7.5 nM [13]	3–5 μM [3]
Docetaxel	Taxane	5–43 nM [14]	~2 μM [3]
Eribulin	Vinca	0.09–9.5 nM [15]	300 nM [16]

[a]sensitive Tb binding sites were traditionally defined based on competition studies, and, because Tb is notoriously difficult to crystallize, few structural studies exist [17–20]. Thus, the binding sites for some compounds are rather poorly defined and this classification can only serve as a rough approximation. For Benomyl the binding site to human tubulin has not been studied and T138067 does not fit into the traditional binding site classification.
[b]Concentration causing 50% of cells to die in a few days, or for causing mitotic arrest in cell culture. These are approximate data for "typical" human cell cultures, and are provided as an approximate starting point for experimental design. Note the reported values may differ widely, and particular cell lines may be much less, or more, drug sensitive.
[c]Maximum concentration in human plasma following a typical therapeutic dose.

the mechanisms of action that have been shown for small molecules binding to Tb, and are commercially available in most cases. For longer reviews that address biochemical and medical mechanisms see [3, 25, 28–30]. A full list of anti-MT drugs would be very long. Here, we focus only on those commonly used for lab research and representatives of classes used clinically (Table 1).

2 Colchicine and Colcemid

2.1 History and Medical Uses of Colchicine

Colchicine (Fig. 1) has been known far longer than any other anti-MT drug, indeed longer than most medicines, and it played a central role in the early development of the MT and molecular

Table 2
Summary of clinical aspects of therapeutically relevant drugs

	Textbook mechanism of action[a]	Clinical use, main indications	Most important side effects
Antiquity			
Colchicine	Inhibits neutrophil migration	Gout, rheumatoid arthritis, familial Mediterranean fever	Diarrhea
1960s—Vinca alkaloids			
Vinblastine	Antimitotic [21]	Germ cell malignancies and some types of advanced lymphomas	Neutropenia
Vincristine	Antimitotic [21–23]	Pediatric malignancies, ALL, blast crisis in CML, lymphomas (Hodgkin and Non-Hodgkin)	Neurotoxicity
1980s/1990s—Taxanes			
Paclitaxel	Antimitotic [24]	ovarian, breast, non-small-cell lung cancer, Kaposi sarcoma,	Neurotoxicity, neutropenia
Docetaxel	Antimitotic	Breast, non-small-cell lung, prostatic (hormone-refractory), head and neck squamous cell and gastric carcinoma	
1995			
Combretastatin A	Antiangiogenic	Cancer—experimental	Pain, cardiopulmonary toxicity
2010			
Eribulin	Antimitotic	Advanced breast cancer	Neurotoxicity, neutropenia

ALL acute lymphoid leukemia, *CML* chronic myeloid leukemia

[a]The mechanism of action according to most textbooks is given. The references given for vinca alkaloids and paclitaxel point to additional suggested mechanisms of action. Information is gathered from Refs. [3, 25–27]. For further references, see main text

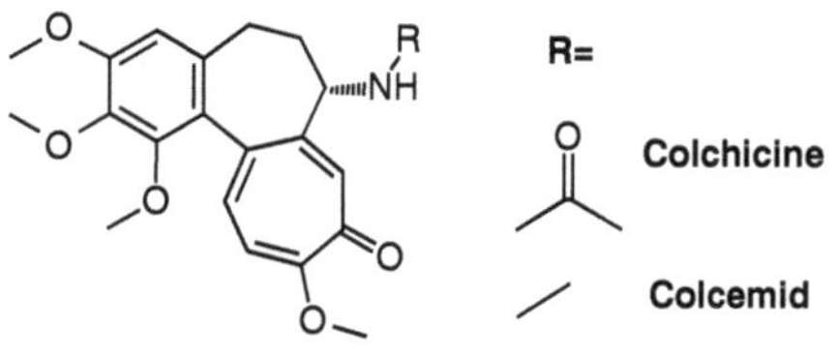

Fig. 1 Chemical structures of colchicine and colcemid

mitosis research fields. Its history and effects on cells were extensively researched by Pierre Dustin and his son Pierre Dustin Jr., and our comments are drawn in part from two seminal books written by Pierre Dustin Jr. [31, 32]. As a natural product extract of the autumn crocus (*Colchicum autumnale*), colchicine was mentioned in ancient Egyptian papyrus scrolls, and was well known to the ancient Greeks. It was the first drug known to arrest proliferating cells in mitosis [33], and this discovery started a whole new field of mitosis research in living organisms in the 1930s, when colchicine was used to estimate proliferation rates in many animals and plants. Tb was discovered, in part, as its binding partner [34] and codiscovered, and named, as a subunit of flagella [35]. Colchicine is still much used as a medicine, and there has been a recent resurgence in interest in its potential, following its recent approval for treatment of recurrent pericarditis [36]. Colchicine is effective treatment for gout, a common inflammatory disease caused by precipitation of uric acid crystals in joints. Taken daily, it also suppresses the symptoms of the relatively common inherited inflammatory disease, familial Mediterranean fever (FMF). Daily colchicine is effectively curative for most FMF patients, with few side effects and no disease progression, which represents a truly remarkable gene–drug interaction. FMF is caused by mutations in Pyrin, a protein that regulates inflammasome activation. It is currently unknown why colchicine is curative in FMF, but has no therapeutic action in apparently similar genetic diseases that cause inflammasome activation by other mechanisms [37]. How colchicine works as an anti-inflammatory drug is, in our view, an interesting unsolved problem. It is known to block only certain types of inflammation, specifically those caused by neutrophils and monocytes. It inhibits accumulation of these cells at sites of inflammation, and also inhibits release of pro-inflammatory cytokines such as IL1β [38]. These therapeutic actions are thought to result from inhibition of MT polymerization in neutrophils and monocytes, but it is unclear how this can occur without antimitotic side effects.

2.2 Lab Uses of Colchicine

Colchicine remains a representative MT and antimitotic drug and can be used for those purposes in research. A major advantage is the huge wealth of published data on its actions at every level, from molecular effects on Tb to human and animal pharmacokinetic data. It is also relatively cheap, water soluble, and is thought to be highly specific, with no known targets other than Tb. Colchicine binds to β-Tb at a site near α-Tb in the heterodimer, and its binding forces the dimer into bent configuration that is not compatible with normal polymerization [39, 40]. The depolymerizing action of colchicine is sub-stoichiometric; only ~5 % of the Tb dimers in the cell need to be bound to colchicine to arrest cells in mitosis [41], and a similar fractional occupancy inhibits polymerization of pure Tb [42, 43]. A very important consideration when using colchicine as a tool is its extremely slow association and dissociation rate constants for Tb binding. The association and dissociation rate constants for colchicine binding to Tb are ~100 M^{-1} s^{-1} and 10^{-5} s^{-1}, respectively [34, 44]. These values represent much slower binding and unbinding rates than those exhibited by most drugs, and they make colchicine binding slow, but effectively irreversible in most experimental contexts, including mitotic arrest of tissue culture cells [41]. If they are taken into account, they can provide experimental advantages. For example, Tb-colchicine complex can be prepared free of excess drug, and is relatively stable. However, mistakes can be made if the slow association rate constant is not taken into account. When adding colchicine to cells, it is important to keep track of the time of incubation as well as the concentration, since it is not safe to assume rapid equilibration of drug with target. The slow binding and unbinding of colchicine was the main reason most laboratories switched to other drugs, notably colcemid and nocodazole, for routine depolymerization of MTs in cells. In our opinion, combretastatin A4 may be the best drug for this purpose since it binds rapidly, and is more potent than nocodazole, and perhaps also more specific. Colchicine can be prepared as a stock solution in water or DMSO. It is stable at physiological pH, but is notably inactivated by long wavelength uv light (see below), so stock solutions should be protected from light.

2.3 Colcemid

Colcemid (Fig. 1), also called demecolcine, was isolated from the autumn crocus in 1950 [45] and commercialized by Ciba. Initially, it was explored as a cancer drug due to its low toxicity. Today, it is only used as a research tool mainly to overcome limitations of colchicine due its very slow association and dissociation rate constants. It binds to Tb at the same site as colchicine, but ~10-fold faster, and it also dissociates faster [46]. Its main use has been to arrest cells in mitosis for cytogenetic analysis, though to our knowledge it offers no special advantages over other drugs in this application. It shares the photosensitivity of colchicine.

2.4 Photo-inactivation of Colchicine and Colcemid

Colchicine and close derivatives are efficiently converted by long wavelength UV light to a series of highly rearranged molecules called lumicolchicines [47]. Lumicolchicines have no known biological activity, and are sometimes used as negative controls in colchicine studies. Local photoinactivation of colchicine and colcemid under a microscope has allowed some creative experimental applications, and should be considered for modern experiments. In one classic example, Hamaguchi and Hiramoto used local photoinactivation of colcemid in echinoderm eggs to control the geometry of the MT aster nucleated by sperm centrosomes during pronuclear migration [48]. This elegant experiment provided the first evidence that a motor protein—now known to be cytoplasmic dynein—pulls on MTs from bulk cytoplasm to center asters.

3 Combretastatin A4

3.1 History and Medical Uses

The combretastatins were discovered, by Pettit and coworkers, through fractionation of extracts of *Combretum cuffrum* [49]. We will discuss only Combretastatin A4 (Fig. 2), which was the most potent derivative, and is also easily synthesized. Combretastatin A4 structurally resembles colchicine. It binds to an overlapping site on Tb, albeit much faster than colchicines, and has similar depolymerizing effects [50]. Combretastatin A4 phosphate ester, a water soluble prodrug, has strong antitumor action in animals, and has shown promise in clinical trials, though its clinical use is limited by high toxicity to multiple organs. The antitumor action of combretastatin A4 is thought to derive mainly from antivascular, rather than direct cytotoxic or antimitotic actions on cancer cells [51]. An enduring mystery is why colchicine and combretastatin A4 have very different clinical effects, given that they bind to similar sites on Tb and have similar effects on polymerization. Part of the answer may lie in their very different association rate constants, and consequent effects of pharmacokinetics and tissue distribution.

3.2 Lab Uses of Combretastatin A4

Combretastatin A4 is a potent and fast-binding drug that cleanly depolymerizes MTs through the same well-characterized mechanism as colchicine. It is easy to synthesize, ~10-fold more potent than nocodazole, has a fast association rate constant and is both chemically and photochemically stable. Combretastatin A4 is soluble in DMSO, and DMSO stock solutions do not tend to precipitate as they do with nocodazole. These properties make it an excellent choice as a routine depolymerizing drug, but for lab studies it is important to purchase the parent drug, and not the phosphate ester prodrug, which is inactive until hydrolyzed.

A "caged" nitrobenzyl derivative of combretastatin A4 has been recently reported that may be useful for spatiotemporal control of MT polymerization, with the opposite properties of colcemid in that it is activated by UV light, rather than inactivated [52]. Photoactivation of caged combretastatin using a confocal

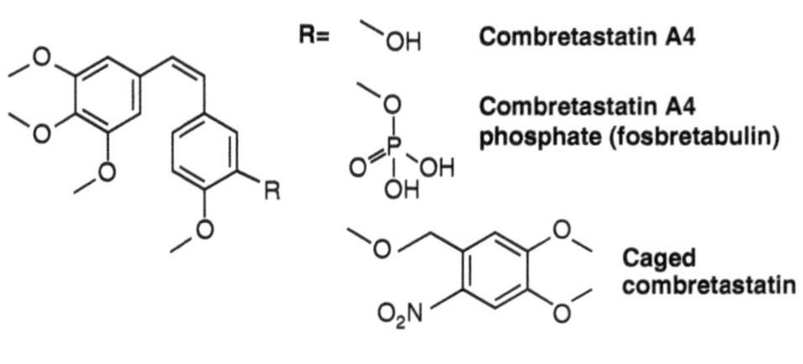

Fig. 2 Chemical structures of combretastatin A4, combretastatin A4 phosphate, and caged combretastatin

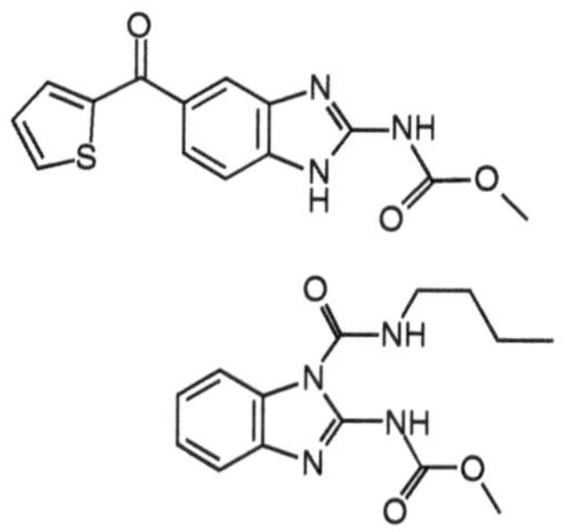

Fig. 3 Chemical structures of nocodazole and benomyl

microscope facilitated the local induction of MT depolymerization in zebrafish embryos, which revealed dynein-mediated pulling forces on centrosomes, in an update of Hiramoto's classic experiment using photo-inactivation of colcemid [48].

3.3 Nocodazole and Benomyl

Nocodazole (Fig. 3) was developed by DeBrabander and coworkers at Janssen in the 1970s [53], and is now perhaps the most used MT depolymerizing drug in basic cell biology research. It is not used clinically. Benomyl, a structurally related benzimidazole, was developed as a plant fungicide at DuPont in the 1960s, and is still widely used for that purpose. Benomyl is less active on mammalian Tb, and has been mostly used in budding yeast experiments. Nocodazole and benomyl bind to β-Tb and block polymerization, but their mechanism is less well understood than colchicine. Nocodazole came to favor in laboratory applications in part because its effects are more easily reversed in washout experiments than colchicine, which reflects faster drug binding and unbinding. Nocodazole addition is often used to arrest mammalian cells in mitosis, and washout to release them into cytokinesis and G1. Nocodazole at concentrations above ~500 nM arrests cells in mitosis with few or no MTs, which causes mitotic arrest, but also mis-regulation of mitotic kinases. For this reason, we prefer to arrest cells in mitosis with the Kinesin-5 inhibitor s-trityl-l-cysteine (STLC) [54] for biochemical analysis. STLC is cheaper than nocodazole, and arrested cells have normal, dynamic MTs. The presence of dynamic MTs significantly altered regulation of Plk1 in cells arrested in mitosis with STLC vs. nocodazole, with the STLC arrested state presumably more relevant to normal metaphase [55]. Nocodazole washout allows progression into cytokinesis and G1, but spindle reassembly tends to be slow and variable, leading to extensive loss of synchrony and high rates of chromosome mis-segregation. For biochemical analysis of cytokinesis, we prefer to override STLC mitotic arrest with purvalanol, a CDK1 inhibitor, which causes rapid and synchronous mitotic exit. The resulting cytokinesis is monopolar, but otherwise an excellent biochemical and morphological mimic of normal cytokinesis [56].

Nocodazole is relatively insoluble in water. It is prepared as a stock solution in DMSO (~10 mM) and stored frozen. Although nocodazole is chemically stable, DMSO stocks tend to precipitate when thawed, and care must be taken to make sure the correct amount of drug is added. Nocodazole precipitates in DMSO can be re-solubilized by warming to ~50 °C. Nocodazole added to tissue culture medium from a DMSO stock also precipitates, and it is important to mix well to ensure it redissolves. Nocodazole is less potent than many natural product anti-MT drugs. This, and its lack of water solubility, are reasons for considering an alternative drug such as combretastatin A4 or colcemid for depolymerizing MTs in tissue culture cells.

Benomyl (Fig. 3) has been much used to probe Tb and mitosis biology in budding yeast, and genetic tests proved it worked by inhibiting Tb polymerization [57]. It is also prepared as a DMSO stock and stored frozen [58].

4 T138067

The structurally simple sulfonamide T138067 (Fig. 4) is of interest because it reacts covalently with Cys-239 of β2- and β4-Tb, causing MT depolymerization and mitotic arrest in cancer cell lines [9]. This reaction was highly selective for Tb, as judged by analysis of whole cell lysate treated with a radioactive derivative. T138067 is useful for experiments where a completely irreversible inhibitor is needed. It was tested clinically, with the hope of being active on tumors that express drug efflux pumps. It exhibited the toxicities expected for an anti-microtubule drug (bone marrow, gut and neurotoxicity), but no useful clinical responses [59, 60]. The question why some anti-microtubule drugs are active in cancer, and others not, remains mysterious, as discussed below.

4.1 On the Mechanism of Action of MT Drugs in Cancer

The next three drug classes, vinca alkaloids, taxanes, and eribulin, are used to treat cancer (Table 2). The vincas and taxanes are widely used in combination chemotherapy, while eribulin has only been recently approved. They have different mechanisms of action on MTs, different pharmacodynamic properties, and different

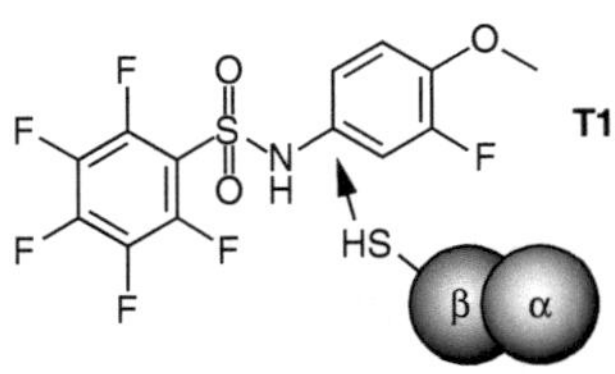

T138067

Fig. 4 Chemical structure of T138067

disease indications. In the last few years, we and others have become interested in the therapeutic mechanism of action of these drugs at the cellular and tissue level, problems that had long been considered solved. The basic concepts about how these drugs work against cancer and inflammation have been developed in the 1960s and 1970s and many of them are based on data obtained in highly artificial cell culture or mouse models with proliferation rates much higher than human cancers. Considering the physiological context in humans, these ideas contain many inconsistencies and contradictions [61–63]. Because mitotic arrest is such a prominent biomarker, especially in tissue culture of highly proliferating cells, mitotic effects of these drugs have been considered responsible for their antitumor efficacy and have been the main focus of interest in basic research [26]. A related concept is that these drugs do not kill cancers by mitotic arrest per se, but by promoting chromosome missegregation [64]. Recent extensive efforts to develop highly selective antimitotic drugs that do not interact with MT have been very disappointing as such drugs hardly show any effects in solid tumors [65]. Thus, it seems likely that interphase effects of MT drugs, which are numerous but have been mostly neglected as a therapeutic mechanism, are at least equally important for their antitumor effect. Interesting examples of such effects are:

- MT depolymerizers like colchicine have been shown to not only induce mitotic arrest, but also cause entry from G0 into the cell cycle, thus *increasing* rather than inhibiting proliferation both in embryos [66] and serum starved cultured cells [67 and references therein]. This effect is antagonized by stabilizing drugs like paclitaxel.
- MT drugs are known to affect a wide range of cellular processes like intracellular transport, signaling transmitted through the adhesion complex and cell migration [68–71] as a few examples. These effects have been regarded as mainly detrimental for therapeutical applications as they are responsible for side effects like neurotoxicity and increased susceptibility to infection, but it is likely that these effects are highly important for drug efficacy.
- MT drugs are known to activate signaling pathways that can cause cell death in particular contexts, such as the Jnk pathway [72, 73]

We believe that interphase effects of MT interacting drugs are therapeutically important and should receive more attention.

5 Vinca Alkaloids

5.1 History and Medical Uses

Vinca alkaloids (Fig. 5) were isolated in the 1950s by Beer, Cutts and Noble at the University of Western Ontario and were the first MT binding drugs to find their way into mainstream chemotherapy [74]. As with most cytotoxic chemotherapeutics, it was a serendipitous discovery. Tea made from the Madagascar periwinkle (formerly *Vinca rosea*, now *Catharanthus roseus*) was used as a diabetes remedy in Jamaica and Noble and Beer originally wanted to test its efficacy more systematically. Instead of finding effects on glucose levels in the blood, they noticed that rats injected with the plant extract suffered from severe depletion of white blood cells and became susceptible to lethal infections. A stepwise fractionation procedure, with depletion of white blood cells as a measure for activity, was used to isolate vinblastine, originally called vincaleukoblastine. Vincristine was also identified, but isolated in larger quantities only later at Eli Lilly. Systematic tests on solid and hematopoietic tumor models revealed that the vinca alkaloids are potent anticancer drugs, and they remain among the most used drugs in chemotherapy, even today. Interestingly, although vinblastine and vincristine differ only by substitution of a methyl group through a formyl group, they have very different clinical efficacy and side effect profiles. Vinblastine is considered effective in germ cell malignancies and some lymphomas in adults and its dominant side effect is white blood cell depletion, the hallmark effect which led to its discovery. Vincristine is mainly used in pediatric malignancies and some adult hematological cancers and its main side effect is peripheral neuropathy.

5.2 Laboratory Uses

The interaction of vinca alkaloids with MTs is very complex. Tb crystals are notoriously hard to obtain and the binding mechanism of vinblastine has only been characterized at the structure level quite recently [19]. Like colchicine, it induces curved Tb assemblies but binds to a distinct binding site between heterodimers (at inter-dimer interfaces). This is an important difference from taxane and colchicine binding sites, which lie within a single heterodimer, on the interior of the MT and at the intra-dimer interface between monomers, respectively (ibid). In the 1960s and 1970s, colchicine

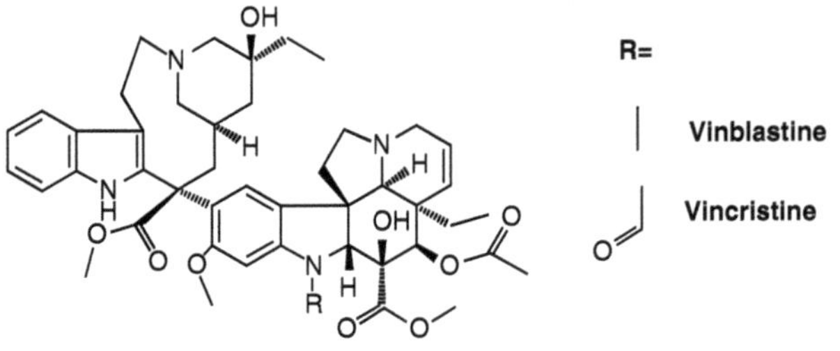

Fig. 5 Chemical structures of vinblastine and vincristine

and the vinca alkaloids were used to determine cell cycle dynamics in human and murine tissues in order to understand how antiproliferative cancer drugs work in vivo [75]. At the time, it was a relatively important field of research, now almost forgotten and superseded by the molecular revolution in cancer research. It seems that, among the classic vinca alkaloids and colchicine, vincristine is the best compound to estimate tissue proliferation by inducing mitotic arrest in vivo ("stathmokinetic agent"—σταθμός—station on a road), because it induces a relatively long lived mitotic arrest without affecting interphase cell cycle kinetics [76].

Three dose regimes have been described for treatment of tissue culture cells with vinblastine, and each has distinct uses. At the lowest active concentration (~2–20 nM) vinblastine causes mitotic arrest and suppresses MT dynamics without depolymerizing MTs [77]. This may be the best tool for blocking plus end polymerization dynamics without gross reorganization of MTs. Paclitaxel, in contrast, tends to promote reorganization, in part due its propensity to promote nucleation. Above ~10 μM vinblastine induces the formation of beautiful paracrystals [78]. In between, it mainly depolymerizes, though the nature of the soluble Tb species in cells in this regime is poorly characterized.

Vinblastine paracrystals have interesting potential as a research tool. They are formed from side-by-side aggregates of protofilaments [79], and thus are structurally related to the ends of growing and shrinking MTs. Comparison of binding affinity for MTs vs. vinblastine paracrystals might be useful to characterize proteins suspected to bind selectively to protofilaments. By forcing Tb dimer into paracrystals, high concentrations of vinblastine very effectively depolymerize MTs, but do not lead to buildup of soluble, drug-inhibited Tb dimers. In this respect, the cellular effects of vinblastine differ from the depolymerizers colchicine, combretastatin, and nocodazole, where soluble Tb bound to drug builds up, and also to the polymerizers paclitaxel and epothilone where most soluble Tb is forced into MTs. These differences between drug classes are useful for studies that probe possible regulatory roles of free Tb dimer, e.g., in experiments showing that translation of tubulin mRNA is regulated by levels of free tubulin dimer [80].

6 Taxanes

6.1 History and Medical Uses

Paclitaxel (Fig. 6) was the first MT stabilizing drug to be discovered and is a hydrophobic compound isolated from the bark of the Pacific or Western Yew (*Taxus brevifolia*). Both clinically and for cell biology, it is a drug of eminent importance. Its unique structure and antitumor activity in multiple preclinical cancer models were discovered in the late 1960s by Monroe Wall and Mansukh Wani [81]. However, for at least a decade, it received relatively

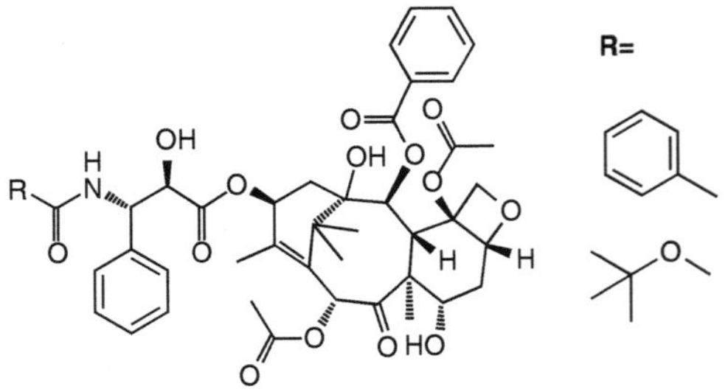

Fig. 6 Chemical structures of paclitaxel and docetaxel

little attention, in part because its complex structure was (correctly) considered an obstacle for production of large scale quantities for clinical use. Interest in paclitaxel as a drug rose again when Peter Schiff and Susan Horwitz published a seminal series of papers showing that paclitaxel was the first known compound to stabilize MTs, and that it allowed in vitro generation of polymerized MTs in absence of GTP or MT associated proteins [69, 82, 83]. The development of paclitaxel for clinical use was difficult. Due to its hydrophobicity, it was formulated with Cremophor EL, a substance known to cause hypersensitivity effects itself, and after one of the first patients treated with paclitaxel died from an anaphylactic reaction, clinical trials were put on hold for 5 years. The problem was alleviated by pretreating patients with anti-inflammatory drugs. A second problem was that a very large number of trees was needed to purify enough paclitaxel from the Pacific Yew and a complete synthesis proved to be difficult and highly inefficient. This problem was solved by semisynthesis approaches, utilizing 10-deacetylbaccatin III, a readily available precursor from the European yew, *Taxus baccata*. While developing this approach, Pierre Potier (who also developed the semisynthetic vinca alkaloid vinorelbine) and colleagues developed a semisynthetic taxane with higher potency and somehow better water solubility, docetaxel (formerly taxotere, Fig. 6) [84]. Paclitaxel and docetaxel seem to have similar antitumor activity characteristics. In the 1990s paclitaxel became one of the most successful cancer drugs to date, gaining initial FDA approval for breast, ovarian, and non-small-cell lung carcinoma. Taxanes have known activity against an impressive number of solid tumors, including sarcoma, melanoma, esophageal, gastric, endometrial, bladder, small-cell lung, hormone-refractory prostate, and germ-cell carcinoma [3]. Even today, in the era of targeted drugs, new clinical indications are approved.

6.2 Laboratory Uses

Taxanes bind to a unique binding site inside the MT lumen, and do not compete with exchangeable GTP, colchicine or vinca alkaloids [17, 85]. Other stabilizing drugs are thought to bind to similar sites [20]. Paclitaxel, which is historically called "Taxol" in the basic science literature (now the commercial name of this drug), is very useful as a tool to polymerize MTs, and to keep them stable in the absence of Tb, GTP, or MT associated proteins. It has been instrumental in studying the function of motor proteins and other MT associated proteins in vitro. A lot of our knowledge about the ability of motor proteins to move along MTs, and to slide them against each other and generate forces in the mitotic spindle, was gained using assays based on paclitaxel-stabilized MTs [86].

Paclitaxel is soluble in DMSO, but only soluble in aqueous buffer below ~100 μM. The ester bond between the taxane ring system and the side chain is quite labile due to hydrolysis, which makes paclitaxel somewhat unstable storage, especially if the DMSO is impure. To obtain reproducible dose–response data is it important to dissolve paclitaxel in fresh, dry DMSO, to store stock aliquots frozen, and to minimize freeze–thaw cycles. Paclitaxel strongly promotes MT nucleation as well as stabilizing polymerized MTs. This property can make it difficult to assemble long, stable MTs from pure Tb, since addition of saturating drug to a concentrated solution of Tb tends to generate very short MTs and protofilament aggregates. To make long MTs we usually start by adding a low concentration of paclitaxel (~1 % compared to Tb) to a solution of Tb + GTP, incubate at 37 °C for ~20 min to induce polymerization, and only then add saturating paclitaxel (i.e., stoichiometric with the Tb) to fully stabilize the MTs. Alternatively, we induce polymerization by adding DMSO to ~5 %, and again add saturating paclitaxel only after the Tb is fully polymerized.

In cells, similar to the vinca alkaloids, the effects of paclitaxel vary with concentration and depend on the binding stoichiometry to Tb. At low nanomolar extracellular concentrations, paclitaxel stabilizes MTs without increasing MT mass. At higher concentrations (1 μM and above) it dramatically increases polymer mass and causes the formation of characteristic MT bundles [69]. The mitotic effects of paclitaxel are heterogeneous and strongly cell type and concentration dependent. Below 10 nM, perturbation of MT dynamics is often too weak to arrest cells in mitosis, but induces chromosome segregation defects which have been implicated in its antitumor effects [87]. Above 10 nM, it induces nucleation of MTs from acentrosomal nucleation centers and leads to formation of multipolar spindles and activation of the spindle assembly checkpoint (SAC). Starting around 500 nM, however, stabilization of chromosome mis-attachments seems to lead to inappropriate fulfillment of the SAC and mitotic exit [88]. Although the binding of paclitaxel to Tb is reversible, it accumulates and is concentrated up to 1000× inside cells [87] and dissociation is very slow. This makes paclitaxel an unreliable drug to synchronize cells in mitosis and its main use in cell biology is to study the consequences of MT stabilization and excess nucleation, and to purify MT binding proteins. Addition of paclitaxel to a cell lysate, followed by centrifugation, is one of the most convenient methods for isolation of microtubule binding proteins [89]. Some motor proteins also co-sediment with taxol-stabilized microtubules, but their binding can be enhanced by adding nucleotide analogs and/or depleting ATP [86].

Paclitaxel derivatives labeled with several fluorescent dyes are commercially available that allow labeling of MTs in living cells from Molecular Probes [90–92]. Since these drugs also affect MT dynamics, they are only useful for correlating drug effects with localization rather than as a tool to observe MTs in living cells. To achieve this, in cells, transiently or stably expressed GFP-Tb has been used, but the effects of overexpression of fluorescent Tb are difficult to determine [93]. Recently, docetaxel derivatives labeled with silicon-rhodamine derivatives have been used to address these problems and look very promising [94]. These far red emitting compounds show a more than tenfold increase in fluorescence intensity upon binding to MT and seem to affect MT dynamics much less than unlabeled taxanes. Thus, they seem to be suitable for long term imaging of MT in living cells and do not require washout of unbound compound.

7 Eribulin

This synthetic analog of the marine sponge natural product halichondrin B was introduced by Eisai as a treatment for metastatic breast cancer in patients who have become resistant to other drugs including taxanes [15]. It is remarkably potent, with a dose in humans ~100 fold lower than a taxane. Eribulin (Fig. 7) binds to Tb dimer and prevents polymerization [95, 96]. This action leads us to question the hypothesis that MT stabilization plays a central role in the actions of MT drugs against carcinomas [26]. Another interesting clinical aspect of Eribulin is its relative lack of neurotoxicity [97]. One notable aspect as a laboratory tool is the free amine, which can be modified with a BODIPY fluorophore with little decrease in potency. BODIPY-Eribulin is a substrate for the MDR/P-glycoprotein drug pump, and has been recently used to probe drug efflux in tumor cells by intravital imaging in living mice [98].

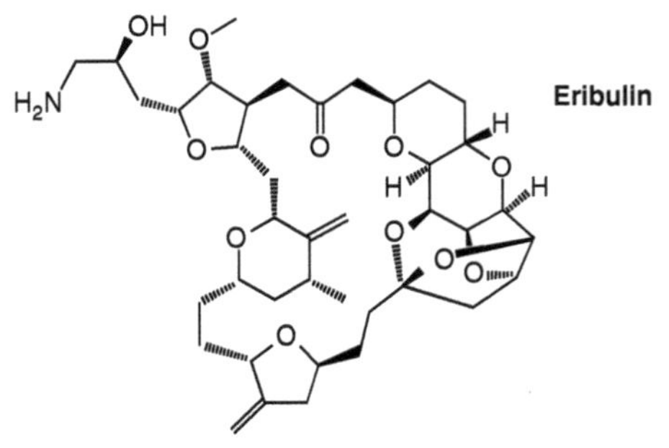

Fig. 7 Chemical structure of eribulin

Microtubule drugs have historically been central to advances both in medicine and basic science. Tubulin was discovered using these drugs, and we owe them a lot of our knowledge about microtubules and mitosis. Recent systematic attempts to develop antimitotic cancer drugs that do not bind to MT have failed rather dramatically [62, 63, 65], making the significant antitumor efficacy of MT drugs seem even more remarkable. It is likely that MT binding is essential for their antitumor effects and besides the induction of mitotic arrest in tumor cells, other, nonmitotic effects on tumor or immune cells are important. Understanding how these drugs kill tumor cells in patients is one of the great challenges ahead, and we expect that work in this direction will continue to result in new discoveries of great relevance both to biology and medicine.

Acknowledgements/Funding

Stefan Florian is supported by a Research Fellowship (FL 820-1/1) from the German Research Foundation (DFG).

References

1. Caner JE (1965) Colchicine inhibition of chemotaxis. Arthritis Rheum 8(5):757–764
2. Xi J, Zhu X, Feng Y et al (2013) Development of a novel class of tubulin inhibitors with promising anticancer activities. Mol Cancer Res 11(8):856–864. doi:10.1158/1541-7786.MCR-12-0177
3. Rowinsky EK (2011) Antimitotic Drugs. In: Chabner BA, Longo DL (eds) Cancer Chemotherapy and Biotherapy: Principles and Practice. Lippincott Williams & Wilkins, Philadelphia, p 848
4. Fujikawa-Yamamoto K, Teraoka K, Zong ZP et al (1994) Apoptosis by demecolcine in V79 cells. Cell Struct Funct 19(6):391–396
5. Lin CM, Singh SB, Chu PS et al (1988) Interactions of tubulin with potent natural and synthetic analogs of the antimitotic agent combretastatin: a structure-activity study. Mol Pharmacol 34(2):200–208
6. Rustin GJ, Galbraith SM, Anderson H et al (2003) Phase I clinical trial of weekly combretastatin A4 phosphate: clinical and pharmacokinetic results. J Clin Oncol 21(15):2815–2822. doi:10.1200/JCO.2003.05.185
7. Xu K, Schwarz PM, Luduena RF (2002) Interaction of nocodazole with tubulin isotypes. Drug Dev Res 55(2):91–96. doi:10.1002/ddr.10023
8. Gupta K, Bishop J, Peck A et al (2004) Antimitotic antifungal compound benomyl inhibits brain microtubule polymerization and dynamics and cancer cell proliferation at mitosis, by binding to a novel site in tubulin. Biochemistry 43(21):6645–6655. doi:10.1021/bi036112v
9. Shan B, Medina JC, Santha E et al (1999) Selective, covalent modification of beta-tubulin residue Cys-239 by T138067, an antitumor agent with in vivo efficacy against multidrug-resistant tumors. Proc Natl Acad Sci U S A 96(10):5686–5691
10. Ngan VK, Bellman K, Hill BT et al (2001) Mechanism of mitotic block and inhibition of cell proliferation by the semisynthetic Vinca alkaloids vinorelbine and its newer derivative vinflunine. Mol Pharmacol 60(1):225–232
11. Nelson RL, Dyke RW, Root MA (1980) Comparative pharmacokinetics of vindesine, vincristine and vinblastine in patients with cancer. Cancer Treat Rev 7(Suppl):117–124
12. Jackson DVJ, Bender RA (1979) Cytotoxic thresholds of vincristine in a murine and a human leukemia cell line in vitro. Cancer Res 39(11):4346–4349
13. Liebmann JE, Cook JA, Lipschultz C et al (1993) Cytotoxic studies of paclitaxel (Taxol) in human tumour cell lines. Br J Cancer 68(6):1104–1109

14. Riou JF, Naudin A, Lavelle F (1992) Effects of Taxotere on murine and human tumor cell lines. Biochem Biophys Res Commun 187(1):164–170
15. Swami U, Chaudhary I, Ghalib MH et al (2012) Eribulin -- a review of preclinical and clinical studies. Crit Rev Oncol Hematol 81(2):163–184. doi:10.1016/j.critrevonc.2011.03.002
16. Goel S, Mita AC, Mita M et al (2009) A phase I study of eribulin mesylate (E7389), a mechanistically novel inhibitor of microtubule dynamics, in patients with advanced solid malignancies. Clin Cancer Res 15(12):4207–4212. doi:10.1158/1078-0432.CCR-08-2429
17. Nogales E, Wolf SG, Khan IA et al (1995) Structure of tubulin at 6.5 A and location of the taxol-binding site. Nature 375(6530):424–427. doi:10.1038/375424a0
18. Lowe J, Li H, Downing KH et al (2001) Refined structure of alpha beta-tubulin at 3.5 A resolution. J Mol Biol 313(5):1045–1057. doi:10.1006/jmbi.2001.5077
19. Gigant B, Wang C, Ravelli RB et al (2005) Structural basis for the regulation of tubulin by vinblastine. Nature 435(7041):519–522. doi:10.1038/nature03566
20. Prota AE, Bargsten K, Zurwerra D et al (2013) Molecular Mechanism of Action of Microtubule-Stabilizing Anticancer Agents. Science. doi:10.1126/science.1230582
21. Madoc-Jones H, Mauro F (1968) Interphase action of vinblastine and vincristine: differences in their lethal action through the mitotic cycle of cultured mammalian cells. J Cell Physiol 72(3):185–196. doi:10.1002/jcp.1040720306
22. Stryckmans PA, Lurie PM, Manaster J et al (1973) Mode of action of chemotherapy in vivo on human acute leukemia—II. Vincristine. Eur J Cancer 9(9):613–620
23. Rosner F, Hirshaut Y, Grunwald HW et al (1975) In vitro combination chemotherapy demonstrating potentiation of vincristine cytotoxicity by prednisolone. Cancer Res 35(3):700–705
24. Rowinsky EK, Donehower RC, Jones RJ et al (1988) Microtubule changes and cytotoxicity in leukemic cell lines treated with taxol. Cancer Res 48(14):4093–4100
25. Jordan A, Hadfield JA, Lawrence NJ et al (1998) Tubulin as a target for anticancer drugs: agents which interact with the mitotic spindle. Med Res Rev 18(4):259–296
26. Jordan MA, Wilson L (2004) Microtubules as a target for anticancer drugs. Nat Rev Cancer 4(4):253–265. doi:10.1038/nrc1317
27. Rowinsky EK (2010) MIcrotubule-targeting natural products. In: Hong WK (ed) Holland Frei cancer medicine 8. People's Medical Pub, House, Shelton, CT, p xxv, 2,021
28. Jordan MA, Wilson L (1999) The use and action of drugs in analyzing mitosis. Methods Cell Biol 61267–295
29. Dumontet C, Jordan MA (2010) Microtubule-binding agents: a dynamic field of cancer therapeutics. Nat Rev Drug Discov 9(10):790–803. doi:10.1038/nrd3253
30. Field JJ, Kanakkanthara A, Miller JH (2014) Microtubule-targeting agents are clinically successful due to both mitotic and interphase impairment of microtubule function. Bioorg Med Chem. doi:10.1016/j.bmc.2014.02.035
31. Eigsti OJ, Dustin P (1955) Colchicine in agriculture, medicine, biology and chemistry. State College Press, Ames, Iowa
32. Dustin P (2011) Softcover reprint of the original 2nd ed. 1984 Edition, 2nd edn. Springer, Berlin
33. B P Sulla cariocinesi delle cellule epiteliali e dell' endotelio dei vasi della mucosa dello stomaco et dell' intestino, nelle studio della gastroenterite sperimentale (nell'avvelenamento per colchico). Sicilia Med 1265–1279.
34. Borisy GG, Taylor EW (1967) The mechanism of action of colchicine. Binding of colchicine-3H to cellular protein. J Cell Biol 34(2):525–533
35. Mohri H (1968) Amino-acid composition of "Tubulin" constituting microtubules of sperm flagella. Nature 217(5133):1053–1054
36. Markel G, Imazio M, Brucato A et al (2013) Prevention of recurrent pericarditis with colchicine in 2012. Clin Cardiol 36(3):125–128. doi:10.1002/clc.22098
37. Ter Haar NM, Frenkel J (2014) Treatment of hereditary autoinflammatory diseases. Curr Opin Rheumatol 26(3):252–258. doi:10.1097/BOR.0000000000000059
38. Cocco G, Chu DC, Pandolfi S (2010) Colchicine in clinical medicine. A guide for internists. Eur J Intern Med 21(6):503–508. doi:10.1016/j.ejim.2010.09.010
39. Ravelli RB, Gigant B, Curmi PA et al (2004) Insight into tubulin regulation from a complex with colchicine and a stathmin-like domain. Nature 428(6979):198–202. doi:10.1038/nature02393

40. Barbier P, Dorleans A, Devred F et al (2010) Stathmin and interfacial microtubule inhibitors recognize a naturally curved conformation of tubulin dimers. J Biol Chem 285(41):31672–31681. doi:10.1074/jbc.M110.141929
41. Taylor EW (1965) The mechanism of colchicine inhibition of mitosis. I. Kinetics of inhibition and the binding of H3-colchicine. J Cell Biol 25(Suppl):145–160
42. Olmsted JB, Borisy GG (1973) Characterization of microtubule assembly in porcine brain extracts by viscometry. Biochemistry 12(21):4282–4289
43. Bergen LG, Borisy GG (1983) Tubulin-colchicine complex inhibits microtubule elongation at both plus and minus ends. J Biol Chem 258(7):4190–4194
44. Fernando Diaz J, Andreu JM (1991) Kinetics of dissociation of the tubulin-colchicine complex. Complete reaction scheme and comparison to thermodynamic measurements. J Biol Chem 266(5):2890–2896
45. Santavy F, Reichstein T (1950) Isolierung neuer Stoffe aus den Samen der Herbstzeitlose Colchicum autumnale L. Substanzen der Herbstzeitlose und ihre Derivate. 12. Mitteilung. Helvetica Chimica Acta 33(6):1606–1627
46. Ray K, Bhattacharyya B, Biswas BB (1984) Anion-induced increases in the affinity of colcemid binding to tubulin. Eur J Biochem 142(3):577–581
47. Chapman OL, Smith HG, King RW (1963) The Structure of β-Lumicolchicine. J Am Chem Soc 85(6):803–806. doi:10.1021/ja00889a031
48. Hamaguchi MS, Hiramoto Y (1986) Analysis of the Role of Astral Rays in Pronuclear Migration in Sand Dollar Eggs by the Colcemid-UV Method. Dev Growth Differ 28(2):143
49. Pettit GR, Singh SB, Boyd MR et al (1995) Antineoplastic agents. 291. Isolation and synthesis of combretastatins A-4, A-5, and A-6(1a). J Med Chem 38(10):1666–1672
50. Lin CM, Ho HH, Pettit GR et al (1989) Antimitotic natural products combretastatin A-4 and combretastatin A-2: studies on the mechanism of their inhibition of the binding of colchicine to tubulin. Biochemistry 28(17):6984–6991
51. Griggs J, Metcalfe JC, Hesketh R (2001) Targeting tumour vasculature: the development of combretastatin A4. Lancet Oncol 2(2):82–87. doi:10.1016/S1470-2045(00)00224-2
52. Wuhr M, Tan ES, Parker SK et al (2010) A model for cleavage plane determination in early amphibian and fish embryos. Curr Biol 20(22):2040–2045. doi:10.1016/j.cub.2010.10.024
53. De Brabander MJ, Van de Veire RM, Aerts FE et al (1976) The effects of methyl (5-(2-thienylcarbonyl)-1H-benzimidazol-2-yl) carbamate, (R 17934; NSC 238159), a new synthetic antitumoral drug interfering with microtubules, on mammalian cells cultured in vitro. Cancer Res 36(3):905–916
54. DeBonis S, Skoufias DA, Lebeau L et al (2004) In vitro screening for inhibitors of the human mitotic kinesin Eg5 with antimitotic and antitumor activities. Mol Cancer Ther 3(9):1079–1090
55. Hu CK, Ozlu N, Coughlin M et al (2012) Plk1 negatively regulates PRC1 to prevent premature midzone formation before cytokinesis. Mol Biol Cell 23(14):2702–2711. doi:10.1091/mbc.E12-01-0058
56. Hu CK, Coughlin M, Field CM et al (2008) Cell polarization during monopolar cytokinesis. J Cell Biol 181(2):195–202. doi:10.1083/jcb.200711105
57. Neff NF, Thomas JH, Grisafi P et al (1983) Isolation of the beta-tubulin gene from yeast and demonstration of its essential function in vivo. Cell 33(1):211–219
58. Straight AF, Murray AW (1997) The spindle assembly checkpoint in budding yeast. Methods Enzymol 283425–440
59. Kirby S, Gertler SZ, Mason W et al (2005) Phase 2 study of T138067-sodium in patients with malignant glioma: Trial of the National Cancer Institute of Canada Clinical Trials Group. Neuro Oncol 7(2):183–188. doi:10.1215/S1152851704000602
60. Berlin JD, Venook A, Bergsland E et al (2008) Phase II trial of T138067, a novel microtubule inhibitor, in patients with metastatic, refractory colorectal carcinoma. Clin Colorectal Cancer 7(1):44–47. doi:10.3816/CCC.2008.n.006
61. Tannock I (1978) Cell kinetics and chemotherapy: a critical review. Cancer Treat Rep 62(8):1117–1133
62. Komlodi-Pasztor E, Sackett D, Wilkerson J et al (2011) Mitosis is not a key target of microtubule agents in patient tumors. Nat Rev Clin Oncol 8(4):244–250. doi:10.1038/nrclinonc.2010.228
63. Mitchison TJ (2012) The proliferation rate paradox in antimitotic chemotherapy. Mol Biol Cell 23(1):1–6. doi:10.1091/mbc.E10-04-0335

64. Weaver BA (2014) How Taxol/paclitaxel kills cancer cells. Mol Biol Cell 25(18):2677–2681. doi:10.1091/mbc.E14-04-0916
65. Komlodi-Pasztor E, Sackett DL, Fojo AT (2012) Inhibitors targeting mitosis: tales of how great drugs against a promising target were brought down by a flawed rationale. Clin Cancer Res 18(1):51–63. doi:10.1158/1078-0432.CCR-11-0999
66. Paff GH (1939) The action of colchicine upon the 48-hour chick embryo. Am J Anat 64(2):331–349
67. Crossin KL, Carney DH (1981) Evidence that microtubule depolymerization early in the cell cycle is sufficient to initiate DNA synthesis. Cell 23(1):61–71
68. Vasiliev JM, Gelfand IM, Domnina LV et al (1970) Effect of colcemid on the locomotory behaviour of fibroblasts. J Embryol Exp Morphol 24(3):625–640
69. Schiff PB, Horwitz SB (1980) Taxol stabilizes microtubules in mouse fibroblast cells. Proc Natl Acad Sci U S A 77(3):1561–1565
70. Zhu ML, Horbinski CM, Garzotto M et al (2010) Tubulin-targeting chemotherapy impairs androgen receptor activity in prostate cancer. Cancer Res 70(20):7992–8002. doi:10.1158/0008-5472.CAN-10-0585
71. Ng DH, Humphries JD, Byron A et al (2014) Microtubule-Dependent Modulation of Adhesion Complex Composition. PLoS One 9(12), e115213. doi:10.1371/journal.pone.0115213
72. Wang TH, Wang HS, Ichijo H et al (1998) Microtubule-interfering agents activate c-Jun N-terminal kinase/stress-activated protein kinase through both Ras and apoptosis signal-regulating kinase pathways. J Biol Chem 273(9):4928–4936
73. Sendoel A, Maida S, Zheng X et al (2014) DEPDC1/LET-99 participates in an evolutionarily conserved pathway for anti-tubulin drug-induced apoptosis. Nat Cell Biol 16(8):812–820. doi:10.1038/ncb3010
74. Noble RL, Beer CT, Cutts JH (1958) Role of chance observations in chemotherapy: Vinca rosea. Ann N Y Acad Sci 76(3):882–894
75. Aherne WA, Camplejohn RS, Wright NA (1977) An introduction to cell population kinetics. Edward Arnold, London
76. Tannock IF (1967) A comparison of the relative efficiencies of various metaphase arrest agents. Exp Cell Res 47(1):345–356
77. Tanaka E, Ho T, Kirschner MW (1995) The role of microtubule dynamics in growth cone motility and axonal growth. J Cell Biol 128(1-2):139–155
78. Bensch KG, Malawista SE (1969) Microtubular crystals in mammalian cells. J Cell Biol 40(1):95–107
79. Starling D, Burns RG (1975) Ultrastructure of tubulin paracrystals from sea urchin eggs, with determination of spacings by electron and optical diffraction. J Ultrastruct Res 51(2):261–268
80. Cleveland DW, Lopata MA, Sherline P et al (1981) Unpolymerized tubulin modulates the level of tubulin mRNAs. Cell 25(2):537–546
81. Wani MC, Taylor HL, Wall ME et al (1971) Plant antitumor agents. VI. The isolation and structure of taxol, a novel antileukemic and antitumor agent from Taxus brevifolia. J Am Chem Soc 93(9):2325–2327
82. Schiff PB, Fant J, Horwitz SB (1979) Promotion of microtubule assembly in vitro by taxol. Nature 277(5698):665–667
83. Schiff PB, Horwitz SB (1981) Taxol assembles tubulin in the absence of exogenous guanosine 5′-triphosphate or microtubule-associated proteins. Biochemistry 20(11):3247–3252
84. Gueritte-Voegelein F, Guenard D, Lavelle F et al (1991) Relationships between the structure of taxol analogues and their antimitotic activity. J Med Chem 34(3):992–998
85. Nogales E, Wolf SG, Downing KH (1998) Structure of the alpha beta tubulin dimer by electron crystallography. Nature 391(6663):199–203. doi:10.1038/34465
86. Vale RD, Reese TS, Sheetz MP (1985) Identification of a novel force-generating protein, kinesin, involved in microtubule-based motility. Cell 42(1):39–50
87. Zasadil LM, Andersen KA, Yeum D et al (2014) Cytotoxicity of Paclitaxel in Breast Cancer Is due to Chromosome Missegregation on Multipolar Spindles. Sci Transl Med 6(229): 229ra43. doi:10.1126/scitranslmed.3007965
88. Yang Z, Kenny AE, Brito DA et al (2009) Cells satisfy the mitotic checkpoint in Taxol, and do so faster in concentrations that stabilize syntelic attachments. J Cell Biol 186(5):675–684. doi:10.1083/jcb.200906150
89. Vallee RB (1982) A taxol-dependent procedure for the isolation of microtubules and microtubule-associated proteins (MAPs). J Cell Biol 92(2):435–442
90. Souto AA, Acuña AU, Andreu JM et al (1996) New Fluorescent Water-Soluble Taxol Derivatives. Angew Chem Int Ed Engl 34(23-24):2710–2712
91. Evangelio JA, Abal M, Barasoain I et al (1998) Fluorescent taxoids as probes of the microtubule cytoskeleton. Cell Motil

Cytoskeleton 39(1):73–90. doi:10.1002/(SICI)1097-0169(1998)39:1<73::AID-CM7>3.0.CO;2-H

92. Diaz JF, Strobe R, Engelborghs Y et al (2000) Molecular recognition of taxol by microtubules. Kinetics and thermodynamics of binding of fluorescent taxol derivatives to an exposed site. J Biol Chem 275(34):26265–26276. doi:10.1074/jbc.M003120200
93. Rusan NM, Fagerstrom CJ, Yvon AM et al (2001) Cell cycle-dependent changes in microtubule dynamics in living cells expressing green fluorescent protein-alpha tubulin. Mol Biol Cell 12(4):971–980
94. Lukinavicius G, Reymond L, D'Este E et al (2014) Fluorogenic probes for live-cell imaging of the cytoskeleton. Nat Methods 11(7):731–733. doi:10.1038/nmeth.2972
95. Smith JA, Wilson L, Azarenko O et al (2010) Eribulin binds at microtubule ends to a single site on tubulin to suppress dynamic instability. Biochemistry 49(6):1331–1337. doi:10.1021/bi901810u
96. Alday PH, Correia JJ (2009) Macromolecular interaction of halichondrin B analogues eribulin (E7389) and ER-076349 with tubulin by analytical ultracentrifugation. Biochemistry 48(33):7927–7938. doi:10.1021/bi900776u
97. Wozniak KM, Nomoto K, Lapidus RG et al (2011) Comparison of neuropathy-inducing effects of eribulin mesylate, paclitaxel, and ixabepilone in mice. Cancer Res 71(11):3952–3962. doi:10.1158/0008-5472.CAN-10-4184
98. Laughney AM, Kim E, Sprachman MM et al (2014) Single-cell pharmacokinetic imaging reveals a therapeutic strategy to overcome drug resistance to the microtubule inhibitor eribulin. Sci Transl Med 6(261): 261ra152. doi:10.1126/scitranslmed.3009318

INDEX

Paul Chang and Ryoma Ohi (eds.), *The Mitotic Spindle: Methods and Protocols*, Methods in Molecular Biology, vol. 1413, DOI 10.1007/978-1-4939-3542-0, © Springer Science+Business Media New York 2016

M

N

P

R

S

T

X

MIX
Papier aus verantwortungsvollen Quellen
Paper from responsible sources
FSC® C105338
FSC
www.fsc.org

If you have any concerns about our products,
you can contact us on
ProductSafety@springernature.com

In case Publisher is established outside the EU,
the EU authorized representative is:
Springer Nature Customer Service Center GmbH
Europaplatz 3, 69115 Heidelberg, Germany

Printed by Libri Plureos GmbH
in Hamburg, Germany